CREATIVE DESIGN OF PRODUCTS AND SYSTEMS

CREATIVE DESIGN OF PRODUCTS AND SYSTEMS

SAEED BENJAMIN NIKU, Ph.D., P.E.
Mechanical Engineering Department, Cal Poly, San Luis Obispo

WILEY

JOHN WILEY & SONS, INC.

Publisher: Don Fowley
Senior Acquisitions Editor: Michael McDonald
Editorial Assistant: Rachael Leblond
Senior Production Editor: Anna Melhorn
Cover Designer: Jeof Vita
Marketing Manager: Chris Ruel

This book was typeset by Thomson Digital and printed and bound by Hamilton Printing Company.
The cover was printed by Phoenix Color Corporation.

To order books or for customer service, please call 1-800-CALL-WILEY (225-5945).

ISBN-13 978-0-470-14850-1

Printed in the United States of America
10 9 8 7 6 5 4 3 2 1

Dedicated to my loved ones, Shohreh, Adam, and Alan
for living with my creations
and for being my source of inspiration through their own creativity

PREFACE

This book is the culmination of years of work, writing, and experience in teaching design to thousands of students at Cal Poly, San Luis Obispo. Like the design process it presents for solving any problem, the manuscript has been edited, modified, added to, and changed numerous times with each iteration. But as one of my students, whose name I have since forgotten, said, "During the design process, there comes a time when you need to shoot the designer and go into production." This book, like any other book, is not perfect or complete. There are always more issues to discuss and present, more information to mention, and other techniques to utilize. But brevity is still an important matter, and expecting one book to cover everything anyone may desire to teach in a class is unrealistic. And to others, it is already too long.

However, this book covers a wide variety of subjects that are usually and customarily covered in design classes. In some schools this material is covered in a Sophomore or Junior class, in others in a Senior, capstone design class. Some schools leave out the engineering calculations and stress analysis, circuit design, and other analysis material to other courses, some cover all in a series of two or three classes, consecutively. This book is intended to be the primary source for most of these classes. It may be used for introductory classes, intermediate classes, or a capstone design class. Only one chapter discusses mechanical design analysis more appropriate to mechanical engineering, and that is because fatigue considerations are so important in so many design decisions. Otherwise, practically all other material is appropriate to the design of products and systems regardless of the type of product or system.

This book is intended for all designers, whether engineers, architects, industrial designers, landscape designers, or any others. It can also be used in all branches of engineering. The detail design of any particular field is taught separately in the rest of the curriculum. But the design process is the same in all fields. As such, mechanical, electrical, architectural, and many other examples are included throughout the book.

At Cal Poly, we teach most of this material in one Quarter-based, Junior-level class. The remainder is covered in a senior capstone class that includes an industry-based design-build-test project with one lecture and two labs per week. Therefore, there should be sufficient material, and enough time, to present most of this book in a one-semester class as well. Our Junior-level class is scheduled for two classes per week, each 1.5 hours long. We have found out that it is much more efficient to have longer classes to cover the material without breaking it into smaller units.

In our experience, we have found the following to work efficiently:

Week	Topic	Reading
1	Introduction Team Assignments, Creativity	Chapter 1 Chapter 1, 2
2	Mental Barriers Mental Barriers, Continued	Chapter 3 Chapter 3
3	Problem-Solving Techniques Design Process	Chapter 4 Chapter 5
4	Patents Visualization, Imagination	Chapter 14 Chapter 6
5	Drawing and sketching Aesthetics	Chapter 6 Chapter 9
6	Design Factors Testing of Midterm Project	Chapter 7
7	Quality in Design Product Liability	Chapter 12 Chapter 13
8	Human Factors Safety, Man–Machine Interfaces	Chapter 8 Chapter 8
9	Economics of Design Entrepreneurship	Chapter 11 (First-half) Chapter 15
10	Final Project Presentations Final Project Presentations	

Due to the present structure of our curriculum, Chapters 10 (Materials and Manufacturing) and 16 (Design Analysis), as well as the second part of Chapter 11 (Engineering Economy), are covered in other classes. Therefore, these chapters are not mentioned in this schedule.

The homework for the first class, due on the second day, is the design of a page-turner for the disabled. This is used as a benchmark for students to realize where they stand before this class, and what their perception of design is. They are also asked to take a personality test at a web-based site such as Humanmetrics. This information is used in the second class for forming affinity groups and assigning team members.

In addition to homework and short projects, we assign two major projects during the quarter, one as a midterm project, one as final project. The midterm project is assigned after the discussion on creativity, and is based on this subject. Students are asked to design some device that will perform a certain job, usually trivial, and make the device for testing in class. Samples of this type of project are given in Chapter 3 (such as a Rube

Goldberg-type project or the garbanzo carrying device). The second project is the design of a product, and follows the design process. It is intended to encompass all the issues discussed in class, from creative design to human factors to economics of the design. Student teams find a need, define it, design it, and at the end of the quarter, present it. Some groups make models of the product for presentation, but we do not require actual manufacture of the product in this class since there is not enough time. The teams are assigned at the beginning of the class, usually the second meeting, and remain in effect throughout the Quarter. Thanks to the National Collegiate Innovators and Inventors Alliance (NCIIA) and the Lemelson Foundation, we introduced the concept of Entrepreneurship Teams (E-Teams) into the class many years ago. The final deliverable is a presentation, a report, and the evaluation of others' projects. However, in our program, senior capstone design is a separate course in which industrially sponsored projects are assigned to teams. Teams of students design, build, test, and present their projects in an exhibit. The contents of this book are the basis for both the senior capstone class and the junior-level creative design class.

We also invite experts, both from the University and local community as well as industry experts, as guest lecturers. These include, whenever possible, entrepreneurs, artists who teach drawing and visualization, quality engineers, safety experts, and so on.

For semester-long classes, more chapters can be covered, for example, material selection, manufacturing techniques, and design analysis, or more free time can be devoted to team work for design projects. If the class does include industry-based projects, more time may be devoted to the design of the product or system while students read the book instead of lectures.

Chapter 2 is informative, interesting, and short. It may be skipped if necessary, or students may read it for their own information. It is intended as an introduction to Chapter 3.

I would like to thank the following individuals for their tremendous help in many different capacities, without whom the project would have been impossible or very different. First, Mr. Joe Hayton, the acquisition editor at Wiley at the time, through whom I received much feedback from colleagues who read the first drafts, as well as Dan Sayre and Michael McDonald, who continued Joe's work. My thanks to the delightful Anna Melhorn, Senior Production Editor. My thanks also go to all the reviewers who made constructive comments that made the book so much better, but (unfortunately) do not know their names to mention. But thank you to all of you. Of course, my thanks to my family who endured all the time I spent working on this, reading the manuscript for errors and providing interesting observations. My colleagues at Cal Poly, especially Jim Widmann, Joe Mello, and Chris Pascual who provided material for different chapters, and my students who allowed me to use some of their homework and projects as examples. Douglass Wilde of Stanford University provided material and valuable review for team formation, Mike Ashby of Oxford University provided graphs and information for material selection, Larry Staufer who provided an example for design for assembly, Bill Bellows of Pratt and Whitney Rocketdyne provided material for quality engineering, Scott Ganaja of Progressive Engineering provided material for design with plastics, Joe Boeddeker and Dennis Fernandez provided invaluable assistance for entrepreneurship, Sina Niku who provided an architectural example, and Rambod Jacoby who provided an electrical engineering example. My thanks also to Shohreh for her own way of breaking the rules, being creative as well helping with the food-related information and testing to Adam and Alan for much helpful editing. My thanks also to NCIIA and Phil Weilerstein for their support to integrate E-teams into our class. My thanks also to my old classmates at Stanford, Dennis Boyle, Pam Kenady, and Dave Kelley, who inspired me in 1975 to look at things differently, as well as James Adams and Jerry Henderson who taught me a lot.

Finally, I am indebted to countless other authors, developers, teachers, engineers, and scientists who have originally created and developed most of the material presented in this book.

I hope you and your students will enjoy this book and will learn from it.

Saeed Benjamin Niku
San Luis Obispo
2008

CONTENTS

PART 3
DESIGN ANALYSIS

CHAPTER 16   *DESIGN ANALYSIS OF MACHINE COMPONENTS*   **533**

CHAPTER 1

CREATIVE PRODUCT AND SYSTEM DESIGN

I am always doing that which I cannot do in order that I may learn how to do it.
—Pablo Picasso

Whether you believe you can, or whether you believe you can't, you are absolutely right.
—Henry Ford

1.1 INTRODUCTION

One of my students, whose name I have long since forgotten, once said, "Design is like getting a hair cut; you do not tell the barber how to cut your hair. You just tell him how you want your hair to look." This implies that the barber (or hair stylist) knows how to accomplish the proper hair cut. No doubt, you or anyone else could also cut your hair. But you still go to a barber or hair stylist and pay him or her to cut your hair. You expect that the barber or hair stylist has learned how to do the job better, quicker, and more safely, even if hair cutting is fundamentally an intuitive project that anyone can perform.

So is design. Most people assume that they are designers. They can invent products, put together plans, and create systems that will do things and perform tasks; and they probably can. However, as with the barber or hair stylist, it takes learning the art and the science of design to actually create products and systems that work correctly and safely, are easy and inexpensive to produce, durable and efficient, and technologically appropriate. It takes much learning in many different fields, including different branches of engineering, sciences, psychology, arts, physical and medical sciences, and mathematics, to become a good designer.

And this is what this book is about. It is about the fundamental knowledge necessary to be a good designer, understand the different steps of the design process, be a more creative and efficient designer, understand the consequences of design decisions, and create products, systems, or plans, that are sound, good, innovative, safe, and useful. The fact that you may have created some products in your garage, or that you have fixed your car, or that you are a hands-on person, does not necessarily mean that you are also a (good) designer. You still need to learn the design process and what it takes to create good designs.

In this book we will try to learn what design means, how to be a more creative designer, and how to look for consequences of design decisions. This book is not a handbook or encyclopedia of all the knowledge you need to be a designer. You still need to learn all the other subjects that are taught in school. But this book will help you to be a good, creative, and forward-looking designer.

Please answer the following questions before you go on. We will refer to these questions and your answers later as we discuss different subjects. However, answering these questions now will help you realize how much difference this book will make in the way you perceive things and how it affects the way you approach design-related subjects:

1. What are your most/least favorite cars, and why?

2. How do you feel about DaVinci's Mona Lisa?

3. What is the last product you bought for your house or for someone else's? Why did you pick that particular product?

4. What is half of thirteen? What is half of 8?

If you have answered questions 1 through 4, please continue.

1.2   WHAT IS DESIGN?

Most people's response to the first question on page 2 is often related to the car's power, looks, price, how easy it is to repair, or how fast (or slow) the car is. Some can clearly describe why they like or dislike the car, and some do not know why. But what is interesting is how different the answers are. A car that is loved by one is hated by another. Some like sports cars. Some like old cars. Some admire unique "concept" cars, and others prefer conservative family sedans. And yet, there are certain cars that almost unanimously everyone likes or dislikes.

It is almost the same with Mona Lisa. Many describe it as dark, sad, common, and not worth anything, while others admire the reality of life it portrays and the extraordinary detailed execution by DaVinci. However, few question its worth as one of the most prized art pieces in the world, even if they do not like it. But why is this so?

This is the nature of design. Design is a subjective endeavor. The design solution one person finds for a problem may be liked by one and disliked by another. There is never a consensus as to whether a particular design is good or bad. Everyone judges a design differently. This is why a particular product (a car, a toaster, or Mona Lisa) is liked by some and disliked by others. Throughout this book, and every time you design something, there will be individuals who will like it and others who will not like it. The same holds true for this book. Some will like this book, some will not. No design can be everything everyone wants. And this is precisely why design is such an exciting adventure. You will never know whether everyone agrees with your decisions or not, and thus, there is a certain mystery in a design that cannot be unfolded until later, when the product is released or the design is finished (regardless of how much market research you perform). There is no safe haven. You will never know whether someone else will do better than you, design a better product or system, and when. But you can imagine that *your* design can be better than anyone else's. The same advantage that others may have over you, you may have over them.

Throughout this book, we will discuss ways to design better. However, we will not be able to present methods on how to design any particular product. That will come from you. Ultimately, the market, your clients, or your teacher will be the judge of your design. What they think about your design will not be known until they are exposed to it. This is the nature of design, and you should expect it, understand it, and live with it. But you should also be open to it, and use it to your advantage.

A fundamental difference between an analytical solution (analysis) and a design (synthesis) is the same point. In analysis, say a dynamics problem, a question is posed which has a clear answer. If you find the same answer as your teacher, your manager, or someone else, then the solution is correct and you get your "reward." If not, you will be appropriately "punished," with the hopes that next time you will do it correctly. In design, however, no single solution exists that will satisfy everyone. Every person will have a different design for the same problem. You may think that you have the best possible solution, but others who judge your design may not agree with you. They may even totally disagree with your assessment. There is also no indication that you will necessarily do better next time. But as mentioned earlier, there is also a good chance that your design may be better than someone else's design (including your teacher's or your manager's design). Mona Lisa is liked by some and disliked by others. Any particular car may be liked by some and hated by others. And yet, we go on designing the best we can. Otherwise, why do you think there have been so many designs for cars throughout the century? And more designs are still coming.

Design is the iterative process of finding a solution for a problem. It is iterative because no solution can be found without evaluations, corrections, modifications, feedback, and implementations. The problem may be of any type, and the solution may take many different forms. A technical problem may have a technical solution: mechanical, electrical, materials related, manufacturing, others, or combinations of these. The problem may be how to spend a fun day in San Francisco. The solution can be a travel plan. The problem may be how to win a war. The solution can be a strategy. The problem may be how to beautify the interior of a house. The solution can be new furniture, moving to a new house, or hiring a decorator. The problem may also be related to a business. The business plan or method of running a business is the solution. Every problem has a solution that is appropriate to its nature. That solution is a design, and finding the solution is a design problem. Planning for a fun day in San Francisco is a design problem; most of the subjects in this book apply to this problem as they do to any other product or system.

In this book, we approach design with a similar attitude. Although many of the examples provided relate to more technical problems, and although many designs are referred to as product or system, any solution is a product or a system, even if of a different nature. The examples mentioned are mostly very common products such as cars or household items that everyone has experience with and can understand. Otherwise, any product, system, or plan may be appropriate.

1.3 WHAT IS A CREATIVE DESIGN?

EXAMPLE 1.1

In October 2001, the U.S. government awarded a contract to an international coalition led by Lockheed-Martin Corporation for $19,000,000,000 (with potential for $200,000,000,000 over its life) to develop the next generation fighter airplanes for the Navy, the Marines, and the Air Force, called the Joint Strike Fighter. It was supposed to have many features, including a vertical take-off and landing (VTOL) capability. The engine was predetermined as a Pratt and Whitney engine.

Both Lockheed-Martin and Boeing had developed proposals and preliminary designs and specifications for the project.

Boeing had taken a common approach to the solution by using air ducts to divert the output thrust of the engine downward and use it for VTOL. To do this with the given engine specifications, they had to move the engine forward of the C.G. (center of gravity) for balance, but the width of the plane became too large, which was bad for stealth (a wider plane can be seen and detected easier). The maximum vertical thrust was barely enough for VTOL.

Lockheed-Martin approached the problem with a new look; they used a drive shaft to run a fan to generate the vertical thrust for VTOL. As a result, they did not have a problem with the size of the plane and the engine location, and could generate as much as 1.5 times as much thrust, which is very good for accelerated VTOL. However, Lockheed-Martin still had to solve reliability problems associated with a new design.

The new approach won them a huge contract. Why? Because it solved the problem with a better solution, yielding better results. This is creative design. It is not necessarily the absolute best design, and it has its own drawbacks too. But it is a better-than-average design. ■

EXAMPLE 1.2

In the 1960s, during the space craze, satellites were being designed for many purposes. One such type of satellite was geosynchronous. *Geosynchronous* satellites are meant to rotate in synch with Earth, always at the same relative position, in order to communicate with a particular antenna. Any change in the position may prove to be unacceptable. However, the problem is that although satellites are in space, there is still a slight amount of particles present in space that create drag, and thus, affect the precise movement of the satellite. There had to be a solution for this.

One commonly considered solution was to design a control system that would measure the precise location of the satellite in space relative to Earth using some sort of sensor, then calculate the error, determine the necessary force or momentum to correct the position, fire a small thruster, and then stop the satellite in correct position. The process would continue forever. Now imagine that you were the designer working on this solution, and you knew everything that one needs to know to do the whole design. Imagine how long it would take you to do this job (conceptual design, calculations, detail design, testing, etc.). Then think about the necessary parts that would need to be manufactured, the sensors, the computers and controllers, the thrusters, and fuel. Also think about the weight of the system, the cost of manufacturing, the cost of sending the system into space, the relative reliability of a system with so many parts, and how many person-hours it would take to design such a device. Finally, think about how much fuel would be required, and what would happen when all the fuel was spent, and although the satellite might be perfectly fine, there would be no more fuel. (In 2008 a stray satellite had to be shut down by a rocket in order to destroy its 1000 lb supply of toxic Hydrazine fuel.)

An alternative solution was to look at the problem in a different way. Engineers and physicists know that if there is absolutely no drag force on an object (in purely empty space), the object will continue with the same velocity forever. In other words, so long as there is no drag, the position of the satellite should remain exactly correct. So why not remove the drag? How? Well, imagine that the satellite is placed inside a bigger sphere and the satellite and the sphere move together (Figure 1.1). In this case, since the satellite is inside the protective sphere, and because they move together, there is actually no drag on the satellite. Thus, it should move perfectly in correct position at all times. The drag is actually on the sphere and not on the satellite, and as a result, it is the sphere that slows down and moves relative to the satellite. All that is needed is to measure the distance between the satellite and the sphere and correct the position of the sphere relative to the satellite. Measuring the relative distance between the satellite and the sphere is a much simpler endeavor to accomplish. The sensing device is much simpler, the control system is much simpler, and since the sphere is lighter than the satellite, it requires much less fuel to move. It is also easier and less time-consuming to design and manufacture the control system. The solution, called "drag-free satellite," is elegant, better fits the problem, and is cheaper, lighter, smaller, and more reliable too.

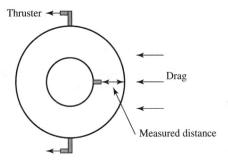

FIGURE 1.1 The concept of drag-free satellite. ∎

When a problem is solved by a solution (design) that fits the problem well, is the best solution possible for the designer, fulfills the needs of the customer (client, user), and is better than an average solution, the design is creative. In other words, a creative design is not necessarily a new invention; it is not necessarily patentable; it is not world shattering. But it is uniquely appropriate for the problem, it is not necessarily the first idea that comes to mind, and it is not an average solution. Can we always find a creative solution? No. There is no guarantee that we will do so every time, or for any given problem. However, creative solutions have been found for countless problems. What we will try to do is learn how to become more creative in our problem solving, and how to design our solutions to the best of our ability. Not only will we look for the most creative and unique solutions for a given problem, we will make sure that each solution has the specifications that are necessary to make it safe, useful, practical, and reliable. A final solution should be creative, but also practical, useful, and safe—that's real creativity.

1.4   INNOVATION VERSUS INVENTION

An innovator is a person who does things differently, who changes a process others use and does it differently, who creates an edge over them, a competitiveness that benefits him or her as well as others. An inventor is a person who finds a way of doing a particular thing that no one else has thought of, that is sufficiently different from others to enable the inventor to obtain the right to exclude others from using it unless they compensate the inventor. These two are related, but not the same. An inventor is an innovator too, since an invention, by its nature, is innovative, novel, and new. But innovations are not necessarily inventions. If you make a table in a form different from others, and because of this difference, create a competitive edge and sell many tables, you are an innovator. But the table is not a new invention.

In a report delivered to President Bush in December 2004[1], the National Innovation Initiative's 21st Century Working Group defined innovation as: "The intersection of invention and insight, leading to the creation of social and economic value." Inventions do the same. Both add value to society, raise the standard of life, and generate new wealth. In this book, we value both inventions and innovations the same. They are both valuable and useful. The goal of this book is to help you be more innovative and more inventive, whether you ever get a patent or not.

We will discuss innovations, entrepreneurship, and benefits to both society and the innovator in Chapter 15. Chapters 2 and 3 relate to inventiveness, creativity, and ideas. The rest of the book relates to the general relationship between these and products we design and use every day.

For a list of 101 innovators who made a difference in America and information about their innovation, see They Made America[2].

1.5   GLOBALIZATION

We live in a global era. Everything in the world is now "connected," both physically and electronically. Not only can we communicate with anyone, anywhere, instantly, through email, cell phones, or the Internet, but we can share data, images, and information as well. You can see the traffic flow at an intersection across the globe from your computer. You can see a person praying at the Western Wall any time of the day. A drawing made in a company in the United States can be manufactured instantly on a rapid prototyping machine in Asia. When you call the information number of a credit card company, you may be speaking with a person in India. The brand of a product is no longer an indication of what it is. The Opel Omega was designed in Germany, was branded as Chevrolet, was made in Australia, and was marketed and sold in Brazil. The German Volkswagen Bug is made in Mexico and sold in the United States, whereas Toyota Camry (like many others) is actually made in the United States. This global life comes with an additional set of requirements with which a designer must cope. Some of these new requirements are:

- Lifelong learning: Graduating from a college is not the end. It is only a beginning of a lifelong learning trip. You should be prepared to continue learning new things and doing new things.

- Communications between teammates, colleagues, and coworkers has changed along with communication among the general population. You may need to communicate with someone across the world who speaks differently, behaves differently, and thinks differently.

- Global technology: a medley of an unprecedented array of possibilities distributed across the globe opens up many possibilities, but also challenges.

- Global markets: where products from all over the world are sold all over the world. Local merchandising is still around, but more as a novelty than as a norm.

- Intellectual property issues are different in other countries: You are not protected unless you have a patent everywhere. Nowadays, there are close to 200 countries. Many of them do not honor patents from other countries and will copy and sell anything.

- Standards can be different in other countries: Your expectations for standards of behavior, success, honor, customs, work ethics, and so on may be different than others. You may be surprised by others people's standards, and you may surprise others with yours.

- Cultures that are as different as people are: The differences in culture are educational, interesting, unique, and honorable. Nevertheless, they are different. You must learn about the culture with which you will deal, but also be open to the differences.

- Teamwork in global environments is somewhat different than local environments. This is driven by the same cultural and behavioral differences that constitute different culture.

1.6 TEAMWORK

The majority of design activities involve teamwork. This is certainly true in most industrial environments and schoolwork. Fundamentally, as we will see in Chapter 3, teams of designers do better with ideas than individuals do. Through their teamwork, as well as by criticizing and complementing each other's thoughts, teams find better solutions. But additionally, technology has become so specialized that it is necessary to combine different expertise in a team in order to integrate many different areas of knowledge. Even planning a wedding requires many "experts." In most cases, on the first day you are in a design class or at work you are assigned to a team.

There are many concerns that must be considered before teammates are assigned to class projects. The following list may be used for assessing team members. Similar characteristics may also be used for professional teams:

- **Interest in a project (based on priority of selection):** People who have an interest in a particular project are more enthusiastic about it and generally perform better. However, in real-world situations, the individual may not have the luxury of choosing an assignment; sometimes you have to work with projects you do not like.

- **Talents and knowledge base (CAD, machining, money management, etc):** A questionnaire helps with assessing this.

- **Background of the individual:** If the individual has experience in dealing with projects, especially the type of project at hand, she or he can be a very effective member of the team.

- **Addresses, whether roommates or members of the same social organizations:** Being roommates facilitates meetings, but it can also create problems.
- **Personality types:** We will discuss this later.

1.6.1 Team Building

A team is as good as its members and their relationship. A team whose members do not trust each other, do not work with each other, are too competitive, and do not communicate properly will fail. In one study[3], the percent of the prizewinning teams of students whose members were matched based on their personalities to complement each other tripled compared to former student teams without matched personalities. Apparently, these teams were superior to the former ones whose members were chosen randomly. There is no doubt that the make-up of a team should be based both on the expertise of the team members and their personalities, complementing each other in more than just their knowledge. Obviously, there is no way to match team members with absolute certainty that they will work well together. However, every attempt must be made to create teams that integrate well, communicate effectively, understand the overall goal of the team, and have a common desire for success of the team.

Parker[4] assigned four styles to team members. They are contributor, collaborator, communicator, and challenger.

- The **contributor** is a task-oriented person who gets things done, has the knowledge and skills to contribute to the team, raises the standard of the team by asking for more and better, completes the assignments, and is the "good old member" of the team on whom everyone can depend. His or her focus is on the details, individual tasks, and assignments.
- The **collaborator** is a goal-oriented person who looks at the big picture, collaborates with everyone, builds relationships, keeps the team focused on the goal, and works behind the scene to make sure the tasks are accomplished. His or her focus is the goal—not individual tasks. She or he is a visionary.
- The **communicator** is a process-oriented person who is the consensus builder and tries to bring the team members together. She or he is an effective listener, a facilitator, a conflict resolution resource person, and a "people person." His or her focus is on the process of achieving the goal, not the goal, and not the tasks and individual assignments.
- The **challenger** is the conscience of the team, the person who criticizes, asks questions, and challenges decisions. This person may disagree with authority, is candid and open, and encourages others to take risks.

Every group needs a balanced range of these roles. I can attest to the same roles in the faculty as well, where different members have different roles they play, and if they work toward the same goals, they will function well and accomplish more.

Varney[5] listed six key factors in building an effective team:

- **Roles:** What is the role of each team member?
- **Clarification of roles:** The assigned roles must be clear to all.
- **Agreed-upon goals:** Everyone should buy-in to the goals of the team.
- **Correct processes and procedures.**

- **Effective interpersonal relations:** Agree on a set of rules of conduct and expectations to prevent future conflicts.
- **Good leadership.**

Parker[4] listed the following characteristics for an effective team:

- **Clear purpose:** The goal of the team should be clear to the team members.
- **Participation:** All team members must contribute their share to the team and accomplish their responsibilities.
- **Listening:** Communication between team members is vital.
- **Civilized disagreement:** Dissent must be allowed for healthy progress toward the goal without fear of retribution.
- **Consensus decision:** When disagreements arise, try to resolve the differences before going on.
- **Open communications.**
- **Clear roles and work assignments:** This creates a sense of order and responsibility.
- **Shared leadership and collegiality:** Shared governance can also help in building trust and consensus.
- **Style diversity:** As in other aspects of life, diversity brings more to the team than uniformity.
- **Self-assessment:** Feedback from the members and attempts to evaluate the team will only improve its work.

Turkman[6] listed the following five stages in the development of a team:

- **Forming:** When the team is formed.
- **Storming:** When disagreements arise, team members get to know each other, and relationships develop.
- **Norming:** When agreements are made, relationships mature and conflicts are settled.
- **Performing:** When the team actually makes progress toward the goal.
- **Adjourning:** When the task is finished and the team disbands.

The following section provides a general overview of the personality types related to team building. These may be used in order to assess individuals' characteristics before assigning members to teams. It requires that each person take a personality test that characterizes the individual into types. If the information is used appropriately, team members may work well together and complement each other in many different ways.

It should be mentioned that there are different types of personality tests. The Meyers-Briggs Type Indicator (MBTI) test is the most common[10]. However, there are many more, including a method based on color preferences, called Luscher Color Test for Personalities[7].

1.6.2 Personality Types

Imagine a person who likes to plan ahead all the details before he embarks on a task, for example, traveling. A person with this personality would tend to arrange for his stay in a hotel or motel in every city, determine how to get to each place, and have tickets for all transportation modes. All schedules would be predetermined accurately, and he would know where he would be on what date, and where he would visit on each day. He would probably read travel books in advance and would figure out what places he would visit, and how much time he would devote to each location. Conversely, imagine a person who does not work with schedules, but prefers to plan "as it comes." Although he might buy a ticket for departure, he would decide where to stay at arrival and how to get there. As he got to the destination, he might talk to a few locals and find out where the best places are to visit, where to eat, how to get to places, and where not to go. If he found the destination exciting, he might stay longer, visit more places, and enjoy the environment. If not, he would quickly move on to the next destination, and depending on the conditions, he would stay there longer or shorter. He would plan each move according to the available resources, the cost, and other variables.

Which one is better? Actually, neither one is better or worse than the other. Each personality has its own advantages and disadvantages. The person who plans all the details will most probably not be stranded in an unknown city, not knowing where to sleep the night. He will not be surprised by the cost of available hotels as he arrives, or whether there is any place available. He will cover all the interesting places he wants to see and will accomplish all the details of his plan. However, his travels may not be too exciting, as there are no surprises. If he goes to a city that turns out to not be so inter-esting, or if the hotel he has reserved is in an undesirable location, or if it turns out to be rainy on the only day he has planned for a specific activity, he will have no flexibil-ity to change things. He is stuck! He will have to stay in an undesirable place because he has already paid for it. He will not be able to visit his desired place during a rain-storm, and he has no choice but to continue with his plans the next day. On the other hand, the person with no plans at all has the flexibility of changing his plans if unex-pected events arise. He will be able to look for different hotels or motels to stay in if the one he finds is not desirable or is too expensive. Depending on whether a particular place has much to offer or is just boring, he can change his plans to stay longer or leave sooner. However, as he arrives at his next destination, he may not be able to find any acceptable place to stay since it happens that there is a large convention or a major bullfight event in the city and all the hotels are sold out. He may end up in an undesir-able place, pay too much for a place to stay, or end up on the street in the cold, rain, or snow.

Now imagine that these two people decide to travel together. There are two possi-bilities. One is for each one to try to force the other to follow his way. In that case, they may be fighting each other all along, the plans may not work, the other person may com-plain about every single situation, and the whole trip may be ruined. On the other hand, they may try to complement each other, by compromising on their ways, and not on their principles, in order to take advantage of the best attributes of each other. In that case, the travel plan may have enough flexibility to be modified, without stranding them in the middle of nowhere. They may find themselves with adequate money, time, and places to visit, and they may return satisfied and fulfilled.

Isn't this what is supposed to happen in a marriage too? Isn't this what is supposed to happen in a team as well? If team members complement each other by compromising for the best attributes of each other, allowing each person who can take care of one need to do so, and cooperating with each other rather than competing with each other, the team may accomplish much more, easier, better, and with more fun. But how are we supposed to do this? What are the personality attributes that are important in this?

A lot of research has been done on profiling people's personalities[8–11]. Most of these are based on standard psychological studies performed by Freud, Jung, and other pioneers in the field. Most profiles are composed of a series of opposing personalities, including the following:

Extraverted	(E)	vs. Introverted	(I)	(where you focus your attention)
Sensing	(S)	vs. iNtuitive	(N)	(how you acquire information)
Thinking	(T)	vs. Feeling	(F)	(how you make decisions)
Judging	(J)	vs. Perceiving	(P)	(how you relate to the outer world)

Extraverted individuals are the type who, among other things, are outgoing, vocal, people oriented, social, talkative, and have many friends. They speak first, then think. They like to deal with people and be noticed. An extravert associate once told me "the worst thing you can do to me is to leave me alone on an island." Extraverts learn best in action and value physical activity, like to study with others, learn better with background sounds, and like faculty who encourage class discussion. They are interested in how others do their jobs.

Introverted individuals are the opposite. They like their own territory, are quiet, think first, then speak, have deeper relationships with fewer people, and are reflective. They like their own privacy, and can work and stay alone for long periods of time. They learn best by pausing to think, prefer reading to talking, prefer studying alone, need concentration to learn, and like faculty who give clear lectures.

Sensing individuals are direct, realistic, factual, and practical. They seek specific information, memorize facts, and follow instructions. They like hands-on instruction, trust material as presented, and like faculty that give clear assignments. They sense their environment, try to understand it as it exists, focus on what works now, and decide based on reality. This attribute relates to the way they collect information.

iNtuitive individuals are more conceptual, future oriented, inspirational, fantasy oriented, and theoretical. They rely on their intuition to guide them in their decision making. They use imagination to go beyond facts, value what is original, and draw conclusions from presented information. They like faculty who encourage independent thinking. This attribute relates to the way they collect information.

Thinkers make up their mind and stick with it, like to follow the rules, are analytical, policy oriented, objective, and just. They like to be treated fairly, but are firm and tough-minded. They like to critique new ideas, easily find flaws in an argument, learn by debating, and like faculty who make logical presentations. They may hurt people's feelings without knowing. They respond to people's ideas, not feelings. This attribute relates to the object-oriented way that they make decisions.

Feelers are subjective, involved, people oriented, and tenderhearted. They express their feelings and can be more emotional than others. Social and personal values are important to them, and they avoid confrontations. They can easily find something to appreciate. They learn by being supported and encouraged. They like faculty who establish a personal relationship with them. They evaluate the effects of their decisions on people's feelings. This attribute relates to the people-oriented way they make decisions.

Judging individuals like to be in control, like to plan ahead, like structures and schedules, and prefer having a deadline, as it will help them set their goals and a schedule to meet it ahead of time. They like formal instructions for solving problems, plan their work well in advance, like to be in charge, and like faculty who are organized.

Perceivers like to wait and see, are more flexible and adapt to the situation as it arises, are open and like open-ended situations, are spontaneous, and do not abide by deadlines, even if they eventually finish at the last minute. They respond to things as they

TABLE 1.1   Sixteen Distinct Personality Types.

E	E	E	E	E	E	E	E	I	I	I	I	I	I	I	I
S	S	S	S	N	N	N	N	S	S	S	S	N	N	N	N
T	T	F	F	T	T	F	F	T	T	F	F	T	T	F	F
J	P	J	P	J	P	J	P	J	P	J	P	J	P	J	P

happen. They value change, work spontaneously, and prefer faculty who are entertaining and inspiring. They feel they never have enough information to make decisions. They tend to keep a list of all the things they want to accomplish in the future.

Each individual has a personality that is a combination of these four pairs. Except in extreme cases, no one is completely one way or another. Rather, one may have more or less of each attribute, even if not equally. In each case, corresponding letters that indicate a preference for each category can describe the personality of the individual. Based on this classification of personalities, it is possible to have up to 16 different types of personalities. For example, an individual described by ENFJ categories has a preference for being extravert, intuitive, feeling, and judging. Table 1.1 shows the 16 distinct possible personality types.

It is vital to recognize that no particular personality type is better than another type. Every personality attribute has its own advantages and disadvantages. You should never brand anyone with a personality, except to understand his or her preferences, and to understand how one may work better with another individual. But the attributes should never be a source for discrimination, ridicule, or harassment.

There are also other types of personality categorizations that differ somewhat from the above-mentioned 16-type personalities. Lumsdaines[9] proposed the four-quadrant thinking preferences A, B, C, D types. Type A is logical, critical, factual, and quantitative; Type B is conservative, structured, sequential, and detailed; Type C is visual, holistic, conceptual, and innovative; and type D is interpersonal, kinesthetic, spiritual, sensory, and feeling. Individuals are categorized based on their preferences (called dominance profiles). Keirsey[11] categories include rationals, idealists, artisans, and guardians, each with its own subcategories. Levesque[12] uses categories such as adventurer, navigator, explorer, visionary, pilot, inventor, harmonizer, and poet, each related to one of the above-mentioned 16 personalities. For example, ISTPs are inventors/adventurers, and INTPs are inventor/explorers. ISFPs are poets/adventurers, and INFPs are poets/explorers. Popular categories, related to the characteristics of our nonhuman friends, include sharks, turtles, teddy bears, tigers, dogs, foxes, eagles, and many more.

It is also popular to assign personality types to groups of people who are in the same or similar profession. For example, most engineers are considered introverted. If college students are about 50 percent extravert, engineering students are 30 percent extravert. If the general population is 70 percent extravert, engineering professors are about 40 percent extravert. In one engineering class assignment, I found 33 guardians, 7 rationals, 10 artisans, 7 idealists, and 1 counselor.

In teamwork, careful assignment of different personalities, as well as simple realization of different personalities of the team members, can be used to complement team members. For example, a team whose members have intuitive abilities and sensing abilities in different individuals, when working together, may complement each other, and as a result, be stronger. Similarly, a team whose members have the attributes of judgers (who work with deadlines) and perceivers (who do not work with deadlines) can complement each other, such that the whole team will set up internal deadlines in order to meet a general deadline. They still may maintain internal flexibility to adapt to situations as they arise. However, in either case, if the team members do not respect each other and believe that another's type is inferior to them, they may experience many disputes among the team members that paralyze the team altogether.

Please refer to references 3, 10, 11, and 13 for available tests in order to determine your personality types. These tests are mostly based on similar principles, but are not entirely the same. The result you get from different tests or from taking the same test more than once or at different times may be different. This shows that as you grow and learn, you may change. It also shows that the result is a function of the type of questions that are asked. This is another reason to always take the personality types with a slight skepticism, and to remember that no one is completely one type. A simple, free test at humanmetrics.com[13] will identify your personality type with corresponding numbers indicating the degree of your preferences.

The following is a simplified version of a method developed by Douglass Wilde, emeritus professor of mechanical engineering at Stanford University. For more information, please refer to his upcoming book on the subject.

1.6.3 Cognitive Modes

Four pairs of "cognitive modes," each combining three out of the eight attributes, may be used in assigning people to teams. They are EN/IN, ES/IS, ET/IT, and EF/IF pairs. Figure 1.2 shows the relationship between these modes and their characteristic attributes.

Assume that you have got a score for your personality in E/I, P/J, S/N, and T/F pairs, for example, through the humanmetrics.com website. Calculate your score for the above four pairs using the following equations:

$$EN = \frac{N}{2} + \frac{E+P}{4} \quad (= -IS) \tag{1.1}$$

$$EF = \frac{F}{2} + \frac{E+J}{4} \quad (= -IT) \tag{1.2}$$

$$ES = \frac{S}{2} + \frac{E+P}{4} \quad (= -IN) \tag{1.3}$$

$$ET = \frac{T}{2} + \frac{E+J}{4} \quad (= -IF) \tag{1.4}$$

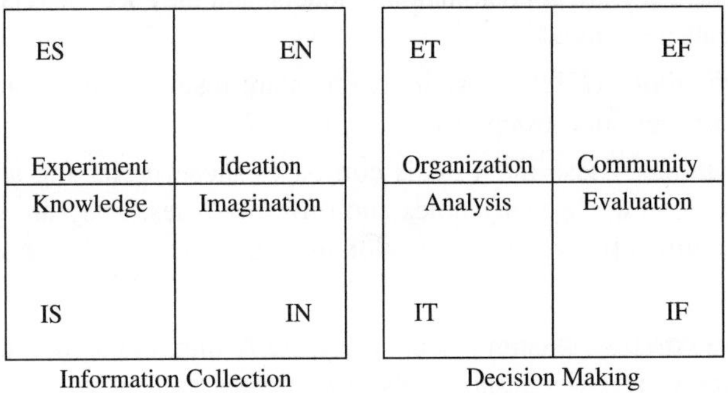

ES	EN	ET	EF
Experiment	Ideation	Organization	Community
Knowledge	Imagination	Analysis	Evaluation
IS	IN	IT	IF
Information Collection		Decision Making	

FIGURE 1.2 Cognitive modes.

In these calculations, if your score is on the I side instead of E, use that for E but with a negative sign. The same is true for the rest. Thus,

If your score is I, use it as $-$ E.

If your score is S, use it as $-$ N.

If your score is T, use it as $-$ F.

If your score is J, use it as $-$ P.

EXAMPLE 1.3

An individual's personality is INFJ with I = 44, N = 38, F = 25, and J = 11. Calculate his cognitive mode values.

Solution Substituting these values into the above equations, with E $= -44$, S $= -38$, T $= -25$, and P $= -11$, we get:

$$EN = \frac{N}{2} + \frac{E+P}{4} = 19 + \frac{-55}{4} = 5$$

$$EF = \frac{F}{2} + \frac{E+J}{4} = 12.5 + \frac{-33}{4} = 4$$

$$ES = \frac{S}{2} + \frac{E+P}{4} = -19 + \frac{-55}{4} = -33$$

$$ET = \frac{T}{2} + \frac{E+J}{4} = -12.5 + \frac{-33}{4} = -21$$

This indicates that the scores for this individual are EN = 5, EF = 4, IN = 33, and IF = 21. ∎

1.6.4 Affinity Groups

Individuals having similar mode scores for some attribute are assembled into one of several "affinity groups." The affinity groups are:

- *Idea individuals (EN):* These individuals are supposedly the ones who are good at generating ideas. Remember that this does not mean others are not good at this task. It also does not mean the "idea individuals" are necessarily good at seeing the idea through implementation. But they have strength in proposing ideas.
- *People individuals (EF):* These individuals are good at keeping the team members together, in forming relationships, and in casting friendships between the team members as well as between the team and others.
- *Action individuals (ES):* These individuals are hands-on experimenters, operators, and model builders. They are good at making models, and they see the relationships through experiments and models.
- *Organization individuals (ET):* These individuals are result driven, deadline oriented, good at schedules, and strong in setting goals.
- *Wild card individuals:* These individuals consist of those preferring any of the other modes. They can fill in many holes and differing roles. They are generally good at complementing others and can function in many roles.

Forming Affinity Groups Assuming that there are N individuals in a class, and that these individuals are to be divided into teams of m members, there will be a total of

$T = N/m$ teams. Obviously, this number must be adjusted such that there will be whole digit numbers of teams, where some of the teams have fewer or more members than m. For example, for a group of 23 individuals, there will be five groups of 4 and one group of 3 ($T = 6$). For a group of 25, there will be five groups of 4 and one group of 5 ($T = 6$), or four groups of 4 and three groups of 3 ($T = 7$). Assuming that the bulk of the teams will have four individuals in it, there should be four affinity groups (EN, EF, ES, and wild cards). Then the class should be divided into four affinity groups with T members in each. So, for the class with 23 individuals, affinity groups will have six individuals each (except the wild-cards with five).

To do this, first list the individuals according to a descending order of their EN scores and pick the T highest scores as idea individuals. Then repeat the same with the remaining individuals for the highest EF scores and pick the T highest scores as people individuals. Then repeat for the highest ES scores among the remaining individuals and pick the T highest scores as action individuals. The remaining individuals are wild cards (whether fewer or more than T). For teams of m members this can also be written as:

- **Idea individuals:** $\dfrac{1}{m}$ top individuals with highest EN scores,

- **People individuals:** $\dfrac{1}{m-1}$ remaining individuals with top EF scores,

- **Action individuals:** $\dfrac{1}{m-2}$ remaining individuals with top ES scores,

- **Wild card individuals:**   the remaining members.

Alternative Approach   You may find out that in many situations, for example in engineering classes, there may not be enough individuals in the first three affinity groups. An alternative method might then be to pick the top EN idea individuals, top EF people individuals, top ES action individuals, and even top ET organization individuals from the whole list, not the remaining. In that case, certain individuals may appear in two or more affinity groups. We will discuss how to assign teams for this alternative in the next section.

1.6.5 Assigning Members to Teams

In general, the size of a team may vary greatly. However, at least in educational environments, many teams are made up of four individual members. The following procedure is also based on a team of four individuals, but it can be adapted to fewer or more members easily. However, with fewer members, there is a chance that not all desired personality attributes are covered. With more, there is a chance that duplicate personalities may exist, sometimes interfering with each other (too many cooks syndrome).

Assuming that teams are made up of four members, four affinity groups should be formed. Each team should ideally have one member from each of these affinity groups. Teams of three individuals should ideally have one member each of the first three affinity groups. Teams of larger size may have additional wild cards. If more than one member is from the same affinity group, it is advisable to ensure that they understand each other's roles and responsibilities in order to avoid conflicts.

Alternatively, if due to the uniformity of attributes in your group, the affinity groups are formed based on the above-mentioned alternative method, an individual may appear in two or more affinity groups. The desire should be to form teams where the four cognitive modes are covered, even if by the same person. So, a team may be composed of an EN, ES, and two wild cards. In fact, if there are not enough individuals to form adequate affinity groups, you may go to the next cognitive mode. We will see an example of this later.

To assign a group of individuals to teams, you may do the following:

- Ask everyone to take a personality test to identify his or her type with related scores.
- Use Equations (1.1) to (1.4) to determine each person's cognitive mode.
- Form affinity groups as mentioned above based on one of the two methods.
- Form teams by selecting one individual from each affinity group. For teams with fewer or more members, select fewer or more members from the wild cards. The best teams will have one individual from each group.

EXAMPLE 1.4

The following is a list of students from one class with their personality scores from Humanmetrics. Column I is the negative of E, etc. After the EN, EF, ES, and ET scores were calculated, the highest six in each group were picked. However, there are not enough high scores in the ES column. Thus, the three individuals with the highest ES scores were included in this affinity group. The remaining individuals were assigned as wildcards. The affinity groups were formed with high EN, EF, and ES/ET scores. Teams had one member from each of these four affinity groups. One team had five members with two wild cards. Alternatively, there can be seven teams, with seven individuals in each affinity group.

	MBTI	E	I	N	S	F	T	P	J	EN	EF	ES	ET	Affinity groups			
1	ENTJ	67	−67	56	−56	−33	33	−33	33	36.5	8.5	−19.5	41.5	x			
2	ENTJ	44	−44	22	−22	−5	5	−1	1	21.8	8.8	−0.3	13.8		x		
3	INTJ	−44	44	22	−22	−56	56	−22	22	−5.5	−33.5	−27.5	22.5			xx	
4	ISTJ	−56	56	−30	30	−56	56	−33	33	−37.3	−33.8	−7.3	22.3				x
5	ESTJ	1	−1	−45	45	−22	22	−22	22	−27.8	−5.3	17.3	16.8			x	
6	INFP	−15	15	6	−6	22	−22	11	−11	2.0	4.5	−4.0	−17.5				x
7	ISTP	−11	11	−11	11	−67	67	56	−56	5.8	−50.3	16.8	16.8			x	
8	ESTJ	33	−33	−1	1	−22	22	−56	56	−6.3	11.3	−5.3	33.3	x			
9	ESTP	56	−56	−30	30	−10	10	44	−44	10.0	−2.0	40.0	8.0			x	
10	ENFJ	1	−1	11	−11	22	−22	−11	11	3.0	14.0	−8.0	−8.0	x			
11	INTP	−22	22	78	−78	−44	44	11	−11	36.3	−30.3	−41.8	13.8	x			
12	ESFJ	33	−33	−40	40	33	−33	−89	89	−34.0	47.0	6.0	14.0	x			
13	ENFP	1	−1	22	−22	11	−11	56	−56	25.3	−8.3	3.3	−19.3	x			
14	INFJ	−22	22	22	−22	11	−11	−22	22	0.0	5.5	−22.0	−5.5				x
15	ENFJ	17	−17	44	−44	33	−33	−33	33	18.0	29.0	−26.0	−4.0	x			
16	ESFJ	22	−22	−28	28	11	−11	−22	22	−14.0	16.5	14.0	5.5	x			
17	INTP	−10	10	33	−33	−30	30	14	−14	17.5	−21.0	−15.5	9.0				x
18	INTJ	−11	11	6	−6	−1	1	−22	22	−5.3	2.3	−11.3	3.3				x
19	ENTJ	11	−11	56	−56	−22	22	−11	11	28.0	−5.5	−28.0	16.5	x			
20	ENTP	11	−11	22	−22	−22	22	34	−34	22.3	−16.8	0.3	5.3	x			
21	ENTJ	1	−1	33	−33	−33	33	−67	67	0.0	0.5	−33.0	33.5			xx	
22	ENTP	17	−17	22	−22	−11	11	24	−24	21.3	−7.3	−0.8	3.8				x
23	ENTP	44	−44	20	−20	−44	44	40	−40	31.0	−21.0	11.0	23.0	x			
24	ENTJ	11	−11	33	−33	−56	56	−33	33	11.0	−17.0	−22.0	39.0			xx	
25	INTJ	−22	22	33	−33	−22	22	−44	44	0.0	−5.5	−33.0	16.5				x

■

EXAMPLE 1.5

The following is a list of students from a class with their personality scores from Humanmetrics. The EN, EF, ES, and ET scores are listed. For this example, we will use the alternative assignment method to form the affinity group. Six individuals are assigned to the EN affinity group with highest scores, followed by six individuals assigned to EF affinity group (from the whole class, not just the remaining), thus individuals 10 and 20 are in both affinity groups. We would have continued to do the same with ES group, but there is really only one individual with a significant positive number. Thus, the third affinity group was formed from highest the ET scores, in which number 22 is in both ET and EN affinity groups. The remaining nine individuals are wild cards. The teams are thus formed by selecting individuals that come from at least three affinity groups. For example, the team with number 22 should include someone from EF and two wild cards.

The advantage of this is that in some cases, the minimum score for teams is raised as a whole and it does not matter which affinity group is formed first.

	MBTI	E	I	N	S	F	T	P	J	EN	EF	ES	ET	Affinity groups		
1	ISFJ	−56	56	−1	1	11	−11	−1	1	−14.8	−8.3	−13.8	−19.3			
2	INFP	−89	89	22	−22	11	−11	56	−56	2.8	−30.8	−19.3	−41.8			
3	INTJ	−11	11	11	−11	−22	22	−56	56	−11.3	0.3	−22.3	22.3			x
4	ISTJ	−67	67	−1	1	−22	22	−33	33	−25.5	−19.5	−24.5	2.5			
5	ISTJ	−67	67	−11	11	−67	67	−44	44	−33.3	−39.3	−22.3	27.8			x
6	ESTJ	1	−1	−22	22	−44	44	−44	44	−21.8	−10.8	0.3	33.3			x
7	INTJ	−22	22	22	−22	−67	67	−33	33	−2.8	−30.8	−24.8	36.3			x
8	INTJ	−56	56	11	−11	−33	33	−56	56	−22.5	−16.5	−33.5	16.5			
9	INFP	−22	22	11	−11	11	−11	11	−11	2.8	−2.8	−8.3	−13.8			
10	ENFJ	56	−56	67	−67	33	−33	−44	44	36.5	41.5	−30.5	8.5	x	x	
11	INFJ	−22	22	22	−22	33	−33	−33	33	−2.8	19.3	−24.8	−13.8		x	
12	ESFJ	1	−1	−11	11	11	−11	−44	44	−16.3	16.8	−5.3	5.8		x	
13	INTP	−44	44	67	−67	−22	22	56	−56	36.5	−36.0	−30.5	−14.0	x		
14	ISFP	−11	11	−22	22	56	−56	11	−11	−11.0	22.5	11.0	−33.5		x	
15	ENFP	22	−22	44	−44	11	−11	11	−11	30.3	8.3	−13.8	−2.8	x		
16	ENFJ	11	−11	11	−11	22	−22	−1	1	8.0	14.0	−3.0	−8.0		x	
17	ESTJ	39	−39	−11	11	−61	61	−67	67	−12.5	−4.0	−1.5	57.0			x
18	ESTJ	22	−22	−1	1	−1	1	−22	22	−0.5	10.5	0.5	11.5			
19	INTJ	−22	22	78	−78	−11	11	−44	44	22.5	0.0	−55.5	11.0	x		
20	ENFJ	44	−44	44	−44	11	−11	−44	44	22.0	27.5	−22.0	16.5	x	x	
21	ENTJ	33	−33	11	−11	−11	11	−17	17	9.5	7.0	−1.5	18.0			
22	ENTJ	11	−11	44	−44	−67	67	−11	11	22.0	−28.0	−22.0	39.0	x		x
23	ENTJ	11	−11	56	−56	−1	1	−44	44	19.8	13.3	−36.3	14.3			
24	INFJ	−44	44	38	−38	25	−25	−11	11	5.3	4.3	−32.8	−20.8			

∎

KOREAN ASSIGNMENT GAME

Koreans play the following simple game for fun and for assigning one set of variables to another set of variables. For example, in Figure 1.3, letters A through E will be assigned to numbers 1 through 5. This game can be played with as many lines as desired and with as many cross members in any format as you wish. The rule is that you start, say with A, and follow down until you get to a cross member. You hop over to the next line that is attached to the cross member. If you have a choice, you always move down. Letter A will end up at one of the numbers which is its assigned number. Next do the same for letter B. Repeat until all letters are assigned. This will always end up with assigning one letter to one number. It can also be used to eliminate individuals until the last person remains.

You may also use this creative, unpredictable, and interesting simple method to randomly assign your team members to individual roles. Substitute letters with team member names and the numbers with team roles.

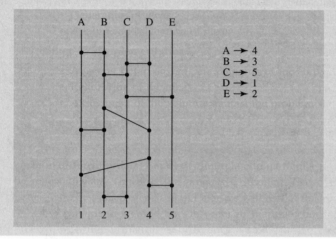

FIGURE 1.3 Korean assignment game.

1.7 THAT'S NOT MY JOB

The following, from an unknown source, summarizes teamwork and personality conflicts in an elegant way:

> This is a story about four people named Everybody, Somebody, Anybody, and Nobody. There was an important job to be done and Everybody was sure that Somebody would do it. Anybody could have done it, but Nobody did it. Somebody got angry about that, because it was Everybody's job. Everybody thought Anybody could do it, but Nobody realized that Everybody would not do it. It ended up that Everybody blamed Somebody when Nobody did what Anybody could have.

1.8 PROJECT SCHEDULING AND MANAGEMENT

In order to meet deadlines, and more importantly, to manage a project, it is necessary to understand the range of activities that is involved, to find out how long each task will take, and to plan ahead in order to meet them and to perform the tasks efficiently and smoothly. For example, imagine that for your project you need to order a motor. Until the motor arrives, you cannot do certain parts of the project. However, even though other parts of the project may still be done, anything that requires the motor has to be delayed until it arrives. If the motor arrives late, the tasks related to it will all be late. If you place the order too late, it may further affect the completion schedule. The planning of the tasks and their relationship is called project scheduling and management.

There are a number of different techniques used for project scheduling and management, including Gantt Charts, Critical Path Method (CPM), and Performance Evaluation and Review Technique (PERT). They allow the designer or manager to predict the lengths of time required to accomplish different tasks or activities within the project, and to control the flow of events. In this section we will briefly discuss these three methods of scheduling and project management.

Commercial programs are available to aid the designer in scheduling and management of the process and may be used wherever available. Learning and using these programs are easy and can be mastered by practice.

1.8.1   Gantt Charts

A Gantt chart is a simple graphical technique of representation of the activities of a project, as shown in Figure 1.4. In a Gantt chart horizontal bars whose length represent the length of time necessary to accomplish each task represent each activity as well as the start and finish dates. The relative position of each event represents the relation between different events. A critical path may be determined from the chart, which indicates the total length that is necessary to finish the project. However, Gantt charts do not show the interdependency of different tasks on each other. To indicate that a task is dependent on another, it is possible to use some convention such as adding a number in front of a task that is dependent on another, as in Figure 1.4. A vertical line may be used to indicate the current date, and filled-in bars may be used to indicate the percentage of the task completed. With these conventions it should be easy to determine whether each task is on time, behind schedule, or ahead of schedule.

In this chart, tasks 1 and 2 can start at the same time, one lasting one week, the other about two weeks. Task 3 cannot start until task 2 has finished (which makes it a critical path). The total length of the project is at the completion of task 6. On the indicated date, task 1 has finished, task 2 is partially done and behind schedule, task 3 has not started since it is dependent on 2, and task 4 is ahead of schedule. Task 5 is on schedule to begin.

Gantt charts are easy to generate, visualize, and follow. They help the manger or the designer keep track of events and required tasks and finish the project on time.

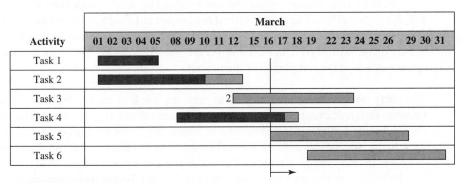

FIGURE 1.4   A Gantt chart.

1.8.2   Critical Path Method (CPM)

The Critical Path Method is also a graphical representation of required tasks, allowing the user to plot the different activities and their relationships on a chart. Activities in a design

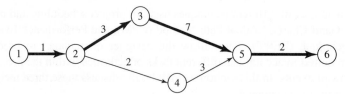

FIGURE 1.5 An example of a CPM chart.

project may range from initial investigation, customer data collection, and concept generation to ordering parts and assembly of subcomponents, to prototyping, testing, redesigning, and delivery. In CPM two alternative methods are used to represent events and tasks (or activities). In the more common method, which we will use for the rest of this section too, a circle represents each event, while arrows or vectors represent activities. The second method is just the opposite, where a circle represents the task and arrows are events. An event can be defined as an instantaneous happening that has no time. For example, the starting of pouring concrete is an event. The task of pouring the concrete, which may take some time to accomplish, is an activity. A number or letter may represent each event. In that case, two numbers or letters, for example, 3-4 or G-H, may represent an activity. Alternately, a single number or letter or name, for example 1, D, or pouring concrete, may represent activities. The length of the arrow may or may not represent the length of the time required to finish the activity. However, the arrangement of the activities represents the logical progression of the activities. The required lengths of activities are generally indicated on the chart, either by numbers or through the length of arrows. Certain activities must be accomplished in the given time without any delay in order to finish other tasks on time and to prevent any delay in finishing the project. The path composed of these critical activities is the *Critical Path*. Other activities may be delayed as time permits without a total delay in finishing the project. Figure 1.5 is an example of a simple CPM chart. The critical path can be indicated by thick lines or by otherwise marking the arrows. After the CPM chart is drawn, it is necessary to do a forward sweep and determine the collection of activities that takes the longest time to accomplish, while considering the relationship between all activities. These activities will constitute the critical path.

In this chart, numbers represent the events. The critical path consists of activities 1-2, 2-3, 3-5, and 5-6. The total length of the project is 13 units. As you can see, activities 2-3 and 3-5 take 10 units of time, while 2-4 and 4-5 require only five units. Thus, 2-3 and 3-5 are on the critical path, but 2-4 and 4-5 are not. They can be floated for five days without any effect on the completion of the project.

Dummy Activities Dummy activities are used to create logical relationships between different events and activities and to remove unnecessary constraints between events. Figure 1.6 shows two situations in which there may be a need for dummy activities. In (a), assume that activity 3-4 is subsequent to 1-3 and 2-3, but 3-5 is subsequent to 2-3 only. In other words, starting 3-5 is not dependent on completion of 1-3, but only on 2-3. Based on the way the chart is drawn, both 3-4 and 3-5 are dependent on the completion of both 1-3 and 2-3, which is not true. To correct this relationship and remove the unnecessary constraint, we can introduce a dummy event 6 and a dummy activity 6-3, shown by a dotted line. This activity does not exist and has no time allotted to it. In this case, we still have the same relationship between 1-3 and 3-4 and between 2-6 and 6-5 as well as the relationship between 2-6 and 3-4. But 6-5 is now independent of 1-3. In (b), there are two activities that have the same start and finish. However, to make it easier to

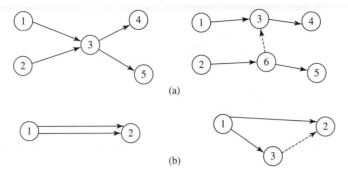

FIGURE 1.6 Application of dummy activities to correct relationships and remove unnecessary constraints between activities.

deal with these two activities, we can introduce a dummy activity 3-2, thus separating the two activities, but still having 1-2 and a similar 1-3.

Earliest Start (ES) Earliest Start time is the earliest time that an activity can start, based on the completion of all other activities to which it is related. The earliest start time for any activity can be found by adding the required length of time of the previous activity to its ES time. If there are parallel paths to an activity, the earliest start time should be calculated based on the longest path. In Figure 1.5, repeated here, the earliest start time for 2-4 is one unit and the earliest start time for 4-5 is three units. So while 4-5 may be accomplished by the end of six units, the earliest start time for 5-6 is still 11 units.

Earliest Finish (EF) Earliest Finish is the earliest time that an activity may end in relation with all prior activities. The earliest finish time is equal to the earliest start time plus the required length of the activity.

Latest Start (LS) Latest Start is the latest an activity may start and still not postpone the final completion date of the project. Latest start can be used in figuring out whether an activity can be postponed without affecting the final completion date, and is important in resource management or in emergencies. We will have a short discussion of this later. Latest start is found by tracing back from the end of the project until the starting event, making sure that the path includes the activity as well. If there are parallel paths, the longest path (critical path) should be chosen. For example, the latest starting time for activity 4-5 in Figure 1.5 is 8, although it could start as early as three units.

Latest Finish (LF) Latest finish is the latest time that an activity may be accomplished without changing the completion date of the project. In Figure 1.5, the latest finish for activity 4-5 is 11. LF may be found by adding the activity time to the latest start.

Total Floating Time (TF) Floating time is the time that an activity may be delayed without delaying the completion date of the project. It is equal to the difference

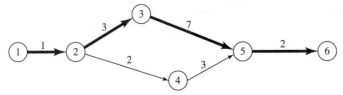

FIGURE 1.5 (repeated).

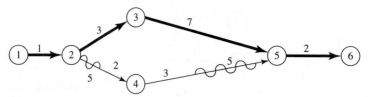

FIGURE 1.7 Representation of the total floating time.

between the latest starting time and the earliest starting time, or the latest finish and earliest finish:

$$TF = LS - ES = LF - EF \tag{1.5}$$

Different conventions may be used to represent TF. Figure 1.7 is one example, in which a wiggly line is drawn on each noncritical activity with the total TF written on it. In this example, each of the two activities 2-4 and 4-5 has the same total floating time of five units, but not both. If three units of floating time are used to postpone activity 2-4, only two units are left to postpone activity 4-5. This is important to remember in order to ensure that the total floating time is not assigned more than once.

Representation of Information on the Chart   The above-mentioned information can also be displayed on the chart in a variety of forms, some better than others. Figure 1.8 shows one technique that is less confusing, but does add to the value of the chart with additional information. Unfortunately, when the chart is very large and complicated, the additional information, although useful, can clutter the chart and become confusing. Obviously, the activities that are on the critical path will have the same ES and LS times.

Determination of the Critical Path   The above method of representation of the information on a CPM chart can be used to determine the critical path. All activities that have the same earliest and latest start constitute the critical path. A similar method is shown in Table 1.2, in which the same information is tabulated. Similarly, all the activities that have the same ES and LS are on the critical path.

Resource Allocation and Management by CPM   There are times when the total required resources to perform the necessary tasks at different stages of a project are not the same. This means that at one time, there may be a need for many individuals to carry on the tasks, while shortly afterwards, there may be a need for only a few individuals. These variations and fluctuations in required resources create many difficulties. On the one hand, the employer may opt to hire the maximum needed number of employees to

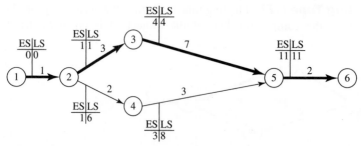

FIGURE 1.8 Representation of additional information on the CPM chart.

TABLE 1.2   Application of a Table to Determine the Critical Path. All Activities with the Same ES and LS Are on the Critical Path.

Activity Name	Activity Designation	Duration	ES	LS	EF	LF	TF	Critical Path
	1-2	1	0	0	1	1	0	*
	2-3	3	1	1	4	4	0	*
	2-4	2	1	6	3	8	5	
	3-5	7	4	4	11	11	0	*
	4-5	3	3	8	6	11	5	
	5-6	2	11	11	13	13	0	*

do the job, but may end up having to pay them unnecessarily during the times when fewer are needed. If the employer opts to hire the minimum number required and tries to hire additional employees during the times more people are needed, basically as temporary employees, then he may not get the quality people he needs, or may not even find enough part timers. The same is true for other resources such as machinery and space.

An alternative solution is to try and optimize the resources available by minimizing the fluctuations. For example, if there are activities that require more resources but have floating time, the task may be delayed or may be spread out to a longer duration in order to spread the required resources. It may also be possible to move the available resources between jobs in order to minimize the fluctuations.

EXAMPLE 1.6

Figure 1.9 shows how the resources were divided differently to reduce fluctuations in resource requirements. ■

1.8.3 Performance Evaluation and Review Technique (PERT)

PERT is generally used with projects that are done for the first time, such as design and development or research projects, where the necessary length of time to accomplish each

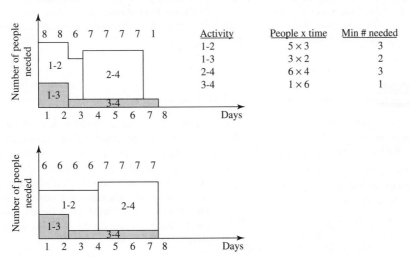

FIGURE 1.9   Resource management through reduction in fluctuations.

activity is not known. In this case, it becomes necessary to somehow estimate the time that may be necessary to finish each activity. Except for estimation of time, PERT and CPM are essentially similar.

Estimated times necessary to accomplish new tasks are calculated from three other estimated times. These are based on a bell-shaped distribution curve and are called:

1. **Most Optimistic Time (T_o):** The time that a task may be accomplished if everything goes well. The probability of the time being any shorter is very low.

2. **Most Probable Time (T_m):** An estimated mean time. This represents the time in which the activity will most probably be finished.

3. **Most Pessimistic Time (T_p):** The time that an activity will take if everything goes wrong. The probability of the time being any longer is very low.

The estimated time is calculated from

$$T_e = \frac{T_o + 4T_m + T_p}{6} \tag{1.6}$$

T_e represents the time that may be safely used in the chart.

REFERENCES

1. "Innovate America," National Innovation Initiative Summit and Report, Council on Competitiveness, 2005, pp. 8.
2. EVANS, HAROLD, "They Made America," Little, Brown, and Co., New York, 2004, pp. 467–472.
3. WILDE, DOUGLASS J., "Team Creativity," Proceedings of the 8th Annual National Collegiate Innovators and Inventors Alliance Conference, San Jose, March 2004, pp. 77–80.
4. PARKER, GLENN M., "Team Players and Teamwork: The New Competitive Strategy," Jossey-Bass Publishers, San Francisco, 1989.
5. VARNEY, GLENN H., "Building Productive Teams," Jossey-Bass Publishers, San Francisco, 1989.
6. TURKMAN, B., "Developmental Sequence in Small Groups," *Psychological Bulletin*, *63*, 1965, pp. 384–399.
7. www.colorquiz.com.
8. KROEGER, OTTO, with JANET THUESEN, "Type Talk at Work; How the 16 Personality Types Determine Your Success on the Job," Tilden Press Books, Bantam Doubleday Dell Publishing Group, Inc., New York, 1992.
9. LUMSDAINE, EDWARD and MONIKA LUMSDAINE, "Creative Problem Solving; Thinking Skills for a Changing World," 2nd Edition, McGraw Hill, Inc., New York, 1989.
10. Myers-Briggs Type Indicator tests, Consulting Psychologists Press, Inc., Palo Alto, California.
11. The Four Temperaments, http://keirsey.com.
12. LEVESQUE, LYNNE, "Breakthrough Creativity," Davies-Black Publishing, Palo Alto, 2001.
13. http://humanmetrics.com/cgi-win/jtypes2.asp.

HOMEWORK

1.1 Determine your personality type by taking at least two different tests provided above. Compare the results between the two. Are they reasonably close?

1.2 Compare the results of your personality test with the results of a similar test taken by a trusted friend. Do the results match your personalities?

1.3 Repeat the same personality test you took a while ago and compare the results. Have the indications about your personality changed? If so, why do you think they have changed?

1.4 When you are assigned to a team for the first time, try to get together and write down a set of bylaws about your group. The by-laws should reflect the nature of your group, the goals, expectations, meeting times, and methods of conflict resolution. If you have not already taken personality tests, do so. Then compare the personalities (anonymously) and see if the make-up of the team is appropriately diverse yet cohesive.

1.5 The following are the MBTI scores for a class. Fill out the table, calculate cognitive mode values, form affinity groups, and form six teams.

	MBTI	E	I	N	S	F	T	P	J
1	ENTJ	67		56		−33		−33	
2	ENTP	44		22		−5		10	
3	ENTJ	44		22		−56		−22	
4	ISFJ	−56		−30		50		−33	
5	ESTJ	10		−45		−30		−22	
6	INFP	−18		6		22		11	
7	ESTP	45		−11		−34		56	
8	ESTJ	32		−1		−22		−56	
9	ESTP	56		−36		−32		23	
10	ENFJ	10		11		18		−11	
11	INTP	−22		70		−44		11	
12	ESFP	33		−40		22		89	
13	ENFP	1		22		33		56	
14	INFJ	−22		22		45		−22	
15	ENFJ	17		44		33		−33	
16	ESFJ	22		−28		8		−22	
17	INTP	−10		54		−30		14	
18	INTJ	−20		6		−1		−22	
19	ENTJ	11		56		−22		−11	
20	ENFP	11		24		25		34	
21	ENTJ	24		33		−33		−67	
22	ENFP	17		22		12		24	
23	ENTP	14		20		−44		40	

1.6 Assign the letters A through F to the numbers based on the Korean assignment game.

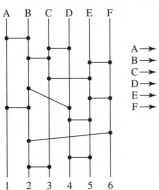

1.7 A class project requires the following steps, each with the associated projected time. The total allotted time for the term is 10 weeks. Select a reasonable start and end time for each task. Draw a Gantt chart for the process.

Find problem	1 week
Define problem	1.5 weeks
Meet with client, finalize definition	2 weeks
Brainstorming, idea generation	1.5 weeks
Selection of final idea	2 weeks
Interim report	Due end of 4th week
Prototyping and testing	5 weeks
Solid model generation	3 weeks
Detail design, material selection	4 weeks
Final report outline	Due end of 7th week
Second interim report	Due end of 8th week
Final report preparation	2 weeks
Final presentation preparation	All along
Final presentation	Due end of 10th week

1.8 You have received an order for your new table design due in four weeks. A total of five tables are ordered. Since this is your second job, you can only devote up to three hours per day to do it. The following must be done. Assign an order to the tasks, and plot a CPM for it. Identify the critical path, ES, LS, EF, LF, and TF for each task.

Design the details	5 hours
Purchase wood	3 hours
Purchase hardware	2 hours
Cut the wood	10 hours
Sanding, drilling, preparation	10 hours
Staining, varnishing (at least three coats)	20 hours
Assembly	5 hours

1.9 You are assigned a new class project (14 weeks total) that requires the following steps. Use a PERT chart to determine how the project could reasonably be managed and how long you might need to accomplish each task:

Preliminary design

Concept development

Testing

Concept refinement

Final presentation

CREATIVITY AND DESIGN

Part One
CREATIVITY AND DESIGN

CHAPTER 2

CREATIVE MIND

The human mind, once stretched by a new idea, never goes back to its original dimensions.

—Oliver Wendell Holmes

A person goes to a brain doctor. After examining him, the doctor says "I have good news and bad news for you. The good news is that like everyone else, you also have a left brain and a right brain. The bad news is that in your left brain, there is nothing right and in your right brain there is nothing left".

—From the Internet, source unknown

So what are the left brain and the right brain?

2.1 INTRODUCTION

A lot of research has been done in order to understand how the brain functions, both physiologically and psychologically[1-10]. What follows is a very simplified description of some simple experiments that show how the two hemispheres of the brain relate to each other and how they perform their own functions independently. These tests are only mentioned here in order to demonstrate the ways we train our brain as it concerns creativity, and how the brain may be retrained to be more creative. There is no intention here to be exhaustive in this presentation or to teach physiology of the human brain.

The human brain consists primarily of two hemispheres, the left and the right. Each hemisphere has a distinct set of areas in it that are dedicated to particular tasks. The left and the right hemispheres of the brain are connected to each other via a thick bundle of white matter (nerves) called corpus callosum (Figure 2.1). Through this bundle, the two hemispheres share information so effectively that we do not even know that the two hemispheres are, in fact, independent.

One particular disease of the human brain is epilepsy, which is caused by the overactivity of the irritated brain cells, causing partial or complete seizures. As a result, the patient may suddenly lose complete control, may be unaware of the environment, may fall, and may violently shake. A very dangerous situation is when this happens while driving, handling machinery, or other dangerous situations. Many patients can be helped by medicines that control epilepsy. However, in severe epilepsy, one method of controlling the seizures is to cut the corpus callosum, the thick bundle of nerves that connects the left and the right hemispheres. Apparently, this allows the patient to continue functioning at an acceptable minimum level even during an attack, because epilepsy generally encompasses only one side of the brain. Thus, by cutting the corpus callosum, at least one side of the brain continues to function even during an attack. Interestingly, although this operation sounds very invasive, except in particular situations, the patient continues to function almost normally after the surgery. By testing the patients under these exceptional

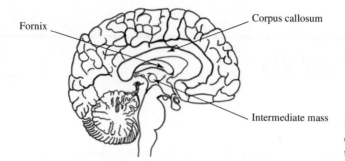

Fornix

Corpus callosum

Intermediate mass

FIGURE 2.1 A simplified cross-sectional drawing of the human brain.

situations, the researchers have discovered many interesting facts about how the brain functions and where the different functions reside in the brain.

The human brain controls the body in a crossover fashion. This means that the left brain controls the right side of the body, and the right brain controls the left side of the body (Figure 2.2). As a result, if the left side of the brain is damaged due to an accident, stroke, or other trauma, the right side of the body will suffer the consequences (paralysis, loss of motion and control, etc.). This has led to the development of many theories about the dominance and the role of each hemisphere. As an example, one theory states that in right-handed people the left-brain is dominant and vice versa (although this in itself is not important to our discussion). What has become clear is that the brain has particular zones that are dedicated to particular functions. For example, there is a section of the brain that is specialized in the development of speech (speech center). In many people, the speech center is in the left side of the brain. If, say due to a stroke, the left-brain is traumatized, the right side of the body may become paralyzed and the person may lose ability to speak

FIGURE 2.2 The two hemispheres of the brain control the left and right sides of the body in a crossover fashion.

well. If the trauma is in the right side, the left side of the body may become paralyzed, but the speech may not be affected.

EXERCISE

2.1. While sitting at your desk, lift one foot off the floor and make clockwise circles. While doing this, draw the number 6 in the air with your opposite-side hand. Your foot will change direction.

Interestingly, the left and the right sides of each eye (nasal and temporal) send their visual information to the same sides of the brain, namely the left hemisphere and the right hemisphere. However the image that forms on the retina passes through the lens, and thus is reversed. As a result, effectively, the left portion of the visual field is transmitted to the right hemisphere and the right side of the visual field is transmitted to the left hemisphere (Figure 2.3). Normally, since the two hemispheres constantly share the information, we do not see the image as two distinct parts. In split-brain patients, the information is not shared, and thus, each hemisphere has its own distinct information that is not known by the other hemisphere[5,6,8].

Through an extensive set of experiments under specific conditions, split-brain patients have been tested in order to understand the way the two hemispheres of the brain work and to understand the strength and weaknesses of each one in performing different kinds of tasks. The explanation of these experiments is beyond the scope of this book. However, the following simple examples are mentioned here to demonstrate the methods used, as well as how the brain functions. For a fascinating full description of these experiments, please refer to the references cited.

Michael Gazzaniga in "The Bisected Brain"[9] describes an experiment that was used to demonstrate the way the two hemispheres process information. In this experiment, visual stimuli are presented to the person by quickly flashing a picture into either the left visual field or the right visual field. Pictures flashed into the right field will only be received by the left hemisphere, while the pictures flashed into the left field will be received only by the right hemisphere. Neither of the two hemispheres would have any information about what the other had received. Objects matching the pictures were placed on a table under cover and out of view (Figure 2.4). Thus, when patients were asked to match the objects they had seen, they could touch the objects, out of view, without being able to see them.

FIGURE 2.3 Optical separation of the two split-brain hemispheres. The nasal part of the right eye and the temporal part of the left eye transmit their visual images to the left hemisphere, while the nasal part of the left eye and the temporal part of the right eye transmit the visual image to the right hemisphere. With central fixation of the eyes, the two split-brain hemispheres will receive the image in each visual field separately.

FIGURE 2.4   Set-up of an experiment to demonstrate the interrelationship between the two hemispheres of the split-brain persons.

Now imagine that the picture of an orange is flashed into the left hemisphere (by projecting the picture into the right visual field). If asked to match the object with the right hand (controlled by the left brain), the patient could easily retrieve an orange from a collection of objects without seeing them. Obviously, since the picture was seen by the left brain, it could use the information through the right hand to retrieve the orange. If asked what the object was, since in most people the speech center is in the left hemisphere as well, the patients could also confirm verbally that the object was an orange. However, if asked to retrieve or match the object by the left hand, the patient would not be able to do so.

Similarly, if the picture of a spoon was projected to the left visual field, and thus only seen by the right hemisphere, the patient would claim he did not see anything (since the dominant, language-strong left brain had correctly not seen anything); but with the left hand, he would retrieve and match the object. If asked what was retrieved, he would reply he did not know.

Ordinarily a split-brain patient can only verbalize information that is received in the left brain (for example, from the right visual field). However, some patients have demonstrated otherwise. In another experiment[8], a picture of a fork was shown to the left hemisphere and a key shown to the right (Figure 2.5). Split-brain patients would normally be unable to tell whether what they saw were the same or different. This is similar to

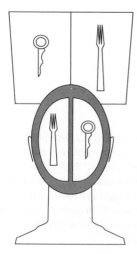

FIGURE 2.5   In patients with capacity to speak out of both hemispheres, stimulus such as this can easily be named because each hemisphere can describe what it has seen. However, if the patients are asked to *not* name the objects, but to judge whether they are the same or not, they are unable to do so because neither hemisphere is aware of what the other one has seen unless it is verbalized.

showing a fork to one person and a key to another. Unless they talk to each other, they would not know whether what they saw were the same objects or not.

However, some patients can verbalize from both sides of the brain, and thus, can mention both the fork and the key. Still, if they are asked to not name the objects, but to say whether the objects are the same or not, they cannot do so. Since some individuals can make speech with both sides, they can verbalize information from both hemispheres, and thus can tell what they have seen. If the hemispheres hear each other, the patient will be able to compare the information. Otherwise, without the verbal response, the hemispheres have no basis to compare their information, and thus, the patient cannot tell whether the objects are the same or not.

In yet another experiment, the researchers were able to demonstrate the domination of the right hemisphere in performing spatial tasks. In this experiment, a patient was given several wooden parts and was asked to arrange them to match a particular design. His attempts to do the job by the left-hemisphere-controlled right hand failed repeatedly. When his right hemisphere tried to help by the left hand, the dominant right hand would knock it away; eventually he sat on the left hand to prevent it from interfering. When he was told to do the task by both hands, the spatially smarter left hand had to shove away the less articulate right hand to keep it from interfering.

Through these experiments, it has become evident that there are sections of the brain that specialize in certain functions, and that the brain is made up of modules that process certain stimuli with certain output. It has also become known that the two hemispheres are, in fact, separate processing units that work independently, but share information so thoroughly that we do not feel that there are two brains at work. Each one processes the information received in its own way. Each one has its own particular strengths and weaknesses, and each one communicates in its own way.

Gardner[8] writes:

> We find, then, an emerging consensus about brain localization. The brain can be divided into specific regions, with each emerging as relatively more important for certain tasks, relatively less important for others. Not all or none, by any means: but with definite gradients of importance. In the same vein, few tasks depend entirely on one region of the brain. Instead, once one examines any reasonably complex tasks, one discovers input from a number of cerebral regions, each making a characteristic contribution. For example, in the case of freehand drawing, certain left-hemisphere structures prove crucial for the providing of details, while right-hemisphere structures are equally necessary for the mastery of the overall contour of the depicted object. Compromise in either half of the brain will result in some impairment, but the kind of impairment can only be anticipated once one knows where the brain injury has occurred.

2.2 WHOLE-BRAIN THINKING

The subject of left- and right-brain was discussed here to demonstrate the fact that we have two distinct processing units that are not similar. One hemisphere is dominant, verbal, logical, linear, and analytic. It is good at reasoning, relating to facts, making conclusions, and drawing upon these to make new assumptions to continue with the processing of the next step. It works in sequential steps based on previous knowledge, methods, and conclusions. It is also good with numbers and facts. It is a one-at-a-time processor, focused on details, and governed by rules. It knows how to do things. It is also a slow thinker. But the other hemisphere is intuitive, imaginative, nonverbal (or verbally weak), spatial, and nonlinear. It can jump from one issue or fact to another in lateral directions, connecting the issues in new ways that may initially seem unrelated. It discovers how to do things. It knows instantly. And it is an all-at-once processor. It can recognize patterns and similarities, and is spatially agile. Since the majority of people are right-handed, and since the left hemisphere controls the right side of the body, it is common to refer to the left brain as the dominant hemisphere. Whether the left or the right hemisphere is the dominant one in right-handed or left-handed persons is not important. What is important is that we tend to nurture the logical, verbal, analytical hemisphere much more than the other one. We tend to favor the side that is logical and can compare information logically and draw conclusions; the side that can verbalize our thoughts; the side that can make decisions based on facts. However, the other side has its own strengths too. It is much more spatial than the dominant side. This means that this hemisphere can perceive spatial relationships much better than the dominant hemisphere. It is also nonlinear; it can draw conclusions without necessarily relying on previous information or facts. As a result, it can "create" new ideas, music, and relationships that are not based on a priori knowledge or facts. Still, through common educational processes used in schools and at homes, where appropriate behavior is rewarded and inappropriate behavior is punished, we strengthen our dominant "left" brain in favor of the *creative, different,* and *wild* "right" hemisphere. In short, we mostly ignore the strengths that our other hemisphere has by not educating it, using it, or relying on it. This is even more prevalent in engineering education, where logic, facts, numbers, and methods are heavily emphasized. As a result, most engineers and engineering students tend to shy away from arts, drawing, colors, and other issues that are not emphasized in their education. This is also emphasized culturally, where the right and the left have their own associated meanings in politics, in alignments and friendships, and in justice. Just think of the word leftist. In fact, the same is true not only in English, but in other languages as well. For example, in Persian, crossed eyes are referred to as left-eyes; in French the left is *gauche,* meaning awkward.

Based on this theory, it is easy to see that by accessing the other hemisphere and nurturing it, we can improve the way our brain functions. Most individuals can remember that as children, they were more creative and could freely associate different facts with each other better than when they became older. In most cases, companies show interest in bringing their projects to engineering schools and having students work on the projects because they are interested in fresh ideas, not limited by the more mature and seasoned engineers who have been trained by the company to behave in a particular way, or to look for standard methods of solving problems. In fact, many instructors agree that seniors are in general less creative than their first-year counterparts. This is because education in colleges is also based on the same reward–punishment system used in schools and homes, where the left brain is favored and dominating. We look for logical answers. We look for verbal descriptions. We want our students to follow procedures. Although these are necessary behavioral patterns for engineers and designers, there is no reason why the right brain should not be nurtured and used as well. Then, it is safe to assume that if we employ the right brain more fully and engage it in our design activities and behavior, and if we nurture it to become stronger, then we should be able to use it more effectively too. Thus, we can assume that no matter how creative one is at any given time, by trying to be more imaginative and creative and intuitive, and by trying to learn how to engage the right brain, the individual may become more creative than before. There is no claim that everyone will be as creative as another, but we can assume that everyone may become more creative if she or he tries to use both sides of the brain; the whole brain has got to be better than half the brain.

2.3   CREATIVITY

So, what is creativity? Is it a gift we have when we are born? Is it a talent? Is it more an attitude toward openness and interest in change, or is it a subtle power? Is it an ability that can be exercised and improved or a trait? Is it learned? Can it be taught? To answer these questions, you would have to read perhaps thousands of books, articles, and journal papers, and still, you would neither find a definitive answer nor a consistent definition about it. In fact, we will need thousands of pages to cover the philosophy and physiology of creativity, well beyond the purpose of this book. However, we do know that it exists, that different people have certain creative capabilities, we know they lose a lot of it as they grow up, that they can gain it again (at least partially), that attitude is an important part of it, and that it is content oriented. Creativity in athletics means a different thing than in engineering or music or business. However, the traits, capabilities, and the fundamentals of a creative behavior are pretty much similar regardless of the content. We also know that it can be cultivated even if not really taught.

An interesting way to try and understand what creativity means to you is to think about its opposite. So, what is the opposite of creative or creativity? Routine? Boring? Not interesting? Old fashioned? The same? Useless? Your definition of the lack of creativity will tell you what creativity is.

When you do something creative, it may be creative to you, even if not to others. What I mean is that you may do something that feels creative to you, but others may have done the same thing before. So for the whole society, it may not be creative. But for your own personal circle, it is creative. This means that creativity may be relative. Your creativity does not have to be at a world record, the ''big league'' level, to be creative. It may be at the ''little league'' level and still be very respectful and appreciated.

EXERCISE

2.2. List the five most creative people you can think of. Then list what they have done that makes them your favorite. Can you identify a common behavioral pattern among them? Would you imitate their behavior if you thought it would make you more creative?

2.4   WHAT MAKES AN INDIVIDUAL CREATIVE?

As was mentioned earlier, creativity is not a trait inherited like eye or hair color, height, or physical looks. Creativity is a function of the brain, and as all other brain functions, it is a natural phenomenon. Everyone is creative when born, but it is estimated that by ages 5–7, most children have lost $\frac{3}{4}$ of their creative abilities. By age 40, most adults demonstrate less than a few percent of their original creative abilities. Thus, it is essential for most adults to reawaken this ability to think creatively. The question is whether it is possible to do so or not.

Torrance, one of the authors of ''The Nature of Creativity: Contemporary Psychological Perspectives,''[11] lists creativity as wanting to know, digging deeper, looking twice, listening for smells (and this is not an error), listening to a cat, crossing out mistakes, getting in, or getting out, cutting holes to see through, cutting corners, singing in your own key, and shaking hands with tomorrow. In fact, each one of these characteristics has deep meanings. Listening to smells indicates that a creative person does things differently, looks for relationships that are out of the ordinary and unusual, relationships that are not common. But it also means that the person uses his or her senses effectively. Crossing out mistakes means that a creative person is not afraid of making mistakes. Mistakes teach him or her. Shaking hands with the future relates also to the fact that a creative person is also an innovator, a person who creates new things, and by doing so, makes friends with the future. We will see much more about these traits in the following chapters.

Similarly, Barron, another contributor to ''The Nature of Creativity: Contemporary Psychological Perspectives,''[11] lists the following as necessary ingredients for creativity: Recognizing patterns; making connections, taking risks, challenging assumptions; taking advantage of chance, seeing in a new way. We will also discuss these in much depth in Chapter 3.

Richards[12] mentions that creative people pay attention to their world, see things differently, challenge assumptions, take risks, are not afraid to fail, and strive to generate multiple solutions to problems. They are passionate about creativity (attitude) and seek opportunities to innovate. We will also discuss these in Chapter 3.

Levesque[13] attributes creativity to all the categories of personalities we discussed in Chapter 1. He asserts that everyone has some form of creativity regardless of the type of personality. Obviously, this is good news. In fact, in a team situation, these different creative abilities can complement each other as well.

According to Eiffert,[14] to reawaken your creativity you must take conscious steps toward a creative mindset. Foremost, you must *want to* be more creative and be open to acting more creatively. Rogers[15] speaks of required internal and external conditions to flourish creativity in individuals. These conditions include:

(a) Openness to experience, flexibility, and receptiveness to changes in conditions and situations.

(b) Self-respect for individual acts of creativity, not looking to others for approval.

(c) Ability to play with concepts and elements, to see relationships between seemingly unrelated elements and concepts, ability to combine them in new ways.

(d) Self-respect for your feelings and abilities.

The emphasis of this book on this subject is concentrated on parts (a) and (c) above. To reawaken our creativity, we will discuss techniques that help you become more receptive to methods that are not as familiar, to be open to different approaches, and to develop the ability to toy with ideas and elements. Certain methods used in design are based on these requirements. For example, we will discuss methods that help you merge many different alternative solutions for each element of a bigger problem into new combinations, thus helping you find new solutions. We will also use puzzles that will point out how we may have lost our creative abilities, but will also open our eyes to new experiences and ways of finding more creative solutions. These discussions will help us realize that there are always other ways of doing things, sometimes better ways than we have thought of. In this country, almost all people tie their shoes the same way; but have you realized that there are other ways of tying your shoelaces that are commonly used in other countries? The same is true for many other simple daily activities, as well as common ways of doing other things.

Over the years, I have heard many teachers express doubt about teaching creativity while others swear by it. So, can creativity be taught, or learned? Most psychologists and educators believe that it can be. The point is that without exception, in every situation, the individuals who love to be more creative, the ones who enjoy creative thoughts and come to class with a positive attitude, do change their behavior and do become more creative. The most important element in it, as mentioned above, is the attitude of the person. After decades of teaching this material to my own students, I firmly believe that the students who want to be more creative, who feel comfortable with change, who are open to new ideas, and who want to be better designers, learn more, become more creative, and design much better. Many who find the material useless and unnecessary are the students who actually need it most; they are not creative, they do not want to change, and they go on doing the same thing for every problem they encounter.

One way to accomplish this is to first identify what holds us back, and in what ways. Then, with the right attitude, we can learn ways to overcome these detriments and win back our lost creativity. This is what we will do in the following chapters.

In the following chapters we will discuss mental processes that result in loss of creative thinking and behavior and the ability to solve problems creatively, as well as methods to overcome these mental barriers.

Please see the additional references (16–23) enumerated at the end of this chapter for more information on the subject of teaching creativity and what it means.

REFERENCES

1. BAYNES, K., J. C. ELIASSEN, H. L. LUTSEP, and M. S. GAZZANIGA, "Interhemispheric Integration Masks Modular Organization of Cognitive Systems," *Science*, 1998, Vol. 280, pp. 902–905.
2. NEAL, E., A. KROLL, A. P. YONELINAS, M. M. KISHIYAMA, KATHLEEN BAYNES, ROBERT T. KNIGHT, and MICHAEL S. GAZZANIGA, "The Neural Substrates of Visual Implicit Memory: Do the Two Hemispheres Play Different Roles?," *Journal of Cognitive Neuroscience*, August 2003, Vol. 15, Issue 6.
3. LONG, DEBRA L., and K. BAYNES, "Discourse Representation in the Two Cerebral Hemispheres," *Journal of Cognitive Neuroscience*, February 2002, Vol. 14, Issue 2.
4. AMISHI, P. JHA, E. NEAL A. KROLL, KATHLEEN BAYNES, and MICHAEL S. GAZZANIGA, "Memory Encoding Following Complete Callosotomy," *Journal of Cognitive Neuroscience*, January 1997, Vol. 9, Issue 1.
5. EDWARDS, BETTY, "Drawing on the Right Side of the Brain; A Course in Enhancing Creativity and Artistic Confidence," J. P. Tarcher, Inc., Los Angeles, 1979.
6. ORNSTEIN, ROBERT, "The Psychology of Consciousness," W. H. Freeman and Company, San Francisco, 1972.
7. GARDNER, HOWARD, "Frames of Mind: The Theory of Multiple Intelligences," Basic Books, N.Y., 2004.
8. GAZZANIGA, MICHAEL S. "The Social Brain; Discovering the Networks of the Mind," Basic Books, New York, 1985.
9. GAZZANIGA, MICHAEL S., "The Bisected Brain," Appleton-Century-Crofts Educational Division, New York, 1970.
10. BENSON, FRANK D., and ERAN ZAIDEL, Editors, "The Dual Brain, Hemispheric Specialization in Humans," The Guilford Press, New York, 1985.
11. STERNBERG, ROBERT J., Editor, "The Nature of Creativity: Contemporary Psychological Perspectives," Cambridge University Press, Cambridge, 1988.
12. RICHARDS, LARRY G., "Everyday Creativity: Principles for Innovative Design," Cutting Ed Online, Spring 2005, pp. 13–15, 30.
13. LEVESQUE, LYNNE, "Breakthrough Creativity," Davies-Black Publishing, Palo Alto, 2001.
14. EIFFERT, STEPHEN D., "Cross-Train Your Brain," American Management Association, New York, 1999.
15. ROGERS, CARL, "On Becoming a Person," Houghton Mifflin, Boston, 1961.

ADDITIONAL READING

16. BROCKMAN, JOHN, Editor, "Creativity: The Reality Club 4," Touchstone Books, Simon and Schuster, New York, 1993.
17. THARP, TWYLA, "The Creative Habit; Learn It and Use It for Life," Simon and Schuster paperback, New York, 2003.
18. NOLLER, RUTH B., S. J. PARNES, and A. M. BIONDI, "Creative Actionbook," Charles Scribner's Sons, New York 1976.
19. PARNES, S. J., R. B. NOLLER, and A. M. BIONDI, "Guide to Creative Action," Charles Scribner's Sons, New York 1977.
20. BAILEY, L. R., "Disciplined Creativity for Engineers," Ann Arbor Science Publishers, Michigan 1978.
21. OSBORN, ALEX F., "Applied Imagination: Principles and Procedures of Creative Problem-Solving," Charles Scribner's Sons, New York, 1979.
22. "Creativity Support Tools," A workshop report sponsored by the National Science Foundation, June 13–14, 2005, Washington, D.C.
23. STERNBERG, ROBERT J., "Successful Intelligence: How Practical Creative Intelligence Determines Success in Life," Plume/Penguin Publishers, New York, 1997.

CHAPTER 3

RECLAIMING YOUR CREATIVITY

To decide which one of his two sons will inherit his fortune, a rich father asks them to race their horses to a distant city. The winner will be the one whose horse arrives second. The brothers wander around for a few days until they receive advice from a wise man. They jump on the horses and race to their destination as fast as they can. What did the wise man tell them?

3.1 INTRODUCTION

In problem solving (or in life, in general), *the same* will get you safety, *different* will get you adventure. They are both valid approaches, they are both good, they are both acceptable, but they are not similar. One will provide consistency, comfort, and safety. You will not have to do something new; you will not have to go somewhere new; you will not have to try a different thing. If you do not have the appetite, the patience, the stamina, the money, or tolerance for taking risk for something new or different, *the same* will be an acceptable answer. If the state of the economy is bad and you will be in real difficulty if you lose your job, or if you have just bought a new house and you cannot lose it, or any other situation in which you cannot tolerate a new difficulty or new situation, *the same* will be an appropriate response. However, *different* will afford you the opportunity to try new things, to see new places, to get into new adventures, to experience something you have not experienced before, and to lose your safety net and land on your back, or your feet. You may lose your house or your job, you may have difficulty with your friends or parents, and you may be scorned for your actions, but in the long run, you will be more experienced, you will know more, and you will be more creative. And in fact there is a good chance you may not have any difficulty, you may not lose your job or house, and you may not experience anything bad, but still benefit from the different situation you experienced, and enjoy it too.

Life is full of decisions we make. Life is full of *different* and *the same* approaches. Our attitude and the way we approach our decision-making will determine our level of involvement in creative activities.

In this chapter, you will learn how and why we lose our ability to think creatively and what to do to counter these barriers and become more creative. Learning about the processes that our brain uses to solve problems will be a tremendous help in restoring our ability to be more creative, although not 100 percent. In addition to this, in later chapters we will also learn how using certain techniques can help us in this endeavor. Collectively, the material in this and succeeding chapters will help us be better, more creative designers and problem solvers.

3.2 MENTAL BARRIERS

Mental barriers are a collection of misconceptions, misunderstandings, biases, mind-sets, predispositions, assumptions, and emotions that prevent a person from understanding, identifying, or comprehending a problem and solving it. Others have referred to these by other names, including mental or conceptual blocks[3], constipated thinking, and many more.

Please read the story at the beginning of this chapter again. What was your answer? Why did the brothers try to speed to their destination although it was contrary to the requirement of the story? If you do not have an answer, it is because you may have assumed a certain limitation in the story, or perhaps you may have approached it as everyone else, assuming the same rule of engagement, and as a result, you cannot solve it. If so, the same assumptions and methods of approaching a problem will prevent you from solving other problems. These are examples of mental barriers that we will study in this chapter.

When NASA first started sending up astronauts, they quickly discovered that ballpoint pens would not work in zero gravity. According to one account, to combat this problem, NASA scientists spent a decade and $12 million developing a pen that writes in zero gravity, upside down, on almost any surface and at temperatures ranging from below freezing to over 300°C. The Russians used a pencil.

So, what did the wise man tell the two brothers? He told them to ride the other brother's horse.

In this chapter, like James Adams[3], we will take advantage of puzzles and simple problems to demonstrate the effect of mental barriers. A puzzle is generally defined as a problem that has an easy solution, but due to some mental barrier, the individual cannot solve it easily. Puzzles are not trial-and-error problems that you solve by repeated application of the same rule. They require a unique and generally different solution.

An essential point about puzzles is that unless you first spend a little time trying to solve the puzzle, it will not provide any benefit to you if the solution is readily provided to you. Thus, it is extremely important that you spend a little time on each and every puzzle before you read the solutions. The solutions to certain puzzles are provided here, immediately after the puzzle, in order to enable us to analyze the mental barrier and see how they affect the solution. In other situations, the presentation of the solution is postponed until the end of the chapter in order to enable you to think about the possible solutions before reading them. Otherwise, as you will see, puzzles can be simultaneously educational and entertaining, and provide a guide to our mental abilities.

It should be added here that many puzzles can be related to multiple mental barriers, all valid relationships. This is because mental barriers can be related to each other, or are affected by each other. Thus, a puzzle may be used to demonstrate different barriers.

Many of these puzzles are common, and thus their original sources are not known; others were created by my former students. Many have multiple acceptable answers, and thus you may think of other solutions or may have heard other solutions for them. That makes them even better. But they are all fun.

3.3   TYPES OF MENTAL BARRIERS

There are many different types of mental barriers, including, but not limited to:

- False assumptions and nonexistent limitations
- Typical solutions
- Making things more difficult than they are: being overwhelmed
- Incomplete or partial information
- Information and sensory saturation
- Associative thinking
- Misunderstanding
- Inability to communicate properly
- Emotions-, culture-, and environment-related barriers
- Fear
- Orderly vs. chaotic; Analysis/synthesis dilemma
- Falling in love with an idea
- Improper methods of solution
- Overabundance of resources

The following puzzles and examples will demonstrate how these mental barriers work and how they prevent the discovery of a solution.

Once again, please spend some time trying to solve the presented puzzles before you read the solutions.

Puzzle 1   Without lifting your pen, draw no more than 4 straight lines to cross all the circles shown in Figure 3.1.

Solution   Unless you have seen this puzzle before, there is a good chance that you may spend a long time trying to solve it without success. What is of interest to us, in fact, is not whether you could solve it, but the approach you took to do so. In most cases, no matter how long you try to solve the problem, unless you change your approach, you cannot find a solution. The approach most people take is to try to draw lines within the confines of the square, left to right, up–down, etc. Additionally, and especially if you are a logical person, you may think that you need to maximize the use of each line, trying to cross as many circles as possible. Thus, you may have tried the above-mentioned approach (Figure 3.2).

However, the opposite is true. First, that abandoning this approach and allowing yourself to draw lines that do not cross as many circles as possible will help you find a solution. Second, because the circles are arranged in a square format, you

FIGURE 3.1   Without lifting your pen, draw no more than four straight lines to cross all the circles.

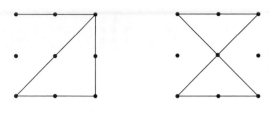

FIGURE 3.2 Common attempts for solving Puzzle 1.

FIGURE 3.3 One possible solution for Puzzle 1.

may have tried to find the solution within the same confines too. Instead, a solution can be found, as shown in Figure 3.3, by abandoning these unspoken, nonexistent, unnecessary assumptions that are not part of our puzzle.

This solution, however, uses the stated maximum four lines. In fact, it is possible to do this with fewer lines too. Can you do it with three lines? There are at least two different solutions for this problem with only three straight lines. In solution one, three straight lines are used, along with two curved lines, to cross through all circles (Figure 3.4). You see, if you read the problem statement, you will notice that there is nothing in the statement that prohibits us from using additional curved lines. It just indicates to use as few straight lines as possible. The apparent assumption that curved lines cannot be used is a false assumption and an artificial limitation that one assumes exists. It does not exist though.

The next solution is also related to another false assumption and nonexistent limitation; that the lines have to go through the center of each circle. The problem statement does not require this at all, and thus, the following solution is perfectly valid (Figure 3.5). Please notice that since the lines do not go through the center, they will intersect at a finite distance (and not at infinity, which is a common solution). To be able to do this more effectively, the size of the circles was increased, but the solution is valid at any size. What about doing the same with even fewer lines? What about just one straight line that will cross all circles?

In fact, there are many solutions for this alternative (Figure 3.6). You just have to make sure you do not make other false assumptions. For example, when we specify small circles, how small? Can they be a little bigger than shown in the figure? Can they be big enough to overlap each other? In that case, one line will cross

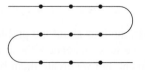

FIGURE 3.4 Another possible solution for Puzzle 1.

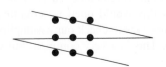

FIGURE 3.5 Yet another solution for Puzzle 1.

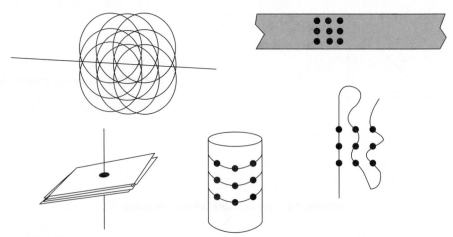

FIGURE 3.6   Other possible solutions for Puzzle 1.

all of them. Alternatively, we know that the mathematical definition of a line is that it has no width. However, in reality, all lines we draw have a width. What if you use a really thick marker pen to draw your line? Could it go through all of them at once? Still another assumption is that the surface on which the circles are drawn is rigid. What if they are on a piece of paper, and you fold the paper three times so that all the circles will be over each other, then a single straight line will cross them all? What if you would cut the paper and place the circles over each other? What if you take the piece of paper and make it into a cylinder and draw one line by circling around the cylinder? What if you draw one line through the first three, go around planet Earth and come back through the second three, and once more around the Earth, and back through the remaining three? Obviously, there is no mention in the problem statement that there is any limit to the length of the line. I suppose one could even argue that drawing one straight line followed by a squiggly one that crosses all circles is also acceptable, since as was stated in the problem statement, we want to draw one straight line that goes through all circles, but did not say that it could not be followed by a squiggly line. As you see, there are many solutions. But each one requires that you break some unspecified rule you may have assumed exists.

One more solution. An eighth grader, using a retractable pen, came up with the idea of retracting the core wherever he did not want to draw the lines although still in contact with the paper, and pushing it out wherever he wanted to draw lines.

3.4   FALSE ASSUMPTIONS AND NONEXISTENT LIMITATIONS

Puzzle 1 shows how we routinely make many false assumptions about the problems we face and place unnecessary limitations on the solutions we seek. In general, these assumptions and limitations are not part of the problem statement, but we create them based on our own expectations and experiences. It is important to remember to eliminate these assumptions when trying to solve a problem, as this would open up countless new approaches for us. We should constantly remind ourselves to check our assumptions and to make sure we disregard the ones that are false. This is a very common, but easy to resolve, problem.

FIGURE 3.7 Possible designs for the tallest structure made of straws to hold an egg.

An example of this occurred a few years ago when I was in a design workshop and, with a team of other faculty members, were given the task of designing the tallest structure made out of 30 plastic straws and needles that could hold an egg. Most groups rushed in to design the structure based on their knowledge of statics and strength of materials, coming up with bracing to prevent buckling of the straws, making sure that the structure was rigid to hold an egg carried in a small pouch at the top (Figure 3.7). However, as you notice, there is no reference in the statement that requires that the egg be at the top. Our group designed a simple structure to hold the egg at the bottom, and we added as many straws as we had on the top to make the structure many times taller than anyone else's. The workshop leader, in fact, argued at first that we were breaking the rules (that did not exist, except in his mind), but finally conceded that we were right (Figure 3.8).

We make similar unnecessary assumptions about real-life problems and limit ourselves too. Consider the evolution of memory technology. Each advancement in the technology was the result of breaching another assumption, from punch cards to tape recorders to 5.25-inch floppy disks to 3.5-inch floppies to Zip disks to CDs and to memory sticks, each one replacing its predecessor. What could be the next medium?

Countermeasure Try to think what assumptions you are making that are not real, necessary, or stated. Break away from these false assumptions. Ask yourself ''what if'' or ''why not...''. Ask yourself ''Do I have to do this in this particular way or can I change my way?''. Unless an assumption is clearly stated, it probably does not exist.

The following additional puzzles relate to the same mental barrier. Please think about possible false assumptions you may be making if you cannot solve them. The solutions are presented at the end of this chapter.

Puzzle 2 (Submitted by Nathan Siegel) Once upon a time, there was a king who wanted to pick a smart advisor to run his affairs. To test the potential applicants, he placed a large gem in the middle of a 20 by 30 foot carpet in his palace banquet room, and asked them to fetch the gem without going over the carpet. Each wise man and woman made elaborate devices, mechanical and such, but none worked.

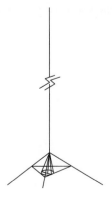

FIGURE 3.8 An alternative design for the tallest structure made of straws to hold an egg.

The dungeon janitor happened to pass by and overheard the puzzle. He solved it immediately and won the King's favor. What did he do?

Puzzle 3 Which one of the U.S. Government buildings is always 540 degrees?

Puzzle 4 You are standing in a room with three light switches in it, all in the off position, each one controlling one of three lights in a room on the floor above. You may turn on any of the switches, but you may only go upstairs once. How can you find which switch controls which light?

Puzzle 5 (Submitted by Tu-Viet Le) A $100 bill is placed inside a wine bottle and the bottle is corked. All that is available is a coat hanger, a hacksaw, a spoon, and a hammer. How would you remove the bill from the bottle without damaging the bottle or removing the cork?

Puzzle 6 (Submitted by Dawn Morrison) Using nine toothpicks, all of the same length, construct a cube without breaking any.

Puzzle 7 (Gene Goodwin, 1996) Divide a circle into 8 pieces using only three lines.

Puzzle 8 A man lives on the 20th floor of a building. Every time he takes the elevator to go to his apartment, he goes to the 15th floor and then takes the stairs for the remaining floors. But when he goes down, he takes the elevator all the way to the first floor. If he is not doing so to visit anyone on the 15th floor or to exercise by walking up five flights of stairs, why does he do so?

Puzzle 9 (Submitted by J. Quebman) Make three squares of the same size by repositioning only three matches and not taking anything away. Can you do it with two matches only? See Figure 3.9.

Now let's discuss another mental barrier with the following question:

Puzzle 10 What is half of thirteen?

Solution Obviously, half of thirteen is 6.5. Right?

Even any elementary-age student would tell you the same, and there is no doubt that 6.5 is the correct answer. But what if I told you that half of thirteen is 7111 or even 8111? In fact half of thirteen is many other things too, as listed next page,

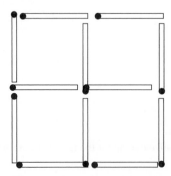

FIGURE 3.9

and every one of them can be proven to be correct. The question is whether you look only for typical solutions, or will you allow for other possibilities and approaches?

Half of thirteen is:

6.5, 6, 7, 1, 3, thir, teen, 8, 11, 2, 10, 3, יכ , יכ , 7111, 8111, ٧/٨ ,٢,١

thirteen , thirteen , ١ , ג , + , 三 , 24/6

To see how this works, let's take the first three answers. Of course, 6.5 is the expected answer. However, suppose that you have 13 people who are to drive to a game in two cars. When you decide that half of the group should be in one car, the other half in the second car, you certainly do not mean to take the real half of thirteen people in each car, you mean 6 in one, 7 in another. Thus, half of thirteen can be 6 or 7. If you cut the number 13 in the middle, half of it will be 1, the other half will be 3. And if you write *thirteen*, and cut the word in half, half will be *thir*, the other half will be *teen*. But what about 8? How can half of thirteen be 8? In this case, we have to switch into another mode of thinking. What if you were to write the number in Roman numerals XIII? If you cut this number in half in the middle, you will get VIII, which is 8 in Roman numerals. Thus, half of thirteen is really 8! If you cut the number vertically in the middle, you will get XI and II or X and III, which are also 11, 2, 10, and 3 (a different 3 than we got before). The next two are the number 13 cut in half horizontally right in the middle. Then what about 7111 and 8111? Once again, we need to shift our thinking, and break another typical rule. What do you think the same half of thirteen we got from cutting the Roman numerals would read in Arabic digits? In fact ٧ is a 7 and ٨ is an 8 and the rest are 1's. Thus, in Arabic notation, half of thirteen will be 7111 and 8111. What about the next three? They are the numbers 6.5, 3, and 1 in the same Arabic digits. If you ask anyone who uses or is familiar with that notation what half of thirteen, is they will write down the same figures. The next two are the word *thirteen* cut in half horizontally. And the next two? Well, they are actually in Hebrew letters. All Hebrew letters have equivalent numerical values. Our two figures are letters *yud* for 10 and *gimel* for 3. Of course you could write many other variations of the previous ones in not just Hebrew, but in Chinese letters (+ and =), Japanese letters, Sanskrit, and many other notations, all valid answers. And what would half of thirteen be if you wrote 13 or its halves in binary numbers (1101)? And what if we wrote 13 in hex numbers? Can you think of other possibilities? According to one student, it is the length of time he will be in college.

Will you now agree that 8 divided in half is 0?

3.5 TYPICAL SOLUTIONS

Typical solutions, as in the preceding example, are perhaps the most common mental barriers. We tend to think of the same solutions, typically arrived at by almost everyone else, every time. It is convenient and safe to come up with solutions that we already know how to do or how to accomplish. Atypical solutions are not safe, are not easily accepted by others, and require breaking rules. We are taught to think based on processes and procedures that everyone else knows and follows. We are asked to follow recipes. We are punished for deviating from the norm, and are rewarded for doing what our peers, teachers, and elders do. What do you think would happen to you if in your math class you would come up with 8 for half of thirteen? Do you believe that your teacher would be amused by your logic? So, we follow the same routine and think of the same solution that everyone else does too. But the reward belongs to the one who breaks the rules and solves a problem in an atypical, different, creative manner. Isn't this what patents are made of?

Edward De Bono in "Lateral Thinking"[2] talks about breaking rules: rules that create mental barriers, rules that prevent finding solutions, rules that are nothing but a hindrance to creativity. Most of these are unwritten, unclaimed rules. Most people are not even aware of them. But they work hand in hand with other rules that we create in our homes, at work, and in the society. Many of these rules are there for our convenience, some necessary, some unnecessary. Certain rules are created to allow us to have more order in our lives. Others are created so that others benefit from them. Many company procedures, store policies, and society rules are created so the people in charge will better control others who work for them. In all cases, these rules are just rules. But since we grow up with rules, we tend to create new ones as we go along (and good managers are good at this too), and as we learn more, we create more. Unfortunately, we go on and create more rules as we try to solve problems, and as a result of following these rules, we lose our creativity, spontaneity, and freedom of thought. We go on and solve our problems based on the rules we know or we create (resulting in typical solutions, as dictated by rules, or in lack of solutions due to new rules we create, namely false assumptions, and nonexistent limitations).

Obviously no one advocates breaking the law. Neither I nor anyone else would tell you to break the law. The laws we have are there to create order in our relationships with each other, and with the authorities. Traffic laws are necessary for our safety. Tax laws are necessary so that we all pay our fair share. Consumer laws are necessary to ensure that products are safe to use. But rules are not law. Make it a rule for yourself to "break a rule a day," and when necessary, even break this rule.

> **Puzzle 11**   (Submitted by Kathie Moore) Without rearranging any of the characters or adding any new characters or leaving any blanks, find as many different numbers as you can in this license plate:
>
> X10BQ3

Following common rules is also the reason why senior engineers are sometimes less creative than engineering students, and why seniors are sometimes less creative than first-year students, and they are sometimes less creative than children. As we learn to abide by rules, we tend to follow them. While working for an institution for some time, the engineer will learn the rules that govern the way the company conducts business, the way things are designed, and the perceived limitations that the company places upon itself. Someone from outside the company who does not have the same assumptions and does not follow the same rules will be free from the same limitations, thus a fresh approach to solving the same problem. A first-year engineering student will also have fewer rules of the "art and science of engineering" to follow, and thus may be more creative. And children have even fewer rules to abide by and fewer common ways of doing things to consider. Thus, they can think of new ways and methods and combinations of elements that adults may not do. This is one of the primary reasons why companies bring projects to design classes. They are interested in the new ideas that students might have that are free from their company rules.

In the mid-1970s, Paul MacCready designed the first human powered airplane and won the £50,000 Kremer prize, something that other teams had failed to do for 18 years before him. He knew that his success was not because other teams did not have the

FIGURE 3.10 A common barometer.

technical know-how. The reason everyone before him had failed was because they all had taken the exact same approach, but were trying to do a better job, which never worked. MacCready called it "digging the same hole deeper."[6] However, he was not handicapped by the same expertise, because he did not have an expertise in the design of airplanes, a mental barrier that could limit his thinking. His out-of-the-box thinking led him to design an airplane that consisted of "6 sticks and 72 wires" and won the prize.

In problem solving, it is essential to consider whether the problem can be solved in ways that are newer, different, or better. Referring to the example mentioned in Chapter 1 about the two approaches taken by two competitors for the Joint Strike Fighter is a good example of the importance of a solution that was not typical. Another commonly used example is the following: You are given a barometer and are asked to find the height of a building. What will you do?

First, a word about barometers. A barometer is a simple device that measures atmospheric air pressure. It is made of a U-shaped tube that is filled with a liquid (Figure 3.10). One side of the U-shaped tube is sealed with no air. As the atmospheric pressure changes, the liquid in the two legs will move relative to each other to maintain balance. Properly calibrated, the air pressure can be measured against the ruler.

Another point of interest is that as the height changes, so does the atmospheric pressure. As you go up in elevation, pressure is reduced. Given this fact, which most engineers know, the most commonly expressed solution for finding the height of a building is to measure the pressure at the bottom of the building, then measure the pressure at the top, and using a calibration table, convert the pressure differential to the height. This is a possible solution, very logical, based on engineering principles, and acceptable. But it is extremely inaccurate. But what about other atypical solutions? Other possible solutions might be to:

- Go to the top, drop the barometer, measure the time it takes to strike the ground, and convert the time to height.

- Same as above, but drop a rock instead, saving the barometer.
- Same as above, but attach the rock to a string, then measure the length of the string when the rock touches the ground.
- Use the barometer as a measuring stick and measure the height based on how many lengths it takes to cover the height.
- Use the barometer as a surveying tool to calculate the height by similar triangles (for example, by comparing the length of the shadows of the building and the barometer, you can calculate the height).
- Do not even use the barometer; use a measuring tape instead. The fact that you are given the barometer does not mean you have to use it.
- Go and ask someone who knows the height of the building (such as in the city hall or the architect).
- Ask someone else to do it for you.
- Sell the barometer, give the money to someone else, and ask the individual to do it for you.

As you see, the less typical solutions are easier, more accurate, and more useful.

As will be discussed in Chapter 14, one of the major requirements for a utility patent is that the invention be novel (new) and nonobvious. This implies that all typical solutions that everyone else thinks of are ineligible for receiving a patent. Is this not an adequate reason to look for nontypical, more creative, solutions?

Unexpectedness of a solution or an idea makes it both valuable and interesting. Almost universally you can detect a certain pleasure in people who are exposed to this unexpectedness, which causes them to smile. You will see the smile, and at times, laughter, in creative sessions; in fact, lack of it in a serious session can be an indication that no new and unexpected solutions or ideas are presented.

> *A man comes to a bar and orders three glasses of beer, sits alone, and drinks from the three glasses one sip at a time. The bartender goes to him and says: "Sir, why don't you order one glass at a time so that your beer will be cool and fresh?" The man says: "I have two brothers that are not here. We used to drink together all the time. I am doing this as if they were here, drinking with me." This goes on until a few weeks later, when the man comes in and orders only two glasses. The bartender later goes to him and says: "I am really sorry for the loss of one of your brothers. You must really miss him." The man replies: "Oh, no. They are fine. It is that I have just decided to quit drinking."*

Jokes are based on the same premise. The punch line is the unexpected information, idea, or consequence. If the punch line were not unexpected, the joke would not be a joke; it would be a story. Due to the unexpected consequence, we laugh.

Magic is the same too. There is no magic, just a consequence that is unexpected, and we do not know how it is done. Finding out how it is done would ruin the magic (or the illusion). The famous Horace Goldin had created the illusion of cutting a person in half with a saw.[5] At the beginning, no one could figure out how he did it. Eventually, in order to prevent others from copying him, he actually patented the idea. In reality, there were two people in the box, curled up, so that the head of one person and the legs of another were visible from the top and bottom of the box, but the center was empty. He sawed the box in half, separated the two parts with each person curled up in each half, and then put them back together. As soon as the patent became public and audiences found out how it was done, its total value was nullified, in fact defeating the purpose of patenting it. No one could copy him, but no one was interested in it either; it was no longer magic.

Puzzle 12 Given that you have a balance scale, what six whole-number weights will give you the maximum choice in weighing any object of whole-pound weights between 1 and the maximum you find?

Mirror image
by Tom Henkemeyer

Bugs Bunny
gets caught

Integrated
bottle opener
by Jim Alves

Square bolt

Easy Squeeze
Glass Bottle
by R. Sommers

Age scale
by N.M. Yazdi

Infinite staircase
Tory Bruno

Curved-neck
guitar
©S. B. Niku

Integrated tooth
brush-paste

Dictionary's
index
Stephane Roussel

Dried paint, ready
to go
M. Stender

FIGURE 3.11 Ideas that defy logic.

Two famous artists, admired everywhere, used in their artwork the unexpected, illogical, and impossible elements that defy logic. One was Maurits C. Escher who lived in the early twentieth century in the Netherlands and, in addition to his extraordinary and intricate designs and woodcuts, drew structures that were physically wrong.[10] The second one was Rube Goldberg (1883–1970), an engineer turned artist who designed impossible devices and machines as cartoons and published them in newspapers throughout the country. His designs became an inspiration for generations of artist that have tried to match his talents. Figures 3.11 through 3.14 are example of works by various individuals inspired by the works of Escher and Goldberg.

Countermeasure The remedy to typical solutions, as already discussed, is to break the stereotypical rules we go by. Usual solutions are just that; usual, common, and bland. Think in other directions, in ways you normally don't. In Chapter 4 we will learn about many different techniques that are devised to help people think in ways they normally don't, laterally, randomly. These techniques are commonly used in many different professional settings as well, from architecture and arts to engineering and advertising.

> **Puzzle 13**   (Submitted by Nancy Stebbins) In Figure 3.15, where do the next numbers belong?

> **Solution**   It is possible to come up with many different patterns and complicated schemes to solve this problem. One could say that the numbers inside and outside follow a pattern of 2 inside, 3 outside, 1 inside, 1 outside, resulting in 8 and 9 inside and 10, 11, 12 outside, and so on. Or you may use some sort of an equation to predict the pattern. However, our solution is much simpler. The pattern is based on the number of letters in the number. Only numbers with three letters are inside (Figure 3.16).

Next, we will discuss how this puzzle is related to a mental barrier.

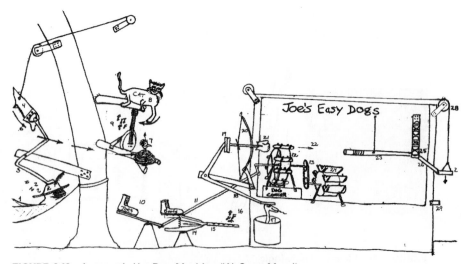

FIGURE 3.12 Automatic Hot Dog Machine (W. Scott Mead).

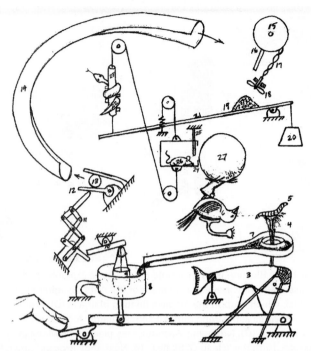

FIGURE 3.13   Automatic Orange Peeler (Richard Goldsmith): Push start button (1) which moves lever (2) that pushes against the whale's belly (3) causing it to spout (4), lifting the worm (5) into the bird's (6) view, which causes it to try to walk toward the worm, thereby turning the orange (27). Meanwhile, the spout is collected by the basin (7), which drains into the tub (8), raising the buoy (9), further moving the lever (2), pushing the lever (10), extending (11), making (12) squeeze the grape (13), shooting it into the trough (14), which directs it against the paddle wheel (16). The paddle wheel winds up the stretchy filaments (17), spinning the propeller (18). The propeller blows away dirt pile (19), allowing the weight (20) to rotate lever (21), lifting snake (22) within reach of climbing tree (23). The snake's weight pulls mouse cage (24) down against springs, opening door (25), exposing starving mouse (26) to the rotating orange (27), which he eagerly begins to nibble on, thus peeling it. You can enjoy your peeled orange.

3.6   MAKING THINGS MORE DIFFICULT THAN THEY ARE: BEING OVERWHELMED

If there is a way to solve a problem more simply, do it. Why should we make it harder? In fact, many problems can be solved more simply if we look for alternatives. This has been shown time and time again in class projects, in product design, and in individual projects. Do you remember the problem of designing a writing instrument for use in space that was mentioned earlier? One solution required millions of dollars and many years of development, another solution was to use a pencil. The team made the problem much more difficult than necessary.

We tend to overestimate the difficulty of many problems by assuming that if there has not been a solution to the problem, it must be a difficult problem. We tend to assume that things are more difficult than they sometimes are. We feel overwhelmed. This even happens in tests. Many students think that if a problem in a test looks simple, that they have misunderstood the problem or that it must have a trick; otherwise it would not have been included in a test. Not true. In many instances, instructors test the understanding of a subject by deliberately including simpler problems in a test and seeing whether it is

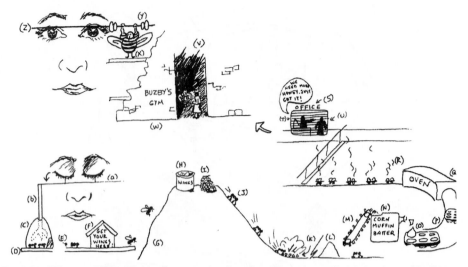

FIGURE 3.14 Why you blink: Blinking is not a simple reflex. There is a pole (a) that is attached to your eyes, and when you close them they move the ant agitator (b). Ants (E) agitated by this highly effective device exit their ant hill (c), which is upon the magical floating platform (D). From there, they have no choice but to rent a pair of wings at the wing outlet (F) conveniently located on the platform, and then fly to the top of the huge mountain (G). Not being wasteful, they place their wings in the recycling bin (H) before donning rollerskates from the "Heap O' Skates" (I). Next, they proceed down the hill (J) until they achieve perfect speed for pulverizing the pile of dried corn kernels (K) at the bottom, making them into a fine mound of flour (L). The conveyor belt (M) transports the flour into the giant mixing machine (N) which then dispenses perfect amounts of sumptuous batter into the awaiting muffin pans (O). Off go the pans on the conveyor belt (P) into the oven (Q) where they are cooked to perfection. As they exit the oven, the aroma (R) rises to the office (s) above where the owner (T) is reminded of his impending need for more honey for the batter. It is the secret ingredient and hard to come by. He has no alternative but to intimidate his employee (U) into desperate measures to attain the honey. He is forced to use illegal means and makes a back-alley (V) trade with the bees from "Buzby's Gym" (W). Honey for steroids! (yes, it is true). Next the bulked-out bee (X) lifts with ease his barbell (y) that is attached to the eye pole (z) and thus raises the eyes open again. This happens every time you close your eyes. It is all invisible but true. (Submitted by N-Stebbins)

FIGURE 3.15 Where do the next numbers belong?

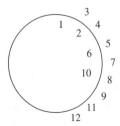

FIGURE 3.16

learned well enough or not. I have encountered countless situations where students design products with so many parts, so many elaborate mechanisms, and so many individual systems that they make the product terribly complicated and difficult to manufacture. By a little guidance, they realize that many subsystems can be combined together, mechanisms can be made simpler, and solutions can be simplified significantly. There is an example of two approaches taken for stabilizing satellites in space so that the antenna would always point to the center of the Earth for proper communication.[3] One, the common approach, is to devise a system with sensors to measure the orientation of the satellite, computers to calculate the necessary motions for correction, thrusters to rotate the satellite to correct the orientation, and an adequate amount of fuel to run the system for the life of the satellite. The alternative solution was to use a retractable hollow beam, similar to a tape measure, that would extend out of the satellite base, and by the action of gravity on the long beam, and the fact that the acceleration of gravity varies with distance from the center of Earth, the satellite would correct itself forever, without any computers, fuel, thrusters, or sensors of any kind. The simpler solution is easier to design, more reliable, more economical, and lighter. Some years ago, one of my students designed a window opener for use by the disabled that was so simple that he practically finished the design and its manufacture within a couple of weeks, and had nothing left to do for the rest of the term. He felt great too.

One of the major ways products are improved is by simplifying their design. Many designers realize that the product can function the same, or better, with a simpler design, fewer parts, or simplified manufacturing processes. However, there are also products that become more and more complicated as new features are added to them, whether necessary or not.

Countermeasure Even if the solution to a problem is difficult and overwhelming, you should try to find the courage and confidence in yourself to tackle it. I am certain that most well-educated individuals have the ability to do so. My evidence is the fact that most of our students get jobs in industry and governmental agencies with high level and sophisticated projects. But with the proper background, they quickly learn the particular trade and become a useful addition to their teams. If they were to feel overwhelmed by the prospect of working on difficult problem at the outset, they would not take the job or would not be able to succeed. Major territorial disputes between countries have also eventually been resolved by creative solutions although they may have appeared as overwhelming at the outset.

Being overwhelmed by a problem is a common trend though, especially in our technologically advanced world. The word ''high-tech'' alludes to this, indicating that the solution is complicated and sophisticated, and that not everyone can solve the problem unless they have the same know-how. However, many problems can have simpler solutions and can be managed with less effort if the right solution is found. The design of a new product may eventually require financing, marketing, finding interested investors, patenting, starting a business, and many others. However, the designer should not be overwhelmed by this prospect at the beginning of the process before the product is designed or developed. Otherwise, the fear of the magnitude of the work may discourage the designer from getting involved in the project.

One of the by-products of being overwhelmed by the seemingly difficult task is procrastination. Postponing the actions required to solve the problem or to take care of the issue, or even ignoring it, is a defense mechanism many choose to respond to a difficult undertaking.

Remember to think positively, have the confidence that you can do it, and do not procrastinate. There is a good chance that there is a neat solution to your problem. Your task is to find it.

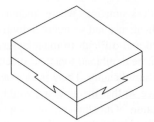

FIGURE 3.17   Is this possible if the two parts are to be easily separated?

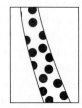 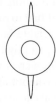 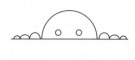

FIGURE 3.18   What are these objects?

The following are some more examples (puzzles) that demonstrate this mental barrier.

Puzzle 14   (Submitted by J. Moak and K. Leven) An engineer has designed a $4 \times 4 \times \frac{3}{4}$-inch machined block to join another $4 \times 4 \times \frac{3}{4}$-inch machined block. They are to be joined with a flanged dovetail on all four sides as shown in Figure 3.17. They are designed to be easily attached and removed. Is this possible?

Puzzle 15   (Submitted by D. Grant) A 1-gallon cylindrical bucket filled with water is given to you along with another cylindrical, empty, 3-quart bucket. Using no other tools, how would you divide the water between the two buckets exactly half and half?

Puzzle 16   (Submitted by Pelling) What are the next numbers in the set below, and what is the pattern?

3, 3, 5, 4, 4, 3, . . .

Puzzle 17   What are the objects in Figure 3.18?

Solution   Obviously, the first one is a giraffe behind a window, the middle one is a bike-rider in a sombrero hat, and the third one is a person peeking behind a wall. Or are they so obvious?

Next, we will discuss how lack of complete information is a mental barrier.

3.7   INCOMPLETE OR PARTIAL INFORMATION

There is a story about three visually impaired individuals who have never seen an elephant, but have been given a chance to touch the animal and describe it. One who touches the trunk describes the elephant as a long, soft, tubular, and flexible animal with a few whiskers. The second one examines the belly and describes the elephant as a big, round,

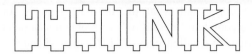

FIGURE 3.19 What is this object?

tough-skinned animal with some hair on it. The third describes it by the legs as a stumpy, hard, flat-bottommed animal. They are all partially right. The problem is that they are only ''seeing'' a part of a whole problem, with incomplete information. Conclusions based on incomplete information are generally erroneous.

Information about the objects in Figure 3.18 is similarly incomplete. Only partial information is provided. As a result, it is easily possible to solve the wrong, inappropriate problem, or come up with an inappropriate solution that only relates to the partial information provided.

Imagine that you are in a room within a building and you are asked to describe the building. That would be impossible. You can only describe the room you are in. The rest is speculation. Similarly, imagine that you are trying to solve a problem that is only partially defined, the information is incomplete, or the problem is misunderstood. Your solution will similarly be incomplete, inadequate, or wrong. It is vitally important to understand the problem correctly and completely. You must try to see the 'whole picture'. However, at the same time, it is also vitally important to not make assumptions that do not exist, or create rules that do not apply.

As we will discuss the importance of clear, concise, and complete problem statements (definitions) in Chapter 5, it is very important to understand what the problem is and what it relates to before we attempt to solve it. Trying to solve only part of a problem will result in an inadequate solution.

The following examples further demonstrate this mental barrier.

Puzzle 18 What is the object in Figure 3.19?

Puzzle 19 The kettle shown in Figure 3.20 has only a spout with no other visible holes on it, and is one piece. Still, it can be filled easily, can pour normally, and can hold a fair amount of water. How?

It is important to realize that there is a difference between partial information as a mental barrier and lack of knowledge about a subject. Obviously lack of knowledge about a subject will hinder its analysis and solution. This is why experts are so valuable in all facets of life, industry, and business. The above barrier is about partial information provided to the individual which causes a lack of understanding or partial comprehension.

It should also be mentioned here that at times, knowing too much about a problem may overwhelm the person and create its own mental barriers. We will discuss this later.

Countermeasure Try to understand the problem as best as you can. Look for missing information that may have been overlooked, eliminated, or not mentioned, but do not create unnecessary limitations. If you work with clients, make sure that you understand

FIGURE 3.20 The kettle for Puzzle 19.

what they want, what they have done, and what the ultimate goal is. In those circumstances, the development of a complete "definition" or problem statement or list of specifications is extremely important. We will discuss this in Chapters 5 and 6.

3.8 INFORMATION AND SENSORY SATURATION

Before you go on any further, please draw a telephone dial and write down the numbers and the letters on it without looking at a telephone.

Actual telephone dials have somewhat changed in the past few years, with more letters and numbers packed on the buttons, and with more buttons on the dials. However, a basic telephone dial looks something like Figure 3.21. Notice that number 1 has no letters associated with it. Also notice that number 7 has four letters, P, Q, R, S, and number 9 has W, X, Y, Z, while all others have only three letters. Number 0 has no letters, but "operator." In fact, until recently, number 7 only had P, R, and S, and number 9 only had W, X, Y. There were no Q or Z letters on the dials. Did you draw the dial correctly?

This is a simple example of sensory saturation. In this case, although we use telephones almost constantly, few of us remember details of it, such as the letters. Few will also remember other details on it (such as which button is the redial button)? Could you remember if the displayed letters were lost?

Now try to remember the details of someone's face that is close to you; perhaps a parent, your spouse, a boy/girlfriend, a child. In most cases, people will certainly recognize all the details when presented, but on their own, they will not remember the details. What about your watch? You see it all the time, but you may not be able to draw the face.

When our brain is repeatedly exposed to the same information for a long time, it tends to ignore it, consider it unimportant, or useless. However, if and when we need to have access to the information, the brain may not easily remember it. For example, if you live on a second floor of a building, you take the steps to your place every day, but you may not remember how many steps there are. Next time you drive on a highway with someone sleeping in the car, notice how they wake up as soon as you stop; the brain wakes the person up to find out why there was a change in the continuous noise of driving. In other words, as soon as the

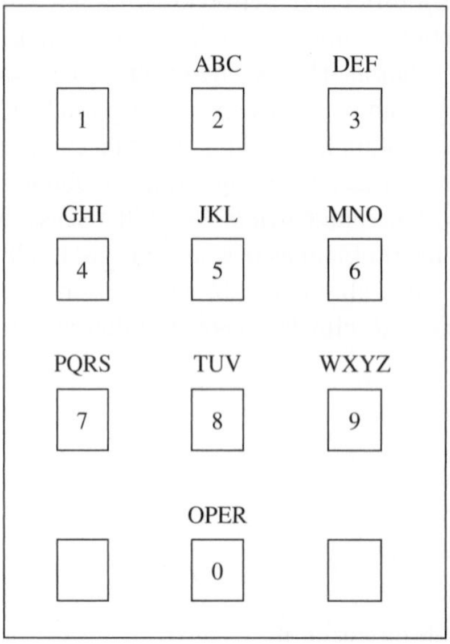

FIGURE 3.21 A telephone dial and the letters on it.

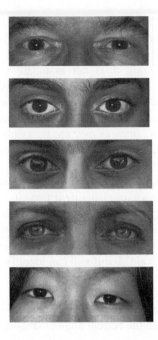

FIGURE 3.22   All eyes are beautiful, but different.

routine of the information changes, the brain will notice it. Have you noticed how your sense of smell also saturates when you are exposed to a particular smell for a short while?

Saturation becomes a more significant mental barrier when it interferes with our perception of data or discovery of a solution. For example, most individuals attempting to draw someone's eye will draw it as if all eyes look the same; they use their own symbolic way of representing an eye that their brain thinks all eyes should look like. In reality, everyone's eye is different (Figure 3.22). So, instead of drawing what they actually see, they draw what they think the eye should look like. In our perception, we see eyes all the time and they all look alike. Or noses or lips or ears. Try them too.

Countermeasure To remedy this problem, you must create a discontinuity in the stream of information that the brain receives. The nonuniform, or unfamiliar, method with which the information is presented will prevent saturation of the brain. In Synectics (which will be discussed later) this is referred to as ''making familiar strange.'' How? Look at Figure 3.23 without turning the book around. Can you tell who this famous

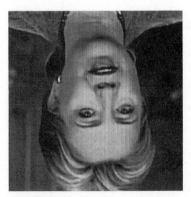

FIGURE 3.23 Can you recognize this famous person or the popular car without turning the book?

person is upside down? How about the popular car? If no, it is because the information is presented in an unusual way; there is an intended break in the usual stream of information, and therefore the brain does not associate automatic information to it. Its components are also not the same, and therefore the brain does not associate symbolic representations to the features of the face. Although this makes recognition of the object more difficult, it makes it easier to draw the features. This is extensively used, and highly recommended, in order to improve the drawing capability of ordinary people. Most individuals who are not trained to draw what they see, draw as they see fit: symbolically. However, when the object cannot be described symbolically, as in an upside-down eye, the brain will draw what it sees.[4]

Next time you want to draw an eye, or a face, or a building, or any other ordinary object, turn it upside down. Turning things upside down, or changing it from its ordinary state by any other method, will change its status from a symbolic object to something that cannot be easily ''categorized'' into known definitions. This will prevent the brain from taking over and drawing what it sees fit, and instead will help draw what it sees in reality.

Here are a couple more puzzles to tease your brain. See if you can identify to what mental barriers they relate.

Puzzle 20   (Submitted by P. Mook) Move only three matches to make the top number a nonnegative number that is lower than the bottom number (Figure 3.24).

Puzzle 21   (Submitted by P. Askins) The mythical country of Zambiziland has a flag as shown in Figure 3.25. One aspect of the flag is physically impossible. Can you find it?

Puzzle 22   If $\lim\limits_{x \to 0} \dfrac{8}{x} = \infty$ then what is $\lim\limits_{x \to 0} \dfrac{5}{x}$?

Solution

$$\lim\limits_{x \to 0} \frac{5}{x} = \text{ǝ}$$

Next, we will discuss how this puzzle is related to another type of mental barrier.

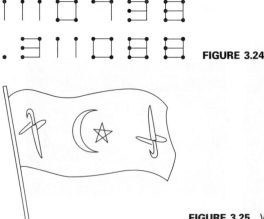

FIGURE 3.24

FIGURE 3.25   What is impossible about the flag of mythical Zambiziland?

3.9 ASSOCIATIVE THINKING

In this puzzle, the ordinary solution associated with the way we think about mathematics forces us to think about the solution in the same mathematical fashion; since the problem looks like a math problem, the solution must also be a mathematical expression. However, instead of this association with a particular method or subject, if we think symbolically about other possible solutions, we will get this solution.

Associating a particular meaning with the common definitions of words, letters, objects, and numbers can limit the way we think about solutions. An expert in mathematics tends to think more efficiently in those terms too. An individual fluent in languages will feel more comfortable analyzing words and thinking in terms of words. An engineer will have a tendency to favor engineering solutions too. Still, we all tend to associate connected letters with words, connected digits to mean a number, and connected parts to mean a whole. We tend to see the space between words and numbers as separators, and we like to disassociate numbers and symbols and letters from each other. For example, can you recognize how many colors are present when big old Bob's car lets out a loud yell owing to the drive up the steep hill? Of course, gold, scarlet, and yellow. But you would only see the colors if you disassociate the particular arrangement of the spaces and letters and combine them in a new form. Read it again.

We also tend to solve problems with the methods we have learned, and associate meanings to the particular subject we see. But there can be many other combinations of elements that we miss. In Hebrew, all letters of alphabet also have an equivalent numerical value. Thus, it is easy to associate numbers and letters together, as equivalents. However, most other languages do not have this association, and thus, words have no numerical value. Based on these equivalencies, countless new meanings and references are created that otherwise, would be missed. You can find similar associative meanings in English and many other languages as well. For example, there are many words that are pronounced similarly, but are written differently (whether, weather), as well as words that have two meanings (train and to train). Victor Borge, the great   Danish-American comedian, had a piece in which he added a number to any words that had the sound of a digit in it. Therefore became therefive. Tonight became threenight. Together became threegether. Talking to someone became talking three sometwo. Think what you can do with this association. Tom Kelley, in the Ten Faces of Innovation, refers to this as cross-pollination. Taking ideas from different fields and combining them together to come up with ideas, new solutions, and new meaning. We will learn more about this later.

Countermeasure Be a cross-pollinator. Take meaning from one thing and associate it with another. Break the common rules of association and the common meaning of things and replace them with meanings from other realms.

Puzzle 23 Change only one of the matchsticks in Figure 3.26 to make a true equality (no unequal sign accepted).

Puzzle 24 Arrange numbers 1–9 (each used only once) in Figure 3.27 such that the first row minus the second row equals the third row:

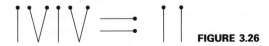

FIGURE 3.26

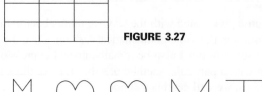

FIGURE 3.27

FIGURE 3.28

Puzzle 25   (Submitted by G. Griffiths) Fill in the blanks to complete the sequence:

$$AB\_\_EZH\_IK\_MN\_O\_P\_T\_\_X\_\_$$

Puzzle 26   What do the following mean?

$$Oicurmt, \frac{UALLS}{NOW}, HOROBOD$$

Puzzle 27   (Submitted by M. Stollenwerk) Find the next symbols in the series (Figure 3.28).

3.10   MISUNDERSTANDINGS

A man had died from drinking milk. "Was the milk spoiled?," he was asked. "No," he replied. "The cow suddenly sat."

Another possible mental barrier is related to misunderstanding the problem concept altogether. In this case, the problem definition or description may be vague or incomplete, it may be written with difficult language, or it may be at a level that is beyond the understanding of the individual. In these cases, the problem at hand is different from the problem understood, and as a result, the individual ends up solving the wrong problem, or worse, is unable to solve the problem. False assumptions, partial information, and unnecessary limitations contribute to this misunderstanding while misunderstanding a problem may lead to false assumptions and unnecessary limitations on solutions. I am certain that you can go back and think of many tests you have taken or problems you have tried to solve where the problem was misunderstood, and you ended up solving a different problem, or were unable to solve it at all. For example, most mechanical engineers know that natural frequency ω_n is the square root of k over m. But ask around, and you will see how many students interpret this as $\omega_n = \frac{\sqrt{k}}{m}$ instead of $\omega_n = \sqrt{\frac{k}{m}}$. This is why it is so important to have a clear and concise definition (or description) of the problem. This point will be discussed in depth in Chapter 4. Synectics, a methodical technique of problem solving discussed in Chapter 4, also emphasizes the importance of correctly understanding the problem by designating "Problem As Understood" as one of its steps.

Countermeasure Try to understand the problem as completely as possible. As was mentioned earlier, we will discuss problem definitions and specifications in other chapters, but it is important to remember to look for the real meaning of a problem statement and what it means. One way to improve your understanding of a problem, even if it is your own, is to try to write it down and communicate it with someone else. You will end up having to ask questions about many important issues as you try to define the problem for others, or as you try to write down what your goal is. I always recommend to my students to write the introduction to their thesis as early as possible. This will force them to answer many questions regarding the problem and reduce the possibility of misunderstandings later on.

Puzzle 28   (Submitted by Sansinena) There are four boats on one side of a river. The times it takes for each boat to cross the river are 1, 2, 8, and 9 minutes. You are to get all four boats to the other side in a maximum of 16 minutes. You may put any faster boat into a slower boat, but one at a time, and only two boats can cross at any time. How would you do it?

3.11   INABILITY TO COMMUNICATE PROPERLY

In the following section, the description of an object is given to you. You are asked to draw the object, within 3 minutes only, as you understand it without going back on the reading. In other words, imagine that the description is given to you on radio, thus there is no chance for you to ask questions, or to review the description again after it is presented. Alternatively, have someone read the description to you, once. Make sure you draw the described object before you continue. In a class environment, the object is placed in a bag. A volunteer, while touching the object in the bag without seeing it, gives the description to fellow students. The following is a typical response by volunteer students:

> *The object is a cube of about 2 inches in size. On one of the surfaces, there is a longitudinal groove in the center of the face, running from one side to the other, semicircular, about $\frac{1}{2}$ inch deep. On the opposite side, there is another longitudinal groove, parallel to the first groove, also somewhat semicircular, about $\frac{1}{2}$ inch wide, but about $\frac{3}{4}$ inch deep. On the face between the first two, and perpendicular to the first two grooves, is a third rectangular groove, $\frac{3}{4}$ inch wide and $\frac{1}{4}$ inch deep. The right far corner is cut off at $\frac{1}{2}$ inch on each side.*

Finish your drawing before you continue.

You will have a chance to check your drawing with the real object later. However, if you feel that your drawing is not very accurate, you are right. It is very difficult to do this effectively without feedback. Now redraw the object (start a new one) doing the same thing, but within 5 minutes, and this time you are allowed to re-read the description as many times as you wish (equivalent of asking questions from the fellow volunteer). If someone is reading it to you, the description may be repeated. When you are done, compare your results with the drawing of Figure 3.29.

As you can see, there is a good chance you may not have gotten the correct drawing even the second time around. In fact, even if you have it correctly, until you see the drawing in Figure 3.29, how can you be certain that your drawing is correct? In other words, the confidence level in your understanding is much below 100 percent. This means that if you do not understand the problem correctly, you will not be able to solve it correctly.

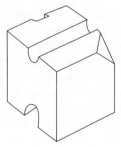

FIGURE 3.29

This is a form of mental barrier that relates to communications between individuals. It is important to understand what is communicated by others and what the problem is before you can solve it.

This communication problem occurs very often in the real world too. There are many instances when one person tries to communicate an idea to another person through words, sometimes adding hand motions to the description. Solutions are described in words, connections between different components are mentioned with gestures, and motions are described with moving arms. However, there is no certainty that the other person has understood the solution (or the problem) at all. Instead, a simple sketch may do the same but much more effectively, more accurately, and with more certainty. Countless times have I had students trying to describe their ideas or problems with words, thinking that I understand their description as they imagine it. Instead I have insisted that they draw the idea or the problem to make sure we communicate correctly. In fact, as we will thoroughly discuss later in Chapter 5, a ''problem definition'' or ''problem specification'' is the exact list of requirements, needs, and goals that need to be accomplished. Without this exact understanding, a wrong problem may be solved and no appropriate solutions may be found for a miscommunicated problem. And this is why the ability to sketch simple drawings of objects and machines and other solutions is so important for a designer. There are thousands of examples of designers who have sketched solutions on ''the back of an envelope'' while eating, or at sports events, or in other environments. You should also develop your ability to sketch your ideas in simple forms and to communicate with others visually. We will discuss graphical representations in more detail in Chapter 6.

Countermeasure Choose the proper way to communicate with others. A picture is worth a couple thousand words, maybe even more. Use graphs, drawings, and pictures abundantly when appropriate or possible. Use models when possible. Be an experimenter and a graphist (not necessarily a craftsman or an artist).

The following are from the Classified ads in papers:

—*Wanted: Widower with school-age children requires person to assume general
   housekeeping duties. Must be capable of contributing to growth of family.*
—*Mixing bowl set designed to please cook with round bottom for efficient beating.*
—*Wanted: Man to take care of cow that does not smoke or drink.*
—*Dog for sale: Eats anything and is fond of children.*
—*For sale: An antique desk suitable for ladies with thick legs and large drawers.*
—*Wanted: Unmarried girls to pick fresh fruit and produce at night.*

Puzzle 29   What is the following recipe for and what is each ingredient?

—*530 cm³ gluten*
—*4 cm³ NaHCO₃*

—4 cm³ refined halite

—235 cm³ partially hydrogenated tallow triglyceride

—175 cm³ crystalline $C_{12}H_{22}O_{11}$

—175 cm³ unrefined $C_{12}H_{22}O_{11}$

—4 cm³ 4-hydroxy-3-methoxybenzaldehyde

—two calcium carbonate-encapsulated avian albumen-coated protein

—475 cm³ fermented, roasted, and ground theobroma cacao

—235 cm³ de-encapsulated legume meats (sieve size #10)

3.12 EMOTIONS-, CULTURE-, AND ENVIRONMENT-RELATED BARRIERS

Please get up and do a chicken dance before you go on, no matter where you are.

If you did so, great. If not, think about why you did not. Was it because you thought this request was unrelated to the subject at hand? Or was it because you thought I must have been joking? Or perhaps because you found it embarrassing to get up and do a chicken dance in the middle of the library, in your room where others are present, or at the beach where other individuals may see you? What if the chicken dance was required for testing a particular motion related to a choreography problem for a play, and you needed to make sure that it would make sense? Would you do it then?

Somewhat interrelated, emotional, cultural, and environmental barriers relate to our personal and societal relationships, feelings, and emotions, and how we feel about things and the environment in which we operate. They include a variety of emotions and our tolerance toward feelings such as fear of failing, fear of being ridiculed, and fear of being singled out, tolerance toward criticism and chaos, jealousy toward others, and falling in love with an idea. They also include emotions that relate to our culture and the environment we live in, what is expected of us to do or not to do, nonsupportive environments, critical colleagues and managers, interruptions in our work, and many more.

Think back to a time when you visited a gallery or a showroom. While you were watching what was showcased, did you feel your creative juices running rampant, ideas flowing in your imagination, wanting to create similar but different things, the way you thought they should have been? That is the effect that a creative environment can have on individuals. Similarly, an environment void of anything interesting, but cluttered with dry facts, tables, and diagrams may induce the same feeling of rigidity and order in people, creating mental barriers. The Science building at Cal Poly is decorated with copies of M. C. Escher's works. Simply walking through the halls makes one's mind wander into a creative mode, wanting to replicate his creativity. Walking into some college dorms does the same thing if the halls are dressed with students' raw, fresh sense of disorder and creative expression. But similarly, walking into a lab, a classroom, or an office void of anything interesting may have the opposite effect on the individual. So, how is your room or office decorated? Do you have items of interest, beauty, and creative thought in your environment? Try to surround yourself with creative, interesting, and thought-provoking articles, artifacts, and images. This will create an environment that will help your mind be more creative too.

Another example of environmental barriers is coworkers. If your colleagues, whether in school, at work, or among your friends, are supportive, you may feel more courageous to take new initiatives, come up with new ideas, or take a different route to finding solutions. If they are not, the negative, nonsupportive environment may cause you to be more conservative in your thinking, more cautious in your decisions, and uneasy about breaking rules. Colleagues who themselves are creative and open to new ways of doing things, and who do what they preach, can be a positive influence on you. But if you meet resistance against, or even punishment for, your ideas and methods, you will certainly think twice before you embark on new ideas and methods. Friends, coworkers, and family members can have a significant effect on your thinking.

In some cultures, if a bowl of fruit or pack of cigarettes or chocolate candies are left on a table, guests automatically assume that it is alright to eat from the fruit, or pick up a cigarette, or have a piece of chocolate; otherwise, it would not be placed on the table. In other cultures, it would be considered almost stealing if the guest takes something from the table before she or he is offered or invited to do so. And would you hesitate sending a friend or lover a beautiful bouquet of yellow roses? Most probably not, except in certain cultures where a yellow rose indicates an intention to break up with the individual. However, red roses would be perfectly fine. In many cultures, yellow is a sign for cowardice, for lack of love. In others, both red and yellow mean life and happiness. Now imagine that you mix these cultural standards with your design and creativity. It may work, or it may create tensions. This is why it is important to learn about other cultures in which you may need to work, travel, or participate in activities.

Interruptions are another source of environmental barriers. Interruptions in our work or in our thinking process, whether intentional or accidental, whether wanted or unwanted, create a disruption in the flow of work or in our thought process that sometimes can be very detrimental. Listening to music while working is used both for concentration and enjoyment, but it can also be disruptive. Phone calls are disruptive too. Others knocking on your door or colleagues talking to you over your shoulder or over your cubicle are also disruptive. However, at times, we also create our own disruptions in order to avoid working on projects and problems we do not enjoy. Have you ever got the sudden urge to clean your room or office, or to call a friend you have not talked to for a long time, or to have a cup of coffee? These disruptions can create the nonsupportive environment that makes it more difficult to concentrate and to be creative. Otherwise, why go to a secluded area to write a book?

Traditional methods of working and problem solving can also be culture-related barriers. Traditions are in fact very useful tools for creating order in societies. Without traditions, we would lose our unique identity. Every country, locality, people, and organization has its own culture. Every household has its own traditions too. We need to respect our traditions and keep them and teach them to the next generation, so long as it makes sense, and so long as it is not a detriment to our problem solving. The fact that a company has used a particular way of solving a problem in a particular pattern does not mean that it must always be done the same way. The company may have a tradition of taking care of customers, and that is a good tradition to keep. But if the same company always uses the same method to introduce a new product, it does not mean that the tradition cannot be changed just because they did it that way in the past. Rules, including traditional rules, are meant to be broken when necessary.

Other examples of culture-related mental barriers are societal expectations. In many circles, daydreaming is a bad thing. You are told to always be mentally present, listen, and not to daydream. But when you are asked to design something, and you start thinking about it, isn't that daydreaming anyway? It actually is daydreaming, except that it is done intentionally and for a good reason, expecting positive results. Otherwise, daydreaming in itself is not bad. So, should you not practice daydreaming?

EXERCISE

Think about how you could use robots to improve the manufacturing of a simple product such as a compact disk, a toy, or a paper cup. Start daydreaming.

Taboos are another example. There are certain words, actions, and behavior that are not acceptable in society, and in general, for a good reason. Taboos are one way we teach our children about acceptable and unacceptable behavior. However, you may need to re-evaluate the value of taboos that are a detriment to your problem solving ability so long as they do not hurt anyone else.

Being serious is another expectation. You may have been told by others in meetings to be serious and to quit joking around, because the meeting is serious. But remember that when there are new and unexpected ideas, the natural tendency is to smile or even laugh. When creativity juices are flowing, it is natural to be less serious and joke around. Being serious in a meeting may mean to not be creative.

Yet another example is the value of intuition in decision making. We are taught to be objective, logical, and scientific in our dealings. Intuition is none of the above. It is a gut feeling about something. Should we always disassociate with it and ignore it? Certainly not. There is no scientific reason why intuition may necessarily be better than logic and an objective approach. But there is no evidence that it is always wrong either. The gut feeling we have about things, the way our experiences make us feel about things, and the way we associate with certain issues may be intuitive, but in fact, that is the way the other side of the brain works too. Intuition means that we cannot describe where the feeling is coming from, why we feel the way we do. But if you remember, since the right brain cannot communicate verbally, many conclusions made by it are also impossible to describe. Malcolm Gladwell in *Blink, the Power of Thinking without Thinking*, describes the power of intuition and gut feeling in an array of situations with wonderful examples, sometimes surpassing scientific judgments. As long as you are careful with your intuition, it should at least be yet another source of inspiration, creative thought, and reasoning for you.

EXERCISE

How did you decide whether to buy or not to buy your last vehicle? Was it based on *Consumer Reports* data and the result of acceleration and slalom tests, or was it based on your gut feeling about the car and how you felt about the way it handled you and the road?

Countermeasure Beware of emotional, cultural, and environmental mental barriers. Create a supportive, creative, fun, and enjoyable environment around your place of living and your place of work, as permitted by others who are involved. Traditions are valuable and play important roles, but they have their place too. Evaluate their role and their effect on your problem solving abilities. Surround yourself with thoughtful and creative artifacts and pictures that induce a sense of adventure and creativity in you.

In the following sections, we will discuss additional emotional barriers that have a strong effect on individuals.

3.13   FEAR OF...(You add the noun as you wish)

Fear is one of the strongest of emotional barriers. Fear stops us from doing things, or can make us do things we do not want to do. It includes fear of making a mistake; fear of failure; fear of being ridiculed by others; fear of not being accepted as a part of a group; fear of losing credibility; fear of rejection; fear of being different from others culturally, physically, or mentally; fear that your solution may not be the best.

Imagine that you are in a group of individuals that is working on a problem, and that you have an idea that is different from others, perhaps unusual. If you feel that by expressing this idea you may be rejected, or others may think you are different than normal, or that you are not "smart," you will most probably keep the idea to yourself and will not express it. If you are afraid that someone may steal your idea or will take credit for it, you may have difficulty presenting it to the group. If you feel that your job may be in jeopardy, you will refrain from participation.

Embarrassments are the same. If you feel embarrassed, if you feel shy, if you have an accent that is different from others (everyone has an accent. What matters is whether your accent is different from the ones around you), or if you are new to a culture, you may have much difficulty dealing with negative emotions that stop you from problem solving.

Fear of failure and getting a bad grade is also prevalent in college work. Students are always conscious of their grades, especially with the emphasis that employers and graduate schools put on grades, with or without regard to "the whole picture" of what the student has done in terms of their extracurricular activities and their total effort in school. This fear of getting a bad grade plays itself harshly in design courses where decisions must be made that may not reflect the teacher's decisions. The result is that students fail in design-oriented courses with open-ended problems and projects.

Competitiveness is another strong emotional characteristic that plays a role in fear. Although competitiveness can help a person in many ways, it can also create a measure of fear that the individual's ideas are not the best among the others involved (the team, friends, classmates, competitors). The brain shuts down in response to this fear to protect the individual.

Mr. Rumbold of the "Are You Being Served?" television series says: "Please don't make a suggestion, Mr. Spooner, unless you are sure it is a good one."

Countermeasure One way to overcome a fear is to look for the worst-case scenario, to consider the consequences of doing something that may have adverse effects. If the consequence is acceptable, and there is a chance that you may be rewarded for the creative approach you want to take, you may decide that it is worth the risk. If it is not acceptable, then by all means, do not partake in the activity. For example, let's say that the worst-case scenario for expressing an idea about a different method of solving a problem is that one may be ridiculed, may be rejected, or be considered less smart than others. Then what? Would that be acceptable to you? If so, then go ahead and express your idea. The worst is that it will be used against you, you may be labeled negatively, and you may be at a disadvantage. No one likes these consequences, but they may be acceptable because you may not care what others think about you, you may have other options that may even be better than the acceptance of your peers, or you may enjoy putting others in the position of having to make a judgment. On the other hand, perhaps the creative approach you have expressed may actually be liked by others, you may get credit for it and advance your position, you may actually be assigned to do it, and thus, you may secure your job for a while, and you may enjoy doing something you suggested. If the result is that you may lose your job, perhaps you can get a better job across the street, and not even care to work for an outfit that does not appreciate creative approaches such as yours. The worst-case scenario could be that you might get a bad grade for solving a problem differently, creatively. If the bad grade is something acceptable to you, if it makes only a small dent in your already prestigious grade point average, and if a low grade is something that will not make you worry about the rest of your life, then go ahead and take the risk. Who knows? Perhaps the teacher will actually like your creative approach, and your reward for being creative may even enhance your grade and your average. If the consequence is unacceptable, the economy is bad and you cannot get a job in this state of the economy, you may lose your house and your car, and you must ensure a high grade point average to get into graduate school, then the risk may not be worth the reward. Your fear is justified, and you should refrain from creative activities. But can you find too many examples like this? Assessing your tolerance for the worst-case scenario is a very good way of overcoming your fears.

3.14   ORDERLY VS. CHAOTIC; ANALYSIS VS. SYNTHESIS

On some occasions, I have included an extra piece of information that is known by students, but is not necessary for solving the problem, in a test. For example, let's say that the value for the Poisson ratio for steel is mentioned in a problem related to the stress analysis of a bar. Although students know about this information, a materials-related constant, and they know what it means and how to use it, many get confused and almost do not solve the problem. Why do I do it? To see if they have any appetite for disorder. To them, this is the worst thing in a test. To have information with which they do not know what to do. However, if you think about it, there are many other known numbers and constants that they know about, but since those are not mentioned, they do not matter. This shows their inability to work in the presence of any chaos. They believe there must be a reason why the Poisson ratio is mentioned. Yes there is a reason, but not what they assume. (Usually the effect of this inclusion on the final outcome of the test is minimal. They do eventually realize that it is not important, but it does occupy them for a while.)

The same is true with other synthesis (design) versus analysis situations. In analysis, there is a problem that the student is trying to solve, to which there is a particular solution that can be found, and if the teacher solves the same problem correctly, their answers can be compared; the student gets the reward if the solution is correct and is punished (with a lower grade) if incorrect. There is a particular *order* that brings everything together, and that order can be defined and followed. Analysis problems abound in high school and in college, in most courses. The other alternative is also writing comparisons (comparative analysis) of literary work, as well as critiquing works of art, plays, cinema, and music. In this case, someone else does the work of design or synthesis; the student or the reviewer only analyzes what already exists and renders judgments.

On the other hand, in synthesis (or design) there is no single answer that the teacher or your boss or anyone else may necessarily find that satisfies all requirements. There is no guarantee that a problem necessarily has any satisfactory solution that anyone in particular can find. A teacher may have better solutions than the students, but the students, too, may have much more clever solutions than the teacher has. Someone who is very smart may have more difficulty (creatively) solving, particular problem, and yet someone with ordinary skills and common intelligence may do even better. There is no particular solution that the teacher can use to compare one design against another. There is no particular solution to point to at all. This *chaotic* situation does not play well with many students and individuals. They like to know that the teacher or the manager can solve the problem, that there is a final particular goal they can look for that renders the problem solved and satisfied, and that it can be achieved by following what has been learned. But the chaotic order of design and synthesis does not guarantee that there will be a solution found, or that one's solution is any better than someone else's. There are many students who do very well in high school and in college until they get to design courses where the order they have learned does not exist. These students either fail or have a miserable time. They feel they have to make decisions that will affect other things, but there is no way to make the right or the best decision. This lack of order and the existence of chaos become a mental barrier and a detriment to their success.

Countermeasure One consideration that can be very helpful in this situation is to realize that the lack of order in synthesis and design works both ways; although someone else may come up with a better solution than you and outperform your creativity, it also means that *you* may do better than anyone else. If you feel that your teacher may have better solutions than you, there is also a chance that you may find a solution that may never occur to him. Any advantage anyone else may have over you in problem solving, you have the same advantage over that individual. And in fact, this is what makes design and synthesis so exciting. Good designs are unique enough to stand on their own feet and be appreciated by others, no matter who is the originator of the idea or the design.

In an unofficial study at Bell Labs, an attempt was made to categorize the characteristics of creative persons. However, the study showed that creative individuals had almost no common characteristic, except a sense of humor and openness for chaos (including a messy work environment).

What about doing something out of the ordinary today?

3.15   FALLING IN LOVE WITH AN IDEA

"I have an idea" can be a dangerous statement, primarily because the word "idea" is singular, meaning that there is just one idea. Although it could mean that the individual has had other ideas from among which the best, so far, has been picked, it usually means that she or he got an idea, fell in love with it, and has stuck to it. The person usually has no interest in other ideas that may be better, more appropriate, simpler, or easier. The person usually falls in love with the idea as if it is their first-born child, to be protected, to be nurtured, and to be defended. But there is always a chance that there may be better solutions if one looks for alternatives.

Countermeasure It takes courage to set aside the strong emotions involved with an initial idea, to abandon it, and look for other ones. If one does not find a better solution after looking, finding, and comparing other ideas, and still believes that the first (or any other idea) is the best, then the idea may actually be the best, and there is no argument against it. But without this search for alternatives, simply taking an idea and falling in love with it is a dangerous situation. Next time you have an idea, try to come up with another three, or five, alternative ideas and compare them before you finalize your solution.

The next puzzle demonstrates another mental barrier.

Puzzle 30   Imagine that you could have as large a piece of paper as you would like to, even larger than a football field. Do you think that you could fold this paper in half each time, 50 times?

Solution   Obviously, one method to solve this problem is to imagine the solution and what it looks like after each folding. It becomes thicker and thicker. It may actually take more than one person to fold it at the beginning, when the paper is really large, but it may also require more than one person to fold it as it gets thicker. But so long as you can have as large a piece of paper as you need, you can keep folding it in half, right? This reminds me of what Archimedes is quoted to have said: "give me a large enough lever and a fulcrum to anchor it to, and I will move the Earth."

Now let's see what this really means. Assume that a stack of 1000 sheets of paper would be about 4 inches high (about the thickness of regular paper), yielding a thickness of 0.004 inches per paper. Every time you fold the paper, the thickness will be twice as much. The total thickness of the paper after 50 folds will be

$$t = 0.004 \times 2^{50} = 4.5 \times 10^{12}(in) = 71 \times 10^{6}(miles)$$

This means that the thickness will approach 71 million miles, or about 75 percent of the distance between the Earth and the Sun. It will take light 6.4 minutes to cross it. Do you think we could do this no matter how large the paper?

There is a story about a ruler of India who fell in love with the game of chess, so much so that he asked the person who invented it to appear before him and offered to bestow upon him anything he wanted. The inventor asked that he be paid in grains of rice, 1 in the first square, 2 in the second, 4 in the third, 8 in the fourth, to the end. The ruler, jokingly, asked for a sack of rice, thinking it would be enough. Eventually, all the rice in India was not enough.

If you assume that on average, 100 grains of rice is 1.5689 g, and that the summation of $1 + 2 + 4 + 8 \ldots 2^{64}$ is $2^{65} - 2$, the total weight of the rice will be 5.79×10^{14} kg. The ruler's lack of vision caused him much grief.

Next, we will discuss how this puzzle relates to the following mental barrier.

3.16   IMPROPER METHODS OF SOLUTION

To solve Puzzle 30, one may choose to imagine the solution, actually try to do it (a favorite method of problem solving is experimentation and modeling of the ideas), or may calculate the result, as we did. In many situations, one may have the luxury of choosing different ways of solving a problem. Depending on the problem, a proper method for solving it will enable the individual to find the best answer in the best way possible. Many individuals who are trained in physics and mathematics have a tendency to use their physical and mathematical knowledge. Others use modeling or experimentations. Individuals with training in arts may use their imaginative powers to visualize solutions. We should thrive to be open to all different possible methods that are available to us for solving problems independent of what our particular training is. In the example above, it is easiest and most convincing to solve the problem using a simple calculation. Experimentation will take too long, and until you fold a large paper many times, this may still mislead you into believing that it could be done. Visualizing the solution may completely mislead you into the wrong conclusion. Still, you must realize that calculations, heavily favored by engineers, are not always necessarily a proper method for solving every problem. Consider the following puzzle.

Puzzle 31   Robert likes to walk around the neighborhood early in the morning, every day, to exercise and to think about his day. He usually takes the same route, going through the same streets, the same side of the park, in the same direction. However, depending on whom he will see on his path, he may stop and say hi to them. He tends to take his time if he finds an animal or a new flower to observe, to smell (the flower), or to pet (the animal). Depending on how tired he is, sometimes he goes slower, sometimes faster. Where it is uphill, he slows down a bit, and when downhill, he goes faster. However, he usually takes less than an hour to make the round.

Today, feeling spontaneous, he decides to do the same, but in the opposite direction. He still takes the same route, the same side of the park, the same street, but he goes around in the exact opposite direction. He still has a varying speed depending on whether he is going uphill or downhill, he still notices the animals and the flowers and the people who pass by him, and he still thinks about what he will do today.

Prove that somewhere along this path, which he takes every day, there is a particular point (other than his house) that Robert happened to be at today and yesterday at exactly the same time of the day.

Since this puzzle contains reference to average velocities, times, and locations, a logical approach to prove that such a point exists would be, especially for someone who knows physics or mechanics, to develop two elaborate equations, one for today and one for yesterday's walk, that describe the relationship between some assumed average velocity, minimum and maximum velocities, time for each portion of the walk, distances related to each portion, and estimated times Robert might take to rest, smell, pet, or observe, etc. By equating these two equations and assigning estimated times and velocities to the variables in these equations, it is possible to come up with the location where he happened to pass at the same time of the day. However, can you think of another to prove this point exists without having to solve such an equation? Imagine that we would call Robert's house at yesterday's departure time his starting point, and his house at return time his destination. Then today, going in the opposite direction, we could call his house at departure time the destination and the house at return time his starting point. Plotting his journey on both days versus time on the same coordinate frame will yield a graph similar to Figure 3.30.

Each path, randomly drawn, represents a variety of speeds, rests, and stops. However, unlike the mathematical representation of the problem, it really does not matter what these velocities are and whether he stops or not. The point is that where the two graphs intersect is the point where he is at the same location at the same point. So, since we are only asked to prove that such a point exists (and not to actually locate it, although with enough detailed information about his walks we could do that too), the simple graph proves the point. However, this is much simpler a solution than the first one.

Alternatively, imagine that instead, we would assume that yesterday Robert and his wife Julia both took a walk, one in each direction, starting at the same time. No matter how fast they walk, how they walk along the way, and whether they stop or not, as long

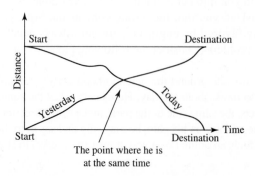

FIGURE 3.30 Robert's journey on both days versus time on the same coordinate frame.

as they follow the same path in the opposite direction, they will meet at some point. That indicates that there is a point where Robert would be on both days at the same time.

Now, which one of these solutions is easier to describe, to accomplish the proof, and to understand? Obviously, the descriptive solution is the best, so far. Unlike the previous puzzle that could mislead an individual into a wrong response, a descriptive solution is a better choice for this problem. Thus, it should be clear that it is our responsibility to look for different possible methods of solving a problem and choose the best fit, independent of what our primary training is. Here is another puzzle to tease your brain.

> ***Puzzle 32***   How much larger would Earth be if its circumference were 1 meter larger?

A simple mathematics problem in the 10th grade algebra book states: ''A plane is observed approaching your house, and you know it is traveling at 550 mph. If the angle of elevation of the plane is $16°$ at one time, and 1 minute later the angle is $57°$, approximate the altitude of the plane.'' How would you attempt to solve this problem if not through drawing a sketch and seeing the relationship between the locations, velocities, and altitudes? My son had difficulty solving it until I drew a simple sketch for him. The rest was easy.

Countermeasure Look for alternative ways and methods to solve a problem. Can you model it? Can you use mathematics, numbers, equations? Can it be graphed? Does it have an equivalent in music? Is there an equal in the animal world? If it is a mechanical problem, is there a solution in the architectural world that applies to it? Chemistry? Food science?

3.17   OVERABUNDANCE OF RESOURCES

The common cliché ''what gift do you buy for someone who has everything'' applies to our ability to be creative as well. When one has everything, all seemingly necessary resources, she or he will not need to come up with new things, new approaches, or new products. This overabundance of available resources creates a mental barrier that prevents the individual from searching for yet other resources, ''I already have it, why do I need another?'' syndrome. This is as bad as underabundance of resources. When your only tool is a hammer, everything needs nailing. But have an abundant number of ways to do anything, and you stop being creative. Evidence of this is the creative, innovative, and sometimes amazing ways people improvise solutions to problems when their essentials are taken away, as in an emergency, or when they are not available, as in many third world countries. The National Collegiate Innovators and Inventors Alliance (NCIIA) has funded many projects specifically designed for third world countries in which only materials and skills available locally are used to produce products or solve problems, from agricultural pumps and vegetable-based fuels to collection of rain water for agricultural purposes and products made from local fibrous woods. When resources are less abundant, you will use them more creatively. This has also been shown repeatedly during times of war, when resources become scarce and people use what is available more creatively.

Countermeasure Eliminate a skill, a resource, or a possibility from your list of possible solutions or available alternatives. Then try to replace that resource or skill or possibility with another, a substitute, another alternative. This will force you to look for better use of what remains available, for new alternatives, or to new uses of the remaining resources. Then try eliminating a second resource.

What we have seen in this chapter demonstrates the variety of different mental barriers that prevent us from solving our problems and how to countermeasure them. Many are learned through the processes that we are exposed to, and by our experiences in life. In

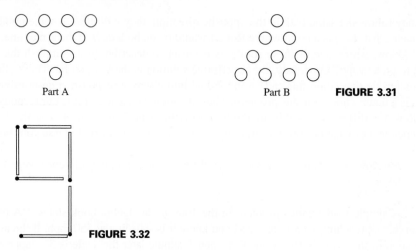

Part A   Part B   **FIGURE 3.31**

FIGURE 3.32

the next chapter, we will learn other ways about how to overcome these barriers, and how to be more creative in everyday life and in problem solving. For more information, please see the additional References (11–23) at the end of this chapter.

The following are more puzzles and the solutions to the previously mentioned puzzles. See if you can identify what types of barriers are responsible for these puzzles as well. These will help you identify your other mental barriers in other circumstances. For other puzzles, please also see Reference (1).

Puzzle 33   (Submitted by Gilpin and Knobf) Move only three of the circles (Figure 3.31) in part A to make it into part B.

Puzzle 34   (Submitted by Steve Wu) By adding one more match and without moving any other matches, change the figure 9 into a figure 5 (Figure 3.32).

Puzzle 35   (Submitted by Cecil Collier) A man in a gray car is speeding at 80 when he notices that a police car is behind him in the left lane, but that the policeman passes and pulls over another red car, going at 75 miles per hour. Knowing that the red car was pulled over for speeding, why didn't the police pull over the gray car?

Puzzle 36   (Submitted by Cecil Collier) A bicyclist is riding down the highway, going south along the coast, on a narrow and winding road. The ocean is only a 100-ft-steep cliff away. There is only a 2-ft-wide bike path with no shoulder, and there are no guardrails. Along comes a car, going in the same direction as the bicyclist, that knocks him off the bike from behind. Assuming that the rider does not stick to the car, what keeps him from going over the edge of the cliff and into the ocean?

Puzzle 37   Prove that $3 + 3 = 0$.

Puzzle 38   (Submitted by Alan Niku) Fill out numbers 1 to 8 in these blocks such that no successive numbers are adjacent to each other.

Puzzle 39 Make four equilateral triangles with only six matches.

Puzzle 40 You are given 10 baskets, each one with 10 balls, and one scale for measurements. Except for one basket whose balls are all 0.9 kg, all other balls in all baskets are 1 kg each. You are allowed only one measurement with the scale, to find which basket has the 0.9-kg balls. What will you do?

Puzzle 41 A long time ago in a small town, a merchant owed a large sum of money to a money-lender. The money-lender, who was old and ugly, fancied the merchant's beautiful daughter. So he proposed a bargain. He said he would forgo the merchant's debt if he could marry his daughter. Both the merchant and his daughter were horrified by the proposal. So the cunning money-lender suggested that they let providence decide the matter. He told them that he would put a black pebble and a white pebble into an empty bag. Then the girl would pick one pebble from the bag. If she picked the black pebble, she would become his wife and her father's debt would be forgiven. If she picked the white pebble she need not marry him and her father's debt would still be forgiven. And if she refused to pick a pebble, her father would be thrown into jail.

They were standing on a pebble strewn path in the merchant's garden. As they talked, the money-lender bent over to pick up two pebbles. As he picked them up, the sharp-eyed girl noticed that he had picked up two black pebbles and put them into the bag. He then asked the girl to pick a pebble from the bag.

What would you have done if you were the girl? If you had to advise her, what would you have told her? Careful analysis would produce three possibilities:

1. The girl should refuse to take a pebble.
2. The girl should show that there were two black pebbles in the bag and expose the money-lender as a cheat.
3. The girl should pick a black pebble and sacrifice herself in order to save her father from his debt and imprisonment.
4. Or would you suggest something else?

Puzzle 42 Count the Fs in the following text:

FINISHED FILES ARE THE RESULT OF YEARS OF SCIENTIFIC STUDY COMBINED WITH THE EXPERIENCE OF MANY YEARS.

Puzzle 43 (Submitted by J. Callan) What is the product of the following expression?

$$(x-a)(x-b)(x-c).\ ...(x-z) = ?$$

3.18 SOLUTIONS AND RELATED MENTAL BARRIERS FOR THE ABOVE-MENTIONED PUZZLES

Solution of Puzzle 2 He rolled the carpet on one side and reached the gem.
A typical example associated with false assumptions: Here, the assumption is that you must reach the gem from outside its perimeter, but there is no mention that it cannot be rolled to decrease the size of the perimeter.

FIGURE 3.33

Solution of Puzzle 3   The Pentagon.
Another example of false assumptions: Here, most people assume that the 540 degree relates to temperature, not angles.

Solution of Puzzle 4   Turn on switch #1 for a few minutes, then turn it off. Turn on switch #2 and go upstairs:

>   The light that is on is controlled by switch #2.

>   The light that is off, but warm, is controlled by switch #1.

>   The light that is off and is cold is controlled by switch #3.

Generally, people assume that only the light can be used to detect the relationship between the light switches and the bulbs, as these are the defined roles for these parts. Another specification, namely heat, is not included in the solution.

Solution to Puzzle 5   Push in the cork into the bottle and remove the bill with the coat hanger.
Here, the assumption is that the cork cannot be moved at all. However, pushing it in instead of pulling it out will solve the problem without breaking the stated requirements.

Solution to Puzzle 6   If you assume that the cube has to be constructed by 12 lines, as in a real 3-dimensional object, you will not succeed. It will only work if you assume that the cube can be on paper, a 2-dimensional reality (Figure 3.33).

Solutions to Puzzle 7   The false assumption here is that the lines must be straight, a common misconception. Lines can come in any shape or form (Figure 3.34).

Solution to Puzzle 8   He is short and cannot reach the buttons for floors above the 15th floor. The long description provided about the actions of the individual makes it hard to concentrate on the person and possible reasons related to him or her.

Solution for Puzzle 9   Most people assume that the geometric relationship between the squares must be maintained, thus trying to keep the squares attached as shown. In the

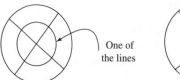

One of
the lines

FIGURE 3.34

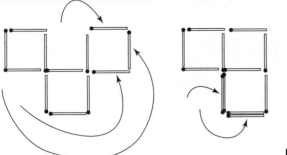

FIGURE 3.35

second part, the assumption is that the sticks must remain active parts of the squares. Nothing was said about not stacking the sticks over each other though (Figure 3.35).

Solution to Puzzle 11   The numbers are: 0, 1, 10, 3, X (for 10 in Roman numerals), X1 (11 in Roman numerals), B (11 in Hex numbers), and 10B (267 in Hex numbers). Additional possibilities, if we allow ourselves to mix different systems, are X10B, X10, and 0B. And if we consider numbers in base-27, Q will be 10 as well.

Typical solutions for this problem consist of only Arabic numbers in base 10.

Solution for Puzzle 12   There are a number of possibilities. As a common, typical solution, for example, six weights of 1, 2, 3, 5, 10, and 20 pounds will allow you to weigh any number up to 41 lbs. The six-weight set of 1, 2, 4, 8, 16, 32 will allow weighing up to 63 lb, much higher than the first set. If you realize that weights can also be placed on both sides (and thus be used as a negative number), the six-weight set of 1, 3, 5, 10, 30, 80 will allow up to 129 lb. For example, 12 lb will be $10 + 3 - 1$, meaning that the 10 lb and 3 lb weights will be on one side while the 1 lb weight will be on the opposite side. Additionally, if you realize that weights can be other odd numbers as well, the set of 1, 3, 9, 27, 41, 162 will let you weigh up to 243 lb. Can you show how? Try to find the right combination for 22 lb and for 87 lb. So, by eliminating limiting factors that do not exist and trying new combinations, you may get a ratio of 243/41 or about a 600 percent better result than the first choice.

Ultimately, if we assume that a lever arm with a moving fulcrum can also be used, theoretically one weight will allow you to weigh any object, including fractional weights.

Solution for Puzzle 14   The dovetails are lengthwise as shown in Figure 3.36 and not crossing each other as normally would be assumed.

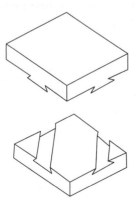

FIGURE 3.36

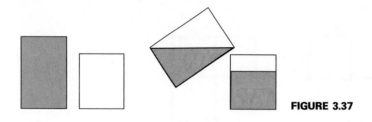

FIGURE 3.37

FIGURE 3.38

Solution for Puzzle 15   The focus here is on measurements, not the actual dividing of the water, a much simpler solution (Figure 3.37).

Solution for Puzzle 16   5, 5, 4, 3, 6, 6. These are the number of letters in digits one, two, three, four, etc. Obviously, many other solutions are also possible based on sophisticated patterns or elaborate assumptions. But this is a simple way to get an acceptable solution.

Solution for Puzzle 18   The elimination of the surrounding box creates incomplete or partial information. A little more detail, and it is an easy problem to solve (Figure 3.38).

Solution for Puzzle 19   Once again, incomplete information about the details of this design makes it difficult to imagine that there can be a "breathing" straw in the center of the kettle providing for air movement (Figure 3.39).

Solution 1 for Puzzle 20   Turn the first stick 90 degrees until it is perpendicular to the paper. It will become a decimal point. You can also move two more of the first sticks to the end. Then .073811 will be smaller than .311088.

Solution 2 for Puzzle 20   Move the sticks as shown in Figure 3.40 and then turn the paper upside down.
   In both cases, we look at the solution in usual ways, whereas a stick normal to the paper or the exchange of numbers into letters will work fine.

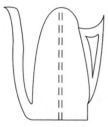

FIGURE 3.39

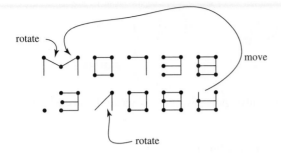

rotate

move

rotate

FIGURE 3.40

FIGURE 3.41

Solution for Puzzle 21   The star cannot be seen behind the moon, as it will block the light. We see the stars and the moon all the time. But we do not notice the relative relationship between them, that the moon is so much closer to Earth that no star can be seen behind it.

Solution for Puzzle 23   If you include mathematical symbols such as square root, you will solve this puzzle, as shown in Figure 3.41. We commonly tend to look for mathematical solutions based on numbers, as given. Turning a number into a symbol requires breaking a rule and converting a characteristic into another.

Solution for Puzzle 24   Obviously, this works if you consider all three digits on each row as one number, thus $873 - 214 = 659$. Individual digits will not yield a satisfactory answer.

8	7	3
2	1	4
6	5	9

Once again, this also requires that we change another rule, that the given individual numbers in each box should be read together and not individually.

Solution for Puzzle 25   Α Β Γ Δ Ε Ζ Η Θ Ι Κ Λ Μ Ν Ξ Ο Π Ρ Σ Τ Φ Χ Ψ Ω. This is the Greek alphabet. The solution requires association of English-sounding letters to the Greek alphabet.

Solution for Puzzle 26   *Oh I see you are empty, all between us is over now* and *ROBIN HOOD*. You have to make a new association between letters and words, or disassociate the order of letters.

Solution for Puzzle 27   Of course, many other variations are possible. In this case, these are mirror-imaged numbers 1 through 9 back to back (Figure 3.42). This requires

FIGURE 3.42

disassociating the whole parts into their elements, especially because they are all symmetrical.

Solution for Puzzle 28   Do the following:

Take $1 + 2$ min boats to other side	2 minutes
Return the 1 min boat to first side	1 minute
Take $8 + 9$ min boats to other side	9 minutes
Return in 2 min boat	2 minutes
Take $2 + 1$ min boats to other side	2 minutes
Total	16 minutes

Solution for Puzzle 29

—gluten is wheat protein.

—$NaHCO_3$ is baking powder.

—refined halite is salt.

—partially hydrogenated tallow triglyceride is oil.

—crystalline $C_{12}H_{22}O_{11}$ is sugar.

—unrefined $C_{12}H_{22}O_{11}$ is brown sugar.

—4-hydroxy-3-methoxybenzaldehyde is vanilla.

—calcium carbonate-encapsulated avian albumen-coated protein is an egg.

—fermented, roasted, and ground theobroma cacao are chocolate chips.

—de-encapsulated legume meats (sieve size #10) are peanuts.

This is a recipe for chocolate chip cookies.

Solution for Puzzle 32   In this case, the best solution is probably mathematics and not visualization, as most people think that adding 1 meter to the circumference of Earth should have a minuscule effect on its size. However, if we add 1 meter to the present circumference and calculate the new radius, we will get:

$$p = 2\pi r + 1 = 2\pi(r + dr)$$

$$dr = \frac{1}{2\pi} = 0.159 (meter)$$

This means that the Earth would be about 30 centimeter (cm) bigger in diameter. Obviously, this number is independent of r. Thus, adding 1 meter to the circumference of a ping-pong ball would add the same amount to its radius.

A wrong language, in this case visualization, can lead to a wrong answer. Another approach, mathematical, can result in a correct answer.

Solution for Puzzle 33   This puzzle can be attributed to unnecessary limitations or typical solutions. We tend to desire to move the circles systematically row by row to accomplish the task (Figure 3.43).

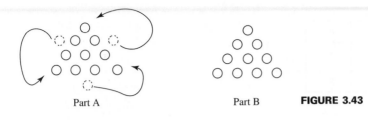

Part A   Part B   **FIGURE 3.43**

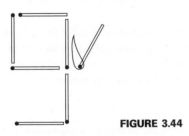

FIGURE 3.44

Solution for Puzzle 34 Add a lighted match (Figure 3.44). This puzzle can be attributed to unnecessary limitations and false assumptions. No one indicated that the match could not be lit.

Solution for Puzzle 35 The car was a foreign car and the speedometer was in metric, showing 80 km/hr, which is about 50 mph.

Another example of false assumptions and unnecessary limitations, we automatically assume that all speedometers are in miles per hour.

Solution to Puzzle 36 The road is in the East Coast. Going South means he is on the other side of the road.

Especially effective in the West Coast, people assume that going South means traveling on the side closer to the ocean. Would individuals from the East Coast make the same assumption?

Solution to Puzzle 37 Start by the known fact that $1 = 1$. Then:

$$-1 = -1 \rightarrow \sqrt{-1} = \sqrt{-1}$$

$$\sqrt{\frac{1}{-1}} = \sqrt{\frac{-1}{1}} \rightarrow \frac{\sqrt{1}}{\sqrt{-1}} = \frac{\sqrt{-1}}{\sqrt{1}}$$

$$\sqrt{1} \times \sqrt{1} = \sqrt{-1} \times \sqrt{-1}$$

$$1 = -1 \rightarrow 3 \times 1 = 3 \times -1$$

$$3 + 3 = 0$$

Yes, we all know enough math to know this is incorrect. But playing with the rules allows you to prove it as shown. Break the rules, and you will get something new.

Solution for Puzzle 38 In fact, this puzzle can be solved by trial and error. Should we even consider it a puzzle? Perhaps, if we break the rule that defines a puzzle.

	6	4	
2	8	1	7
	5	3	

Solution for Puzzle 39 This can only be done if you think in three dimensions, as shown in Figure 3.45. Typical assumptions of 2-dimensional solutions will not work.

Solution for Puzzle 40 Pick 1 ball from basket one, 2 balls from basket two, 3 balls from basket three, etc. Then weight all the picked balls together. If they were all 1-kg balls, you would have had a total of 55 kg. The difference between this number and the measured value ×10 is the number of the basket whose balls are 0.9 kg each.

Solution for Puzzle 41 This story is supposed to make us appreciate the difference between lateral and logical thinking. The girl's dilemma cannot be solved with traditional logical thinking. Think of the consequences if she chooses the above logical answers.

What did she do? She put her hand into the bag and drew out a pebble. Without looking at it, she fumbled and let it fall onto the pebble-strewn path where it immediately became lost among all the other pebbles. ''Oh, how clumsy of me,'' she said. ''But never mind, if you look into the bag for the one that is left, you will be able to tell which pebble I picked.''

Since the remaining pebble is black, it must be assumed that she has picked the white one. And since the money-lender dare not admit his dishonesty, the girl changes what seems an impossible situation into an extremely advantageous one.

Mental barriers such as typical assumptions and being overwhelmed by the cruel conditions make it difficult to see that there is an easy solution available. Imagine that similar solutions would be applied to other global and international conflicts as well.

Solution for Puzzle 42 There are actually six. If you missed all of them, it is probably because the brain finds the F in OF negligible, a sign of sensory saturation. We see certain common words so often that we tend to ignore them.

Interestingly enough, a seventh grader found 20. In his mind, there is a hidden F under every E as well.

Solution for Puzzle 43 The product of the series will be zero, since one of the factors will be $(x - x) = 0$.

FIGURE 3.45

Because of information saturation, we focus on the shear number of parentheses we need to work with rather than the fact that since one of the later ones is zero, the result is simply zero.

REFERENCES

1. FUJIMURA, KOBON, "The Tokyo Puzzles," edited by Martin Gardner, translated by Fumie Adachi, Scribner, New York, 1978.
2. DeBONO, EDWARD, "Lateral Thinking: A Textbook of Creativity," Ward Lock Educational, London, 1970.
3. ADAMS, JAMES L., "Conceptual Blockbusting: A Guide to Better Ideas," W. H. Freeman and Company, San Francisco, 1974.
4. EDWARDS, BETTY, "Drawing on the Right Side of the Brain," J. P. Tarcher, Inc., 1979.
5. BROWN GARY, "Sawing a Woman in Half," *The American Heritage of Invention and Technology*, Vol. 9, No. 3, Winter 1994, pp. 34–39.
6. CIOTTI, PAUL, "More with Less; Paul MacCready and the Dream of Efficient Flight," Encounter Books, San Francisco, 2002.
7. KELLEY, TOM, "The Art of Innovation," Doubleday Publishing, New York, 2001.
8. VON OECH, ROGER, "A Whack on the Side of the Head; How You Can Be More Creative," Warner Books, New York, 1983.
9. VON OECH, ROGER, "A Kick in the Seat of the Pants," Harper Perennial, a division of Harper Collins Publishers, 1986.
10. LOCHER, J. L., Editor, "The World of M. C. Escher," Abradale Press, H. N. Abrams, Inc. Publishers, New York, 1971.
11. BAILEY, ROBERT, "Disciplined Creativity for Engineers," Ann Arbor Science, The Butterworth Group, 1978.
12. PARNES, S., R. NOLLER, and A. BIONDI, "Guide to Creative Action," Charles Scribner's Sons, New York 1977.
13. LUMSDAINE, EDWARD, and M. LUMSDAINE, "Creative Problem Solving, Thinking Skills for a Changing World," McGraw-Hill, Inc., New York 1993.
14. HANSON, T. F., "Engineering Creativity," T. F. Hanson, 1987.
15. RUBENSTEIN, MOSHE, "Patterns of Problem Solving," Prentice-Hall, Inc., 1975.
16. GARDNER, HOWARD, "Frames of Mind; The Theory of Multiple Intelligences," 2004.
17. GLADWELL, MALCOLM, "Blink, the Power of Thinking without Thinking," Little, Brown, and Company, 2005.
18. KELLEY, TOM, with JONATHAN LITTMAN, "The Ten Faces of Innovation," Doubleday Publishing, New York, 2005.
19. STERNBERG, ROBERT, Editor, "The Nature of Creativity; Contemporary Psychological Perspectives," Cambridge University Press, 1988.
20. LEVESQUE, LYNNE, "Breakthrough Creativity," Davies-Black Publishing, 2001.
21. OSBORNE, ALEX, F., "Applied Imagination; Principles and Procedures of Creative Problem Solving," 3rd ed., Charles Scribner's Sons, 1979.
22. STERNBERG, ROBERT, J., "Successful Intelligence; How Practical and Creative Intelligence Determines Success in Life," Plume, a Penguin Co., 1996.
23. RAUDSEPP, E., "Creative Growth Games," Perigee Trade, 1980.

DESIGN PROJECTS

The following are examples of projects we have assigned in the past. Obviously, they may be changed to suit your purpose.

3.1 C3 (CAL-POLY CABLE CAR) DESIGN PROJECT

The objective of this project is to design and build a vehicle that travels down a catenary cable between two points (Figure D.3.1), with the following limitations and definitions.

- The vehicle to be designed and built by you will travel down a catenary cable, hung in a designated area, as shown below. The vehicle should be released from rest, move down the cable, and eventually stop.

- The objective of your design is to maximize the following equation:

$$G = \left(\frac{(D - 50)}{M} \right) \times (1 + CF)$$

- D is the horizontal distance of travel in inches (to where the vehicle stops), as shown in the diagram, M is the mass of your vehicle (or the average of the mass of the vehicle at the beginning and end of each test) in grams, and CF is Creativity Factor, solely at your instructor's discretion, awarded for exceptional creativity and resourcefulness.

- You must realize that the actual shape of the cable is related to the weight of your vehicle.

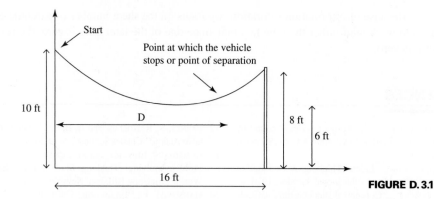

FIGURE D. 3.1

- If any part of the vehicle (or the vehicle itself) falls off of the cable, we will measure the distance D to the point of first separation.

- The device may not be made of separate portions that are connected together by a cable or other similar flexible elements. Therefore, a falling weight that converts stored potential energy to kinetic energy is not allowed.

- There can be no stored energy in your vehicle when it is released, except potential energy of the weight of the vehicle. This means that if the device is placed on a taut horizontal cable, or on a horizontal plane, it should not move at all.

- The distance D will be measured from the point the vehicle stops, not the max distance traveled.

- Your vehicle may not mar or damage the cable, or otherwise leave it dirty.

- You may work in groups of 1 to 3 as assigned by your instructor.

- The rules are subject to my interpretation if disagreements arise.

- You must **construct** your own vehicle. As such, a single chain-link will not do. You may use off-the-shelf items as your budget allows.

- You may perform minor repairs to the vehicle, but no design changes are allowed.

- The vehicle must be mounted on the cable within 10 seconds and be released. You may not unhook the cable for mounting your vehicle. The vehicle may hang down the cable.

- The catenary cable is nominal 0.127-inch-multistrand steel cable, and will be available for your inspection a few days prior to the test.

- We will start testing at the beginning of class time. You must be present and ready to test. You may not repair, assemble, or build your vehicle while we test the other vehicles.

3.2 CP-TRADE (CAL POLY TRANSPORT DEVICE) DESIGN PROJECT

The objective of this design project is the following, with the given limitations and definitions:

- A transport device is to be designed and built by you that will transport a garbanzo bean along the path shown in Figure D.3.2. The path will be marked on a paper or on a tabletop. The device (and the bean) should be released from rest, the bean should travel along the path, and should stop at the end of the path.

- The objective of your design is to maximize the following equation:

$$G = \left(\frac{D_1 - D_2}{\sqrt{M}} \right) \times (1 + CF)$$

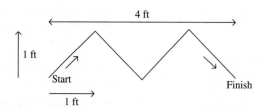

FIGURE D. 3.2

where D_1 is the distance the bean travels in inches along the path. D_2 is the distance the bean will go beyond the finish point, in inches, as well as any portion of the path that it deviates from as described below. M is the total mass of your transport device (or the average of the mass of the device at the beginning and end of each test) in grams. CF is Creativity Factor, solely at the discretion of your instructor, if at all given, awarded for exceptional creativity and resourcefulness.

- The path will be drawn at a 1-inch width (1/2 inch on each side of the actual path). The bean must stay within this 1-inch width if the distance is to be counted as traveled. Any portion of the path over which the bean does not traverse will be counted as D_2.

- You may use a battery, rubber band, balloons, etc. as a power source. Using a battery will cost you 5 percent of the final grade before CF is applied. Only one AA battery may be used.

- Absolutely no microprocessors are allowed. No remote control is allowed.

- You may not make an artificial path over which your device or the bean will travel. However, other accessories are allowed to assist in your navigation so long as it is not a continuous "assister." An assister is defined as an element that assists the bean's travel. It may not be a continuous element.

- You must construct your own device. You may use off-the-shelf items as your budget allows. Using a toy car chassis is allowed if the rest is constructed by you.

- You may perform minor repairs to the device, but no design changes are allowed.

- Read the rules and understand them. Ask if you have questions. I will not judge your solution before testing.

3.3 Design a letterhead for a company with a specific business such as in the food industry, airline industry, manufacturing, distribution, etc.

3.4 Design a poster for a cause such as a blood bank, driving under the influence, global warming, etc.

3.5 Design a modular chair for movies that is economical, comfortable, and versatile.

3.6 Design a better briefcase with more efficient utility for professionals.

3.7 Design a new sport or sport equipment.

3.8 Design a new nonelectric toy.

3.9 Design a Commercial for a Phony Product.

- You are to consider a phony, nonexistent, ridiculous product or service that makes no sense, and make a 45–60 second commercial for it.

- You or anyone else you choose may act in it. Commercials with scripts and acting are much preferred to announcements with written material only.

- You may edit, mix, add music and special effects, etc. to your commercial. Check the final quality of your recording. Beware of wind noise if you record outdoors and of lighting if you record indoors.

3.10 For the lack of any better name, this project is called QWERTY, which also stands for "*Qan We Ever Really Top Yt*?" It is a machine that will receive a starting trigger from you and will work for at least 30 seconds (but not more than 60 seconds) to perform a task that you have decided, and stop.

"Working" has the following meaning: You will consider a certain task that you wish your machine would do. The task does not have to be a serious task at all. It is intended to be a "Rube Goldberg"-type machine, where the machine as well as the task it performs can be trivial, but the machine should do the job completely. Still, it should be very creative and fun. The design should have at least 15 separate and distinct motions that can individually be defined. Repeated similar motions will be counted as one motion. A set of cause–effect will constitute one motion. An effect alone is not a motion. There can be as many additional motions as necessary to do the job or to make the project work or to make it fun. No more than five types of electrical parts (batteries, light bulbs, wires, switches, etc.) will be counted (if you decide to use any). No single motion may take more than 5 seconds. Explosives and rocket projectiles are not allowed. You may procure parts from any available resource. You may use the facilities on campus for manufacturing your machine. You will have to clean up whatever mess you make. You have the option of dressing appropriately for the occasion!

HOMEWORK

3.1. If Ernst, who lives in Amsterdam, is in the Netherlands, in what land will your neighbor be?

3.2. Find the feminine/masculine version of the following words. Then write as many others as you can think of:

e.g.,   heritage   hisitage or sheritage

           shenanigan

           hero

           women's shelter

           hemisphere

           sheep

           hermit

           shelf

3.3. Guess what the following mean. Then make five of your own:

3.4. Rube-Goldberg Type design

You are asked to make a drawing of a Rube-Goldberg-type device of at least 26 elements (A to Z) that will perform a task of your choice. The task can be as trivial or significant as you wish. However, the main purpose of the assignment is creativity. Be creative and inventive, and try to do things without the usual limitations.

You must turn in a sketch of the device. You may not use a computer to create the sketch, but you may use cutouts of pictures or other ''things'' to complete the sketch. Along with the sketch, you must have a text, preferably humorous, that describes how the system works.

A L L   N I G H T

D model T
D model T   pen cil   ning fork
D model T   ning fork
D model T

Story bus   hwallowing8
Story bus   youIcan'tdecideyou

partisanship   clomonsterset   saurne
partisanship   cyclo2pedias   dicherokeeans

     b e
g bush a   o__er__t__o__   ri
   n i t   poorch

$\dfrac{\text{Fire}}{\text{High}}$ ELO	$\dfrac{\text{Morning + Morning}}{\text{NPR News}}$	$\dfrac{1}{1000}$ tary	dons	
			dons	g
			dons	n
			dons	i
$\dfrac{\text{bed}}{\text{monster}}$	$\dfrac{\text{my mom is an}}{\text{lord}}$	$\dfrac{2}{\text{takers}}$	sore dons	w
			dons	o
			dons	r
$\dfrac{\text{he is an}}{\text{achiever}}$	$\dfrac{\text{I'm}}{\text{internet}}$	$\dfrac{\text{cray}}{\text{boxes}}$	dons	h
			dons	t
			dons	

$\dfrac{0}{\text{M.D.}}$
Ph.D.
D.D.S.

$\dfrac{1}{n}\sin x = ?$

3.5. Puzzle (Submitted by Louis Mann): Draw two lines perpendicular to each other that will cross all five dots in Figure P 3.5.

• • •

• •

FIGURE P 3.5

3.6. Puzzle (Submitted by Brian Zeller): Touch three points on a circle, drawing only one straight line with an ordinary pencil.

3.7. Puzzle (Submitted by Dmitri Yegoshin): Connect points A and B with a straight line without crossing CD in Figure P 3.7.

• A

C ——————————————— D

• B

FIGURE P 3.7

FIGURE P 3.10 There is no end to it.

3.8. Puzzle: A farmer plants 10 trees in five rows with four trees in each row. How did he plant them?

3.9. Puzzle: Write the next few digits for these series:

112000222001332002 and 191419391950

3.10. "There is no end to it" homework is based on Figure P 3.10. As shown, four starting and four ending lines are given. You are supposed to connect these four "inputs" and four "outputs" together, in any form, pattern, or shape that you see fit, so that if we place everyone's paper next to each other, we will have a continuous, un-ending set of lines that will do something different on each sheet. No other limitations are given.

3.11. If you play backgammon, try to play it with a new set of rules and see how it works. You will notice that as the rules change, you will have to come up with new strategies and new plans in order to win the game. You will also have to continually update your rules as you encounter new situations.

3.12. As the game is played traditionally, a checker can hit (bear off) an enemy's checker if it is single, but not if there is a pair of checkers in one point. Now assume that one single checker can hit and remove a single enemy checker, a pair of checkers

can hit a pair of checkers occupying the same point, etc. A simple rule change like this will force you to rethink all strategies for defense, for attack, and for reaching your home board.

3.13. To make medical braces for wrists and ankles, steel or aluminum wires are covered with a variety of medical grade foams and straps and tied together. To further mechanize the manufacturing process, it is desired to design an automatic wire cutter to cut the wires at specified lengths. The steel or aluminum wires may be up to 1/8-inch thickness. The device should be able to cut the wire from 6 to 18 inches of length in whole numbers. It should be fully automatic. The wire is wrapped on a spool 1 ft in diameter. Pressurized air is available. You may assume that a programmable logic controller (PLC) and encoders, as well as electric motors and pneumatic cylinders may be used. Write a complete set of specifications for this device, and then design a machine to accomplish the job.

3.14. Find new uses for empty aluminum soda/beer cans. List as many diverse uses as you can. Consider the different characteristics of the can and find applications for each one.

3.15. Find as many uses for cut-out pieces of leather which is left over during the process of

FIGURE P 3.17

shoe-making. These pieces may be of any shape, mixed colors and thicknesses, and different softness grades.

3.16. *Radius of Gyration* is defined as the square root of the moment of inertia divided by mass of the object. Write an equation describing this relationship, then investigate what the actual equation is that describes this formula. Write the description again in such a way that will leave no doubt as to what the correct equation should be.

3.17. Explain why the arrows on this parking lot entrance are as shown in Figure P 3.17.

3.18. How many seconds are in a year?

3.19. Puzzle (Submitted by Schmitt et al.): Identify the people in Figure P 3.19.

3.20. Puzzle (Submitted by Brangham et al.): A truck's odometer displays 151517. Which digit had been in the visible position the longest?

3.21. A horse travels the same distance every day. Two of its legs travel 10 miles every day, while the other two travel 9 miles. The horse is normal and does not have any natural defects. How is this possible?

3.22. Write a six-letter word such that every time you remove one of the letters, the remaining letters still spell a correct English word.

3.23. Puzzle (Submitted by Robia Choi): What do the following letters stand for?

HIJKLMNO

FIGURE P 3.19

CHAPTER 4

CREATIVE PROBLEM SOLVING TECHNIQUES

When there is nothing else left to do, let x = 8.

—Unknown source

The following is a crude translation of an old Persian poem. Unfortunately, I no longer remember the poet's name, but the meaning is still fresh:

If you know, and you know that you know, you can accomplish anything.

If you know, and you don't know that you know, someone should wake you up from your sleep.

If you don't know, and you know that you don't know, you can still lead your lame horse to destination.

If you don't know, and you don't know that you don't know, you will forever remain in your eternal darkness.

4.1 INTRODUCTION

In Chapter 3 we discussed how mental barriers may potentially prevent us from creatively solving problems and from finding new and better solutions to challenges we encounter. As was discussed, being aware of the mental barriers has a tremendous effect on our ability to overcome them and in being more creative. The more awareness, and the more practice, the better we will be in our problem solving. Still, for many, this awareness is not enough impetus to set aside the old habits and learned methods of approaching problems. In order to circumvent these problems, and to help the practitioner with overcoming the mental barriers, a number of different techniques have been developed—some formal, some informal. These techniques are supposed to help us overcome our mental barriers by subconsciously subverting them. In order for them to be effective and to help us in our endeavors, we must follow their basic rules of practice. Some of these techniques are simple and straightforward; some are more complicated and require much practice. You will eventually see that many of these techniques may be used for the same problem, and still at other times, one particular technique may lend itself much better to particular types of problem. In this chapter we will discuss a few of these techniques.

Tom Kelley[1] and Lynne Lavesque[2] talk about the types of personalities involved in innovative and creative activities. Kelley lists 10 types of personalities, including anthropologist, experimenter, cross-pollinator, hurdler, collaborator, director, experience architect, set designer, caregiver, and storyteller. For example, a cross-pollinator is a type of person who takes from different disciplines and puts them together to solve a problem, a multidiscipliner. This type of person may also cross-pollinate among different people. An experimenter is the type of person who jumps into making things to see the solution, one who experiments with ideas, situations, and strategies. Lavesque speaks of the eight creative talents of an adventurer, a navigator, an explorer, a visionary, a pilot, an inventor, a harmonizer, and a poet. In fact, each personality mentioned above can in itself be considered a method of problem solving. Therefore, a cross-pollinator uses the method of cross-pollination to solve problems. The experimenter uses the method of experimentation, and so on.

One may find an impressive number of methods of problem solving in the literature. Some references list a large number of sometimes minor methods with varying degrees of usefulness and applicability. For more information on these, please see References 3 through 15 at the end of this chapter. However, we will consider the following methods, which can be applied directly to most engineering design problems. These methods also vary in their ease of application, flexibility, and usefulness. We will discuss the following methods:

- Brainstorming
- 6-3-5 method
- Morphological attribute lists (menu matrix)
- List of alternative actions
- Analogy (Case-Based Reasoning) method
- Random attributes
- Asking questions
- Do your own slogans
- Incubation
- Synectics
- TRIZ (TIPS) method

4.2 BRAINSTORMING

Brainstorming is a very simple, but effective, technique for overcoming a number of mental barriers. In everyday use, it has become a common verb. You may have heard phrases such as "we were brainstorming and..." expressed by people who generally do not know how to use it, what the rules are, and how it should be conducted; it has come to mean that we were just generating and expressing ideas. However, brainstorming is a technique with certain rules that should be followed if it is to be effective and useful. Without these rules, its effectiveness will be questionable.

There are at least four fundamental rules that govern a brainstorming session:

- Delay and postpone judging the ideas until later.
- Quantity matters. Go fast, generate as many ideas as possible.
- Build on previous ideas to improve them and jump to new ones.
- Stretch your imagination. Be creative. Freewheel.

The first rule is very important in overcoming a variety of fears. As you may remember from Chapter 3, one of the most significant barriers we discussed is fear; fear of being ridiculed, fear of being criticized, fear of not being accepted, fear of others thinking less of you, fear of not being considered smart, and many more. The fact that in brainstorming one may not criticize any ideas allows the participants to freely express their ideas without being subjected to ridicule, or being judged, or being rejected. The fact that brainstorming allows one to overcome this mental barrier makes it very desirable for people who are shy, may have an accent, or may not be very assertive. Notice that the rule requires that we delay the judgment until later, not that we forgo judging or criticizing the ideas altogether. In short, delaying judgment shifts the brain from analysis mode to synthesis mode; the analysis is delayed until later, allowing the brain to focus on synthesis. Delaying judgment of ideas also has other benefits that tie this rule to the rest, as will be discussed shortly.

One easy way to implement this is to use "*yes, and*" instead of "*yes, but.*" In the latter, you criticize someone else's idea by offering an alternative or a modification or addition to it. In the former, you praise the idea by adding to it, by modifying it, or by offering an alternative to it.

Unlike the popular assumption and as stated by the second rule, in brainstorming it is the quantity that matters, not quality. This is because, theoretically, as more ideas are expressed, the chances of finding a more suitable solution for the problem and finding an elegant solution that is hopefully different and superior to typical solutions increases. In other words, as more and more solutions and ideas are expressed, chances of variations in the solutions are better, and we can assume that we have considered a larger variety of solutions. In reality, many ideas expressed during a brainstorming session are not necessarily useful.

The third rule for brainstorming is to build new ideas on the old ones and to combine ideas into new possibilities. It is assumed that as a participant expresses an idea, even if the idea is not good, it may trigger new ideas in other participants' minds, and therefore, help in finding better and more varied solutions. The power of free association is why both the quantity of the ideas as well as the deferring of judgment help the session; as a result of both rules, we intend to increase the participation of everyone in the discussion, and therefore, increase the chances of a better and more elegant solution. If participants are not judged, and if they come up with as many solutions as they can, including through building ideas on others' ideas, we may have better, more varied, solutions.

The last rule of brainstorming is to be creative. This rule in itself is an oxymoron; we are using brainstorming to be more creative in finding a solution, but brainstorming requires us to be more creative. This shows how important it is for us to consciously try to be more creative at all times, even during brainstorming. We should try to set aside our mental barriers. We need to be aware that we gravitate toward the usual, the mundane, the typical, and the stereotypical. But in brainstorming, we need to think freely, in different directions, be free of our fears, be free of our usual limitations, set aside judging of our own ideas even before they are expressed, even if they are not judged by others. Being more creative also increases the quantity and variety of our ideas, and as was mentioned earlier, this will improve our chances of finding a better, more elegant, and more appropriate solution.

Brainstorming is a group exercise, best accomplished with 4–12 individuals; fewer people will make it less effective due to lack of diversity in thinking; more people will make it excessively chaotic and difficult to manage. However, brainstorming may even be used individually, even if less effectively, as a simple and useful technique to make you more creative. The major difference between a group session and individual brainstorming is that the variety of solutions will be less, the total quantity of ideas will be smaller, and the judgments may be harsher (we are our own worst critic). This is one of those cases when 1+1 is more than 2. The group may be composed of anyone who is involved in the problem, including the customer (if any), experts (if necessary), and designers. The group usually has a leader whose role is to record and write down all the ideas that are expressed by the group members. The leader is in fact not supposed to lead the session, but has more the role of a secretary than a leader. Generally, it is recommended that *all* ideas be recorded to ensure that nothing worthy of later consideration is missed, but also to allow free association between ideas. It is also recommended that all ideas be written in large print, on a large paper or large board to allow the participants to readily see all the ideas, and to be able to read the ideas at any time. This will encourage the triggering of new ideas as a result of the previous ideas. In practice, good leaders are very prolific and even artistic, and record the ideas creatively and humorously, allowing a more relaxed and fun session. In many cases, drawing the ideas or developing and making models may be even more helpful than writing the ideas. Some expert brainstormers actually take their "play kit" of Lego parts, play dough, duct tape, Styrofoam, and other similar parts to the session in order to quickly make a model of their ideas. Your drawings and models do not have to be artistic, perfect, or even nice looking; sketches and simple models will do.

A better definition of the problem helps sharpen the focus. Without a good understanding of the problem, the solutions may not be focused enough to yield good results. On the other hand, limiting your problem may also limit your solutions and ideas. In fact, in another technique which we will discuss later, the problem is thoroughly analyzed with the client to completely understand the background of the situation, the solutions already tried, and the capabilities and limitations on the solution, before any attempt is made to solve it. However, in brainstorming, it is desirable to have a good understanding of the problem, but we should try not to limit ourselves.

It is recommended that a brainstorming session be relatively short and to the point, perhaps 20–30 minutes. After the session is over, you may start analyzing the solutions you have found, or you may continue with other problems. If you find that the session has not produced enough solutions, or that the solutions are not satisfactory, you may have another session later, but not immediately. It is assumed that a short session, followed by a break and other activities and then a second session at a later time, is more effective than a long session or multiple consecutive sessions. We will discuss the reason behind this later when we discuss incubation.

Variations on Brainstorming   At times, certain problems arise during a brainstorming session that may reduce the usefulness of the activity. As an example, consider a situation where a very assertive, talkative, and creative person may express countless ideas to the extent that he may negatively affect the others in the group. As a result, others in the group may become too quiet, overwhelmed by his assertiveness or creativity, and withdraw from participation. On the other extreme, it is possible to have someone who is quiet, an introvert, or shy who will not contribute to the session. It is also possible that, perhaps due to someone's influence, the ideas may become so "wild" that they will become useless, or they may all be typical solutions without any novelty. In all these cases, the leader of the group may try to exercise some control over the session and correct the situation. As an example, the leader may invite the person who is not actively participating to present ideas, providing a chance for him or her to speak. However, it is not recommended that participants speak in turn, as this will kill instantaneous creativity. The leader may also want to interject either novel or typical ideas to steer others into particular directions. This should not become a method to exert individual likes and dislikes into a session, but only as a resource to correct problems.

Another problem with the free-flowing format of brainstorming is that it is very hard to give credit to individuals when credit is due. For example, if someone improves another person's idea by building upon it, eventually who should be given credit? The person with the original idea or the person who improved it? Due to this sharing of credit, participants may withhold their ideas. To remedy this, a variation of brainstorming called Programmed Invention may be used. In *Programmed Invention*, the session is conducted as usual until an original idea is expressed by someone. At that point, with a signal from the leader, all idea generation stops except on improving this idea. The originator of the idea records all improvements until done. The session continues from there. In this way, the originator of the idea will get credit for the idea.

Killer Phrases   Any comment, sentence, word, gesture, or behavior that may negatively affect the participants is a "killer phrase"; it kills the atmosphere that brainstorming tries to create, in which ideas flow freely without the fear of ridicule, criticism, or reprisals. Examples are:

- Oh no!
- Won't work.
- Too expensive.
- Are you sure?
- I don't understand you.
- Where do we find people to do this?
- A disapproving look.
- Why would you want to do it that way?
- Old idea!
- Ignoring the idea.
- That's crazy.
- We are already too much into it to change it.
- It is not in the budget.
- That's not what we want.
- I am tired of hearing this.
- Really?
- How will it do that?
- You think managers will like this?
- A sigh.
- We did this before and it did not work.
- That requires too much work.
- We don't have enough time to do it.
- Yes, but.
- Too wild.
- You don't understand the problem.
- Let's bring that up some other time.
- I don't understand it.
- What about our image?

In all these cases, the phrase conveys a negative feeling toward the person who expresses the idea. As a result, that person may feel threatened, rejected, or ridiculed, and consequently may no longer fully participate in the session, sometimes even unconsciously.

EXAMPLE 4.1

Every year, Cal Poly's Food Science Product Development Team participates in a national contest for new food products, in which a new food product or combination is created, tested, packaged, and presented. This competition was used as an impetus for an exercise in an engineering class. Figure 4.1 is the result of a brainstorming session regarding possible new foods. When multiple groups participate in the exercise, you will notice that there are always some ideas that are common between them, although these groups are at different locations and they do not hear each other. These ideas are usually very common and typical. Whether they are new or not and directly useful or not, these ideas are indirectly useful in providing seed ideas for other ones. In addition to these, there are always some ideas that are totally useless or impractical. These ideas are also indirectly useful in providing a platform for others' ideas. And hopefully, there is a good chance that within the list there are a few that warrant further consideration and analysis. Those ideas are the main benefit of brainstorming and the fundamental value of it.

One of the interesting ideas from the example in Figure 4.1 is "beer concentrate," perhaps warranting consideration by a food scientist.

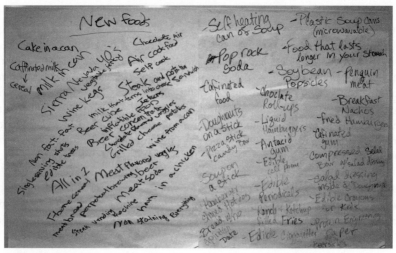

FIGURE 4.1   An example of the results of a brainstorming session.   ▪

4.3 THE 6-3-5 METHOD

The 6-3-5 method is a variation of brainstorming, in which 6 participants, seated around a roundtable for continuity, sketch (or deliver) 3 solutions within 5 minutes without being interrupted (some use the number 5 as the number of solutions picked from the list, or the number of times the sketches (solutions) are passed to the next person who will add to them within a certain time limit).

As in brainstorming, ideas are supposed to trigger new ideas. The added ideas should be improvements, not criticism. This process is void of verbal communication. This provides a chance for everyone to participate in the process and have an opportunity to contribute to the solution. Similar to other variations of the brainstorming technique, 6–3–5 circumvents many participant-related shortcomings of brainstorming, where one may be too apprehensive about speaking up or too aggressive in talking.

4.4 MORPHOLOGICAL ATTRIBUTE LIST (MENU MATRIX)

Morphological attribute list or menu matrix is a very easy-to-use, but fun, method that has the potential to generate countless alternative solutions. In fact, you will find that you may have to limit the scope of it before you can pick a useful solution from among the alternatives.

Morphological attribute list technique consists of picking several attributes or subcomponents of the whole system, listing alternative solutions or ideas for each attribute or subsystem, and then randomly connecting the entries on the lists to each other to create a whole solution. In order to understand how this technique works, we will apply it to the previously mentioned problem of developing a new food item for a competition. We want to find new ideas for foods such as snacks, packaged foods, etc. Although many different attributes may be considered, we will choose the following attributes:

• Main ingredient
• Flavoring (added) ingredient

- Form of the food
- Packaging

In this list, the *main ingredient* will be the essential part of the new food item, while the *added ingredient* (flavoring) will be a second major ingredient or a flavor that will be added to change or modify the main ingredient. *Form of the food* refers to the state in which it will be presented to the customer, and *packaging* will be the form in which the food is stored for sale. Next, we will list as many different alternatives as we can think of for each attribute. The following lists are by no means exhaustive or complete. You certainly may come up with other alternatives that are not included in the list, and therefore, achieve other solutions.

Main Ingredient	Added Ingredient	Form of the Food	Packaging
meat	chocolate	sandwich	tablet
ice cream	vanilla	powder	can
cereal	strawberries	liquid	box
hamburgers	almonds	dehydrated	bag
soda	nuts	vapor	cup
rice	cumin	soup	on a stick
beer	spices	jerky	on the cob
alcoholic beverage	cheese	processed	on a plate
		frozen	
		instant	
		concentrate	

If we randomly connect a different alternative from each of the lists of attributes to each other, every set of connected alternatives will give us a solution. First, let's look at the following:

Main Ingredient	Added Ingredient	Form of the Food	Packaging
meat	chocolate	sandwich	tablet
ice cream	vanilla	powder	can
cereal	strawberries	liquid	box
hamburgers	almonds	dehydrated	bag
soda	nuts	vapor	cup
rice	cumin	soup	on a stick
beer	spices	jerky	on the cob
alcoholic beverage	cheese	processed	on a plate
		frozen	
		instant	
		concentrate	

As you can see, we have a combination of one item from each of the lists, giving us a regular cheeseburger sandwich served on a plate. Obviously, there is nothing new in this solution; it is something that has existed for a long time. And of course, there are many other combinations that are similarly familiar foods. Now let's take another combination, as shown below:

Main Ingredient	Added Ingredient	Form of the Food	Packaging
meat	chocolate	sandwich	tablet
ice cream	vanilla	powder	can
cereal	strawberries	liquid	box
hamburgers	almonds	dehydrated	bag
soda	nuts	vapor	cup
rice	cumin	soup	on a stick
beer	spices	jerky	on the cob
alcoholic beverage	cheese	processed	on a plate
		frozen	
		instant	
		concentrate	

Obviously, this combination does not work. Here, we get a liquid mixture of meat and chocolate in the form of a tablet. But then again, who knows? Market research may show that in fact there may be people who like chocolate covered meat of some sort, in liquid form. What do we do for the tablet? Well, maybe instead of a tablet, we use a small capsule (or add capsule to the list and use that instead). What this shows you is that there may be ideas in the list that may be totally useless, at least for now.

Next, let's consider the third combination as shown:

Main Ingredient	Added Ingredient	Form of the Food	Packaging
meat	chocolate	sandwich	tablet
ice cream	vanilla	powder	can
cereal	strawberries	liquid	box
hamburgers	almonds	dehydrated	bag
soda	nuts	vapor	cup
rice	cumin	soup	on a stick
beer	spices	jerky	on the cob
alcoholic beverage	cheese	processed	on a plate
		frozen	
		instant	
		concentrate	

In this combination, we get a nut-flavored meat in the form of jerky on a stick. I do not know whether such a product exists or not, but it is a new product I have not seen. So, as you see, in this method it is possible to get known combinations of attributes that already exist, combinations that do not make sense, and combinations that may be new and worthy of further investigation. Because so many new combinations are possible, there is a good chance that we may find solutions that are atypical, random, and "lateral" in nature rather than "vertical." By the way, do you see that we could also find the beer concentrate through this method? In this case, you may in fact skip the added ingredient (if you do not like spiced beer) and come up with a beer concentrate in a can.

How many combinations are possible from this list? The total number of possibilities, if we combine only one from each of the four lists, will be at least $8 \times 8 \times 11 \times 8 = 5632$ ideas (scores more if we include combinations with fewer or more than one item from each list). We can get additional combinations if we combine more than one from each list, or skip any list. What this also means is that we will have to check thousands of ideas from this simple list to decide how many are good, new, old, useless, or really worth further analysis. Some researchers have tried to use a computer program to check the combinations. However, until we learn how to create computers that are capable of creative thought and imagination, it will be fruitless to try and program a computer to decide whether or not a new combination is useful and creative. In the absence of creative comprehension on the side of computers, the designer will need to check all combinations. This is why it may eventually be better to go back to the list and eliminate any possibility that is unacceptable.

Main Ingredient	Added Ingredient	Form of the Food	Packaging
meat	chocolate	sandwich	tablet
ice cream	vanilla	powder	can
cereal	strawberries	liquid	box
hamburgers	almonds	dehydrated	bag
soda	nuts	vapor	cup
rice	cumin	soup	on a stick
beer	spices	jerky	on the cob
alcoholic beverage	cheese	processed	on a plate
		frozen	
		instant	
		concentrate	

Obviously, it is possible to skip certain lists (or attributes) or to add other ones and try new combinations too. For example, you may skip the flavoring attribute and connect the meat, jerky, and on-a-stick together. Alternatively, it is possible to combine hamburgers and soda, in the form of a sandwich, placed in a box-and-cup combination, and offer them together.

This technique also emphasizes the trivial nature of having "an idea." As you see, in a matter of a few minutes, it is possible to come up with a basis for thousands of ideas. So, having an idea in itself is not important; you can come up with countless ideas very quickly. What matters is an idea's value. This technique helps us look in new directions, laterally, that we may normally fail to look. It allows us to have a large number of ideas from which we can search for new directions and inspirations.

One important note about this technique is its potential to create a design that is not wholly integrated, but a combination of individual elements in which each element relates to an attribute. However, a good design is a whole; this means that every element of the design is useful, but that all the elements work together to accomplish the task. But when the whole purpose of a problem is artificially divided into attributes and a separate solution is found for each, the combination may or may not be optimized. For example, the solution for one attribute may in fact address concerns about multiple other attributes too. In that case, there may not be a need to have elements from other attributes added. The designer should always go back and simplify the solution and integrate all the individual elements into a whole solution.

4.5 LIST OF ALTERNATIVE ACTIONS

In this technique, a list of alternative actions is used in order to help the designer look into new directions and possibilities for solutions. Imagine that you have prepared a list of random verbs. When you are trying to come up with new ideas to solve a problem, use the verbs on the list to help you come up with a solution for each term. As an example, consider the following list:

> Drop, rotate, alternate, shoot, copy, multiply, eliminate, change, increase, decrease, compress, hide, add, lengthen, wrap, transfer, shorten, camouflage, select, fold, expunge,...

Each of the above may now be used to create a new solution, even if not immediately useful or feasible. These new ideas may eventually help the designer think in new and unexpected directions, or may become impetus for new ideas.

In fact, each of the above action verbs may further be broken into other actions, each providing a new alternative. For example, you may use the following additional alternatives with the verb "change":

> Change lengths, color, texture, elements, components, frequency, image, timing, initial conditions, destination, means of achieving it,...

As you see, hundreds of new directions may be looked at easily by simply considering each action and trying to find a solution based on the action. Since these actions are randomly chosen, they provide a means to look laterally in directions you may normally not look. You may prepare a new list by simply looking into a dictionary and choosing new action verbs.

EXAMPLE 4.2

Let's apply some of the above-mentioned verbs to the idea of energy sustainability. We want to generate a few new ideas on how to further sustain our energy sources.

- For "alternate," the idea was to alternate heat transfer during the day and night. At night, when outside temperature during the summer months is low, the water from a large tank may be cooled through a cooling tower. The colder water can be used later during the day to cool down a building. This system is implemented at Cal Poly (Figure 4.2).

FIGURE 4.2 A cooling tower used to cool the water during the night. The cooled water is used during the day.

- For "rotate," the idea was to integrate a hand-crank recharger to a cell phone similar to the now common flashlight (no more need for an electric recharger).
- For "compress," the idea was to use wind energy to compress different gases into capsules for later use, thus reducing the need for electricity.
- For "hide," the idea was to bury a network of pipes underground in order to warm the fluid in it for home heating during winter.
- For "wrap," the idea was to wrap the cold water pipe of a shower around the drain in order to warm it before going to the showerhead. ∎

As you see, each one of these verbs yields an idea, some new, some variations of other ideas. How many of these ideas might you have thought of without these stimuli? Can you suggest an idea for some of the other verbs?

EXAMPLE 4.3

Until the mid-1860s, train engineers had to engage the brakes to stop the train. The problem with this was that if the braking force was not large enough, the train would not stop. This used to happen frequently in down-slopes, where the force needed to stop a large train was tremendous. Additionally, when a pneumatic brake was used, if the engine would shut off due to any problem, or if the air pressure was lost due to a leak, no breaking force would develop in the brakes. As a consequence, there used to be many accidents in the railway industry. In one estimate, 30,000 people died each year due to derailments and from brakemen jumping between cars to turn the hand-operated mechanical brakes on each car.

Imagine that you are involved in this problem and you need to find a solution for this problem. Let's imagine that you are using the above-mentioned technique, and one of the words is reverse. What would you find as a solution to this problem based on the word *reverse*? One solution is to reverse the action of the brakes; instead of applying the brakes when needed, what if we would reverse the direction and have the braking force applied at all times, but remove (disengage) the braking force when we do not need it? The patented solution was a reverse-acting fail-safe braking system that is always engaged through the force of large springs. However, when we want the train to move, we apply a force to the system to disengage the brake, thus allowing the train to move. In this case, if the engine power or the air pressure is lost due to a problem or an accident, the train will automatically stop. This simple reversal of the application of the forces has saved countless lives and has abated countless accidents. Nowadays, the same system is standard in buses, trains, and trucks. In fact, the same idea is utilized in robot arms in order to prevent the arm from falling and creating a hazard if electrical power is lost. For this, the electric joint-brake is normally on until the arm needs to move, at which time, the brake is disengaged.   ∎

Although there is no indication that Westinghouse employed this technique, you can imagine how easily it could have helped him. Can you come up with other ideas for the same problem using other words?

George Westinghouse was traveling from Albany, New York, to his home in Schenectady in 1866 when his train came to a halt as the one ahead derailed. Westinghouse and others took over two hours to bring the train back on track using crowbars and force. As he got back on the train, he thought about the accident and how to prevent it in the future. He invented a system comprising a short piece of rail that could be run off the track at an angle with the derailed car. A locomotive could then pull the cars back on the track by getting close to them on this accompanying rail. He continued to invent other equipment for the railroad industry, including automatic rail switching equipment and air brake systems.

Until Westinghouse invented the air brake system, brakemen had to stand between cars. On signals from the engineer, they would engage the brake on each car by turning a wheel, and then jump to the next car. Lack of a coordinated action between the brakemen would result in derailment. Additionally, due to the lack of sufficient braking power, the trains had a limited speed.

Westinghouse first invented the air brake system in which an air compressor was placed in the engine cab. The engineer would apply the pressurized air onto the brakes installed at each car through a set of pipes between the engine cab and each car. However, if air pressure was lost, there would be no braking power. He later reversed the action; the brakes were normally applied, but they would be released by applying the pressurized air. This way, if air pressure was lost, the brakes would automatically be applied and the train would stop. Westinghouse received his first patent on the air brake in 1869. He eventually received 103 patents related to the air brake systems alone. This system is still in use on all air-brake vehicles.

Later, Westinghouse got involved in the production and distribution of electricity. His rival, Thomas Edison, had started small electricity generating plants in New York. He produced DC power and distributed it through short distances into neighborhood houses. However, since he could not transform DC power into higher voltages, and due to electrical losses of higher currents at lower voltages, he could not distribute the power over long distances. Westinghouse built alternating current power plants and transmission lines, proving the practicality and feasibility of AC power. In his case, since the AC power could be transformed into very high voltages and therefore very low currents, the electrical loss was minimal. Thus, he could transmit the power over huge distances without much loss. Edison invented the electric chair to demonstrate the danger of AC power, but could not convince the users against using it.

Westinghouse received close to 400 patents and started 60 companies that had a total of 50,000 employees. At his peak, he was the largest private employer in the United States.

EXAMPLE 4.4

A new *ciclobus*, a bus that is modified to accommodate about 30 passengers with their bikes that board it by going up a ramp, has been introduced in Havana, Cuba. A similar list of alternative verbs, as well as the other techniques mentioned so far, can easily result in this new idea.   ■

4.6   ANALOGY (CASE-BASED REASONING) METHOD

This method[16] is based on reasoning by analogy, where knowledge about previous experiences may be used to find a solution for the present problem. Many experienced designers rely on their "expertise" in solving problems. Although at times this expertise and a priori knowledge creates its own mental barriers, it can still be an asset.

One example is when experience with nature helps a designer find solutions based on natural designs (called bio-mimicry). In bio-mimicry, nature is used as a database of solutions for problems. Examples include artificial kidneys, artificial joints, cameras, radar, sonic sensors, stereo vision system, depth measuring device (which follows the way salamanders fix one eye on the object in flight while the other eye ranges from side to side to calculate depth), bridge design (see Chapter 16) and others, where natural designs were used as a guide to human designs.

To use the analogy method, apply the following four steps:

1. **Recall** cases similar to the current problem, and how they were solved. Successful implementation of a procedure in a previous experience may help solve the current problem. As an example, imagine that you want to design a cart that will slide down a cable at an angle of 10° between two points with minimum friction. The cart may not leave any residues on the cable. You recall that lubrication has been effective in reducing friction, but it leaves the cable dirty.

2. **Reapply** the process to the current problem by adapting it to the current situation. These may not be exactly the same situations, and therefore, changes may need to be made in order to match the process to the problem. In our example, even though lubrication can help, it may not be used. Instead, you decide to attach a small piece of ice between the cart and the cable. As the ice melts, it provides local lubrication and reduces friction, but will evaporate without leaving anything behind.

3. **Revise** the process in order to match it with the needs of the current problem. This may involve testing, simulation, or experimentation. In our example, you may test it to see whether this idea works properly or not, and if necessary, you may still revise further.

4. **Store** the new solution for future use in case a similar problem arises.

4.7   RANDOM ATTRIBUTES

This technique is in fact somewhat similar to the above-mentioned list of alternative actions technique. However, in this case, instead of limiting the words to action words or verbs, you may choose any random attributes. For example, imagine you pick a color (e.g., green), a temperature (e.g., hot), or an object (e.g., chair). You may then force yourself to come up with an idea for the problem at hand that relates to these attributes. In this technique, too, we try to force the mind to look in new directions that it may ordinarily not do so on its own, going laterally into uncharted territories and trying to find new solutions.

EXAMPLE 4.5

To create a new food item for a class assignment using the colors of Cal Poly (green and yellow), the participants created a banana ice cream with kiwi swirls.

Can you think of a new food item based on the content of a poster in your room? Would you have otherwise thought of the same idea?   ■

4.8   SCAMPER METHOD

The SCAMPER method by Bob Eberle[16] is similar to other techniques we discussed above. It relies on a lateral thinking approach to finding other alternative solutions. SCAMPER stands for:

S	*Substitute*	who else, what else, what other ingredients, other materials, places
C	*Combine*	units, components, elements, needs, solutions
A	*Adapt*	what else is like this? what else can be done with it?
M	*Modify*	change meaning, color, motion, sound, odor, or shape, add, subtract, make it bigger, smaller, simpler, eliminate, divide, etc.
P	*Put to other uses*	new ways to use as is, other uses if modified, etc.
E	*Eliminate*	what to subtract, what to remove, smaller, condensed, lower, shorter...
R	*Rearrange*	interchange components, other patterns, change layout, other
	Reverse	sequence, rearrange positive and negative, opposites, upside down, reverse,..

EXAMPLE 4.6

Consider the evolution of cell phones. The design of cell phones has changed over the years into the flip-open type, sliding type, cameras were added, calculators, alarms, and many other features included, antennas were eliminated, new uses were found, and so on. Although there is no evidence that any of these improvements or changes occurred as a result of SCAMPER, it is clear that the method could help in coming up with the ideas. Can you guess what might occur next?   ■

4.9   CREATE YOUR OWN SLOGANS

Torrance[17] has listed creativity as wanting to know, digging deeper, looking twice, listening for smells, shaking hands with the future, cutting corners, and many more. In fact, you may use these and many more of your own inventions as slogans or guides for creative thought. For example, listening for smells relates to someone who looks for new and unrelated relationships, while cutting corners may mean to change the shape, to reduce, or to change the process. Like the methods discussed above, this is yet another way of forcing your brain to think in directions it may normally not. So, make your own slogans and use them. Here are a couple more: listen to the wind, feel the sun, eat when you are full, tic with a tac.

Barron, too, lists additional ingredients for creativity, including recognizing patterns, making connections, taking risks, taking advantage of a chance, and seeing in a new way. Add these slogans and more to your list too. When you are trying to solve a problem, look at the list and try to use them.

4.10 ASKING QUESTIONS

What is your definition of the three words *smart*, *intelligent*, and *clever*? For most people, these three words have different meanings. Smart means someone who knows a lot, has a good memory, someone who can remember everything she or he sees, and understands what everything means. Intelligent means scholarly; it is a more formal definition of knowledgeable, someone who studies a lot. Clever is someone who is ''street-smart,'' someone who can scheme plans and respond to others quickly. However, if you look at any dictionary, you will see the definitions of the three words similar to the following:

Smart *(adj)*   mentally alert; bright; clever or skillful.
Clever *(adj)*   showing mental quickness; skillful; alert.
Intelligent *(adj)*   possessing intelligence or intellect; having the power of reflection or reason; having a high degree of intelligence or mental capacity or power of perception.

As you notice, these words mean the same thing. But the perception that most people have is that someone who is considered smart must have much knowledge, and vice-versa. To be considered smart, the person must have (or show that she or he has) much knowledge. As a result, people assume that if someone appears to not have as much knowledge as someone else, then that person must not be as smart. You probably have heard people say ''Oh, she is so smart she can remember everyone's name at the first meeting'' or ''he is so smart he can solve this math problem without using a calculator.'' However, consider a very young child, perhaps one year old. The child has very little knowledge. She or he can barely say any words at all, does not know any math or physics or engineering or history, cannot perform any significant operation or task, and yet can understand what others tell him or her, can follow instructions, and learns at an amazing rate. Just think how quickly a child learns to understand and speak a language. How many of us adults can learn another language that quickly, even though we know a lot more? Isn't the child very smart? She or he definitely is very smart to be able to learn at least one language, social skills, a variety of functions, and still have fun growing up. Then clearly, it is not how much you know that makes you smart, but how much you can learn, how strong your intellect is, and how you can process the information you receive.

Why is this brought up? Because one of the ways that children learn so much, so quickly, is that they ask questions. They are not afraid of asking question after question, until, as the popular joke goes, their parents and teachers go crazy! By having a questioning attitude they learn more. However, as we grow up, and especially at the age of puberty when we become more aware of our environment and other people, and when we notice how others perceive us, as we do them, we also become more aware of how others judge us. And since everyone wants to be thought of as being smart, and since we think that being smart means we know much, and since asking questions means that we do not know, we stop asking questions. Have you been in a situation, say in a meeting or classroom, where many are happy when someone else asks the same question they have, so that they do not have to ask? Have you ever been in a situation where someone says something that you do not understand, but you go on listening and nodding and never ask a question? A while ago, in a television show, this was used to create a situation comedy. In a party, the host would strike up a conversation with one of the guests while candidly they were filmed and the conversation was recorded. The host would talk for a long time about subjects that made no sense while the guest listening to him would nod constantly as if he understood and agreed with the guest, while clearly it was impossible to make any sense out of the conversation. Later, another person would strike up a conversation with the guest and ask him about that conversation with the host, and whether he understood what was being discussed, and the guest would confess that he did not, but never asked any clarifying questions.

Ask and you will learn more, you will know more, and you will be "smarter." You will also be able to solve problems better when you ask questions about the problem, and when you know more about the problem and possible methods of solving it. Ask the users, the experts, your colleagues, your superiors, and anyone else who can help.

Saul Steinberg once said: "There are no foolish questions and no man becomes a fool until he has stopped asking questions."

4.11 INCUBATION

Have you ever been in a situation in which you try to solve a problem, remember someone's name, or recall a song, but you cannot, and later, at an unexpected time, you will know how to solve the problem, recall the name of the individual, or remember the song? This may have happened to you in taking a test, when you think of the method of solving the problem after the test is finished while you are watching television or while playing sports. It may have also happened to you while in a gathering place, where you could not remember the name of an acquaintance until after you left the party. Or perhaps, when you remember a song before you go to bed a day after the time you tried to first remember it. This happens under a particular set of conditions, which will be discussed shortly, and is called *incubation*.

Incubation is based on a theory that if one works hard on a problem for a while, and then leaves the problem alone and works on other matters, the "other side" of the brain will continue working on the problem, even if subconsciously. Then there is a chance that it may find a solution to the problem, and since the "other side" of the brain is generally the more holistic processor, it may find ways of solving a problem that are different, more creative, and less common. In many cases the person may not even be able to describe where the solution came from since the speech center is usually in the more logical dominant half-brain. In incubation there is no control over if or when a solution is found. So, the realization of the solution is usually sudden, at unexpected times when the brain is relaxed or is doing other things, and sometimes too late. Still, incubation is responsible for many things we remember or solve without really trying. However, it is necessary that you first really try to solve the problem by almost frustrating your dominant brain before the "other one" will do its best.

EXAMPLE 4.7

There is a famous story about Archimedes, the great Greek philosopher who lived about 200 B.C., who was asked by the contemporary ruler to find a method to ensure that the gold brought to his court as tax was pure gold. Archimedes tried very hard to solve the problem for a long time, but could not find any solutions for it. One day, while relaxing in a public bathtub, he suddenly noticed how he was floating in the water and realized the buoyancy effect of objects in water and the relationship between the density of a material and its buoyancy. He was so happy and excited about his discovery that he ran into the streets yelling "Eureka! Eureka!"; "I found it! I found it!" He had found the solution to measuring the density, and thus purity, of gold by floating the object in water. This, in fact, is incubation at work. He had worked hard on the problem, then relaxed and left it alone. His brain continued to work hard on its own, found a solution, and spurted it out without words. ∎

This frequently happens to students too. The student tries hard to find a solution for a problem during the test, perhaps even unsuccessfully. This exercises the brain and forces it to think hard. However, after the test, although the student is not actively thinking about the problem, the ''other side'' of the brain continues to process the problem, and if it finds a solution, as if a light is turned on in the brain, the student suddenly realizes that he could have done the problem, usually followed by swearing and disappointment as to why he could not do this earlier during the test.

Since there is no guarantee whether incubation works, and when, it is not something that you should rely on for your problem solving or design activities. But since there is no cost in trying it, any solution found this way is a bonus. However, remember that you must first do your best in trying to solve the problem before relying on incubation.

If you remember, when we were discussing brainstorming, one of the requirements for effectiveness of the technique is to work hard in the session for awhile, and then stop the session and go back to analyzing the solutions and working on other problems. The theory is that while you analyze the already suggested solutions, or as you work on other problems, the ''other side'' of the brain will do its job. So, if you go back to a new brainstorming session, there is a better chance that you may find new solutions than if you were to spend extra time during the first session; the period that you leave the problem alone will help your brain process its information without your interference!

4.12 SYNECTICS

Synectics is a research-based technique originally developed between the 1940s and 1960s. One of the principal inventors of the techniques was William Gordon, who has described the technique in a book called Synectics,[3] and later founded a company by the same name. Synectics is a Greek word meaning joining together different and seemingly irrelevant elements. You will shortly see how the technique is actually based on this premise, lending it to the same notion as we have mentioned before, that is, to help the designer think in ways and directions that he may not normally think (lateral thinking), and thus increase the possibility of a new, creative, atypical solution. The technique is based on much research into the ways individuals think creatively, and the process of creating new ideas. To do this, team members were recorded on tape and their thought processes were inscribed and analyzed to determine the thought processes that they had employed in order to arrive at their solutions. Later, the common processes were summarized into the principal steps of Synectics and were employed to design solutions to problems in all industries.

Research in Synectics was based on the premises that (1) the creative process in humans can be concretely described and that this description should be usable in teaching the methodology to increase the creative output, (2) the cultural phenomenon of invention is the same whether in arts or in sciences and that they are characterized by the same psychic processes, and (3) the process of creative enterprise is analogous between an individual and a group. These assumptions were necessary due to the common thinking at the time that creative processes only belonged to the arts and literature, and that group-thinking dynamics were different than an individual's.

As a result of the research into creative thought processes, the following hypotheses were developed, on which Synectics is based[3]: (1) the creative efficiency in people can be much increased if they understand the psychological process by which they operate, (2) in creative process, the emotional and irrational components are more important than the intellectual and rational components, and (3) it is the emotional and irrational elements that must be understood to increase the probability of success in problem solving.

Synectics defines the creative process as "the mental activity in problem-stating and problem-solving situations where artistic or technical inventions are the result." First notice how both stating the problem and solving the problem are included in the creative activity. This means that understanding and defining the problem in a creative manner are in fact as important as solving it. Next also notice how artistic and technical solutions are equated in the creative process. In other words, creative technical solutions and artistic solutions are essentially similar as far as the creative process is concerned. This belief in the creative process results in the following basic principles for Synectics:

(i) making the strange familiar

(ii) making the familiar strange

Making the Strange Familiar   This, in a nutshell, means that the problem is analyzed, the customer is interviewed, and all aspects of the problem are considered. In doing so, the problem is scrutinized, all previous solutions and attempts to find a solution are investigated, existing products are dissected and looked at, all limitations are discussed, and "what-if" questions are asked to understand the problem completely. However, it is important to ensure that the designer does not become so involved in this phase that it becomes the final goal. Although this investigation may yield some superficial solutions, it is important to find a new *viewpoint*, a new way of looking at the problem, in order to find novel solutions.

Making the Familiar Strange   This is to distort, change, invert, turn upside down, and transpose the familiar ways of looking at the problem that give the designer a sense of security, thus making the familiar strange again. This is an attempt to develop the new viewpoint which we need to create new solutions. In the familiar world, everything is in its right place, in correct orientation, and in logical order. In the new viewpoint, when we make it strange again, things are not in their expected order, they are not right-side up, and they do not follow the familiar logic. In fact, if you remember, one way to overcome the problem of sensory saturation is to turn it upside down, to look at the object in an unfamiliar orientation in order to notice the details we normally miss. This is also the reason to talk to someone who is not familiar with the problem to get feedback from a different point of view, a new viewpoint.

Operational Mechanisms   Synectics has identified four operational mechanisms for accomplishing the task of making the familiar strange. These mechanisms are all metaphorical analogies that help the designer see the problem with a new look, and in different ways. These analogies are characterized as *personal*, *direct*, *symbolic*, and *fantasy* analogies.

Personal Analogy is the personal identification with the elements of the problem. In this analogy, the designer may put himself into the problem, as an element, in order to see what may be happening to the element, to feel what it does, and to sense what the element can do to solve the problem. The personal identification with the problem is related to the personal experiences the designer has had, and therefore, is helped by how imaginative the designer is. In one example, Gordon relates the personal analogies of the team members when they were working on the problem of designing a machine to deliver a constant angular velocity output for variable speed input. One team member described how he would grab the input shaft, the output shaft, and let the input shaft turn in his hand if it were going too fast, and how he felt the heat of the shaft as it turned in his hand.

Direct Analogy is the actual comparison of parallel facts, knowledge, or technology. In this case, the individual relates the elements of the problems to real technological, physical, or worldly examples. It is related that Alexander Graham Bell compared his telephone with the human ear and the relative disparity between the masses of tympanic membrane and the three bones of the middle ear (hammer, anvil, and stirrup), and how his telephone could similarly function with a relatively light membrane and a heavy steel piece. This direct analogy between a physiological system and his device made the telephone possible. Many other similar analogies have been found between physiological and human systems, including jet engines and octopi, human inner ear and vibratory tachometers, human eyes and camera systems, and honeycombs and lightweight structural elements.

Symbolic Analogy uses impersonal and poetic images to relate to the problem. Gordon relates the example of an Indian rope trick to find a solution for a 4-ton jack that could lift the load 3 feet into the air, while fitting into a $4'' \times 4''$ box. The Indian rope is soft and collapsible at the start, but through "*magic*," it becomes hard so that the magician can climb on it. The symbolic analogy was first reduced to the idea of a jack consisting of a rubber tube inside a rounded tape measure with hydraulic pressure build-up in the tube, and eventually to a pair of bicycle chains linked together on their sides so that each one would only fold in one direction. Initially, the chains would individually fold into the box, but when extended and linked, they would be stiff. The pair would prevent collapse in directions perpendicular to each other.

Fantasy Analogy is the fantastic fulfillment of our wishes, whether in art or in technology. When the designer fantasizes possible solutions, no matter how impossible, he is in fact using fantasy analogy.

In one example, when designing a vapor-proof space suit, Gordon [3] relates the following analogies (with many others in between), including fantasy analogies:

—"Let's imagine you could will the suit closed (fantasy analogy)."

—"You wish it closed and invisible microbes, working for you, close hands across the opening and pull it tight."

—"A zipper is kind of a mechanical bug (a direct analogy), but not air tight, or strong enough."

—"I feel like a Coast Guard insect (personal analogy), worst storm in the winter, vessel on the rocks, can't use lifeboats, some impatient hero grabs the rope in his teeth and swims out."

—"If you used a spider, he could spin a thread, and sew it up."

—" Spider makes thread, gives it to a flea. Little holes in the closure, the flea runs in and out of the holes closing as he goes."

Eventually, the suit was designed with small steel rings protruding out of it, with a steel cable that could be threaded into the rings pulling the suit together. This way, the suit would be air tight, but also very secure. Notice how a few fantasy analogies were eventually reduced to a practical solution.

Synectics theory can be summarized into the three following phases and their associated steps. These steps are used to reduce the theory into practice:

• Examine Phase

(a) **Problem as Given:** This is the statement of the problem as given to those who are responsible for solving it. The statement may be an accurate account of what the problem is, or it may be somewhat vague and unclear, where the actual problem is hidden in other requirements that are not as important or as basic. Although the title implies that the problem is to be given, the same applies when the problem must first be found or developed, except that the process will be longer and more complicated.

(b) **Making the Strange Familiar:** Any problem that needs a solution, no matter how old or familiar, must become clear in the sense that any analysis can uncover elements that are not yet revealed. Since we are not yet looking for solutions, it is not important that these elements may be contrary, but it is important that they are made clear.

(c) **Problem as Understood:** In this phase, the different individual pieces of information revealed in the previous analysis are isolated for examination, and a clear understanding of the problem is achieved. This is similar to *problem definition* or *problem specification*, in which a clear understanding of the problem and what the goals are is sought. We will discuss this in detail in Chapter 5.

- **Stretch Phase:**

  (a) **Operational Mechanisms (Personal, Direct, Symbolic, and Fantasy Analogies or Metaphors):** These metaphors or analogies are developed in relation to the problem as understood, or are caused by it. In this phase, the problem as understood is pushed and pulled and is deformed until its rigid form is broken, allowing new conceptual ideas. Free association of words and concepts are used to enhance the process.

  (b) **The Familiar Made Strange:** As a result of the analogies, the problem as understood is seen in a somewhat different light, and the familiar is made strange again.

  (c) **Psychological States:** According to the theory of Synectics, the mind goes through the states of involvement, detachment, deferment, speculation, and being commonplace in order to arrive at a solution. These psychological states help the mind with the creative process.

- **Return Phase**

  (a) **States Integrated with the Problem:** At this point, when the mind has been helped by the mechanisms to reach this state, the most pertinent analogy is compared to the problem as understood.

  (b) **Viewpoint:** Each time an analogy is compared to the problem as understood, a viewpoint is possible. If the comparison can lead to a technically feasible insight, the viewpoint is actual (it can be implemented, it is real). If not, a viewpoint may only be potential (it is not feasible as is, but may help in other insights).

  (c) **Solution or Research Target:** In this phase, the viewpoint is implemented, tested, and put into practice, or it may become the subject of further research (to arrive at a feasible solution). The activities in this phase depend on whether the viewpoint implies the reintegration of known material, or new material has to be developed. One requires a solution, the other requires further research.

As is clear from this description about the phases of Synectics, this method is highly technical, requires substantial training and insight, and its success depends on effective leadership by knowledgeable personnel. However, with proper training, it can be effective, and can lead to creative solutions.

4.13 TRIZ

TRIZ in Russian, TIPS (Theory of Inventive Problem Solving) in English,[18, 19, 20] is a method through which solutions to problems with inherent conflicts can be developed systematically. It was developed by Genrich Altshuller and his colleagues starting in 1946, but is practiced in many parts of the world today. TRIZ is a very systematic method based on the hypothesis that creative innovations follow universal principles, which, if identified and codified, can be followed to predictably find new solutions or improve existing ones. Over 2 million patents have been studied and classified by their level of inventiveness in order to recognize common principles of innovation.

Altshuller divided his findings into five categories:

1. Apparent solutions, 32 percent of cases, where solutions are based on personal knowledge. This may require consideration of perhaps 10 solutions.

2. Minor improvements, 45 percent of cases, based on knowledge within the company, with consideration of perhaps 100 solutions.

3. Major improvements, 18 percent of cases, based on knowledge within industry and requiring consideration of approximately 1000 solutions.

4. New concept, 4 percent of cases, based on knowledge outside the industry, and requiring consideration of 100,000 solutions.

5. Discovery, 1 percent of cases, based on all that is possible to know, requiring consideration of 1,000,000 solutions.

This led to his assumption that perhaps 90 percent of the problems that engineers face are already solved. Therefore, following a path, starting with their own knowledge, and advancing to the higher levels, most problems may be solved with knowledge within the industry or outside of the industry.

To use TRIZ, you must first identify your problem, understand it, and recognize what each characteristic of the problem means. Next, the problem is defined (please see Chapter 5 for discussion about this), including all contradictory requirements. For example, as the size of a shaft is increased to increase its strength, its weight increases as well, requiring more strength. Similarly, as the speed of a device increases, the forces acting on its rotating parts increase, requiring more strength (or stiffness), which may require more weight, further increasing the forces on the rotating parts. The Quality Function Deployment (QFD) method (which will be discussed in Chapter 5) can assist the designer in recognizing conflicting requirements when product characteristics are analyzed.[21] Although QFD is not a problem solving method, it may be used for this step. The next step involves searching for previously well-solved problems that might have similar characteristics, even if the relationship is not apparent. TRIZ suggests 39 engineering parameters that create conflicts, including weight of moving and non-moving objects, their lengths, their area and volume, speed, force, tension, strength, durability, power requirement and energy waste, convenience of use and repairability, and many more.

Inventive Principles   The following 40 Inventive Principles are identified for use in TRIZ, and may be used to search for previous solutions. Table 4.1 is the list of TRIZ principles with some possible actions:

TABLE 4.1 TRIZ Principles with Some Possible Actions

Principle	Possible Actions
1. Segmentation	Divide the object into independent parts
2. Extraction/removal	Remove the part from the object, or extract the necessary part from the object
3. Local quality	Change from homogeneous to heterogeneous, allow different parts of the object to do different things
4. Asymmetry	Change symmetry into asymmetry
5. Combining	Combine parts that are homogeneous into an integrated section
6. Universality	Combine different functions into one, or let one object do multiple tasks
7. Nesting	Place objects within each other to reduce number of parts, or pass one object through another object's open space
8. Counterweight	Counterweight an object's weight with other lifting forces
9. Prior counter action	Provide a counter action or anti tension in advance
10. Prior action	Provide an action or tension in advance
11. Cushion in advance	Compensate for low reliability by countermeasures in advance
12. Equipotentiality	Change the condition to avoid having to raise or lower an object
13. Inversion	Implement the opposite of what is needed. Make a moving part stationary and a stationary part moving
14. Spheroidality	Change linear motions to curvilinear, use rollers and bearings
15. Dynamicity	Make parts move or adjust relative to each other
16. Partial or excessive action	If you cannot achieve 100%, try a little less or more
17. Moving to a new dimension	Increase the object's degrees of freedom, use multilayers instead of a single layer
18. Mechanical vibration	Make the object vibrate, change its frequency, use resonance
19. Periodic action	Change a continuous action into intermittent action, change its frequency
20. Continuity of a useful action	Remove idle or intermittent actions, make it continuous, make all parts perform all the time
21. Rushing through	Increase the speed of harmful or hazardous actions

(Continued)

TABLE 4.1 (Continued)

Principle	Possible Actions
22. Convert harm into benefit	Use the harmful factor to create a positive effect, combine with other harmful effects to reduce the impact, increase it until it no longer remains harmful
23. Feedback	Introduce feedback, or if it already exists, reverse it
24. Mediator	Use an intermediary object to perform a task
25. Self-service	Let the object service and repair itself
26. Copying	Use a simple and inexpensive copy, use an optical copy of an object
27. Inexpensive vs. expensive	Replace a long-lasting and expensive object with a short-lived and inexpensive object
28. Replacement of a mechanical system	Replace a mechanical system with an optical, electric, acoustic, etc. system
29. Pneumatic/hydraulic construction	Replace solid parts with fluid or gaseous parts
30. Flexible membrane or thin film	Replace solid parts with flexible, thin-film, or membrane parts
31. Use of porous material	Make an object porous, or add porous material to it
32. Changing the color	Change the color of an object or its translucency
33. Homogeneity	Make interacting object of the same material
34. Rejecting and regenerating parts	Remove an object after its task is performed. Parts that become needed should automatically be generated
35. Physical and chemical transformation	Transform characteristics of an object
36. Phase transformation	Use phase transformation phenomena
37. Thermal expansion	Use thermal expansion properties of the material
38. Use strong oxidizers	Use oxygen, ionized oxygen, ozone, etc.
39. Inert environment	Use inert environment or vacuum instead
40. Composite materials	Replace homogeneous material with composite material

EXAMPLE 4.8

Principle number 19 is the application of periodic action. In fact, applying this principle to the design of an internal combustion engine will yield the concept of a hybrid engine. The engine will only start and run when needed, and otherwise it shuts down. Number 33 is homogeneity, requiring that objects which interact with other primary objects be of the same material or close to it. Similarly, diamond is cut with diamond, because one behaves similarly to the other (primary) material. Number 22 suggests converting harm into benefit. Nike Corporation recast their waste rubber into sandals for the victims of the December 2004 Tsunami.

Can you relate any of these principles to the conversion of vinyl records to cassette tape, to compact disks, and to iPods™? ■

FP International's parent company, Safe-T Pacific Co., used to manufacture paper soda straws by pulling two strips of paper around a mandrel and cutting it to length. However, this process produced a large volume of "straw ends," which were costly to dispose of. To address this problem, Safe-T Pacific purchased a compacting machine; however, the compactor did not work as air was trapped in the "straw ends" upon compaction. Realizing that these leftover, waste pieces would probably make good cushioning material, Safe-T Pacific established FP International to produce and sell a new product they called Paper Flo-Pak. Flo-Pak was first used for shipping ammunition, and later for other products. As a product improvement effort, FP International eventually developed the first polystyrene foam "peanut" for packaging.

One problem with all packing material is the high cost of shipping to users. Since most of the volume of any shipping material is air, even though weighing little, the material occupies a large volume. To remedy this, FP International Inc. came up with a new process to add the air to the protective material at the point of application, where products are packed for shipping. They created a continuous roll of polyethylene film configured into rectangular pouches that when blown up with air, form air pillows. A small machine adds the air to the pouch and seals the open end where the air is blown into the pouch. Perforations in the film allow as many air pillows as are needed for shipping to be torn from the strip and used for packing. This innovative process saves money and space, is clean, and the polyethylene film can be recycled as well.

TRIZ suggests that most innovations may be derived from the application of one of these inventive principles, discounting the role of right-brain insight. For validation, one may reverse-engineer any innovation and, with hind sight, find a relationship with one of the above-mentioned principles. But whether or not the same innovation may occur to anyone else cannot be proven.

Discussing each of these inventive principals in detail is beyond the scope of this chapter. For more information about TRIZ, please refer to http://www.triz-journal.com.

REFERENCES

1. KELLEY, TOM, "The Ten Faces of Innovation," A Currency Doubleday Book, New York, 2005.
2. LEVESQUE, LYNNE C., "Breakthrough Creativity," Davies-Black Publishing, Palo Alto, Calif., 2001.
3. GORDON, WILLIAM J. J., "Synectics; The Development of Creative Capacity," Harper and Brothers, New York, 1961.
4. ADAMS, JAMES L., "Conceptual Blockbusting: A Guide to Better Ideas," W. H. Freeman and Company, San Francisco, 1974.
5. KELLEY, TOM, "The Art of Innovation," Doubleday Publishing, New York, 2001.
6. VON OECH, ROGER, "A Whack of the Side of the Head; How You Can Be More Creative" Warner Books, New York, 1983.
7. VON OECH, ROGER, "A Kick on the Seat of the Pants," Harper Perennial, a division of Harper Collins Publishers, New York, 1986.
8. HANKS, KURT, and JAY PARRY, "Wake up Your Creative Genius," Crisp Publications, Menlo Park, Calif., 1991.
9. ADAMS, JAMES L., "The Care and Feeding of Ideas: A Guide to Encouraging Creativity," Addison-Wesley Publishing Company, Massachusetts, 1986.
10. OSBORN, ALEX F., "Applied Imagination," Charles Scribner's Sons, New York, 1979.
11. JONES, JOHN CHRIS, "Design Methods," 2nd Edition, Van Nostrand Reinhold, New York, 1992.
12. BAILEY ROBERT L., "Disciplined Creativity for Engineers," Ann Arbor Science, The Butterworth Group, Michigan, 1978.
13. PARNES, SIDNEY J., R. B. NOLLER, and A. M. BIONDI, "Creative Actionbook," Charles Scribner's Sons, New York, 1976.
14. PARNES, SIDNEY J., R. B. NOLLER, and A. M. BIONDI, "Guide to Creative Actionbook," Charles Scribner's Sons, New York, 1977.
15. KOBERG, DON, and JIM BAGNALL, "Universal Traveller," Crisp Publications, Los Altos, Calif., 1991.
16. EBERLE, ROBERT F., "Scamper, Games for Imagination Development," D.O.K. Publishers Inc., Buffalo, New York, 1972.

17. STERNBERG, ROBERT J., Editor, "The Nature of Creativity: Contemporary Psychological Perspectives," Cambridge University Press, Cambridge, 1988.
18. ROGER SCHANK, "Dynamic Memory: A Theory of Learning in Computers and People," Cambridge University Press, New York, 1982.
19. http://www.mazur.net/triz/, Glen Mazur, 1995.
20. ALTSHULLER, HENRY, "The Art of Inventing," translated by Lev Shulyak, Worcester, Massachusetts, Technical Innovation Center, 1994.
21. REVELLE, JACK B., J. W. MORAN, and C. A. COX, "The QFD Handbook," John Wiley and Sons, New York, 1998.

HOMEWORK

4.1 Design a new sport. Describe the activities, scoring, and the equipment needed. You may design the sport for age-specific audiences, or alternately, no age limits.

4.2 Design a new toy. Specify the specifications of the toy and what it needs to be used or manufactured. Describe the activities involved and why the users will enjoy playing with it, or what they will learn from it. Specify the age limits too.

4.3 Design a new food item. Design the recipe and specify the ingredients needed, its appeal, and advantages or disadvantages. Specify for what users the new food item is intended.

4.4 Design a new movie or TV series. Describe the situation, locale, the content, and the appeal of the series. Describe why this is a good idea, and specify the age group and the type of audience.

4.5 Design a new form of entertainment. This may be a game, an activity, a process, or another event that has a specific intent. Specify for whom it is designed and where and how it will be implemented.

4.6 Design a new type of terminator. Describe the subject of termination, and how it will be accomplished. Also specify who will do the actual termination.

4.7 Design a new type of pump for an artificial heart. Let's assume that the pump should work by the person doing some activity. This will encourage the recipient to exercise more to receive more blood and feel better (although this is medically incomplete, countless ideas can be generated, and perhaps one will someday be used).

4.8 Find new uses for the left-over leather scrap resulting from making leather shoes.

4.9 Find new ways of sustaining our energy sources longer.

4.10 Find ways of making driving safer.

CHAPTER 5

THE DESIGN PROCESS

"It is hard to do something right when you do not know what right is."
—Unknown Source

5.1 INITIAL BENCHMARKING: A DESIGN PROJECT

Throughout this chapter, in addition to other examples, we will follow the design of three different products, namely a page-turner for the disabled as our primary example, and an electronic alarm clock and a house as our secondary examples for the design process. Before you go on with this chapter, it is absolutely necessary for you to benchmark your abilities by engaging in a design exercise. Depending on your field of study and your interests, please pick at least one of these projects and first try to come up with a solution for the problem(s) you picked on your own. You must think of a solution for the choice(s) and make a drawing of it before you continue. Without first attempting to design the product(s) on your own, you will think that much of this chapter is pointless and trivial, whereas after having tried to find a solution yourself first, you will better appreciate the points that are made. This benchmarking will help you see where you stand in understanding the design process.

So, please stop here and spend a little time finding a solution for your choice problem(s) and prepare a sketch of your design (for each one). Then continue with the rest.

5.2 THE ITERATIVE DESIGN PROCESS

As was mentioned in Chapter 1, design is the iterative process of finding a solution for a problem. The problem may be of any type, and the solution may take many different forms. A technical problem may have a technical solution: mechanical, electrical, materials related, manufacturing, many others, or combinations thereof. A nontechnical problem may have a nontechnical solution. The solution for winning a war is a strategy. The solution for spending a fun day in San Francisco is a well-planned travel plan. The solution for beautifying the interior of a house may be new furniture, moving to a new house, or hiring a consultant. The solution for succeeding in business may be a business plan and strategy. Every problem has a solution that is appropriate to its nature. That solution is a design, and finding the solution requires a design process.

Obviously, there are countless cases and examples of "a light going off in someone's head with an idea" that eventually becomes an invention, a new product, or a system. In these cases, it appears that the product or the invention is conceived instantly, and it is finished quickly. In reality, this is not true. Most products go through the design process with many iterations before they are finalized. It is true that some products are conceived instantly as an idea, at least initially. However, in general, when solutions are sought for problems as a systematic practice, such as in industry, instantaneous solutions are not common; designers work hard through the design process until they arrive at a satisfactory solution. So, although it is possible that you may have an inspirational idea that results in a new solution or a new product or a new idea, there is still an undeniable value to the design process and what it accomplishes in real life. In fact, our expectation should be that in general, it is as part of the design process that inspirational ideas come about and creative solutions are found.

Design is a journey, not a destination. It is a process, not a single step. It takes planning and requires many activities that eventually bring us to a destination. In between, we may encounter good and bad experiences, exhilaration and boredom, enjoyment and frustration, satisfaction and anger, and many other factors. This also means that in order to design, one has to go through a number of different steps that collectively enable the individual to find an appropriate solution for a problem.

There are many different design processes defined in literature, most of them with similar steps, even if they are defined differently. For example, Harrisberger[1] defines the design process as advanced planning, feasibility studies, preliminary design, production, and marketing. He also defines the nine stages of design, jokingly, as confusion, frustration, despair, resentment, anger, discovery, elation, disappointment, and compromise. Henderson[2] uses the four stages of definition, generating schemes, picking best schemes, and implementation. Kemper[3] lists definition, invention, analysis, decision, and implementation. Hewlett Packard Corporation at one time used business/portfolio planning, product definition, conceptual system design, product/process component development, initial production, and mature production as their design process. George Polya[4] includes understanding the problem, devising a plan, carrying out the plan, and looking back in the design process. Jones[5] does not even list a set of steps, but includes phases that are iterated when necessary. Edel[6] lists steps such as exploration of needs, synthesis of possible solutions, preliminary design (which includes selection), detail design for production, and experimentation and testing. Love[7] lists different design processes for mass production, construction, military system procurement, and book development. For mass production, he includes concept phase, feasibility phase, preliminary design, detailed design, pilot run, and use. A more detailed process originally suggested by Koberg and Bagnall[8] includes all steps defined by others, but in more detail. In this book we will use stages of **initiation** (including finding a need or a challenge and accepting the challenge); **specification** (including initial study, the development of specification and design requirements or problem statement); **ideation** (including generation and selection of ideas); **implementation** (including detail design and planning, realization of the idea, and assessment or verification); and **iterations**. This can be applied to the design of almost any type of problem too. We will discuss this process in more depth shortly.

Design process is an iterative process. In most cases, the first time a solution is developed, it may not completely satisfy all the specified requirements, or it may not be totally satisfactory, or it may not even work. In that case, the designer will have to reiterate the design steps in order to resolve the problem. This evaluation may be between any two steps, and it may be necessary to reiterate many times until a satisfactory and acceptable solution is finally achieved. We will discuss this in more detail later.

Design can also be defined as *change*.[9] Engineers create change from one state to another, from one way of doing things to another, and from one state where there is no solution for a problem to another state where there is a solution. As an example, consider humans; they are capable of walking from one place to another, from home to work, the bottom of a hill to the top, and from one side of a mountain to another. However, in this state of affairs, how large a distance any individual can cover is much related to the person's capabilities, stamina, time, and patience. Some can cover long distances over a long period of time, and we know that in the past, armies of soldiers walked from one country to another to fight. However, in this state, many other situations arise that make it impossible to do things on foot. For example, it is impossible to cross the ocean on foot, without other devices or machines such as a ship or airplane. It is impossible for someone to live in the suburbs and work in a city, miles away, and still go to work and return home on foot. To create a *change* in this state of affairs, a design engineer may create other devices that facilitate the desired objectives easier, faster, better, or even make them possible. For example, one design may be something like a bicycle, which will enable a person to cover longer distances, easier, faster, and with less effort. This change in the state of affairs will benefit mankind in many ways, but still does not provide a solution to crossing the ocean or covering a large distance (city to city, state to state, etc.). Another device to create a change may be a car, which creates even more possibilities. One can drive long distances and go far with a car. It is more comfortable too, whether in heat, cold, rain, or snow. Even individuals with disabilities can go to places with a car. But still it is impossible to get from one side of the country to another in a day, or to drive through oceans. Still another mechanism for change is an airplane, allowing mankind to fly large distances in a very short time, fly over oceans, and carry large loads. All these designs create a change from one state to another, each with its own benefits and consequences. A bicycle is cheap, easy to learn to ride, and can be mass-produced relatively easily. A car requires much more training, safety is an issue, it creates pollution, is much more expensive than a bicycle, and requires much more engineering know-how to design and manufacture. An airplane is significantly more expensive (in many cases unaffordable by individuals), requires extensive training, needs airports and significant infrastructure to fly and land, is noisy, and pollutes the air. The point is that the design engineer has a large number of choices in the design of each of these solutions. Additionally, the design engineer, although very careful about consequences, may not be able to predict all ramifications of a particular design. The first designers, producers, and users of the automobile did not know, nor could they have predicted, the social, economical, environmental, and physiological consequences of driving cars. And perhaps this is why design improvements go on forever. There is no single solution for any problem that can be achieved and be done. Any new or improved design or product can still be improved and made better. The role of the designer is to come up with *the best* solution that can be found for creating this change from one state to another, for the given conditions, at a given time in history. Have we found the best design for a car yet? Have we found the best design for transportation yet? In 2001, a new concept called the Segway emerged. The Segway is a device with a simple platform on which the rider stands and holds a steering column with controls on it. The platform rides on a pair of wheels on a single axis. The motion of the Segway is controlled by an inertial guidance system that keeps it upright. As the steering column is pressed forward, the Segway goes forward, and if pressed sideways, it goes sideways, etc. This device is something between a car, a bicycle, and simple walking. As in all other devices, Segway has its own benefits, but comes with its own problems. The design created a change in the state, but is still not the ultimate answer.

A scientist works toward finding what exists. That is discovery. A designer may find the best contemporary solution that satisfies the need with the most advantages and least disadvantages. That is invention. The designer may never find the ultimate solution to a problem because the ultimate solution may not exist.

Next, we will discuss the design process with the following stages and steps:

Stage I. Initiation

- Finding a need or challenge
- Accepting the challenge

Stage II. Specification

- Initial study
- Specification and design requirements: development of a problem statement

Stage III. Ideation

- Generation of ideas
- Analysis and selection of ideas

Stage IV. Implementation

- Analysis, detail design, and planning of the converged solution
- Realization of the idea
- Assessment of the solution against the problem statement and verification

Stage V. Iterations

Figure 5.1 shows a schematic representation of the design process. These steps may occur in different orders, some may not relate to particular problems at all, and some may be repeated many times as the process is iterated. However, before we discuss these issues, let's first discuss each step. Later we will revisit this representation.

As was mentioned earlier, in order to facilitate our discussions about the design process, we will consider three different examples throughout this chapter, a page-turner for the disabled, an electronic alarm clock, and a house. We will look at the design of the page-turner in detail as our primary example. The other two will be used as secondary examples to show that the same design process can be applied to other problems, whether mechanical, electrical, architectural, or in fact, anything else. So, even if for the sake of brevity we do not follow all three examples fully at each step, the hope is that you can extend the same ideas to those two, or in fact any other problem you will encounter.

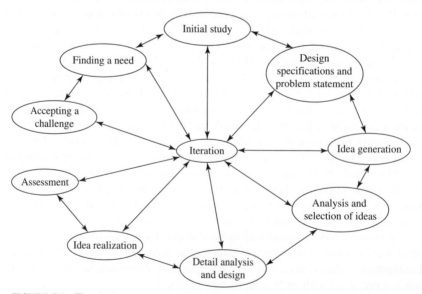

FIGURE 5.1 The design process and its steps.

STAGE I. INITIATION OF THE DESIGN PROCESS

One of two things must happen before a design project gets under way. One is finding (or being presented with) a challenge or a problem on which to work. The other is accepting the challenge of solving a problem or fulfilling a need. Either of the two may happen first, followed by the other. But without both happening, the design process will not initiate. I am certain that you have many ideas in your mind, but if you do not decide to work on them, nothing will happen. On the other hand, in many cases one decides to solve a problem for fulfillment, to make money, or to affect a change in the world, even if the problem is not clear. Unless she or he finds a problem to work on, that remains a dream too.

In the following sections we will discuss these two steps which may happen in any order.

5.3 FINDING A NEED OR A CHALLENGE

In life, one may be faced with a personal problem, a need, or a challenge that must be solved or attended to. Many products, systems, or plans are the result of a personal need, or an awareness of a problem in someone else's life. One may also be presented with a problem to solve, a need to fulfill, or a challenge to overcome that relates to others' lives. Alternately, one may search and look for a need to fulfill, a challenge to overcome, or a problem to solve. How might these happen?

If you are in a design class, your teacher may give you a problem to work on. In many cases, the problem is well defined, or the teacher can clarify the difficult points of the problem. Generally, the problem is sized to match the time available in school for design projects, so the scope is also defined. In the previous section, I challenged you to design a page-turner for the disabled. At work your manager, your boss, or your client may present the problem to you. You will be told what is needed, when a solution is expected, and what resources are available. As in school-related projects, the scope of the problem is generally clear, or it can be clarified. The due date is also given, even if very short. In these cases, you will need to understand the problem or the challenge or the need, but you need not find it. It is provided to you, and in these cases, you will have to work on the problem and find a solution.

In other cases, you may have to actually find a problem to work on, a challenge to overcome, or a need to fulfill. Here, the problem is much more open-ended. There are no guidelines for what the need should be, how big a problem you should look for, or what an appropriate challenge is. You may find a need that is great and a good solution for it may be very significant, or you may choose something that is helpful to only a few, with little effect in the world. It may be a need applicable only to a few, yet with much social value, or it may be a mediocre need with a lot of potential for commercial success. Still it is you who has to decide what need, problem, or challenge you will work on. Your impetus for finding a problem to work on may also be based on different reasons. Perhaps you need to find a new product or service for your employer, and by doing so, be able to get ahead on the corporate ladder. Perhaps it is to help others improve their lives. Or perhaps it is for your own betterment, socially, mentally, emotionally, or financially. Fulfilling a great need may help many people, but will make you financially successful as well. Designing a page-turner for the disabled will probably not make you rich, but may give you much personal satisfaction. All these reasons are valid, but they may provide different levels of encouragement to the designer.

There are a number of different ways to find a need to fulfill, a problem to solve, or a challenge to overcome. The following is a simple list of some suggestions to consider.

Many companies do the same kind of market research to improve their products, to come up with new products, or to branch into new ventures. You may add to the list as you wish:

- Do a patent search in the areas in which you are interested. A patent search allows you to see what others have done in the same area, recognize areas of industry in which elevated levels of activity are common, and choose devices and products that can be improved and patented. It may also steer you in a direction that you may not otherwise explore. Most large companies maintain complete sets of copies of patents along with a patent department just to do this. Please note that as in brainstorming, a patent search may also provide a nucleus idea from which you may branch out. However, you should be aware that, as we discussed in the previous chapters, there is always a danger that this may create a mental barrier in your mind that will make it harder to come up with new ideas.

- Look into your own life and see what bothers you, what products need improvement, what you can use to make your own life easier, what needs are not fulfilled.

- Look into the lives of other groups; children; home-makers; mothers with babies; truck drivers; teachers; a disabled person. See what they need, what can make their lives easier and better.

- Become a consumer in a particular group of products; imagine you are a child, visualize what games or new toys you would enjoy playing with; be a gardener, imagine what you would like to have to make your gardening easier and more enjoyable; become a truck driver, imagine that you have driven for hours in the desert, you need to rest in the middle of nowhere, and you are tired. Think about what you will need to make resting possible.

- Talk to members of different groups to find out what they need. I have sent many a student to the Disability Resources Center on our campus in order to talk to their personnel or to get in touch with individuals with disabilities to see what they need, what they can use to improve their lives, and what can be done to make them more independent. We have designed automatic seat belts, a device to mount a person on a horse from a wheelchair, a swing for the disabled, an artificial hand, and countless other products. Talk to children, mothers with babies, or teachers. Ask physicians about what they need in their practice (an excellent source of innovative product needs). They will give you insights which you may never have yourself. Go and talk to your professors and teachers. They also have many ideas that they do not have time to work on. You may get a chance to do those projects.

- Go and study other people's works. Consider looking into other people's designs and products and see whether you can improve them. If you find significant and worthwhile improvements to existing products, it may be possible that the original manufacturer may be interested in your improvement idea as well. Alternatively, if you use an existing product in another idea, because your idea is dependent on their product, the manufacturer may have a greater interest in adopting your idea and in producing it.

- Study trade journals and technical magazines and manufacturer's catalogs. These are excellent sources of information, and as was discussed in Chapter 4, a great way to plant new ideas into your mind for other products.

- Keep a notebook or a bug-list. Write down what annoys you and what delights you. Use the ones that annoy you as problems to be solved. Use the ones that delight you as a jumping board to do others like them, improve them, and learn from them.

- Keep a wish-list, a cool-list, an idea-list. Write down about things that you like, ideas that you think others may like, any idea that comes to your mind. Otherwise, you may not remember them later.

The following is a list of some interesting needs in everyday life. Add yours to the list:

- When it rains, even if you have an umbrella, your shoes and pants can get wet.
- Butter, when cold, cannot be easily spread over bread.
- It is hard to aim an eye-drop into your eye.
- During allergy season, a runny nose needs constant wiping.
- When napping on a train, your head falls over to the next person.
- The neighbor's cat uses your child's sandbox for his natural needs.
- When a wet car is parked in a garage after rain, the water runs to other stuff in the garage.
- Holding a cell phone while driving is dangerous.
- A disabled college student needs temporary door-openers for a rented house.

At this point, let's assume that you have been given, have previously had, or have found, a need, a challenge, or a problem to work on. Either the next step has already occurred and the present step was succeeded by it, in which case you will continue with the initial study, or you will need to consider the next step. We will assume that you have worked on at least one of the three challenge projects that were presented at the beginning of this chapter too.

Case Study #1: Page-Turner for the Disabled   Through discussions with disability experts, we have identified that there is a need for a page-turner for the disabled. We plan to come up with a solution for this problem. We will continue with this case study throughout this chapter.

Case Study #2: Electronic Alarm Clock   We have identified that college students (as well as others) have a tendency to shut off their alarm clock and go back to sleep, and therefore miss their work, classes, or appointments. We seek to solve this problem. We will continue with this case study throughout this chapter.

Case Study #3: Design of a Family House   A family of five (parents plus three children) have decided that they would like to build a house. In arriving at this decision, they have considered alternatives and associated costs, and their consequences. For example, their present house is in a desirable neighborhood with good schools, is close to the place of their employment, and is close to shopping, recreation, and other necessities. Selling this house and buying a new one will end up being more expensive than demolishing and building their own "dream" house. Any new house they buy will still require some remodeling, which adds to their final cost too. Consequently, they consider building a new house in place of the present building.

We will continue with this case study through throughout this chapter.

5.4   ACCEPTING THE CHALLENGE

Accepting the challenge of solving a problem or fulfilling a need is either the first step of the design process or the second, after identification of a need or challenge. You must either be motivated to take on the challenge, be asked to do so, or because you face a need, you decide to do something to solve it.

If you are employed by someone to do design work, you have already accepted the challenge. The fact that you are paid as a designer implies that you have taken on the challenge of trying to solve problems, in this case for an employer. Your supervisor may assign you to solve certain problems, or you may be asked to find new problems to solve (innovation, new product development). In both cases, you may not refuse to work on the assignment without retribution. The fact that you are employed determines your acceptance of the responsibility to attend to the problem and its solution, that you will go to work instead of to the beach, and that you will do your best to solve the problem.

Another scenario in which you automatically accept the challenge by the virtue of participation in an activity is a design class. The fact that you take a design course establishes that you will work on problems to which you are assigned. Your acceptance of the responsibility is automatic. In this case too, you may still have to find an appropriate problem to work on.

On the other hand, imagine that you are not in a class or employed as a designer. Then why would you want to solve a problem or fulfill a need? For example, why would you ever want to design a page-turner for the disabled? In doing so, you will have to spend your own time, effort, resources, money, and brainpower to accomplish the task. Instead of going to fun places, to the beach, spending time with your family or friends, or working out, you spend your time in a garage, workshop, or studio, working on a project. Obviously, you do this because you want to gain financially, emotionally, or personally. But isn't it true that everyone would like to gain financially, personally, or emotionally? Then why is it that only a select few engage in design and problem-solving activities?

Why would anyone decide to solve a problem or fulfill a need before a need or problem is identified? The motivation usually comes from financial, social, or emotional interests. You may be interested in inventing a product or system to solve a problem or fulfill a need in order to sell it and make money. This is a very strong motivation for many people: to be an inventor and live off of the royalties generated by it. In fact, many people have done so and are still doing it. One of the most prolific inventors in U.S. history, Jerry Lemelson, made a fortune by licensing his inventions to others (mostly after suing them). Your interest in partaking in design activities may be due to your social interests in helping others. For example, I have known many designers who design for the disabled, not because they make money, but because they like to be a positive force in the society by helping others. Society and its members are important to them. They like the satisfaction that they get from helping others in need, whether the need is related to their own life or not, and whether they gain any financial benefit from it or not. And finally, your reasons may be emotional, wanting to make a difference in the world, regardless of financial rewards. Over the years, I have met many people who have spent their energy designing devices for use by farmers in third world countries, by the Eskimos, or by refugees in African countries. These individuals never made any money from their useful, interesting, innovative, and simple solutions to world problems. But they had the utmost satisfaction of knowing that they had helped thousands of others in their daily lives.[10,11]

The heart of the issue is that one needs to accept the responsibility of engaging in problem-solving activities for some reason, devote time, energy, resources, and money to it, and sacrifice one thing to gain another. This acceptance must occur sometime in the process, either as a result of having a problem that one needs to solve or to look for a problem to solve in order to gain other advantages. Finding a problem to solve and accepting the responsibility to solve it go hand in hand. You must have both before the rest of the design process materializes.

Case Study #1, continued: Page-Turner for the Disabled We have decided to continue with the design of this project in order to learn about the process. In reality, it may be assigned to you as a project, or you may have decided on your own to work on it. In either case, it will require your time, energy, attention, and resources. Next we will continue with the other steps of the design process.

Case Study #2, continued: Electronic Alarm Clock We have decided that this will be taken up by a student as his senior project. He knows that he will have to devote time, energy, and resources in designing and building this clock.

Case Study #3, continued: Design of a Family House After careful consideration of the costs, advantages, and disadvantages, the family has decided to go ahead with this project. They have also decided to talk to an architect.

Obviously this means that the family is prepared to temporarily move to a rental house for the period of the time their present house will be demolished and built. They will have to spend much time choosing the materials, design, and specifications of their house, must be prepared to take out a loan and spend as much money as required to have their house built, and must be prepared to deal with differences in taste, likes, and dislikes of each other and the tension it causes.

STAGE II: SPECIFICATION

In this part of the design process, we develop the necessary specifications and requirements that define the scope of our problem, expectations, limitations, and what the final solution must accomplish. This requires that we first analyze the need, and subsequently, define and specify the requirements and expectation. We will continue with the discussion about these two next.

5.5 INITIAL STUDY: ANALYSIS OF THE REQUIREMENTS OF A DESIGN

Before one learns to follow a design process, there is a tendency to jump right in and come up with a solution based on the *title* of a problem without understanding what the problem really requires. For example, when you were asked to design a page-turner for the disabled, did you just go to work and design a device without first thinking about what a page-turner is and what it is supposed to do? If so, you missed one of the most important steps in the design process. In this section we will try to understand how important this step is and how we can make the final result better by applying certain methods such as Quality Function Deployment to it.

Case Study #1, continued: Page-Turner for the Disabled So, did you design a page-turner for the disabled? Can you explain what a page-turner is? You may say that it is a device that will turn pages of a book for the disabled. But what kind of a book? Should it be able to turn pages of a textbook, which is usually hardback, or should it be able to also turn pages of a paperback book? "What difference does it make?" you may ask. Well, when you open a hardback book, in general, it stays open and the pages lay relatively flat. But a paperback book, when opened, will close again, unless you hold it or you do something else to prevent it from closing. As a result, if you intend to have your page-turner turn pages of a paperback book, then you must have designed your device with appropriate mechanisms to handle this task. Otherwise, it will not accomplish what you thought it would do. Additionally, should it be able to handle a magazine too? What do you think? Would that be a desirable feature? And what would be different for a magazine? What about a newspaper? Would we want pages of newspapers turned? Obviously, in the case of newspapers, not only are the pages larger (and perhaps at different sizes and with flaps), they are not bound together either. And what about a bundle of papers stapled together on the upper left corner? Should we consider that too? What about the pages of a notebook in a 3-ring binder? What do you think the additional requirement would be that your design would need to accomplish? Would it make the design too hard or actually simpler?

Perhaps the next question to ask is what size book (or magazine or papers) should the machine handle? Many textbooks vary around a nominal 7×9-inch size, but they may easily vary as much as ± 2 inches in width or length. Paperback books may be as large as 8.5×11 inches (such as instruction booklets, lab manuals, tutorials), but also as small as 4×6 inches for pocket books used in popular novels. In fact, I have many books that are bound on their shorter side, rendering them 8.5×22 inches wide when opened. If you want your page-turner to be able to handle a variety of different size books, in addition to the above mentioned requirement of different types of books, it should also be designed to allow it to handle different size books.

The next question can be related to the thickness of a book. How thick a book should we allow? Or does the thickness of the book matter anyway? Well, in fact, it does. First of all, a thicker book will be heavier, which means that your design should be able to handle more weight and larger stresses, and it may require more power to operate. But also as the thickness of the book increases, not only does the manner in which it opens and stays open differ, the center seam of the book moves a larger distance as the pages are turned, and their height relative to the base of the device also change (try a thick book and see how the seam moves as you go though the book). Your machine should be able to also handle this movement and accommodate the change in the location of the center of the book. This may require additional features or adjustability.

Another consideration is to decide whether it is necessary to turn pages back and forth, or in one direction only. In fact, this depends on the type of book or magazine. For a magazine or a novel, chances that one may need to go back may be lower. In a textbook or a reference, it is very probable that the reader may need to go back and forth. Here, too, it will be necessary to make your design capable of turning the pages back and forth. This also means that you would need two types of signals to the machine, one to turn the page forward, one to turn it backward.

The next issue is whether the machine should be able to turn single and multiple pages at once or not. If you assume that the reader may need to move to another part of a book to refer to another subject, to look for an equation, or to check the index, it will be necessary to be able to turn multiple pages at once; if not, it will be too distracting and unnecessarily time consuming to turn a large number of individual pages back and forth. The same would be necessary in a magazine, if the article is continued on another page elsewhere. You must also consider how many pages to turn at once if this feature is required. What about increments of 1 and 10 only? Or perhaps 1, 5, 10, and 20? If so, then you will need to include the necessary detail in your design to do so, as well as additional signals for each increment.

We also have not yet discussed what possible range of weights for the book or magazine to allow. Obviously, as the weight of the book or magazine changes, the power requirements and methods of turning the pages might be affected.

So far we have only discussed the page-turner. The next question is who is a disabled person? Or should we call it design of a page-turner for a disadvantaged person? Or physically challenged? You will shortly see why this simple distinction is more than just a politically correct statement, and rather can become important and even help us. So, who is a disabled person? A rather common assumption is that a disabled person is someone who has lost the use of a limb, and perhaps is in a wheelchair. Such an individual would, indeed, be disabled (or disadvantaged) and may need this device to be able to read a book. But who else might our audience or user be?

In fact, this narrow definition of a disabled person can severely limit the capabilities of your design, its value and utility, its applications, and your solutions. If you consider the definition of *disability* more closely, you will notice that disability means that an individual is in a condition that makes performing a particular desired task difficult or impossible, either temporarily or permanently. Taking this approach, you will see that, for example, a sighted person in a dark room (or in total darkness as a result of, say, a blackout) may be more disabled than a blind individual, since the blind person is used to working in total darkness and can manage to perform many tasks better than a sighted person (one of the reasons why "visually challenged" has been suggested to replace the word blind). Then who is more disabled in the dark room or in total darkness, the blind or the sighted individual? This particular disability may have no effect on the design of a page-turner, but it demonstrates how a definition may change the result. Now take a pianist. Not only are a pianist's hands not disabled, they are probably more articulate than average. Still, while playing a piece, the pianist may need someone next to her to turn the pages as she plays, because the particular circumstance she is in disables her from turning the pages. In that case, the page-turner will be a useful device to turn the pages of the score for the pianist (Many comedy situations have been created through personal page-turners, and perhaps the same situations can be created by your machine.) The same is true for secretaries or typists, a person holding a baby, and one with a broken arm. It is also true for older individuals with certain diseases such as arthritis, when even a simple turning of the pages of a book may be impossible. Another possibility is for a person who is laying supine with his or her arms temporarily unavailable for turning pages, such as in a hospital bed with broken arms, a person in an artificial lung (formerly called iron-lung, in which the whole body except the head is inside a sealed cabinet, where the changing internal pressure will induce breathing). In this case, in addition to all the above-mentioned requirements, the page-turner may need to work with the book upside down, where not only must the pages be kept up against gravity, they must be turned in this configuration. Obviously, the disabled person may be an individual who is totally paralyzed, or totally disabled due to stroke or degenerative diseases, in a wheelchair, that may have the most pressing need for this device. However, the range, the duration, and the need for this machine vary among the different "disabled" individuals with varying disabilities. Therefore, do you need to go backwards or turn multiple pages for the pianist or the typist or a person holding a baby?

By the way, should we worry about rare books? Should we assume that the person may need to read such a book? In that case, what happens if the deteriorated pages tear during the operations?

Next, we need to consider what is allowed and/or available, and what is not. For example, should we specify the availability or lack thereof of any source of energy, e.g., vacuum? Sometimes these limitations and requirements are set by a client, or are known by the employees of a firm who are working on a problem. Other times, the capabilities and limitations may be spelled out by authorities (the military list of requirements for fruitcakes is about two pages long). In either case, it is important to know what is permissible and what is not.

Another issue to consider for this design is the type and range of available signals. The page-turner will need to receive a signal from the user to turn the page(s). If it is a simple, one-way, single-paged device, one signal may suffice. Otherwise, there may be a need for multiple signals to operate the machine in forward and backward modes, in single and multiple-page modes, or even in other modes. But what can be the source of these signals, assuming that the person is "disabled"? Can we assume that a person will be able to provide some sort of a signal no matter how disabled? Perhaps a blink signal? A puff signal? A lip signal? A head signal? A simple signal provided by one of the fingers? A signal from a keyboard?

Another issue is the cost. If there were no concerns about the cost, you could probably design a fantastic machine with impressive capabilities, but your machine might cost a million dollars. Or conversely, if you assume that most "disabled" individuals may not be financially comfortable and believe that no one would buy this machine for more than $50, you might unnecessarily limit the device's features to make it inexpensive, whereas there is a good chance that the government agencies may be impressed with a good machine and may offer to help the disabled individuals with the purchase of your machine if it really is helpful. If you do not specify either an upper limit, or better yet, a range of cost or price for the product, the ultimate outcome may be completely useless.

You also need to determine the needed time to accomplish the task. This is extremely important when you sign a contract with another party to develop a product in return for provided funds. In that case, as in any other legal contract, it is necessary to know the available length of time, the expectations, and the probable penalties for not accomplishing what was promised (called breach of contract).

There are other issues to consider as well. For example, you may need to do a market study to figure out what customers may need, want, or prefer, what the market potential is for your product or system, and what the potential size of the market is for such a product or system. You may also want to research the market for similar existing products, their features, and their limitations. You may also want to do a patent search to find out what has been patented or suggested by others, whether already in the market or not. The only danger in doing the latter analysis and search is that as a result of knowing what has been done in the past, you may develop unnecessary artificial limitations and other mental barriers that hinder your creativity in later stages of the design process. Therefore, it is imperative that you either postpone analyzing the market, finding what already exists, and patent searches to a later time, or be aware of the danger this poses to your creativity and try to overcome the barrier.

This was a long study of the possible issues related to this problem. However, as you see, it is vital that this initial study and analysis be performed in order to understand what the problem really is.

Case Study #2, continued: Electronic Alarm Clock   As mentioned earlier, this alarm clock is supposed to be designed in such a way as to force the sleeper to wake up and not go back to sleep. A snooze button allows the user to shut down the alarm quickly, and even if it goes off again later, it still allows the user to go back to sleep again. One option is to not have a snooze button at all. In that case the alarm can simply be shut off, which is worse. Others have tried to add additional requirements to the snooze button to make it more difficult for the user to deactivate the alarm (e.g., placing it inside a compartment with a simple lock, placing it further away).

In this alarm, the intention should be to force the sleeper to actually wake up by requiring some higher-level mental activity before the snooze button can be activated. Therefore, we should force the user to do something that will wake him up. We also need to be wary of associated cost, the power source, and many others.

Case Study #3, continued: Design of a Family House   The family and their chosen architect first try to understand the requirements of the family for this house. Questions such as the number of bedrooms and bathrooms, whether one story or two, and how many cars in the garage (or carport) must be answered. Would the family like to have an office too? Is the kitchen supposed to be large enough to have an island and breakfast area as well? Would they like to have an open structure style, where the living, dining, and family room are open to each other, thus creating a large open area, or would they like to have separate areas? Is there any preference about what faces the street and what faces the back yard? How big a back yard do they desire? A major question will be about the eventual cost. In California, the current average cost of new construction per square foot ranges around $200. The cost of construction for a house with 3,000 square feet can easily be around $600,000, excluding the cost of the land, demolition, fences, etc. Is this affordable for the family? Do they want cathedral ceilings? What style do they prefer, modern, Victorian, western, or others? Each one of these specifications can have a significant effect on the final outcome of the plan.

As you see, the purpose of this initial analysis is to thoroughly understand the need, determine what needs to be accomplished with what resources, and by when. This initial study will allow the designer to decide about important factors that will determine what the product will look like, its functions, its characteristics, its users and clients (audience), its price, its reliability, and many other factors. It is also extremely important in understanding the needs and the market value in the future. This analysis is an invaluable tool and an essential step in determining the *definition* of the problem and in developing the specifications and requirements of the problem and its characteristics. In short, the questions you ask will determine the answers you will look for.

While going to a service, Sol and Mort were wondering whether it would be ok to smoke while praying.

Mort says "why don't you ask the Rabbi?" So, Sol goes to the Rabbi and asks, "Rabbi, is it ok to smoke while praying?" to which the Rabbi replies with a resounding "No. That is an utter disrespect." Sol goes back and tells his friend. But Mort says, "I am not surprised. You asked the wrong question. Let me try." Mort goes to the Rabbi and asks, "Rabbi, may I pray while smoking?" to which the Rabbi replies, "By all means, my friend, by all means."

At times this initial study may seem to be a trivial analysis. Many designers think that there is no need for it, because "of course I already know this, why waste time?" they think. But how many of these issues did you consider? Or would you have gone back later and redesigned everything when you learned about the other issues?

5.6   QUALITY FUNCTION DEPLOYMENT (QFD), HOUSE OF QUALITY

In the following section we will look at a method called Quality Function Deployment (QFD), which can be used to pin down customer requirements and preferences and integrate them into the list of design specifications, requirements, and definition (which we will do in the next step). Normally, QFD is used for mass-produced consumer products where customer surveys and user feedback can be collected and used to improve the product or system. However, it can also be used for specialty designs and new product development. In order to use this method, you will need to collect customer information and preferences, likes and dislikes, and complaints. If your product competes with competitors' products, you also need to collect information about those products, their advantages over yours, and their shortcomings. These data are then used in QFD to incorporate customer specifications into the product and make it more competitive in comparison with other available products.

Quality Function Deployment[12,13,14] was developed in the late 1960s and early 1970s by Mizuno and Akao. It is applied through a table called a *Product Planning Matrix* that integrates customer requirements and perceived priorities with product features to allow the designer to design quality products that satisfy the customer. QFD provides a method for understanding customer requirements, thinking about the relationship between individual system components and the system as a whole, psychology of the customer, maximizing positive qualities of a product that customers want, customer satisfaction, and a strategy to stay ahead of competitors. It provides for analyzing the needs of customers and deciding what features to include as well as linking the needs of customers to the design of the product and its functions. QFD helps the designer focus on *what* needs to be designed, *not* how. Therefore, it should be used along with the development of the problem definition and design specification (the next step).

Quality Function Deployment is the translation of equivalent Japanese words. As in many other instances, the actual meaning is somewhat lost in translation, or at least the words could have been translated differently. So, more than the meaning of these words, please concentrate on the actions and the results.

QFD is based on the assessment of the needs, desires, and preferences of customers and users. The spoken and unspoken needs and requirements of the customer are somewhat subjective and related to the psychology of the customer as well as how these needs are assessed. Have we identified the real customers? How do we know what customers want? What are their priorities? How much do they value lower prices versus aesthetics or performance? What qualities are positive, and which are not? These questions need to be answered accurately before QFD can be implemented. In general, this is accomplished through market research, feedback from users, feedback from retailers, dealers, and maintenance workers, and from focus groups. However, the returned data must be sorted and normalized for accurate assessment and evaluated for relevancy. There is also the danger that public trends may change too quickly, creating backlash. For example, a recent trend in the automobile market is toward larger cars and SUVs. However, as soon as the price of gas increases, the demand for large cars falls significantly, faster than the manufacturers can respond.

5.6.1 Product Planning Matrix

QFD is a methodology that is accomplished with the application of a table called Product Planning Matrix. Due to the shape of the matrix, it is also referred to as the *house of quality*, although to some, house of quality includes the rest of the product's life as well. Figure 5.2 shows the basic components of this matrix. Different users and commercial programs use variations of this table, which is perfectly fine. You may draw your own table, use a commercial table, or make your own matrix using common databases such as Excel very easily. The following is a description of these areas:

- **Area 1**, in the left side of the table, is used to list customer wants and needs as WHATs. This area may be divided into categories and specific needs for better understanding of these needs.

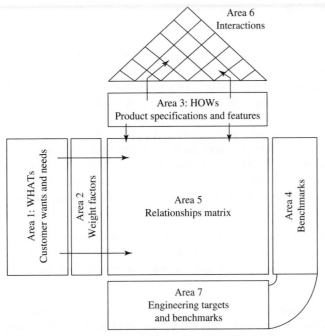

FIGURE 5.2   The general configuration of a QFD product planning matrix.

- **Area 2** is used to quantify each WHAT with a weight factor that specifies the importance of each customer desire or need.

- **Area 3** is used to list product specifications and features as HOWs and may be qualified as "More is better," "Nominal is best," and "Less is better." For example, if customers desire a powerful device, more power may be more desirable. If they want a quiet device, less noise is more desirable. In this area, a list of the product's specifications and features are listed and quantified. For example, if customers desire a powerful device, the word powerful is quantified into a specific number as a specification. Through these features and specifications, it is hoped that the customer needs and wants (WHATs) will be satisfied.

- **Area 4** is used for benchmarking the present product (if there is one) as well as competitors' products (if any). Your goal should be staying ahead of your competition. This information may help you to plan against your competition.

- **Area 5** is the relationship matrix which details the relationships between the WHATs and the HOWs. In each cell, we may indicate this relationship with symbols such as $+$, $-$, or N (neutral), or we may indicate the strength of the relationship with weight factors such as

9   (for strong relationship)
3   (for medium relationship)
1   (for weak relationship)

- **Area 6** is used to denote interactions, correlations, trade-offs, or compromises between different product specifications and features. The following notations may be used to symbolize these interactions:
  ⊖   strong negative
  −   moderate negative

⊕   strong positive

+   moderate positive

In reality, negative interactions are compromises. When there is a negative interaction, one specification may need to be compromised or even sacrificed in order to achieve another one that is more important. In other cases, these interactions provide opportunities for improvements and for setting your product apart from the competition. Assuming that the competition has similarly compromised between different specifications, if you find a way to reduce this compromise, your product will be more competitive and better received by the users. For example, more power requires more fuel, a trade-off. But if you can deliver more power with less fuel, your product will be more competitive. Similarly, longer shelf-life for a cookie and the use of hydrogenated oils are compromises. If you can increase shelf-life but also reduce reliance on hydrogenated oils, your cookies will be superior.

- Finally, **Area 7** is used for engineering targets and benchmarks. This area is used for technical evaluation and deciding on target values that will be used in the design of the product, the final result of this exercise. It provides answers to the "how much" questions.

Customarily, one line of area 7 is devoted to units. If you cannot include a unit for the desired specification, it means you will not be able to measure it either. Although not always possible, you should strive to set your targets in measurable ways. For example, instead of specifying low cost, try to define what your goal is with a number. Instead of specifying "efficient," try to specify how much power to use. On the other hand, how will you define comfortable or aesthetically pleasing? The lack of units in this area indicates that either the specification is not well understood, or that it cannot be measured, and is therefore difficult to achieve objectively.

Area 7 also includes the summation of weight factors (area 2) times relevancy factors (area 5). Larger numbers indicate where opportunities exist for improving the product.

Customarily, another line in this area is dedicated to showing how difficult it is to achieve a particular specification (1–5 range). If it is more difficult to achieve, there is an increased chance that a compromise may be required. But it also indicates that there is a better chance that you may surpass your competition if you can accomplish it.

5.6.2 Customer Requirements

In QFD you must first determine what it is that you want to achieve before you decide how to accomplish it. In a QFD program too you must first enter your knowledge of the customer desires before entering how it will be accomplished. This allows you to understand the relationship between the customer and your product and how to better set limits and specifications for it.

QFD relates to three types of customer requirements: Performance quality, Basic quality, and Excitement quality. Figure 5.3 shows Kano's Model of Quality, which relates customer requirements and reactions.

- **Performance quality** relates to customer satisfaction. Customers indicate their preferences, what they would like to see in a product, and what may be considered as niceties. If these requirements are provided or delivered by the product, they will be satisfied or delighted. If not, they may become disappointed or disgusted. For

FIGURE 5.3   Kano's Model of Quality relating customers' requirements with customer reactions.

example, more power, lower cost, and modern look are in this category. If the price is lower than what customers expect, they will be delighted. If it is higher than expected, they will be disappointed or disgusted.

It is important to realize the difference between a feature and a requirement. A customer requirement is a need; a feature is only one solution to it. The type of solution delivered may create a varying degree of delight or disgust. For example, the customer may ask for safety (a need) in an automobile. However, the features that deliver additional safety may vary. A system of sensors that senses the weight of a passenger and his or her height and signals whether or not the passenger will be adequately protected by the air-bag system may create an additional degree of delight. But this is only one possible feature to accomplish additional safety in an automobile.

- **Basic quality** relates to specifications that are normally not even expressed, but automatically expected. For example, it is expected that a product performs its basic function, or that it is safe to operate. No one will specify these requirements in a customer feedback form or during a focus group meeting; they are expected. Delivery of these specifications by the product is neutral and will not create additional satisfaction or delight in the customer, but their lack will create disappointment or disgust. Thus, it is imperative that these requirements be delivered, even if they do not create competitive advantages.

- **Excitement quality** relates to features or specifications that the customer may not have even requested. These are unexpected quality features and specifications that add to the quality and value of a product. Their lack will not create disappointment or disgust, as they are not even expected. But their presence will generate additional satisfaction or delight for the product. They are a bonus.

5.6.3   Application of QFD Method

QFD allows strategic competitiveness by improving product quality. This is accomplished in a methodic way by integrating customer data with product features and specifications. In general, a QFD Product Planning Matrix is formulated by following these steps:

1. Identify the real customers (area 1). List the customer requirements from market analysis and feedback data. This should be done at two levels, general requirement categories (functional, aesthetic, safety, etc.) and specific requirements within each category.

One of the most important factors in doing this is to first determine who the real customers are. In most cases the real customers are the end users who actually purchase the product and use it. However, there are times when other constituents may be the customers, including the manufacturers of the product, other assemblers who use the product as a subassembly, management, etc. Additionally it is important to also consider who orders the product even before the end user finally purchases it. In fact, in many cases the corporate buyers, dealers, decorators, architects, etc. will select products based on what they think will sell or how much money they will make on it and not what customers may actually want. In that case, the first (real) customer is not the user (for example, the page-turner for the disabled will most probably be selected by others on behalf of the disabled).

2. Collect information about customer wants, needs, desires, and preferences (area 1). Clearly, for products that are already in the market, collecting customer data is easier. It can be accomplished through dealers, direct questionnaires, warranty data, returned products, customer comments, focus groups, market studies, surveys, and others. However, if the product or system is new, and especially if it is not patented yet, getting these data may be much more difficult. In some cases, the product is kept secret until introduced in the market in order to keep the competition from gaining on the product and its market share. In a case like this, it may be very difficult to collect customer data, except by estimation or from very limited groups.

Additionally, make sure that the data are current. As mentioned earlier, customer preferences can change quickly if economic or societal conditions change. Also remember that at times, customer desires may be undesirable in the long run. For example, kids like more sugar in their cereal; parents don't.

The customer requirements and desires can generally be divided into categories such as price, quality, functions, aesthetics (including modernity, high-tech features), lifetime issues (life expectancy, maintenance, recyclability), and features.

3. Assign priorities (importance, weight factor) to each requirement (area 2) based on customer feedback, customer psychology, or estimates. This will allow the designer to allocate resources to the most important requirements later. Distinguish between customer "musts" and customer "wants."

Some suggest that the total of all weight factors should add up to 100, but this is not required. One important issue in selecting importance factors is "whose importance?" The weight factor for a customer may be different than that of management or corporate buyers. For example, during the design of the Ford Pinto, the final cost of the product was extremely important to the management, even more so than the customer. The customer might have been perfectly willing to pay an extra few dollars for a safer car, but the decisions of the designers on safety issues were overridden by management for cost. Their priority was different than the customers'. The designer must decide whom to please. We will discuss methods for assigning weight factors to competing ideas shortly. Please refer to that discussion as well.

4. List the design specifications and features of your product (area 3). This list (usually) includes all the measurable features, design specifications, and important details of your design. You may consult your initial study to decide what features and specifications should be included in the product. These specifications, features, and characteristics are expected to satisfy the customer requirements as discussed above.

QFD can be used to help set the values and features that will constitute the problem definition. For example, during the initial study stage, the designer may consider many features and specifications, not all of which may need to be included. Based on customer data and information (and competitors' products) QFD will allow the designer to decide what features are important, at what level, and where opportunities lie. This will help the designer select appropriate specifications for the product.

Some suggested specifications to consider include time (duration, cycle, how long), dimensions (distance, areas, volumes, width, length), value (precision, reliability), and power (efficiency, currents, voltage, miles per gallon, etc.).

5. Benchmarking your competition (area 4). Determine how well each requirement is satisfied by the current product (if any) and by competitors. This is called benchmarking. It allows you to determine where you stand in comparison with your competition, and how to move beyond them. It also allows you to determine where opportunities exist for competitive advantage. Obviously this can only be done if there are competitive products on the market, and if you have already had previous models in the market. If this is a new product, or if you are new to this market, you will not be able to benchmark competitors or benchmark your own product. You may use a scale of 1–5 for benchmarking, with 1 indicating that the competition does not meet the requirements, to 5, that competition fully meets the requirements.

6. Determine the relevancy of each feature to each customer requirement (area 5) by filling out the table. Through this analysis, the relationship between each feature and each requirement is determined. Use the following weight factor between each pair of customer requirement and design feature as follows:

For strong relationship use   9
For medium relationship use   3
For weak relationship use   1
For no relationship leave blank

7. Determine interactions and compromises (area 6). Each cell is related to a pair of features or specifications. You must determine whether fulfilling each specification may have a positive or negative effect on the other specification, and if so, to what extent. You may use the following symbols:

⊖   strong negative
−   moderate negative
⊕   strong positive
+   moderate positive

Negative and strong-negative interactions indicate where compromises and trade-offs must be made, but also show opportunities to surpass your competition.

8. Set engineering targets and benchmarks (area 7). This area will assist you in determining what to do, what your targets should be, what needs to be accomplished, and what your benchmarks should be.

You should look at the benchmarks, your competition, degree of difficulty, and the summations of weight factors multiplied by relevancy factors, in order to draw conclusions about what should be done, what targets you should strive to achieve, and what specifications you must set for your product. These results will assist you in completing your problem definition and setting your engineering specifications. For example, for the page-turner, if users indicate "easy to hold book," or "quick turn of pages," through QFD we can assign measurable values to these customer requirements. You should look for

- Engineering specifications that do not relate to customer requirements
- Engineering specifications that relate to multiple customer requirements (this is an opportunity)
- Customer requirements that are not adequately addressed by engineering specifications
- Benchmarking results that may provide opportunities for advancement
- Specifications that are easier to achieve while having the most positive effects on customer requirements

The following is an example of the application of quality function deployment.

Case Study #1, continued: Page-Turner for the Disabled   We will now apply the QFD product planning matrix to the page-turner project. In our initial study we raised a number of questions and came up with answers, but no decisions were made about most of those issues. Here, we will try to identify which of these issues are more important, and if so, how important. The information developed in this section will be used later to specify the exact nature of the page-turner.

Please note that this is only an exercise in the application of QFD, and it is very subjective. Therefore, a different set of issues can easily result in different conclusions.

Step 1:   Due to the lack of access to a large group of disabled individuals as customers, the students in a design class were selected as our representative customers. Their reaction to our study is only their perception of a page-tuner, and therefore, it may not be as accurate as a real group of customers in the future.

Step 2:   Since this is a new product, we selected the following issues for further consideration and collection of customer information. In our case, a group of 65 students were asked to rank the list based on their perception of what is important and how much. They were also asked to identify what else might be important, but no statistically significant answer was received from the group. The issues included reliability, going back and forth, multiple pages, weight, ease of use, price, aesthetics, hardcover versus paperback, magazines, and "what else?."

Step 3:   Table 5.1 shows the result of this customer data collection. Each factor was ranked in importance between 1 and 10. The number in each cell is the summation of how many students chose that level of importance. The summations of the importance levels multiplied by number of people who chose them were divided by each total to obtain the weight factors.

Another method of assigning weight factors is to perform pair-wise comparisons. We will discuss this method in more detail later when we discuss the process of selecting ideas. But for now let it suffice to say that each of two characteristics can be compared

TABLE 5.1 Customer Preference Data for the Page-Turner

Importance→	1	2	3	4	5	6	7	8	9	10	Total A	B	B/A	Rank	Weight factor
1. Reliability					1	2	5	10	19	25	62	553	8.92	1	9
2. Back/Forth	1	1	2	4	7	11	10	10	9	8	63	437	6.94	4	7
3. Multiple pages	5	4	8	10	9	11	7	3	3	1	61	298	4.89	7	5
4. Weight	7	6	12	9	8	2	10	6	1	1	62	280	4.52	8	4
5. Ease of use	1		1	2	1	5	3	9	17	24	63	533	8.46	2	8
6. Price	2		4	4	9	9	11	12	5	5	61	397	6.51	5	7
7. Looks	12	12	10	8	5	5	5	3	2	1	63	240	3.81	9	4
8. Hard cover/ paperback		1	2	4	10	7	13	11	8	9	65	457	7.03	3	7
9. Magazines	1	3	6	7	7	14	11	7	5	3	64	380	5.94	6	6

A: Total number of people in the row.
B: Summation of importance levels multiplied by each number.

based on the perceived level of their importance. The summation of the number of times that each characteristic is chosen can be used as a weight factor, as follows:

Between	Choose	Between	Choose	Between	Choose	Between	Choose
1–2	1	2–4	2	3–7	3	5–7	5
1–3	1	2–5	5	3–8	8	5–8	8
1–4	1	2–6	2	3–9	3	5–9	5
1–5	5	2–7	2	4–5	5	6–7	7
1–6	1	2–8	8	4–6	6	6–8	8
1–7	1	2–9	2	4–7	7	6–9	9
1–8	1	3–4	3	4–8	8	7–8	8
1–9	1	3–5	5	4–9	9	7–9	9
2–3	2	3–6	3	5–6	5	8–9	8

resulting in $7[1] + 5[2] + 4[3] + 0[4] + 7[5] + 1(6) + 2[7] + 7[8] + 3[9]$. Thus the weight factors based on this analysis could have been, respectively, 7, 5, 4, 1 (instead of 0 for #4), 7, 1, 2, 7, and 3. These are different from the ones we obtained from our survey. Since this is based on the opinion of one person, we will choose to continue with the results of the survey.

Steps 4 and 5:   Figure 5.4 shows the QFD table with the associated characteristics and specifications in it.

As you see, the customer requirements are listed along with the weight factors. The competitor products are also listed as Competitors 1 and 2. These are two student projects at Cal Poly and University of Rhode Island.[11] Neither of these two projects is commercialized and therefore limited information about customer satisfaction is available on either one. However, the information shows our perception of the advantages and disadvantages of these two competitor products. Lower numbers in this area indicate opportunities to outperform the competitors' products, whereas higher numbers indicate areas where we must at least meet the same level of competence in our product.

Step 6:   Next, we will fill area 5 of the matrix with our estimation of the relative relationship between customer requirements and design specifications. Blank cells indicate no relationship.

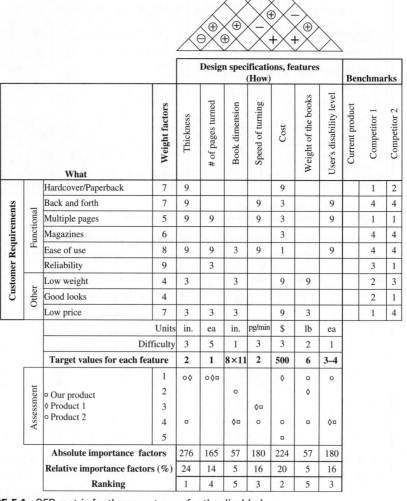

FIGURE 5.4 QFD matrix for the page-turner for the disabled.

It should be mentioned here that, in general, it is much better if this analysis of the relationship between these groups is done in a team with people who have knowledge about them. In many cases people's views change as they discuss these issues, and therefore the results may change significantly.

Step 7: Next, we will fill out the roof of the matrix with our estimation of the interactions between different design specifications. There seems to be only one strongly negative relationship between the thickness of the book and the number of pages turned. Obviously, the thicker the book, the more important it is to be able to turn larger number of pages at once. This requires a compromise between the desired values. There are also a number of strongly positive relationships, indicating that as one is increased, the other will also increase (such as thickness and weight of the book).

Step 8: Lastly, we will fill out the bottom area of the matrix. This includes units, the level of difficulty of achieving the desired specification (for example, achieving multiple pages is difficult, whereas a larger dimension for the book is not too difficult), and target values. We have also included a comparison of how our specifications compare with the competition in meeting customer requirements. This is another indication of how good our specifications are. Absolute importance factors and relative percentages (out of 100%) are also included. These values are the summations of each weight factor multiplied by the relationship numbers for each column (for example, $7 \times 9 + 7 \times 9 + 5 \times 9 + 8 \times 9 + 4 \times 3 + 7 \times 3 = 276$). These factors are also ranked in importance.

Considering the weight factors, where the competitors stand, the levels of difficulty of achieving different customer requirements, and the relative importance factors, we can decide about our target values. Based on these, let's say that our page-turner will be designed for books up to 2 inches thick and 6-lb maximum weight, with pages as large as 8×11, one page at a time, at least 2 pages per minute, costing less than $500. We will also assume that the user will have enough movement and control in a body part to be able to provide three to four distinct signals.

We will use these targets in the next section to define and specify the project.

5.7 SPECIFICATIONS AND DESIGN REQUIREMENTS: DEVELOPMENT OF A PROBLEM STATEMENT

As you may have noticed, a number of questions were raised in the previous section regarding the three case studies, but we never answered any of them. This is because the answers depend on what is really needed or sought. In many cases, the answers come from the client, although the questions are posed by the designer. In many other cases, both the questions and the answers come from the designer. But in all cases, someone who has the authority (a company representative, a client), the knowledge (a consultant, an expert), or the need (a customer, a user) will provide the answers to the questions that are asked. It may also come from the applications of methods such as QFD. The collection of the answers to the posed questions will constitute the actual *definition of the design project*, the *problem statement*, or *problem specifications and requirements*.

The problem definition or list of specifications and requirements is the "what" of the design. The solution we will find is the "how" of the design. Before we know how to do something, we need to know what we need to accomplish.

Case Study #1, continued: Page-Turner for the Disabled   The first question we posed was "What kind of a book?" A hardback book, a paperback book, a magazine? Notes? A binder? Based on the information we collected from our customers and the QFD table, let's say that we would like the machine to work with both hardback and paperback books (weight factor 7). We believe that it will also be able to turn pages of a magazine with the same features, but we will not actively design it to do so. We assume that the page-turner will not be used to turn pages of notebooks in a 3-ring binder. We decide that the books may be as large as 8.5×11 inches and as small as 4×6 inches. We assume that the books may be as thick as 2 inches. We decide that it is necessary to turn pages back and forth (weight factor 7). After considering the added cost versus the utility of turning multiple pages (weight factor 5), we decide that the page-turner will only turn one page at a time in either direction (with the capability to do multiple pages as a bonus). The maximum weight of the book will be about 6 lb. We will not handle the book upside down.

Now assume that we decide that the audience of our design (the user) is a disabled individual who for one reason or another does not have the capability of turning the pages of a book. The disability may be temporary, such as a broken arm, or permanent. However, we will design the device for a person with permanent need for the page-turner. Such a person will probably be willing to pay more money for a device that has more utility for them compared to someone with a temporary use. However, if the person with a temporary need can either afford the higher cost or can rent it for a period of time, they should be able to use it. We also assume that others who may have less severe disabilities should be able to use the machine effectively, even if they do not need all of its features. This includes the elderly, someone confined to an awkward situation that prevents him or her from turning the pages, a musician, etc. This will become a determining factor in our cost analysis. We will also assume that the person will be able to provide at least two types of signals to the machine, one for forward motion, and one for backward motion. The signal may come from any type of switch that the individual can operate, including a simple switch, a head switch, a tongue switch, a blink switch, a puff switch, a keyboard switch, etc., and that the page-turner will be able to use the simple on-off switch to operate.

At this time, we will not make any restrictions on the type of materials, methods of manufacturing, or other design elements. If necessary, we may amend our assumptions later.

Let's also assume that we expect that the selling price of the final product will be about $500 (weight factor 7). This means that we should design the product to not cost more than about $150 to manufacture. This cost needs to be adjusted based on the total number of page-turners manufactured, the financial ability of the users and the agencies that assist the users, market conditions, and other factors. But for now, in order to move on, we will make this cost assumption. We also want a reliable device (weight factor 9) that is easy to use and set up (weight factor 8). The weight of the device and its looks are also important, although not as much as the rest.

And finally, we would like to have a working prototype in about 6 months from the starting date. The final date of a product release will also depend on a multitude of factors, but as designers, we want to finish our design work in 6 months.

Now that we have made decisions on all of our questions from the previous section, we are ready to define design requirements for our project. The design requirements (specifications or definition) for any design project are the collection of all the answers to all the questions that are posed. Please notice that this definition *is not* the same as the analysis of the problem; rather, it is the collection of the answers and the decisions that are proposed in response to the questions and concerns posed during the initial study. Otherwise, analyzing the problem without offering a decision is incomplete, and in many cases, useless. Therefore, our design requirements (definition) for the page-turner for the disabled project can be summarized as follows:

> *We intend to design a page-turner for the disabled that will turn the pages of a hardback or paper-back book (no 3-ring binders) as large as 8.5 × 11 inches and as small as 4 × 6 inches, up to 2 inches thick, not weighing more than about 6 lb, back and forth, but one single page at a time. Turning multiple pages will be an added bonus. The page turner is not recommended for rare books. It will be designed for disabled individuals who for one reason or another do not have the capability to turn the pages of a book. The user will be able to provide at least two distinct signals to the machine. The selling price of the final product will be about $500. We want the page-turner to be reliable and easy to use. We plan to have a working prototype in about 6 months from starting date.*

Please note that other decisions such as the use of the machine for magazines, lack of restrictions on materials or components, etc. are not mentioned in this definition of the design requirements. This is because if something is not a part of our requirements, it does not have to be mentioned either. Thus, what we have specified will be what we will have to do.

Case Study #2, continued: Electronic Alarm Clock   We intend to design an alarm clock that requires the user to wake up before the alarm can be turned off. It should require a higher level of mental awareness, motor skill, and visual signal from the user that cannot be memorized or be learned as a habit. The clock should have regular expected features of showing time and a buzzer. Easy adjustments for time display and setting the alarm will be provided. The final price of the alarm should be less than $10.

Case Study #3, continued: Design of a Family House   We would like to design a house with five bedrooms (a master, three for the kids, one for guests), four bathrooms (master, guest, powder room, kids bathroom), a family room, kitchen with an island, and an office for mom. She prefers the office to be closer to the entrance door. The living and dining rooms should be adjacent. The kitchen should be adjacent to the dining room. A breakfast area is a bonus. A two-story building will be acceptable if necessary. A two-car garage will be acceptable. The size of the yard is dependent on the other requirements. Bedrooms should be larger than 12×12 ft. California style with tile roof is preferred. The cost of construction is to remain below $200 per square foot.

Many people ask engineers (or car enthusiasts) about what car they should buy. Of course, a logical answer can only be provided if you first ask the person what his or her requirements are. If you just tell the individual about your favorite car, the answer may be totally irrelevant. Someone with a family of four will probably have a very different utility for a car than a single person. The need for fuel economy versus size may be different for a college student and an executive. Then it is appropriate to first ask the person what she or he wants in a car, what his or her utility is for the car, what price range she or he has in mind, and what is expected from a car. If the utility is to go around town, go camping, shopping, to work and school, and have a safe, low-maintenance, and relatively inexpensive car, you would not recommend a Ferrari. Therefore, it is very important that you ask the person to define his or her design requirements before answering the question.

Now imagine that you want the exterior of your house painted. Would you ask a painter to just paint the house? If so, will you know what color, type of paint, or process the painter will use? How many coats of paint? Will the painter first wash the walls? Will the wood trims be sanded and cleaned first? Will the trim and the walls be the same color? How long will it take? How much will you be charged? You will tell the painter what colors you want, what type of paint you prefer, and when you would like to have it started. He will tell you about the process he will use, the details of preparation, how long it will take, the cost, workmanship, warranties, etc. Then the mutually agreed upon details (the definition of the design problem, or the design requirements) will be listed in a contract. After signing the contract, the painter can go ahead and do the job, and in return, expect to get paid. You do not expect that the painter will do things differently than agreed upon, and the painter will expect that he will be paid in due time. The painter cannot, on his own and without your agreement, change the details of the contract, and neither can you. Otherwise, either party may be breaching the contract and creating a cause for a complaint or a lawsuit. Similarly, imagine that someone asks you as a consulting design engineer to design a page-turner for the disabled. After discussing the issues, you will have a definition of the design requirements, which will be written in a contract. In return for your deliverance of a design or prototype, you will get paid an agreed-upon compensation. Neither you nor the customer can change the definition without the other's agreement. Thus, in this kind of a situation, the definition and the design requirements become a contract that must be fulfilled. As a painter should not be allowed (or would not want to) paint a house without a contract, neither should a designer design a product or a system, nor solve a problem, without first understanding it, whether or not it becomes an actual contract.

In fact, design drawings are contracts (definitions, specifications). When a drawing is submitted to a shop or to a manufacturing facility for manufacture, it specifies what is expected. If the manufacturing facility makes the parts differently from the drawing, the parts will be unacceptable. And similarly, as I remember the story of a large company that had ordered 10,000 brackets made from a drawing they had submitted, if the drawings are incorrect, the submitting institution is responsible. In this example, it was possible to interpret the specified tolerances in two different ways. The shop had interpreted them differently than what the company had assumed. The latter was left with 10,000 useless brackets and had to correct the drawings and order again.

Defining the design requirements will also help us know our goals and how to proceed toward achieving our goals. When we understand the requirements we are to achieve, we can start working toward getting to our goals with a much clearer destination. In fact, a good way to make progress toward a solution in any design problem is to ask the designer or the customer to write down what they want. This goes a long way toward defining the problem and specifying its requirements because the only way to write about something is to think about it, and thinking about a design project forces one to look into its requirements.

Additionally, we will also know to stop designing when we can measure our success against a specified goal. A carefully defined list of design requirements will enable us to know when we have reached our destination. Otherwise, we may continue designing forever and never stop if we do not know when we have reached our goals. Another important benefit of defining the design requirements is that during the process of designing and finding a solution, especially (and hopefully) if there are multiple solutions available, the definition will assist us in sorting through the alternative solutions and selecting the best one. We will discuss this in more detail later, but as an example, imagine that we will have two alternative ways of turning the pages, one which requires a vacuum, and another which is electric. If a part of our definition is that a vacuum is not available, then we will eliminate the solution, or we will try to solve the problem of the lack of availability of vacuum. We will only know this if it is already a part of the design requirements. This attempt at defining the requirements will be invaluable in later stages when we need to pick the best solution and when we evaluate our final design.

STAGE III: IDEATION

The next two steps involve the generation of ideas and the analysis and selection of the best idea (or combination of ideas).

5.8   GENERATION OF IDEAS

Now that we have determined the design requirements and created a problem definition which specifies exactly what we are trying to accomplish, it is time to find, or generate, as many alternative solutions, plans, or strategies, as possible. These ideas and solutions will be analyzed later in order to select the best possible solution. However, in this step we intend to create the solutions, even if we do some analysis and comparison at the same time for better understanding or for creation of even more solutions. In other words, the primary intention is the generation of solutions that satisfy the problem definition or specifications. So far, we had concentrated on the "what" of design. At this point, we will concentrate on the "how" of design; how will we accomplish the task?

This is a very important and crucial step in the process. Many people skip some of the previous steps and jump right into this step, and by taking the first idea as the solution, assume they are finished. However, the previous steps are the building blocks for getting us to this point, and this step is the one that will launch us into a creative, appropriate, and hopefully the best solution for the given problem. Chapters 1, 2, and 3 directly relate to this step, as we should strive to generate as many creative solutions as we can. Chapter 4 was devoted to a complete discussion of methods and techniques of idea generation and how to overcome mental barriers, which is also directly related to this step in the design process. Obviously, you may use any of the techniques discussed in the previous chapters, or use any method that you know or learn. The intention here is to generate ideas, from which we will subsequently select the most appropriate one(s).

Case Study #1, continued: Page-Turner for the Disabled During brainstorming sessions, as well as a response to assigned homework, the following ideas were suggested for consideration:

- A puff of air under the page and an arm to turn it.
- A puff of air over the page with the arm to rotate it.
- Consider a narrow belt stretched over the pages, which is operated by a motor. As it turns one way or another, it will turn one page forward or backward. Continuously running the belt (perhaps with additional force exerted on it at the same time), multiple pages may be turned (Figure 5.5).
- A belt with friction knobs on it is used to lift and turn a page. Changing the direction of the belt will reverse the process.
- A sticky arm lifts a page, and an arm slides under the page and turns it (Figure 5.6).
- A cutter cuts the seam of the book or magazine to separate all the pages. Then a machine like a copier will feed each page and even flip it to allow the user to read (B. McLuen).
- A stick in the user's mouth with sticky tip to lift and turn the page (Figure 5.7).
- The same as above but attached to the head with head gear.
- I have a toy in my room to show the first law of thermodynamics. It is a bird who's beak goes into a cup of water and then comes out as it gets cooled. This can go on forever. We can use it to turn the pages (K. Patterson).
- Suction to lift the page and an arm to turn it (Figure 5.8).
- For a person with an amputated arm, an artificial "arm" that is attached to the arm with a friction ball at the tip (M. Knorr).

FIGURE 5.5 Idea #I.

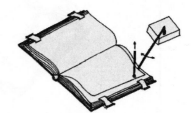

FIGURE 5.6 Idea #II.

FIGURE 5.7 Idea #III.

FIGURE 5.8 Idea #IV.

- A mechanical arm attached to the chest can be used to turn the pages.
- The book is placed on a turntable that rotates and turns the pages! (T. Robinson-Carter).
- A pair of two-link arms on both sides, each with a roller at the end. The roller lifts a page, and as the arm extends toward the seam it moves the page to the other side. The second arm grabs the page by the roller and completes the turn and holds on to it too (Figure 5.9).
- A hook-like hand with a sticky finger, lifting the page, turning $\frac{1}{2}$ turn under the page, and then sweeping the page to the opposite side (Figure 5.10).
- Instead of using clips to hold the book, use rollers. These rollers can lift the page, a sweeping arm will turn the page, and the rollers on the opposite side will grab the page and hold it (Figure 5.11).
- A rotating arm is lowered near the edge of the paper to come into contact with the paper. As it turns, due to friction, it will curl up the paper, eventually moving under it. Then it can be swept to the other side.
- A rotating wheel that lifts the page and then moves sideways to the other side to turn the page (Figure 5.12).
- An electronic book.
- A bug is ordered to go under the page and fly to turn it (J. Jiang).
- The book could be scanned into an electronic file. The computer file can then be manipulated easily (Figure 5.13).
- A rotating wheel on the page to separate it from the rest, with an arm to rotate it.
- A monkey could be trained to sit next to the user to turn the pages whenever the user sends a signal to the monkey (Figure 5.14).
- A small disk with two short arms attached to it along with two sets of rollers that hold the book in place. The rollers roll to lift one or more pages. The disk rotates to

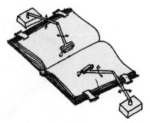

FIGURE 5.9 Idea #V.

FIGURE 5.10 Idea #VI.

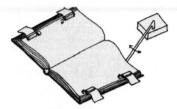

FIGURE 5.11   Idea #VII.

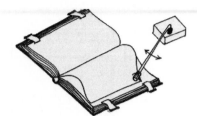

FIGURE 5.12   Idea #VIII.

FIGURE 5.13   Idea #IX.

FIGURE 5.14   Idea #X.

let the short arm under the page, which turns it. The opposite-side rollers grab it. The opposite can be done to go backwards.

From among all of the above, let us consider the following few ideas for the page-turner to enable us to discuss the remaining steps in the design process. The ideas relate only to the page-turning function. Other requirements of the design will be considered separately. It should be mentioned here that the following list is by no means a complete set of possible solutions, and is only used here for the sake of discussing the design process. There are many more that are not listed.

Case Study #2, continued: Electronic Alarm Clock   A number of different solutions are suggested, including

- A joystick should be used to move a pointer within a window through a maze and get it to home position before the alarm can be turned off.
- Start with a nice tune and move into obnoxious sounds. A sequence of numbers in a keypad must be pressed to stop it.
- Use a few LEDs that light up in a random order. The user must press them in the same sequence to turn off the alarm.
- The Off switch is activated by a steel ball that must be moved within a labyrinth to home position.
- Fresh water must be added to a small cavity within the alarm to stop it. If the cavity is filled at night, it should evaporate by morning time.

Case Study #3, continued: Design of a Family House   Four alternative plans were developed by the architect. They are shown in Figures 5.15 through 5.18. As you notice, each plan satisfies some of the requirements. Only the first floor is shown. The bedrooms are on the second floor.

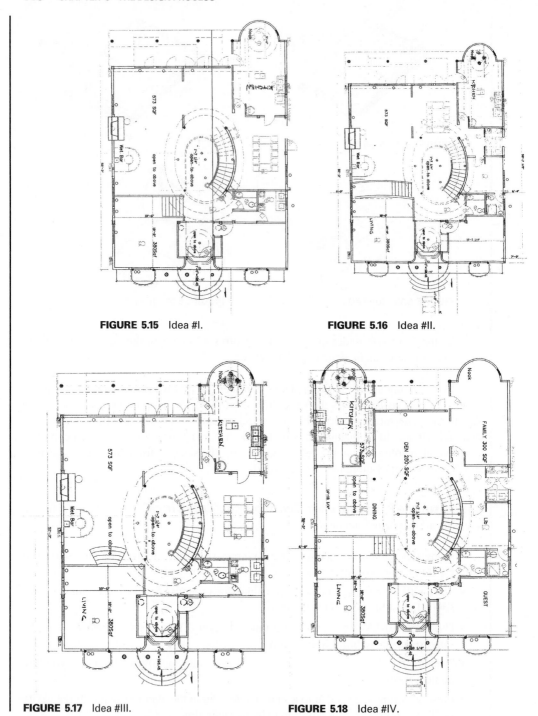

FIGURE 5.15 Idea #I.

FIGURE 5.16 Idea #II.

FIGURE 5.17 Idea #III.

FIGURE 5.18 Idea #IV.

5.9 ANALYSIS AND SELECTION OF IDEAS

The next step in the design process is to select one (or multiple or a combination of) idea(s) from the above-generated ideas. You must realize that although in this step we will compare the characteristics of each idea, we are not bound to pick one whole idea from this list; rather, we are interested in picking the best idea possible, including combinations of ideas or combinations of parts of different ideas. We will discuss this a bit more later.

To analyze our ideas and to select the best one, we will refer back to our specifications and design requirements or problem statement. All the decisions we made earlier about the product requirements will now be used as a benchmark to decide whether or not any particular idea is advantageous and useful. Without the problem statement (definition), we would not be able to determine whether or not any particular idea is useful. But with the specifications list completed, we can refer to the list for our comparisons and decision making. This is why that step in the design process is so crucial.

Different methods may be employed here to facilitate our selection, ranging from methodical approaches such as the decision or Pugh matrix, to the Decision-Based Design method[15] which is primarily an economic decision, to heuristic and instinctive decision models. Many of these methods have common elements too. We will employ the following approach. First, we will analyze the ideas in order to understand the ramifications, consequences, advantages, and disadvantages of each idea, and whether or not the idea may satisfy our needs, and to what extent. Eventually, using the methods we will discuss later, we will use this deductive reasoning analysis in order to select the best idea. This analysis is necessary because in many methods, including brainstorming and the morphological list of attributes, there is no analysis or critique of the ideas during the session. Therefore, it is necessary to analyze these ideas for their merit.

5.9.1 Analysis of Ideas

In this analysis, for each individual idea, we should consider all the consequences, advantages, and disadvantages of the idea and compare their characteristics with the list of requirements to see whether the idea can satisfy the needs, and if so, to what degree. Eventually, this deductive reasoning will lead to the selection of the best idea we have had so far, an idea that will be used for implementation and testing. In this analysis, you may consider any number of the following methods to enhance your analysis:

- **Personal opinion:** This constitutes your personal opinion about the idea and its characteristics. If you have had much experience in design, and if you have adequate expertise in the area, your personal opinion may be very useful and trustworthy. However, all designers also have a "gut feeling" or intuitive feeling about ideas which must be trusted as well. We gain this experience and expertise by our knowledge and understanding of engineering, physical and biological principles, and by repetition. All design projects you do in school help you gain experience in judging ideas. And obviously, the more you design and analyze ideas, the better your judgment will be. However, you should not shy away from judging design ideas based on lack of experience alone, as all judgments can be valuable.

- **Ask your colleagues, your boss, your teachers:** Other good sources of information and judgment about different ideas are your colleagues, your boss, or your teachers. Your colleagues are generally intimately involved in the same type of work, they are familiar with the strengths and weaknesses of your industry and your institution, and may have worked on similar ideas in the past. They generally possess a wealth of knowledge and expertise that you may lack, and may be able to help you analyze an idea and determine its advantages and disadvantages. The "in-house" knowledge developed by working in a place for many years can be very valuable and may not cost you at all.

One thing you should be aware of is that the motivations of these individuals may be different. For example, in education, the ultimate goal is for you to make decisions and learn from your mistakes. Therefore, a teacher may refrain from directly expressing opinions about an idea in order to entice you to decide for yourself. In industry, the ultimate goal is making profit. Thus, your colleagues and your boss may more readily respond to your request for their opinion, sometimes even before you ask. However, there may also be ulterior motives involved with colleagues and coworkers that may influence their judgment.

• **Ask an expert:** There are times when there may be a need to ask experts about ideas and their consequences. If the idea at hand is complicated, and if you or your colleagues lack the knowledge, the judgment, or the expertise to analyze the idea, you may need to seek the opinion of an expert. Experts, by definition, have the understanding and the experience to be able to do what you need, usually in exchange for money or other favors. An in-house expert (not necessarily your colleague or coworker) may not cost the institution much, but an external expert may be expensive and time consuming.

• **Ask your clients, customers, or the users:** Another source of information to form opinions about ideas is to ask your clients, customers, or the people who will use the product. You may be able to gain different knowledge and varied opinions about their personal preferences from these entities to help you decide about ideas and their characteristics.

The client, the customer, and the user may or may not be the same entity. In fact, you may or may not even have a client or customer. Your client may be the institution that will use the idea to make a product or offer a service. The client may be internal or external. It may be an institution that wants to see ideas developed so that others will be able to use it to produce products (such as a funding agency). The customer may be the individual who will purchase the product or service for personal use or for others to use (a parent who buys for a child, a purchaser buying products for retail sale, etc.). The users are the individuals who use the product or service. Thus, the client, the customer, or the user may be different entities and individuals with different opinions and points of view, or the same individual or entity.

It should be pointed out here that if the idea you are considering is novel enough to be patentable, you may want to be careful about seeking others' opinions that are not directly associated with your team. There is always the matter of divulging too much proprietary information to others and of disclosing important information that may jeopardize the future value of an idea. We will discuss this issue in more depth in Chapter 14.

• **Use models to test the idea:** If the idea you are analyzing is complicated, untested, or unfamiliar, if you cannot decide whether it works or not, and if you cannot predict its appropriateness to the problem at hand, you may gain useful insight by modeling it. The model may be simple, quick, and cheap, or may involve sophisticated set up and much expense. But this may be necessary if the idea has the potential to be very useful and you do not want to dismiss it unnecessarily. Many designers have a slew of parts, actuators, toys, and other "junk" at their disposal to make a quick model of an idea to visualize it, to test it, and to understand it. Foam blocks, Lego pieces, toys, electronic parts, and computer simulation programs are very common tools used to model ideas. It should be mentioned here that as you may have noticed, we are already testing ideas, and based on the result, we may change our decisions or approach. This, in fact, is an iteration in the design process and will be discussed in more detail later. Suffice it to say that, as mentioned before, design is an iterative process that involves many repetitions before a final solution is achieved.

- **Use scientific (statistical, engineering, . . .) analysis to assess the feasibility of the idea:** Yet another approach to analysis of ideas is to use scientific analysis to determine the feasibility and virtue of an idea. This includes engineering analysis, statistical analysis, computer simulation and modeling, and many more, depending on the idea and the issues involved. For example, statistical data available on weather patterns may be used to decide whether a June wedding should be held outdoors or indoors. Similarly, a statistical analysis of the risks involved in using a nuclear battery to power an artificial heart may determine whether or not it should be pursued. A mathematical model of a dynamic system involving a drive system based on a swinging pendulum can predict whether or not a cylindrical robot will work before it is made. The scientific analysis may or may not be applicable to every idea, and at times it may not be worth the time. But when the system is complicated and the designer is not able to predict the feasibility of the idea, this method can be very effective and useful.

5.9.2   Combine Ideas

It should be mentioned here that we are not just interested in picking a single idea out of our list of proposed ideas; rather, we are looking for the best idea possible. Therefore, if combinations of ideas, or even combinations of the best features of different ideas, can give us a better solution, we should do so. Our intention is to find the best solution without considering whose idea is picked and whether or not ideas from different people are combined, all without emotional mental barriers of ownership and pride.

5.9.3   Compare to the Definition

During, or after, your analysis of the ideas, you should compare the characteristics of the ideas with the problem statement or the list of specifications and requirements set earlier. This will allow you to eventually decide whether any of the ideas can be used for your purpose, and whether they are adequate and appropriate. If there are minor inadequacies that can be overcome by redesign or modifications, the idea is worth retaining. If there are major shortcomings that cannot be easily overcome by redesign or by improving the idea, then the idea is probably not worth keeping. Each redesign or improvement can be considered as an iteration in the design process (and we will discuss this later), and each time you try to overcome a shortcoming, the process can be a new round of design challenge. However, having analyzed the problem at the beginning and understanding the requirements and specifications will enable us to compare our ideas with the requirements and needs and decide what idea is worth pursuing.

5.9.4   Decision Matrix (aka Pugh Matrix, Screening Matrix, Scoring Matrix)

This method can be used as a culminating step to make a final decision. In most cases, picking the best idea from among a large list of ideas can be daunting and very confusing. Many ideas may have both advantages and disadvantages, and comparing all these ideas and their features simultaneously can be confusing and difficult. In order to be able to compare all ideas in an orderly fashion that allows us to simultaneously evaluate all the ideas to our stated specifications and requirements, we will use a table called a Decision Matrix. The decision matrix contains all the required specifications, through which all ideas are evaluated.

TABLE 5.2 An Unweighted Decision Matrix

	Idea #1	Idea #2	Idea #3	Idea #4	Idea #5	Existing Product
Characteristic 1	+1	+1	+1	+1	−1	0
Characteristic 2	−1	+1	0	−1	0	0
Characteristic 3	+1	−1	+1	−1	+1	0
Characteristic 4	0	+1	0	−1	−1	0
Characteristic 5	−1	+1	−1	+1	−1	0
Sum of +1	+2	+4	+2	+2	+1	0
Sum of −1	−2	−1	−1	−3	−3	0
Total	0	+3	+1	−1	−2	0

To do this, we will form a table with one column containing the specifications required by the problem statement, followed by columns relating to each of the solutions we want to consider (Table 5.2). In general, there are two possibilities. One is that a product already exists in the market, with which we intend to compete or upon which we intend to improve. In this case, every characteristic of each one of our solutions are either the same (0), better (+1), or worse than (−1) the characteristics of the existing product. The other possibility is that there is no similar product in the market with which we can compare our ideas. In that case, each characteristic is evaluated based on its own merit and whether it is a positive trait (+1) or negative trait (−1). In other words, in this case, there are no benchmarks to which we can compare our ideas. Thus, we will only have an advantage factor of +1 or a disadvantage factor of −1. Since we have already analyzed the ideas, we can compare their characteristics against the problem statement and fill out the table. The table is filled out, and the positive and negative characteristics are added. The summations will indicate the strength of each idea and how closely it relates to our problem statement.

For example, imagine that we are designing a new bookcase. Based on the specifications set by the marketing department, we have chosen aesthetics (A), cost (C), ease of assembly (E), long life (L), and weight (W) as our five important characteristics. Further, imagine that we have five alternative solutions which include metal, plywood, particle board, softwood, and hardwood cases. Our benchmarking target is a common white particle board bookcase sold in stores. (The following table is formed only as an example. The comparisons are not accurate, since we do not have an actual design.) Applying these characteristics and the ideas to the decision matrix, we will get Table 5.3.

Based on this simple analysis, both the hardwood and softwood have an advantage over the existing bookcase, a plywood case would be similar in virtue, while metal cases would be less useful.

As you notice, there can be much disagreement about this result. For example, the ease of assembly is deemed superior for hardwood or softwood cases over the existing white particle board cases because real woods do not crack and split as easily as particle boards; and therefore, there are fewer chances of damaging the case during assembly. All products, except for the bare particle board, have been marked as superior to the white particle board as well, because all these have a longer life, and they do not break or bend under the weight of books. However, it is clear that hardwood and metal are far superior to plywood. None of these are taken into consideration in this table; it is strictly whether a characteristic is superior or inferior that gets a +1 or a −1. This clearly is inadequate. An

TABLE 5.3 An Unweighted Decision Matrix Applied to a Bookcase Design

	Metal Case	Plywood Case	Particle-Board Case	Softwood Case	Hardwood Case	Existing Product
Aesthetics	−1	−1	−1	+1	+1	0
Cost	−1	−1	+1	−1	−1	0
Ease of Assembly	−1	0	0	+1	+1	0
Long Life	+1	+1	−1	+1	+1	0
Weight	−1	+1	0	+1	+1	0
Sum of +1	+1	+2	+1	+4	+4	0
Sum of −1	−4	−2	−2	−1	−1	0
Total	−3	0	−1	+3	+3	0

alternative to this is to not just consider whether a characteristic is superior or inferior, but also to consider how much more (or less) important it is. To do this, we will assign a weight factor to each characteristic. Each +1 or −1 is multiplied by the weight factor, and the summation is compared for the best result.

However, the question that arises is how do we assign an importance value (weight factor) to a characteristic? Obviously, it is possible to simply consider the importance of a characteristic and just assign a number to it. You may improve your decisions by attempting to normalize all the weight factors toward a particular value, such as 1, 10, or 100. In other words, you may try to assign weight factors so that the summation of all weight factors will equal 1 or 10 or 100. However, this is still a difficult and subjective task. To improve on this, we will use a pair-wise comparison technique. As an example, with the five characteristics of aesthetics (A), cost (C), ease of assembly (E), long life (L), and weight (W) we will do the following:

- First, compare the design characteristics against each other and write down which of the pair is more important. Then add the chosen characteristics, normalize, and assign to the decision matrix.

Compared Between	*Chosen*
A & C	A
A & E	A
A & L	L
A & W	A
C & E	C
C & L	L
C & W	C
E & L	L
E & W	W
L & W	L

Total values =10; normalization factor =1/10.

- Add the above preferences to get a weight distribution formula $3A + 2C + 0E + 4L + 1W$.

The total number of comparisons for N characteristics is

$$n = \frac{N(N-1)}{2} \tag{5.1}$$

It should be mentioned here that as a result of this technique, it is possible that one (or more) of the weight factors may end up being zero, indicating that the characteristic is completely unimportant, which is not necessarily true. To remedy this, change the 0 to another value, indicating that all characteristics with which we are concerned are of value. We will choose to increase the weight factor for E to 0.5. As a result, the total will change to 10.5, and the normalizing factor will be 0.095, and our weight distribution formula will change to $3A + 2C + 0.5E + 4L + 1W$. The weight factors will be

	Normalized to 10.5	Normalized to 1
A	2.9	0.29
C	1.9	0.19
E	0.5	0.05
L	3.8	0.38
W	1	0.1

To simplify the task of remembering the pairs of characteristics and to ensure that they are not unnecessarily repeated, you may use a table similar to the following. You may increase or decrease the size as necessary.

A	C	E	L	W

W	A	C	W	L
L	L	L	L	
E	A	C		
C	A			
A				

Applying the normalized weight factors to the decision matrix shown in Table 5.3, we will get Table 5.4.

Comparing the totals, in this case too, the best choices will be either the softwood or hardwood cases, followed by a plywood bookcase and metal bookcase. Comparing these results with the result of Table 5.3 shows that although the best choices are the same, the values of metal cases and particle board cases have switched.

TABLE 5.4   A Weighted Decision Matrix Applied to the Bookcase Design

	Weight Factor	Metal Case	Plywood Case	Particle-Board Case	Softwood Case	Hardwood Case	Existing Product
Aesthetics	0.29	−1	−1	−1	+1	+1	0
Cost	0.19	−1	−1	+1	−1	−1	0
Ease of Assembly	0.05	−1	0	0	+1	+1	0
Long Life	0.38	+1	+1	−1	+1	+1	0
Weight	0.1	−1	+1	0	+1	+1	0
Sum of +1		+0.38	+0.48	+0.19	+0.82	+0.82	
Sum of −1		−0.63	−0.48	−0.67	−0.19	−0.19	
Total		−0.25	0	−0.48	+0.63	+0.63	

As you can see in our latest comparisons to decide about the weight factors, it appears that, in fact, ease of assembly was not really an important factor as compared with all other factors, and that aesthetics is much more important than weight. (This can be true, since, although ease of assembly is important, it really is done just once, and as a result, it is much less important than, say, long life, which is a characteristic that relates to the whole life of the product and its quality.)

On the other hand, as you go through these comparisons, you will notice that there are times when not only is one characteristic clearly more important than another, it is much more important, or not as important. In other cases, it may be that two characteristics are equally important. Selecting one characteristic over another may still be inadequate. To improve this, you may choose to assign a simple weight factor to the selected characteristics in the above comparison and emphasize or de-emphasize one in comparison to another. As an example, we may come up with the following formula:

Compared Between	Chosen (not weighted)	Chosen (weighted)
A & C	A	0.5 A
A & E	A	2 A
A & L	L	1.5 L
A & W	A	A
C & E	C	3 C
C & L	L	L
C & W	C	2 C
E & L	L	2 L
E & W	W	W
L & W	L	2 L

Total value =16; normalization factor =1/16

Adding the above preferences results in $3.5A + 5C + 0E + 6.5L + 1W$. As before, changing the weight factor for E to 0.5, we will get $3.5A + 5C + 0.5E + 6.5L + 1W$. Notice that in this case, the cost became more important than aesthetics, while long life remained as the most important. If you do not agree with this notion, you should change it now.

The weight factors for each will then be

	Normalized to 16.5	Normalized to 1
A	3.5	0.21
C	5	0.3
E	0.5	0.03
L	6.5	0.39
W	1	0.06

Applying these results to Table 5.4, we will get Table 5.5.

As you can see, the results here are different too, but they show the same tendency. This may not be true for other cases, and as the weight factors change, the final result may be different.

It can be argued here that in assigning the above weight factors, when one characteristic is chosen, the other one is actually ignored (since it will not be mentioned), whereas we know both are of value. Additionally, it is not possible to indicate that two characteristics are actually of the same value. Another alternative for assigning weights to the characteristics is to always mention both terms, but with different values. In this

TABLE 5.5   A Revised Weighted Decision Matrix Applied to the Bookcase Design

	Weight Factor	Metal Case	Plywood Case	Particle Board Case	Softwood Case	Hardwood Case	Existing Product
Aesthetics	0.21	−1	−1	−1	+1	+1	0
Cost	0.3	−1	−1	+1	−1	−1	0
Ease of Assembly	0.03	−1	0	0	+1	+1	0
Long Life	0.39	+1	+1	−1	+1	+1	0
Weight	0.06	−1	+1	0	+1	+1	0
Sum of +1		+0.39	+0.45	+0.3	+0.69	+0.69	
Sum of −1		−0.6	−0.51	−0.6	−0.3	−0.3	
Total		−0.21	−0.06	−0.3	+0.39	+0.39	

TABLE 5.6   Weight Factors for the Bookcase Example

Compared Between	Chosen (not weighted)	Chosen (weighted)	Chosen (both characteristics, weighted)
A & C	A	0.5 A	1.5A / C
A & E	A	2 A	3A / E
A & L	L	1.5 L	2.5L / A
A & W	A	A	2A / W
C & E	C	3 C	4C / E
C & L	L	L	2L / C
C & W	C	2 C	3C / W
E & L	L	2 L	3L / E
E & W	W	W	2W / E
L & W	L	2 L	3L / W

Weight factor formula = 7.5A + 9C + 4E + 10.5L + 5W; total value = 36, normalization factor = 1/36.

way, you ensure that no characteristic is completely ignored, and that they can be assigned equal values. Table 5.6 shows the result. To do this, increase the previously mentioned value by 1.

	Normalized to 36	Normalized to 1
A	7.5	0.21
C	9	0.25
E	4	0.11
L	10.5	0.29
W	5	0.14

As you notice, once again, the importance has remained the same, but the emphasis has changed. Still, long life is the most important trait, followed by cost, aesthetics, weight, and ease of assembly. But in this case, all characteristics have value. Table 5.7 shows the result of these weight factors used for comparing the presented ideas.

You may wonder which one of these revised methods is better. In fact, it depends on the product you are analyzing. The best thing is an engineering judgment. However, the last variation is perhaps the most thorough, and thus, the most representative.

TABLE 5.7 Twice-Revised Weighted Decision Matrix Applied to the Bookcase Design

	Weight Factor	Metal Case	Plywood Case	Particle-Board Case	Softwood Case	Hardwood Case	Existing Product
Aesthetics	0.21	−1	−1	−1	+1	+1	0
Cost	0.25	−1	−1	+1	−1	−1	0
Ease of Assembly	0.11	−1	0	0	+1	+1	0
Long Life	0.29	+1	+1	−1	+1	+1	0
Weight	0.14	−1	+1	0	+1	+1	0
Sum of +1		+0.29	+0.43	+0.25	+0.75	+0.75	
Sum of −1		−0.71	−0.46	−0.5	−0.25	−0.25	
Total		−0.42	−0.03	−0.25	+0.5	+0.5	

However, you should always check the final values and see if they represent the true desires of the designer or not, and if not, they should be modified.

A close inspection of the numbers used in all the preceding tables shows another shortcoming. When you look at the comparison between softwood and hardwood bookcases with the existing white particle board cases, you will notice that they are both marked as better (+1) for aesthetics and worse (−1) for cost. Obviously, whether or not we use a weight factor associated with these two characteristics, the final sum of these characteristics will always be the same. However, we know that hardwoods look differently than softwoods, and that the cost of hardwood is much more than softwood. But these are not affected by a weight factor that only indicates their importance and not their different values. To further improve our assignment of values to these characteristics we can also include variations in the level of advantages and disadvantages by assigning numbers larger or smaller than +1 or −1. Assigning values with varied emphasis, together with the weight factors, will enable us to compare the different ideas much more accurately and more truly. We will apply this to our comparison in Table 5.8 and will compare the results.

In this case, the best choice is hardwood bookcase, followed by softwood and plywood, which make sense for our chosen weight factors. Obviously, if we were to change our weight factors or advantage factors, the results might change accordingly.

TABLE 5.8 A Revised Set of Weight Factors and Variable Advantage Factors Applied to the Bookcase Design

	Weight Factor	Metal Case	Plywood Case	Particle-Board Case	Softwood Case	Hardwood Case	Existing Product
Aesthetics	0.21	−0.5	0	−2	+1	+2	0
Cost	0.25	−3	−1	+1	−2	−3	0
Ease of Assembly	0.11	−1	0	0	+1	+1	0
Long Life	0.29	+1	+1	−1	+1.5	+2	0
Weight	0.14	−1	+1	0	+1.5	+1	0
Sum of +		+0.29	+0.43	0.25	+0.965	+1.25	
Sum of −		−1.105	−0.25	−0.71	−0.5	−0.75	
Total		−0.815	−0.18	−0.46	+0.465	+0.5	

Case Study #1, continued: Page-Turner for the Disabled   In the following sections we will analyze our 10 chosen ideas developed for the page-turner and then try to select an idea to pursue. The design requirements and specifications statement is repeated here for convenience.

> *We intend to design a page-turner for the disabled that will turn the pages of a hardback or paperback book (no 3-ring binders) as large as 8.5 × 11 inches and as small as 4 × 6 inches, up to 2 inches thick, not weighing more than about 6 lb, back and forth, but one single page at a time. Turning multiple pages will be an added bonus. The page-turner is not recommended for rare books. It will be designed for disabled individuals who for one reason or another do not have the capability to turn the pages of a book. The user will be able to provide at least two distinct signals to the machine. The selling price of the final product will be about $500. We want the page-turner to be reliable and easy to use. We plan to have a working prototype in about 6 months from the starting date.*

Idea Number I   (Figure 5.5): *A narrow belt stretched over the pages, which is operated by a motor. As it turns one way or another, it will turn one page forward or backward. Continuously running the belt (perhaps with additional force exerted on it at the same time), multiple pages may be turned.*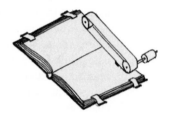

This idea can work nicely and can be adapted to many different situations and requirements. It can work with both hardback and paperback books so long as we ensure that the books are secured in place. However, since the paper used with paperback books is generally stiff, as the belt rotates it may wrinkle or tear the paper. Additionally, as one page is bent and turned, before the operation is finished, the belt may start on another page prematurely. We also need to improve on the holders on the sides to allow the page to go under the holders (we will need to do this for most of the ideas anyway). This idea may be more appropriate for a magazine, but this is not what was required in the specifications. It may have difficulty working on a 3-ring binder, but this is not required by our problem statement either. Since the belt sits on the top pages, as the thickness changes, it remains in contact with the pages, and thus it may require less adjustment. However, as more and more pages are turned to either side, the contact with the pages may be lost and the system may not work any more. This method can work for turning pages back and forth by changing the direction of the motion of the belt, and there is no need to move the belt into a particular starting position; it is always ready to turn pages. Although this is not required in our specifications list, it is possible to turn multiple pages either way by continuously running the belt and by adding to the force of friction, even if the number of pages turned cannot be controlled well. The set maximum allowable cost can also be met by this design, as it is relatively simple to make the device.

One disadvantage is that the belt may cover the very top portion of the page, and thus, may interfere with titles, chapter information, or page numbers. If the book is large, handling the page only at the top may also make it difficult to turn without jamming. We will need to improve the design of the holders to accommodate different sizes of books as well. This design is relatively bulky and large and requires much space and may be heavy. However, the design is simple, and requires only one motor. Overall, although this idea has potential, it is not very well suited for our purposes.

Idea Number II (Figure 5.6): *A sticky material attached to a pen-like arm will lift one page, while another arm will rotate under the page turning it to another side.*

This is also a nice idea and, with proper adjustments, can work for a variety of book types and sizes. As in idea number I, the holders need improvement, and we will need a better way of pushing the paper under the holders as pages are turned. This idea can also work for magazines and newspapers, although this is not required.

This idea cannot be used for turning multiple pages, but that is not required. However, with proper design and with two lifters, it is possible to go back and forth. The idea can be implemented with relative ease and simple parts. There will be a need for multiple actuators or motors to operate the lifters and the sweeper arm. There will also be a need for control circuits to reverse the motion.

The sticky material used with the lifters may also create a problem. Many different materials may be used, including Post-It ™ notes. However, we will need to change the sticky material or the paper once in a while when it becomes ineffective.

Idea Number III (Figure 5.7): *A sticky material is attached to an arm. The arm is held in the mouth of the user. By pressing the sticky material against the page, the user will lift it from one side and will move it to the other side by moving the head.*

This is one of the simplest ideas to implement. It requires no external power, can be used to turn pages back or forth, but not for multiple pages. The same limitations exist with the use of a sticky material as in Idea number II. It can work for almost all types of books, different sizes, magazines, and even 3-ring binders. There is no need for vacuum, motors, or controls. A major concern is whether the disabled person can control his or her head with enough dexterity to be able to pick up the stick, hold it, and try to turn the pages or not. Depending on the type of disability, the idea may work or not. For example, for someone with a temporary disability in the arms (e.g., someone with a broken arm) this may be easily implemented. However, someone with a more debilitating disadvantage may not be able to use this system. We also need to consider the cultural acceptability of the idea. A pianist at home may not mind using this, but during a concert, it will look unacceptable for an individual to move pages of notes by this system. The effectiveness of the idea can be tested very easily.

Idea Number IV (Figure 5.8): *Use a vacuum (suction) cup to lift the page, while a rotating arm sweeps under the page turning it.*

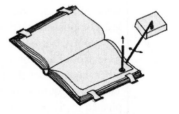

This idea has a lot of similarities with Idea number II. It requires a similar source of power for the arm, similar actuators and motors, and similar controls. However, it does also require a source of vacuum or a vacuum pump. Both the vacuum pump and vacuum cups are noisy, including when the paper is separated from the cup. Of course, it is possible to release the vacuum in the cup before the paper separates, which will also make the pages safer, but this requires an additional control valve. With proper control, the pages can be turned both back and forth, but only one page at a time. Since more parts are needed in this design, the cost will be more than other ideas too. On the other hand, vacuum cups can be more reliable than sticky materials, and they do not need to be replaced as often. Since the vacuum cups are lowered into the page, this system can handle a variety of book types and sizes.

By the way, a variation to this idea is to use a pressurized puff of air to lift the page instead of vacuum. A puff of air is perhaps not as reliable as vacuum, but it may be easier to produce the puff of air, either with a simple pump, or a pressurized bottle of air (including a simple balloon that is inflated once in a while or a hand pump that fills a simple tank).

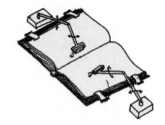

Idea Number V   (Figure 5.9): *Use a pair of two-link arms with rollers at their tips rotating to lift the page, and the linkages extending to turn the page. The opposite pair grabs the pages and holds it. The opposite process can be used for turning backwards.*

This is an interesting idea. The functions of both lifting the page and turning the page are accomplished by the same part. As the roller rotates and lifts the page, the arm extends and turns it. This means that the system is somewhat simpler and more reliable. Unless the rollers become dusty or glazed, they can be effective in lifting the page and in moving it (similar rollers handle thousands of pages in copy machines and printers). However, there is still a need for motors to roll the rollers and to rotate the arms, on both sides. It is possible to design the system such that all motions on each side could be accomplished by one actuator, but the system becomes more complicated with more parts. Notice that unlike other ideas, this system needs to be duplicated on both sides, but it can turn pages back and forth easily. It is also possible to force the roller to lift multiple pages at once, a bonus. However, it might be difficult to control how many pages are lifted. One other nice feature about this design is that since the rollers are always in contact with the pages at either side, there is no need for holders or clips, which helps in lowering the cost and lessens worries about fitting the pages under the holders. The motion of the pages is coordinated and smooth too. This system can work with variable heights too, and thus does not need to be adjusted.

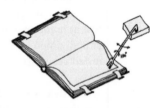

Idea Number VI   (Figure 5.10): *A hook-like hand with a sticky finger lifts the page, turns $\frac{1}{2}$ turn under the page, and then sweeps the page to the opposite side.*

This is also a nice idea. It is actually simpler than some of the other ones. In this idea, the lifting of the page is accomplished by friction between the hook, which rotates 180° to lift, but also simultaneously getting the opposite ''finger'' on the hook under the page. To do this for a single page, the arm has to lift slightly while rotating to allow the other finger under the page. The arm moves to the left to turn the page as in other ideas. This idea requires one motor to turn the page, but the motion of the hook can be accomplished by a simple solenoid. Multiple 180° rotations of the arm may lift multiple pages, an added bonus. By including proper controls, the arm can start on the opposite side to turn the pages backward as well. This idea can also work with a variety of sizes, thicknesses, and book types.

Idea Number VII (Figure 5.11): *Replace the holders with rollers. These rollers can lift the page, a sweeping arm will turn the page, and the rollers on the opposite side will grab the page and hold it.*

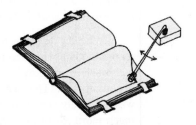

This idea is similar to the ones before as far as turning the page is concerned. It requires an arm to turn the pages back and forth. However, the difference is in the way the pages are lifted and held. Both tasks are accomplished by the same set of rollers. The rollers on both sides turn in the same direction. This causes the page on one side to lift, while the arm is inserted under it and turned. The rollers on the opposite side grab the page, pull it in, and hold it securely. Because the rollers hold the pages in place, there is no need for additional holders, and since these rollers adjust with each new page (removed or added), there is no need for adjusting holders either. One thing that is nice is that, in fact, these rollers can replace the holders in all other ideas too, solving one of the concerns we have had with the holders previously. We might want to actually replace the holders in all other ideas with these rollers anyway. By the way, these rollers can be used to lift multiple pages too. To accomplish this, the roller should be left on a little longer to lift additional pages after each page is lifted. And while we are discussing the replacement of holders with rollers in all these ideas, it should be added that the rotating arm in all these can also be modified to function along with a 4-bar mechanism that, instead of a simple rotating motion, makes the arm go forward under the page, turn, and then come back out.

Idea Number VIII (Figure 5.12): *A rotating arm is lowered near the edge of the paper to come into contact with the paper. As it turns, due to friction, it will curl up the paper, eventually moving under it. Then it can be swept to the other side.*

This idea is similar to some of the previous ones, but in this case we have combined the roller into the arm. Thus, the roller lifts the page, the arm moves under it and turns the page. The same roller can be turned on again on the opposite side to assist the page in turning and the holders in grabbing the page and holding it. In fact, the roller can be used to lift multiple pages too. This idea requires two motors, one for the roller and one for the arm unless the motions are combined. This idea can turn pages in both directions. It can work with a variety of book types and sizes as well. As long as the rollers are clean and not glazed, the system can work reliably, and it does not require additional resources such as vacuum or pressurized air. Additionally, we can combine the previous idea with this by replacing the holders with rollers as well.

Idea Number IX (Figure 5.13): *The book could be scanned into an electronic file. The computer file can then be manipulated easily.*

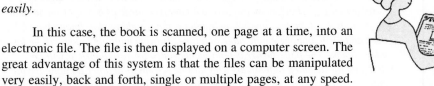

In this case, the book is scanned, one page at a time, into an electronic file. The file is then displayed on a computer screen. The great advantage of this system is that the files can be manipulated very easily, back and forth, single or multiple pages, at any speed. It requires a simple input from any pointing device used in computers. Additionally, many books are becoming available in digital form, and thus there may be no need to even scan the pages. The book could be purchased in digital form to begin with. Additionally, either an electronic book (there are several tens of patents on these books) or a laptop computer could be used for this purpose. No doubt, with a large enough market, an electronic book display could also be designed for use by the disabled as well. Another advantage is that if necessary, the print size could be enlarged for individuals with impaired vision. On the other hand, there are several disadvantages to this system too. One is that for any book that is only available in print form, whether old or new, and for all other printed material that does not come in digital form, the user will have to scan all the pages. For a large book of several hundred pages, this can be a daunting task to accomplish. Additionally, unlike with a mechanical page turner, you cannot just take a new book and read it. It must first be scanned. This can be a huge detriment to someone who likes to pick up a new book and read it. On the other hand, this is light, quiet, and free of most other limiting physical specifications (paperback, small books, large books, thickness, etc.).

One major point about this idea is whether or not we require a mechanical device. The idea is different from most of the other ones expressed, and in itself, this is a positive virtue. There is also no indication in our definition that we cannot design a completely electronic solution either. However, we still need to clarify our intention here as to whether or not we want a more mechanical device. Assuming that the problem of scanning all the pages of a book is not unacceptable, this idea solves most of our other concerns, and then some. Imagine a pianist who has a simple screen located on the piano displaying the notes. It may be a refreshing and revolutionary idea to some and a sacrilege to others.

Idea Number X   (Figure 5.14): *A monkey could be trained to sit next to the user to turn the pages whenever the user sends a signal to the monkey.*

Whether we can actually train a monkey to sit around and do our bidding when required remains to be seen. Curious animals such as monkeys are not known for sitting still for long periods of time. (The idea of using a dog as a guide for the blind was not easily accepted either. However, dogs can be trained much more easily than monkeys.) As with all other animals, there is also the question of the well-being of the animal, need for food, rest, and other daily requirements, and whether owning a monkey is allowed by the community or State laws. Still, it is conceivable that there may be a possibility to train an animal to turn pages of a book when told.

Now that we have analyzed the ideas, we will use our problem statement with our chosen specifications and characteristics and will compare the ideas through a decision matrix. We will list our requirements in the table and will assign a positive or negative number to each characteristic for each idea. To be able to assign weight factors, we will first do a comparison analysis of the characteristics and will use the weight factors as well. Tables 5.9 and 5.10 show the results. The problem statement is repeated here for your reference.

> *We intend to design a page-turner for the disabled that will turn the pages of a hardback or paperback book (no 3-ring binders) as large as 8.5 × 11 inches and as small as 4 × 6 inches, up to 2 inches thick, not weighing more than about 6 lb, back and forth, but one single page at a time. Turning multiple pages will be an added bonus. The page-turner is not recommended for rare books. It will be designed for disabled individuals who for one reason or another do not have the capability to turn the pages of a book. The user will be able to provide at least two distinct signals to the machine. The selling price of the final product will be about $500. We want the page-turner to be reliable and easy to use. We plan to have a working prototype in about 6 months from the starting date.*

TABLE 5.9 Weight Factors for the Page-Turner

	L	R	P	C	D	S	T
T	T/L	T/R	2P/T	2T/C	T/D	S/T	
S	S/L	S/R	2P/S	2S/C	S/D		
D	D/L	D/R	2P/D	2D/C			
C	2L/C	2R/C	2P/C				
P	2P/L	2P/2R					
R	2R/L						
L							

TABLE 5.10 The Decision Matrix for the Page-Turner

	Weight Factor	I	II	III	IV	V	VI	VII	VIII	IX	X
T	0.13	−1	+1	+1	+1	+1	+1	+1	+1	+2	+1
S	0.11	−1	+1	+1	+1	+1	+1	+1	+1	+2	0
D	0.13	+2	+1	+2	+1	+2	+1	+2	+2	+2	+1
C	0.11	+2	+1	+3	+.5	+1	+2	+2	+1.5	+3	−3
P	0.22	−2	+1.5	+1	+2	0	+2	+2	+2	+2	−2
R	0.17	−2	+1.5	+1.5	+2	+.5	+2	+1.5	+2	+2	−2
L	0.13	+1	+1	+1	+1	+1	+1	+1	+1	+1	−2
Sum of +		+0.61	+1.2	1.44	1.34	0.83	1.5	1.55	1.58	1.98	0.26
Sum of −		−1.02	0	0	0	0	0	0	0	0	−1.24
Total		−0.41	1.2	1.44	1.34	0.83	1.5	1.55	1.58	1.98	−0.98
Ranking		9	7	5	6	8	4	3	2	1	10

We will consider the following characteristics and concerns from our problem statement. Notice the similarity as well as differences between these and the issues we considered to define the problem. Since we decided that certain characteristics are not as important and therefore not included in the definition, they are not listed here either (such as weight, or multiple pages):

- (T) Works with different types of books
- (S) Works with different sizes and thicknesses
- (D) Direction of turning pages back and forth
- (C) Cost
- (P) Performance (how well it works, ease of use)
- (R) Reliability (how reliable is the system)
- (L) Long life (whether or not components last)

From Table 5.9 we get

Weight factor formula $= 7T + 6S + 7D + 6C + 12P + 9R + 7L/(\text{TOTAL} = 54)$

Normalized weight factors $= 0.13T + 0.11S + 0.13D + 0.11C + 0.22P + 0.17R + 0.13L$

Based on this formula, performance is the most important characteristic, followed by reliability, which makes sense. We want to make sure that the device works well and

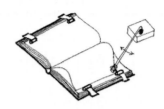

FIGURE 5.19 The two chosen ideas for the page-turner.

is reliable. Next, type of books, going back and forth, and long life are about as important as each other, followed by cost and size of books. In fact, these are similar to the results we got from our survey. In the design of a device such as this, the cost is not as important as the function because the product is limited in its numbers (not mass produced for the general public), and the success of a specific design like this is more dependent on whether it works properly than how expensive it is. However, we need to ensure that what we choose can be made within the specified price range. If you do not agree with these numbers, you may change them. Table 5.10 compares these ideas relative to each other and based on whether or not each idea may satisfy the need. These numbers are usually subjective. Experience helps, as does a team assessment.

Based on this analysis, the best fitting idea is number IX, the idea of scanning the book into a digital display unit. The second best is idea VIII, which also makes sense because it can satisfy most of the requirements, especially if holders are replaced with rollers. Obviously, it is possible to combine the rollers idea with the other components of other ideas to see what the result will be, but we have not done this here. Also, it should be clear that the numbers chosen above are subjective and are influenced by the designer. Thus, you may do the same analysis and come up with a different result. However, the point of this exercise is to learn about the process.

In order to continue with the design process, we will consider both ideas IX and VIII with the modification. At this point, nothing in the design specifications prohibits us from using idea IX, but eventually, our client should clarify whether or not this solution is acceptable. For now, we will continue with the development of these two ideas, shown again in Figure 5.19.

Case Study #2, continued: Electronic Alarm Clock A similar analysis for the ideas expressed for the alarm clock will result in our ability to select one idea and pursue it. For example, the idea that a joystick may be used to move a pointer within a display through a maze and get it to the home position before the alarm can be turned off is very interesting. The maze can be designed to change every time to prevent the person from memorizing it. However, the disadvantage of this idea is the cost of a microprocessor, the display, and the joystick. The idea of using a steel ball within a labyrinth that has to be moved to the end to turn off the alarm is similar in nature, but mechanical. It does not need as much electronics and can be used anywhere, but it cannot easily be changed to prevent memorization (unless it is designed to be excessively difficult). Fresh water needed to switch off the alarm will force the person to get the water and fill up the alarm's funnel with water, but a small glass of water next to the alarm will defeat the purpose. Besides, this is easy to get used to and do without really thinking. A sequence of sounds from nice music to obnoxious sounds and a set of numbers pressed on a keypad to turn off the alarm is good and practical. But it may be unacceptable if there is more than one person present. Therefore, we assume that we will continue with the use of a series of LEDs that light up in random order, whereby the user must press them in the same sequence to turn off the alarm. We will continue with this idea.

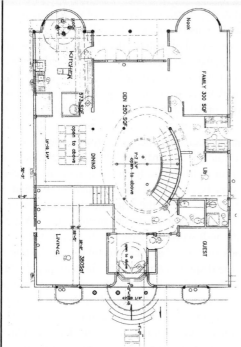

FIGURE 5.18 Repeated.

Case Study #3, continued: Design of a Family House A similar analysis can be applied to the four suggested plans for the house. For example, idea IV (shown again here) provides an office at the entrance, a dramatic staircase to the second floor, a kitchen facing the yard (quiet, nice view) that is close to the dining room, open space between the family room and the dining area, a kitchen with an island, and so forth. This preliminary plan satisfies many of the requirements, except that the living room is not integrated with the open space of the dining area and the family room. We will select this idea as a first choice.

STAGE IV: IMPLEMENTATION

5.10 ANALYSIS, DETAIL DESIGN, AND PLANNING OF THE CONVERGED SOLUTION

As a result of the last step in the design process, the designer should have selected an idea to pursue. This pursuit involves the detail analysis of the idea, detail design of the components and parts, and planning of the events that may be necessary to later implement the design. The level at which this analysis and detail design occur is dependent on the type of project, plan, or strategy. The design of the page-turner requires a different amount of analysis and detail than an automobile or a fun day in the city. All you may require for planning a fun day in the city may be the addresses and names of a few interesting locations to visit or dine. The detail design of a new automobile may take many months with a large group of engineers and designers. The detail design of a page-turner may require many days of work with your group. However, the designer must decide on the details of all the necessary parts and components that must be designed before they can be implemented.

The level of detail in this step also depends on whether it is the first iteration of the design process or later iterations. We will discuss iterations later, but at this point it should be mentioned that during most first detail design iterations, the level of detail may be lower than later iterations. For example, if we decide to use the idea of the rollers and a rolling/swinging arm for the page-turner (Figure 5.19), we may start with the detail design of the whole system first, and afterwards, go into the details of the subassemblies (the arm)), and then the components (the motor, the roller, the friction element) until all parts are completely designed.

This detail design includes calculations, detail drawings of all parts, detail plans of assembly, bill of materials, economic considerations, safety considerations, performance and quality requirements, and selection of components and other parts that will be purchased from other vendors. For example, consider the page-turner idea we are pursuing. The rotating arm and the rollers must rotate in a coordinated fashion with enough frictional force to lift the page, but not to tear it or to lift multiple pages, unless this is desired. You need to estimate what the necessary force is, design the rollers to deliver the required force, choose motors that are powerful enough to rotate the arm and the rollers, calculate the forces and stresses in the load-carrying components to ensure that they do not bend or break under load, choose bearings that will work properly for the life of the product, and so forth. The idea of scanning the book requires that you determine how the books will be scanned, the memory requirements, the display requirements, the way pages are selected and displayed, the way pages are turned (and the type of pointing devices), the loading and saving of the digital information, and so forth. In each case, all these must be designed in detail. Engineering design involves engineering details. Architecture design involves architectural detail. Design of electronic devices requires electronic detail design. Planning a wedding requires detail design of the ceremony, the banquet, food, and venue.

For an engineer, one of the most important steps in the design process is the detail design of a system based on engineering calculations and engineering judgment. You may spend most of your career in college learning just that. In mechanical engineering, you will learn about forces, stresses, motions, materials, thermodynamics, mass and heat transfer, energy, controls, vibrations, and much more. In aeronautical engineering, you learn many similar subjects, but also about the design of aircrafts. In chemical engineering you learn about the design of reactors and chemical processes. In electronics, you learn to design circuits, digital processors, signal processing, and other similar subjects. In architecture you learn about designing building and required codes, aesthetic elements of buildings, and human functional requirements. Each profession has its own detail analysis methods and knowledge base that the students of the field must learn in order to perform this step of the design process. However, as important as this is, it is still just one of the steps in the design process. You must perform the other steps to be able to do this step too. Jumping into this step at the beginning without having completed the other steps leaves you with an incomplete solution that may or may not be adequate, and is certainly not exhaustive.

Case Study #1, continued: Page-Turner for the Disabled

(A) The Mechanical Solution: Think about the page-turner idea with the rollers and a rotating arm. What do you think needs to be designed now? Once again, it depends on whether this is the first attempt at the detail design or a later iteration. However, we eventually need to design the details of the rollers, the arm, and their relative positions and coordinated motions. We also need to ensure that the rollers are adjustable to work with the required sizes and types of books. We must calculate all the loads, stresses, and deflections. We need to choose the proper materials and appropriate dimensions for the rollers, the arm, and all other components to ensure that nothing will yield excessively or break under load. We also need to select motors, bearings, switches, and other components. Additionally, we need to design the wiring diagram and control circuits. When all calculations are done and we know how every part is to function, we need to make drawings of all the parts and

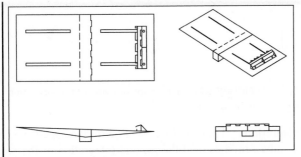

(*a*) The base of the page-turner.

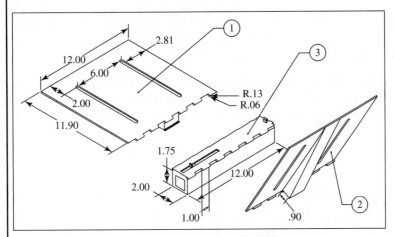

(*b*) Base of the page-turner with detail.

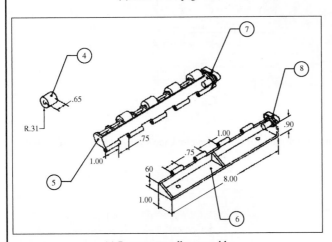

(*c*) Page-turner roller assembly.

FIGURE 5.20   A sample of the preliminary drawings for the page-turner (drawn by M. Welch, D. Sanatana, R. Sommers, and J Seaman).

components. Figure 5.20 shows a sample of preliminary drawings that describe some of the details of the page-turner. Most dimensions are not shown. Similar drawings must be drawn for all components that will be manufactured, as well as an assembly drawing showing the way these components will be assembled together.

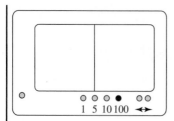

FIGURE 5.21 Schematic drawing of the electronic page-turner.

(B) The Electronic Solution: The electronic solution for the page-turner involves the design of the circuits, subassemblies, and the general integration of the parts. This detail design is very involved and will not be shown here. However, design engineers with expertise in the design of circuits should be able to design such a system. The actual detail design of a system such as this is beyond the scope of this book. Figure 5.21 shows a schematic drawing of what the interface of such a system may look like. The electronic interface would be located on the back or the side.

Case Study #2, continued: Electronic Alarm Clock Figures 5.22 through 24 show the design of the circuits for the alarm clock. The drawings show the detail design of the printed circuit board, the logic circuit, and the light assembly diagram.

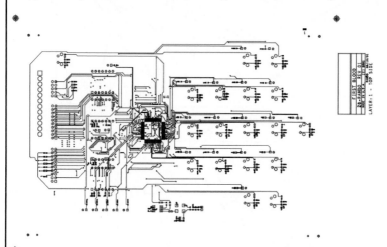

FIGURE 5.22 The circuit diagram for the alarm clock (courtesy of Rambod Jacoby).

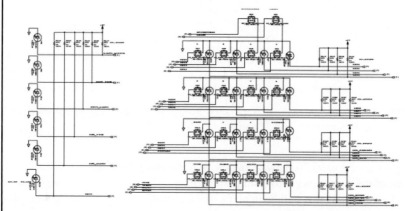

FIGURE 5.23 The light assembly diagram for the alarm clock (courtesy of Rambod Jacoby).

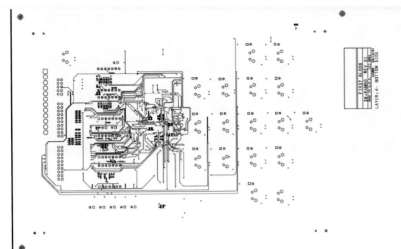

FIGURE 5.24   The printed circuit board design for the alarm clock (courtesy of Rambod Jacoby).

Case Study #3, continued: Design of a Family House   The actual detail designs of the house are too large and extensive to be shown here. Suffice it to say that the architect spends a substantial amount of time preparing these detail drawings according to the architectural requirements and codes as well as the local structural codes. Similarly, a civil or structural engineer will design the structure, sheer walls, the foundation, and other structural elements of the building. Additionally, a wiring diagram, plumbing and sewage diagrams, and other utilities have to be prepared by the architect and his or her associates.

5.11   REALIZATION OF THE IDEA

In this step, the product or the plan will be implemented and realized. This means that if the idea is to make a product, we will make it based on the detail design that we have developed. If it is a plan, we will execute the plan. If it is a strategy, we will implement the strategy. During the first attempts, this realization may be as simple as a model or a prototype in order to test the feasibility and effectiveness of the idea, it may be to learn about the specifics of the idea or plan, or it may be the final product. In the final realization of the product or plan you may need to set up the process, create necessary casts, jigs and other implements that will be used for mass manufacturing, procure machinery and install it, and set up a process to manufacture the parts and assemble them. Nonetheless, in the realization step we execute our plan and make the product. For example, in the case of the page-turner, we will make a prototype to later evaluate the idea and its effectiveness. In the example of spending a fun day in the city, we will execute the plan by actually going to the city and spending the day there doing what we planned. In the example of redecorating our living room, we will implement the plans we have made by perhaps hiring a decorator and seeking his or her advice.

This step may require the designer to make models in a shop or in a garage. It may also involve the designer's cooperation with shop personnel to make the product from the drawings. Or it may involve hiring an outside entity to make the product or execute the plan. Many designers and engineers keep boxes of sometimes seemingly unnecessary parts in their possession in order to be able to make models of their ideas quickly and inexpensively.

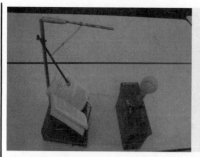

FIGURE 5.25 This prototype simple page-turner was made out of an army surplus lunch box, scrap metal, and a tennis ball. The model was used to test the feasibility of a particular idea (Courtesy of M. Rodriguez, et al.).

Many others have access to cardboard, foam, wood, and other building materials that they use to bring their ideas into reality and to see whether or not the ideas are feasible and useful.

Case Study #1, continued: Page-Turner for the Disabled Figure 5.25 is a preliminary model of a page-turner that was made based on the idea of an arm, controlled by the chin, turning the page with the help of a sticky material on its tip. To make this model, an army surplus lunch box, scrap metal, and a tennis ball were used. Although not the final product, it helped in evaluating the idea.

Case Study #2, continued: Electronic Alarm Clock Figure 5.26 shows the fabricated circuit boards of the alarm clock. The fabrication of this prototype was limited due to the availability of mass production facilities on campus, but a more integrated manufacturing approach could have been taken for mass production.

Case Study #3, continued: Design of a Family House After the plans for the house were completed, a contract was signed for its construction. In architecture and construction there is not much chance for testing ideas other than models. In most cases, models are used for visualization. Although this house is actually built, pictures cannot show all its details at once.

FIGURE 5.26 The fabricated printed circuit boards with mounted electronic components for the alarm clock (courtesy of Rambod Jacoby).

5.12   ASSESSMENT AGAINST PROBLEM STATEMENT AND VERIFICATION

When an idea is brought into reality by making it, when a plan is executed by doing it, or when a strategy is implemented by performing it, the next step in the design process is to assess the results against the problem statement and our list of requirements and specifications and decide whether the idea is appropriate, the plan is useful, and the strategy is feasible. This assessment will help us determine whether we have finished the job at hand or need to do more. If the result indicates that we have achieved what we set out to do, and if all the requirements have been met and all specifications have been achieved, we are done; there is no more to do. We have satisfied the problem statement by reaching our goals as set in the problem statement. However, if any of the requirements or specifications are not met, or if the need is not fulfilled, or if we feel that the solution is either inadequate or that it is not the best we expected, then we will need to iterate the process by going back to the drawing board and repeating the steps that are necessary to achieve our goals. The cliché of going back to the drawing board is not literal. What this means is that we need to go back to whatever step that is necessary and repeat what is needed in order to achieve our goals.

Assessing the implemented idea, whether a manufacture, a plan, or a strategy, may require testing, collecting data, and analysis. The testing may be simple and quick, or it may be extensive and costly and require much time. For example, testing an airplane will require hundreds of hours of wind tunnel testing and evaluations as well as hundreds more hours of flying time. During these tests, data are collected and analyzed in order to make sure that everything functions properly and adequately, that the airplane is safe, and that it satisfies all the internal and external requirements. In contrast, testing a simple device like a one-way valve for natural gas may be quick and easy. Testing machines and products that are subject to fatigue loading may require millions of repetitions, while a static device may be tested quickly. A medical device may need both engineering and clinical evaluations and approval from the Food and Drug Administration (FDA) before it can be marketed and used. The need for the type of assessment is determined based on the requirements of the design, governmental regulations, and safety needs. What is learned during this assessment should enable the designer to decide whether the design has achieved all the requirements and specifications that were specified in the problem statement or not and whether the process has ended or more needs to be done.

If there is a need to repeat any part of the design process in order to achieve the final goals set in the problem statement, there needs to be iterations of the design steps, as discussed next.

STAGE V: ITERATION

Iterations in the design process are not a single step. Each iteration may require the repetition of many steps in the design process. The following section ties all the above-mentioned steps of the design process together into a repetitive, iterative process called design.

5.13   ITERATIONS

Figure 5.1, repeated here, schematically shows the relationship between different steps of the design process and how the iterations may work. As you notice, there can be numerous iterations between different steps, when necessary, and not in any particular order. For example, imagine that we have made a prototype product and we have assessed the

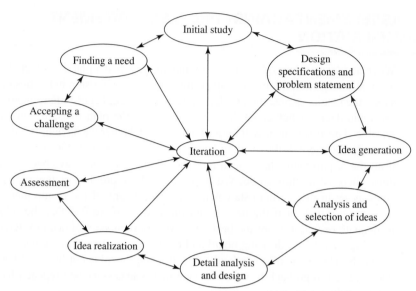

FIGURE 5.1 **Repeated:** Iterations can be between any two steps of the design process and may be repeated as many times as necessary to achieve the final goals of the design project.

product against the original problem statement and we have decided that the result is not yet satisfactory. In that case, we will need to iterate the process, but to where should we iterate? In fact, to any step. For example, imagine that you have worked hard on the process of coming up with a solution for the page-turner and you have so far spent much time, money, and energy to make a prototype, but you feel it is not satisfactory. This expenditure of time, energy, and money has exhausted you. While you originally had decided to do this to perhaps invent a new product, patent it, and make a fortune, now you have come to believe that this will not happen. In that case, you may decide that this endeavor is not worth it, and you decide to quit. This means that your iteration is back to the beginning of the process; you change your first step (accepting the challenge) by quitting. That would end the project too. This iteration, although severe, is a valid iteration. On the other hand, you may think that although you find this particular need to be a difficult task, you are still interested in designing a new product that will satisfy another need, one that will make you rich and famous. In that case, your iteration may be to the second step, finding a new need to satisfy. This means that you still are committed to work on a challenge, but perhaps a new challenge. That is an iteration to the second step.

Now imagine that you have decided that the final realization of the idea was not satisfactory, but that you are not ready to quit, and you still want to continue on this project. You may then proceed by iterating to other steps, for example to the initial analysis in order to determine if your specifications were accurate and adequate and whether or not you missed anything. You may also go to your problem statement and change it (if allowed), or generate more ideas, consider (select) other ideas, take the same idea and change it, come up with a new converged design, make it again, and assess it once more. As you see, you may go to any other step in the process that you, as the designer, deem appropriate. Sometimes, iterations happen quickly and repeatedly between two steps. At other times, many iterations are repeated between two consecutive steps, back and forth, until the result is satisfactory. And eventually, there may be no reason to iterate any more. That is when the project is finished. It is also possible to continue indefinitely with the process and never be satisfied. In that case, I always recall what one of my students said a long time ago, that "in the life of any product, there comes a time when you have to shoot the designer and go into production." Although this sounds like a joke, it is true. Otherwise, as perfectionist as some designers are, the process may never end. On the other hand, when the result is satisfactory, it may be possible to go on to production, while the design process continues toward improvements (after over a century of cars made by countless manufacturers, new ones are still being designed).

It should also be mentioned that iterations may be necessary between other steps of the process, even before the process is finished for the first time. For example, imagine that you have worked on the initial analysis of a problem, and that you have developed your problem statement and design specifications. After you generate some ideas, you may decide that the definition is either too narrow, or not complete enough, prompting you to return to the problem statement step to change or modify it. Or imagine that after you accept the challenge, as soon as you look for a need and you encounter difficulty finding an exciting project, you may change your mind and quit. All these are variations of the same iteration between different steps.

The point is that design is an iterative process with the above-mentioned steps. All these steps eventually happen, hopefully many of them repeatedly, whether you like it or not. As a designer, you will do better, and you will increase your chances of success, when you systematically follow the process and ensure that all steps do occur. I have had countless students who come to me with an idea (a solution) and want to see whether it is a good idea, say for their final project. What I ask them is "What is the problem you are trying to solve?" They usually realize that they do not really know what it is they are trying to solve, the problem statement, what it is that must be achieved. Their idea (solution) is only one solution, one of many other possibilities that they have not, and would not, consider unless they look at the problem statement first. When they realize the difference, they approach design not as an activity, but as a process.

Writing a book follows the same design process. The author realizes the need for the book and takes the challenge of writing it (and all the consequences thereof). The scope of the book is defined, it is written, reviewed, and edited (sometimes many times) until the objectives of the design are met. The book is printed and used. Evaluations and feedback are implemented in the later editions, and the process continues until the book is no longer valid or used.

It should be mentioned once again that certain iterations may not be allowed due to prior agreements (such as in a contract). For example, quitting the project may be feasible if you are working on a project on your own, but it may not be an option if you are hired by a company and have a contract to do it, or if you are registered in a class and need a grade. Quitting (iterating to the acceptance step) may not be an option. Similarly, you may not be able to change the project and start working on another project because you now deem the project too difficult, impossible to achieve, or boring. If you have committed to someone that you will design a page-turner, and especially if you have a contract, quitting or refusing to work on it can be a breach of contract, a serious issue (as is changing the project without your instructor's consent). It may also be impossible to change your problem statement or its specification if you have a contract, if others are involved, or if you are doing this for a specific reason or for a particular group. In these cases, you must make sure that the other entities agree to the changes.

Application of iterations in the design process is also important in other ways. For example, imagine that you are designing an automobile. In order for you to be able to select the correct size engine in relation to the required acceleration rates, you will need to know the weight of the car. In practice, automobile manufacturers have classes of engine sizes already designed, where the appropriate size engine is selected for the desired specifications. Each engine will be capable of satisfying acceleration requirements for a range of weights. However, since the engine is not yet selected, the designer does not know the final weight of the car, and thus, cannot select the engine. This apparent discrepancy is very common in engineering design. To remedy this problem, iterative design cycles are used, where the design engineer estimates the weight of the car (with a hypothetical engine, engine mounts, and other accessories), selects an engine appropriate for that weight and the desired acceleration rates, recalculates the weight of the car with the selected engine (and its required engine mounts and other accessories), and checks to see if the selected engine is still adequate. If it happens that the selected engine in this particular class and range is the same as what is necessary, great. Otherwise, the first iteration has probably given the design engineer a starting point from which she or he can make a second guess at the final possible weight, select a new engine, and check again. The process will be iterated until the selected engine, the total weight, the accelerations, and the required engine all match.

Similarly, in high-cycle fatigue design of machine elements, it may be necessary to use iterations in order to calculate the correct size of the load-bearing element. The idea of high cycle fatigue loading is discussed in more detail in Chapter 16. Here, suffice it to say that the strength of a particular machine element in fatigue (varying) loading, also called the *endurance limit* of the element, is a function of the strength of the element's material and modifying factors,[16] one of which is size factor. This means that the fatigue strength of the material is a function of its size. Therefore, we need to know the size of the element to be able to design its size—a contradiction.

As in the previous section, this is impossible to do unless the design engineer assumes (estimates) a size, calculates the size modification factor, calculates the fatigue strength for the part, and using this strength, calculates the size of the machine element. If the calculated size is close to the initially estimated size, great. If not, the calculated size becomes a new estimate for this iterative process until the estimated size and the calculated size are approximately the same.

Many other similar situations arise in engineering and design, where the iterative process must be used to arrive at a final conclusion. You will see examples of these as you study other subjects as well.

Another very important point to make here is that there may be a significant number of secondary iterations and design processes within the main process. For example, when an idea is selected, and while details are worked out, there may be a need to go through additional secondary design processes to design subassemblies, meet additional requirements, or address other details that were not included in the original idea. For instance, for the page-turner, we need to design the roller assemblies and the arm assembly. Each one of these can in itself be a new secondary design process, where we need to develop new ideas for each subassembly, select ideas, and develop the details. Each one of these design processes can have their own iterations, assessments, and idea realizations. The larger the project, the larger the number of secondary design processes and the need to repeat these steps. Figure 5.27 is a symbolic representation of these secondary processes.

Case Study #1, continued: Page-Turner for the Disabled

Previously, we selected two ideas for consideration for the page-turner problem, one involving scanning the book into an electronic display system, another involving a pair of rollers on each side (to assist in lifting, grabbing, and holding pages) and a rotating arm for turning the page, as shown in Figure 5.19. We also discussed the need to develop these in more detail, come up with detail solutions and drawings for all required parts, and to make the page-turner for assessment. A sample device that was used previously to assess another idea was also shown.

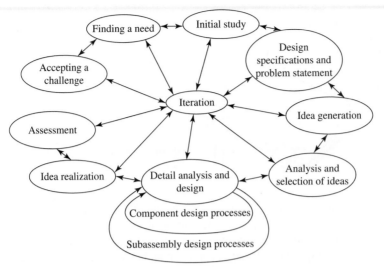

FIGURE 5.27 Schematic representation of secondary design processes within the original design process, as needed for subassemblies and development of details.

In reality, there may be many other scenarios happening at this juncture, and as a result, it is impossible to discuss every possible scenario and possible iterations here. For example, as was mentioned earlier, one may iterate to any step of the process, from acceptance to detail design and realization at this point. A couple of possible scenarios on what may happen next are discussed below. These are used only as examples, and not that they need to be followed exactly in this order, as the order of remaining events entirely depends on what is perceived as necessary by the designer.

One possible scenario is to assume that the prototype design does not completely fulfill the needs of the problem. For example, let's assume that it turns out that as the roller rotates to lift the page, it lifts more than one page. In that case, we need to find a solution for this problem, for example, by stopping the roller exactly as it lifts one page. (Depending on the circumference of the roller, this will be an exact number of rotations, and the controller can rotate the roller only that much before stopping it.) For multiple pages, the roller may be turned multiple times. This should then be tested to verify that it does work.

Another possible scenario is to consider another solution (other than the arm) for turning the page. In our selected solution, we used the rollers to assist in lifting the page while the rotating arm lifted the pages and swept them from one side to the other. Instead, we may use a small fan to turn the page. In this case, the rollers lift the page farther up (above the roller) while the fan is turned on, pushing the page to the other side. The roller on the opposite side will grab the page and will tuck it in and hold it. However, referring to the first scenario, in order to prevent multiple pages from being lifted and still lift the top page to a point above the roller, we might need to have two actions on the rollers, first to turn the roller to "select" the top page, and instead of stopping, to lift the roller while it turns. This will continue to lift the page up and over the roller. The fan can then push it to the other side (Figure 5.28). Is this a better idea? On the one hand, this eliminates the need for the arm and all its associated needs. The fan is simple and needs no other parts. On the other hand, the additional motions needed by the rollers will add to the complication of the design. Should we do it? Should we make a prototype and assess it?

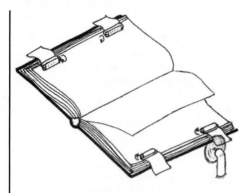

FIGURE 5.28 Another idea for the page-turner. Is this better?

Another scenario would be to consider a change in the problem statement and list of specifications. As you probably noticed, during the analysis of the previous ideas, we considered whether or not the idea was capable of turning multiple pages backward and forward, although this was not specified. Perhaps then we should redefine our problem statement and add the multiple-page requirement to the list, since many of the ideas discussed are in fact capable of this additional requirement, and if added, this can be very beneficial to many users. In that case, the list of requirements and the numbers in the decision matrix might change as well, resulting in a different choice and a different final product.

An actual final product may involve a microprocessor-controlled digital display unit, or a microprocessor-controlled mechanical device with a collection of subassemblies mentioned above which will enable the device to manipulate the pages of the book. The application of the microprocessor will allow us to integrate sophisticated decision-making algorithms related to the operation of the machine.

Case Study #2, continued: Electronic Alarm Clock Like the page-turner, in this case too, many iterations may have occurred already between different steps. The detail design of each subassembly and circuit requires many iterations, testing, and redesign. The final testing must verify that the design specifications are all met. Otherwise, more iteration will be necessary before the process is finished.

Case Study #3, continued: Design of a Family House In this case too, many iterations occur before the design is finished and before construction commences. Even so, many details and subsequent design changes are possible while construction continues. So long as these changes are not substantial, it is possible to affect changes in order to meet the customers' satisfaction.

In this chapter, we discussed the design process and its related iterations. In the next chapters we will discuss other related subjects that enable us to design products and systems that are better, cheaper, safer, and more useful. We will discuss issues such as human factors, patents, product liability, and many others that are all an integral part of the design process.

REFERENCES

1. HARRISBERGER, LEE, "Engineersmanship . . . The Doing of Engineering Design," Second Edition, Brooks/Cole Engineering Division, Monterey, California, 1982.

2. HENDERSON, JERALD, "Engineering Systems Design Class Notes," University of California, Davis, California, 1979.

3. KEMPER, JOHN DUSTIN, "The Engineer and His Profession," Holt, Rinehart and Winston, New York, 1967.

4. POLYA, GEORGE, "How to Solve It," Princeton University Press, New Jersey, 1957.

5. JONES, JOHN, C., "Design Methods," Second Edition, Van Nostrand Reinhold, New York, 1992.

6. EDEL, HENRY JR., Editor, "Introduction to Creative Design," Prentice Hall, New Jersey, 1967.

7. LOVE, SIDNEY, "Planning and Creating Successful Engineered Designs: Managing the Design Process," Advanced Professional Development Inc., North Hollywood, 1986.

8. KOBERG, DON and JIM BAGNALL, "The Universal Traveler, a Soft-Systems Guide to Creativity, Problem Solving, and the Process of Reaching Goals," Crisp Publications, New Horizons Edition, Los Altos, California, 1991.

9. KOEN, BILLY VAUGHN, "Definition of the Engineering Method," American Society for Engineering Education, Washington D.C., 1985.

10. PAPANEK, VICTOR, "Design for the Real World," Bantam Books, New York, 1970.

11. National Collegiate Innovators and Inventors Alliance (NCIIA), Proceedings of the 8th Annual Meeting, San Jose, California, March 2003.

12. REVELLE, JACK B., J. W. MORAN, and C. A. COX, "The QFD Handbook," John Wiley and Sons, New York, 1998.

13. COHEN, LOU, "Quality Function Deployment: How to Make QFD Work for You," Addison Wesley, Mass., 1995.

14. ULLMAN, DAVID, "The Mechanical Design Process," McGraw-Hill, New York, 1992.

15. FREY, DANIEL D. and KEMPER LEWIS, "The Deciding Factor," *Mechanical Engineering Magazine*, March 2005, pp. 20–24.

16. SHIGLEY, JOSEPH and C. R. MISCHKE, "Mechanical Engineering Design," 6th Edition, McGraw Hill Co., New York, 2001.

17. CANNIZZARO, MARIE, "Smooth Operator," *Stanford Magazine*, Nov.–Dec. 2006, pp. 42–43.

HOMEWORK

5.1 Design an electric kettle. What is needed for safety?

5.2 As discussed in Reference 17, an inventor found out that many customers complained about two major aspects of common household irons. Develop a questionnaire and distribute it to a group of users (e.g., your classmates) and see if you arrive at the same conclusions. Use these complaints as "needs" and develop a problem statement.

5.3 As discussed in Reference 17, an inventor found out that many customers complained about two major aspects of common household irons. One was the fact that an iron must be placed in a side location when the material is arranged, and picked up again for ironing. The other major complaint was that the iron must be turned on its side every time it is placed aside. The inventor created a new iron that solves both problems by integrating three small lifters in the soleplate of the iron. When the handle is gripped, the lifters retract and the iron may be used on the fabric. When the iron is to be left waiting, the touch of a simple switch pushes out the lifters, enabling the iron to stand on its own feet less than an inch above the table. Can you solve this problem in another way?

5.4 Most curricula have a capstone design project in which a system or product is designed and developed. In many institutions the need for a project is presented by industrial associates too. Apply the design process in your next design project.

5.5 Among other things, customers may require performance (P), quiet operation (Q), low energy requirement (E), long life (L), utility of features (F), and size (S) as important characteristics of a dishwasher. Assign weight factors to these characteristics.

5.6 Write a problem definition and list of specifications for a door opener for a wheelchair user. Assume that the user is a college student that may have to temporarily live in a dormitory room, an apartment, or a house.

5.7 Write a problem definition and specifications list for a hair-washer for the disabled. Assume that the user is quadriplegic with limited capability, but capable of providing 2–3 distinct signals.

5.8 Develop a list of problems that a quadriplegic individual may face in daily life. Select at least five problems that can be used as a capstone design project.

5.9 Develop a list of problems that a surgeon may face during surgery. Choose one problem that in your opinion has the potential for a new product development.

5.10 The decision matrix developed for the page-turner was based on the opinions of one team. Redo the matrix based on your opinion and compare the results.

5.11 In a second attempt at surveying customers for a page-turner, the following data were collected. Calculate weight factors for these customer requirements.

Customer Preference Data for the Page-Turner, Attempt #2

Importance→	1	2	3	4	5	6	7	8	9	10	Total A	B	B/A	Rank	Weight Factor
1. Reliability			1				8	11	17	26					
2. Back/forth	1		3	1	10	11	12	8	9	10					
3. Multiple pages		7	4	9	4	10	7	3	5	14					
4. Weight	4	6	10	14	12	2	8	6	1	1					
5. Ease of use	1			1			13	19	17	14					
6. Price	2		4	8	9	15	11	6	3	3					
7. Thickness	6	10	12	7	4	6	5	3	2	8					
8. Hardcover/paperback			2	5	9	8	13	10	9	8					

5.12 Using the following table, assign weight factors to the same customer requirements as in Problem 5.11. Compare the results.

Between	*Choose*	*Between*	*Choose*	*Between*	*Choose*	*Between*	*Choose*
1–2		2–3		3–5		4–8	
1–3		2–4		3–6		5–6	
1–4		2–5		3–7		5–7	
1–5		2–6		3–8		5–8	
1–6		2–7		4–5		6–7	
1–7		2–8		4–6		6–8	
1–8		3–4		4–7		7–8	

5.13 Fill out the following QFD matrix for a dishwasher. Use a stainless-steel lined dishwasher as your first competitor and a double-decker dishwasher for your second competitor.

			Weight factors	Design Specifications, Features (How)							Benchmarks		
		What		Heating element	Washing jets	Motor assembly	Programming	Cost	Lining material	Compartment size	Current product	Competitor 1	Competitor 2
Customer Requirements	Functional	Performance											
		Quiet operation											
		Long life											
		Energy use											
		Utility of features											
		Size											
	Other	Low cost											
		Good looks											
		Easy installation											

Units									
Difficulty									
Target Values for Each Feature									

Assessment	□ Our product ◇ Product 1 ○ Product 2	1							
		2							
		3							
		4							
		5							
Absolute Importance Factors									
Relative Importance Factors (%)									
Ranking									

CHAPTER 6

IMAGINATION, VISUALIZATION, GRAPHICAL REPRESENTATIONS, AND COMMUNICATION

That which I have not drawn I have not seen.

—*Unknown Source*

Imagination is more important than knowledge. For while knowledge defines all we currently know and understand, imagination points to all we might yet discover and create.

—*Albert Einstein*

6.1 INTRODUCTION

Imagine two chess players playing a game. While they plan their strategies, their movements, and their countermovements, the players are not allowed to touch any of the pieces, or they will have to move them. So how do they do it? They visualize all the possible or probable movements and countermovements that either player might make, they plan it in their head without actually moving anything, and they imagine what the consequences might be. Everything happens in their mind, not in reality, until they make the move.

Design is the same. Imagine that I ask you to design something, say a page-turner for the disabled. Most probably you will think about the requirements, how it will work, what is needed, and how it will operate, all without a physical system at hand; everything is in your mind. You are visualizing a solution; you are imagining the operations of the machine in your mind's eye. This is not daydreaming. In this case, you actually see and compose alternative solutions and techniques of making and operating the product in your mind. You may consider many different solutions, different approaches, and different consequences, all in your mind. This is why imagination and visualization are such necessary and important skills for a designer. You must see it before it exists. Einstein once said, "Imagination is more important than knowledge." Many have the knowledge. Not all of them have the imagination. A designer must imagine things.

Not everyone is good at visualization. Some people can visualize much more easily than others; some can see things in their minds much better too. This is a necessary skill for designers. Learning the fundamentals of imagination and practicing them will help you gain the experience that is needed to be an effective designer. In this chapter, we will see what this involves and how one can learn to imagine more effectively.

Tesla,[1] an imaginative engineer who invented the AC motors, has been credited as a powerful visualizer, able to design machines in his mind, run them for a period of time, check for wear and tear, and redesign. Whether this is in fact true or not remains to be proven; no one can read another's mind. But the story is supposed to emphasize the importance of being able to visualize what does not yet exist. Engineers have to do this quite often. Every time an engineer solves a dynamics problem, she or he has to visualize motions and forces and accelerations based on a static, lifeless, simple schematic. Every time a strength-of-material problem is solved, the engineer has to visualize the load path, forces, stresses, and strengths. And every time an electric circuit, a building, a mechanism, or a process is designed, the engineer visualizes it first. Solving problems in the kinematics of robots requires that the individual visualize three-dimensional motions of the robot in space, with perhaps up to six joints moving simultaneously, all from a static picture. In fact, most problem-solving in engineering and design requires visualization and a good amount of imagination to see the requirements, synthesize the solutions, and solve the problem before they become real objects.

Michael Gazzaniga, in the Social Brain,[2] states that visualization is actually not a function of the visual system of the brain, but a separate function. To test this, split-brain patients who were tactually split but not visually split were asked to perform a series of functions which showed that visualization and vision are two independent systems. In these patients, tactual (touch) information is not shared by the two brains, but visual information is. As a result, for example, if an object is placed on the right hand (hidden from view), since the information goes to the left brain, and since the left brain in most people contains the speech module, the individual could name the object. An object in the left hand could not be named, since the left brain would not be aware of the object in the left hand. However, if a picture of the object were presented in either visual field, since the brain was not visually split, the person could name the object. These individuals are able to easily perform a visual-tactile match. This means that if an object, say a pencil, is presented to either visual fields, since both half brains have the same information, the person should be able to retrieve the same object by using only touch. Thus, the brain "sees" the object in both halves. To test the relationship between visualization and vision, the patients were given an object in the right hand and were asked to not name the object, but instead, to form a mental image of it; just visualize the object in their mind's eye. However, unlike the actual vision, the image formed in the mind could not be transferred to the other brain, thus the patients were unable to retrieve the object with the left hand. This indicates that visual imagery is not really a function of the vision system, but of computations taking place elsewhere in the brain in another module. Other similar tests have confirmed this finding.

Thinking visually may involve one of the following three activities:[3]

1. **Seeing** involves observing the environment, learning through visual sensory information gathered, and using the information for solving problems.

2. **Imagining** involves the power of imagination and seeing it in the mind's eye to solve problems.

3. **Drawing** involves sketching and drawing to communicate with others, to conceive ideas, and to find solutions.

Seeing, rather observing, is key to learning through the received visual information, and then using the information to solve problems as they arise. Observant individuals do see things in more detail, and since they actually pay attention to the details, they also remember them better. This does not really require any extra effort or learning. All it takes is to consciously be more detail oriented, look more carefully at the surroundings and notice the details, and to pay attention to what goes on. This is a significant trait in photographers, who see the interesting elements of the environment, and capture the moment forever, whether it is news, nature, the good or the bad, the beauty or the beast. Designers need to be observant as well. The sensory information that is gathered and remembered through observing the environment and seeing the details is an additional weapon in the arsenal of techniques, solutions, and possibilities.

Imagining is a level higher than observing what is already in existence. Unlike seeing what already exists, imagination requires inventing what does not yet exist and seeing it in the mind's eye. For some, this is natural and it comes easily. For many, it is not an easy thing. These individuals have a difficult time visualizing or imagining what is not in front of them. Most people can practice and learn how to visualize better. Many life experiences and educational activities involve imagination, and the more we do them, the better we will be. What the designer needs to do is to engage in activities that help him or her learn to visualize better.

Drawing pictures, sketches, or engineering drawings is another level of visualization. If you have learned to draw orthogonal views of an object, if you have drawn an isometric pictorial of an object from its orthogonal views, if you have done any descriptive geometry project, and if you have sketched any pictures to get your idea across, then you have already learned to communicate with pictures or drawings and you know how effective this can be. Drawing sketches or engineering drawings is the culmination of the visualization process, both to help you visualize better, and to communicate to others what you have visualized. Drawing is the process of converting an intangible image into something that can be seen by others.

6.2 IMAGINATION

Figure 6.1 shows a truss and a mechanism. The truss is made up of two links, attached to another link (in this case, the ground), creating a 3-bar structure. The mechanism is made up of three links attached to a 4th bar (in this case the ground), forming a 4-bar mechanism. However, perhaps you noticed that one was called a structure, the other a mechanism. What is the difference? The truss (a structure) is rigid and does not move; the mechanism is a machine, and it does move.

The schematic drawings shown in Figure 6.1 are static; they do not move. However, you may be able to imagine the bars moving relative to each other. In the case of the truss, imagine that a force is applied to the contact point between the two bars. Will the force cause the bars to move relative to each other? Now take the mechanism. What if the crank (the bar on the left) of the mechanism is rotated to the left? What will happen to the mechanism? To figure out what will happen to the mechanism, you must be able to

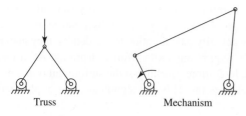

Truss   Mechanism

FIGURE 6.1 A truss and a mechanism; one moves, one does not.

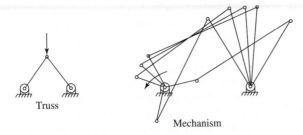

Truss

Mechanism

FIGURE 6.2 The mechanism, as it moves to other positions.

visualize all connected links moving relative to each other while the integrity of the geometry is maintained. In other words, the lengths of the links in this mechanism remain constant, but the angles relative to each other change as the mechanism moves. Can you imagine the full range of motion of the mechanism? As an engineer and a designer, you must be able to imagine all this and see the whole range of motion in your mind. Figure 6.2 shows the mechanism in four other positions.

As you can see, there are many instances in our daily lives where we visualize events and objects, sometimes without even noticing. In many games we play, when we find a solution, move objects, decorate a place, plan a meeting, decide about the route we take to a particular place, and when we hear a sound from the engine of our car, we imagine or visualize. The following simple examples are presented as an exercise to see how well you can visualize certain things.

EXAMPLE 6.1

Figure 6.3 shows three multilink structures. Determine whether each one is a rigid structure or a mechanism. Visualize the range(s) of motion of the mechanism(s).

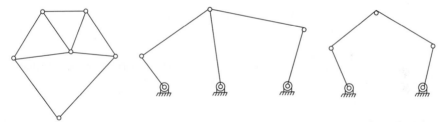

FIGURE 6.3 Which structure is rigid, which one is a mechanism? ■

EXAMPLE 6.2

Can you imagine an object that would look like an E on one side, like a B on another side, and a G on the third? Can you imagine whether it is possible to construct such an object (Figure 6.4)?

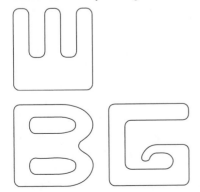

FIGURE 6.4 Can an object be constructed to look like an E and a B and a G on three sides?

This is actually not a puzzle. Here, you are asked to visualize an object with particular requirements although it does not exist. You need to construct it in your mind to see what it would look like. For an answer, please refer to the end of this chapter. ∎

Engineering drawings are also based on visualization. We visualize the object, with a set of a priori rules, by projecting it onto three orthogonal planes. Conversely, by looking at the three views of an engineering drawing, we try to visualize the object. Obviously, this requires much training in order to develop the ability to read drawings and understand them. To do this, we need to relate the details of each view to the similarly corresponding details of the other views in order to understand the shape of the object. If this is so hard compared to visualizing a three-dimensional pictorial, why do we bother? Because it is more difficult to draw pictorials than engineering drawings, and it is easier to relate and specify information about dimensions on an engineering drawing.

EXAMPLE 6.3

Figure 6.5 shows the three orthogonal views (in architecture, views are called elevations) of an object. Can you visualize the object? The solution is presented at the end of this chapter. However, assess how difficult (or easy) it is for you to visualize the object. Obviously, this can be made much more involved as soon as other details are added to the object. A designer must learn to visualize objects as represented by engineering drawings (or other methods of graphical representations).

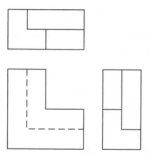

FIGURE 6.5 Visualize the object represented by these orthogonal views. ∎

EXAMPLE 6.4

Figure 6.6 is another representation for an object, except that only two views are given. Try to visualize the object, assuming that there are no missing hidden lines in these views.

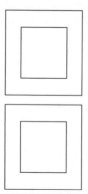

FIGURE 6.6 Can you visualize this object?

If you have difficulty visualizing this object, it is probably due to the lack of a third view that can contain much additional information. Consequently, it is more difficult to relate the details of the two remaining views to each other. The solution for this object is presented at the end of this chapter. You will see that if the third view were given, it would have been much easier to visualize the object. ■

EXAMPLE 6.5

Figure 6.7 is another representation of an object where the front and side views are given. Assuming that there are no missing hidden lines, can you visualize the object?

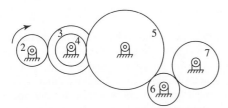

FIGURE 6.7   Visualize this object represented by front and side views with no missing hidden lines. ■

EXAMPLE 6.6

In this example, we deal with motions. Assuming that in the gearbox of Figure 6.8, gear 2 is rotating clockwise, what is the direction of rotation for gears 6 and 7? Gears 3 and 4 are concentric and attached to each other.

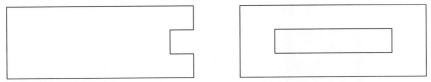

FIGURE 6.8   Determine the direction of rotation of gears 6 and 7 based on the direction of rotation of gear 2.

Following the directions of motion, you should find that gear 6 rotates counterclockwise and gear 7 rotates clockwise. ■

EXAMPLE 6.7

In the machine shown in Figure 6.9, pulleys 2 and 3 are connected together through a belt. Gears 4 and 5 are bevel gears. Gear 8 is a right-hand worm gear. If pulley 2 is rotating as shown, determine whether gear 9 rotates clockwise or counterclockwise.

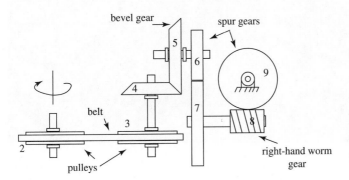

FIGURE 6.9   Determine the direction of rotation of gear 9 based on the given rotation of gear 2.

The correct direction of rotation for gear 9 is clockwise. If you did not get the correct answer, you need to trace the motions again, remembering that a right-hand worm gear moves along the direction of your thumb if the worm rotates in the direction of your curled fingers (on the right hand). Visualizing the motion, will allow you to determine the correct direction of rotation. ■

EXAMPLE 6.8

Figure 6.10 is a schematic of an automobile differential. Can you visualize how it works?

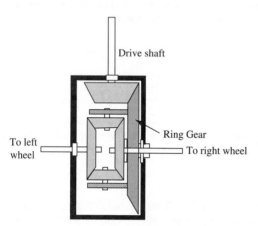

FIGURE 6.10 A schematic of an automobile differential gearbox. ■

A differential gear is used to allow the tires to rotate at different rates as the car turns to the right or left. In the differential, the ring gear is attached to a box-like set of arms to which the differential gears are attached. When going straight, the drive shaft rotates the ring gear. Since the differential gears are attached in a series format, it is impossible to rotate the two opposite gears in the same direction (as would be the case when going straight). Thus, there is no motion in the differential gears. Then, the rotation of the ring gear is directly transferred to the arms and the differential gears as a unit, rotating them all at once with the axles, rotating the tires. When the car turns to the right or left, the differential gears rotate in opposite directions relative to the neutral axis along the axles, providing for a differential movement between the two tires. If only one tire is lifted off the ground, it will rotate twice as fast because the total relative motion between the two tires is constant. This bicycle was designed by Jarrod Sheetz for a class project at Cal Poly, San Luis Obispo. He used two differentials and a wrench with swivel u-joints to drive the bike. Although the bike was too heavy to ride more than a few feet, it looks exceedingly interesting.

EXAMPLE 6.9

A cube is shown in Figure 6.11 with a set of markings on it. Determine which one of the other ones can be the same cube. Notice that there is no information about the other faces of the cube.

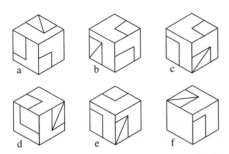

FIGURE 6.11   Which of the cubes shown can be achieved from rotating cube a?

The answer is cubes b, d, and e. You must be able to visualize multiple steps to get the result. ■

EXAMPLE 6.10

Visualize the necessary rotations about the reference frame x-y-z that are needed to transform the cube shown in "a" into the other cubes (Figure 6.12).

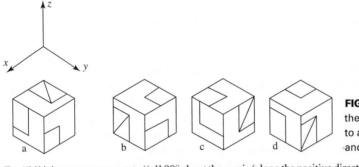

FIGURE 6.12   Determine the necessary rotations to achieve states b, c, and d from a.

For "b" it is necessary to rotate "a" 90° about the z-axis (along the positive direction, where rotation about the curled fingers of your right hand will be about an axis along your thumb), followed by a rotation about the y-axis for 90°, for "c" it is a rotation of $-90°$ about the x-axis. For "d" we need a rotation of 180° about the z-axis, followed by a rotation of $-90°$ about the x-axis. ■

EXAMPLE 6.11

In 3-dimensional kinematic analysis of machines, including robotics, frames are used to represent objects, positions, and motions (called transformations). This analysis requires the representation of transformations between frames, enabling one to move from one frame to the next, for as many frames as necessary. One particular technique uses a representation of transformations based on the Denavit-Hartenberg representation of frames.[8] Figure 6.13 is an example of this representation. Imagine that there are two nonparallel, nonintersecting, axes AB and CD. Imagine that there are also two reference frames, denoted by $x1$-$y1$-$z1$ and $x2$-$y2$-$z2$. The z-axes are collinear with AB and CD lines. The x-axes are along the common normal line between these two lines. The y-axes are perpendicular to x- and z-axes. The angle between AB and CD (or between $z1$- and $z2$-) is shown as α. The angle between the two x-axes is denoted by θ. We plan to go from frame "1" to frame "2." To do this, imagine that you are on frame "1." To go to frame "2," do the following:

- Rotate about the $z1$-axis an angle of θ to make the two x-axes parallel.
- Move up along the $z1$-axis until the two x-axes become collinear.

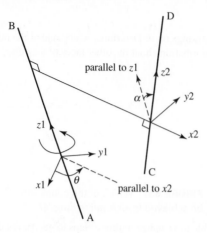

FIGURE 6.13 Three-dimensional representation of movements with frames.

- Move along the $x2$-axis until the origins become the same.
- Rotate about the $x2$-axis an angle of α to make the two z-axes collinear.

The point of doing this exercise here is to see whether you can visualize all these motions in your mind. Practice will enable you to imagine these motions with ease. ■

In order to visualize effectively, one has to be able to visualize clearly and be in control of the images, as discussed below.

Image Clarity relates to one's ability to visualize objects, events, or other phenomena with clarity.[3] If the image is very clear, with all the details of the subject clear in the mind of the individual, when you ask a question about it, the individual should easily be able to respond to the question. Some can visualize very clearly, while others may have vague images and be unable to respond to questions about details, but still capable of seeing the subject. And some may not be able to visualize particular images, especially in the absence of any prior experience with the subject.

Robert McKim[3] suggests the following exercises in order to see how clearly you can imagine different situations, events, and objects. This exercise can be best done if someone else reads the following to you, while you try to visualize them without interruption. However, if alone, you may read the list yourself and try to visualize each one. For each subject, make some note as to whether the image was clear or vague, or whether you had trouble visualizing the subject matter (such as with V, H, N for vivid, hazy, and nothing). The list should not be used to label anyone, and should not be regarded as a weakness or strength. It is also not important whether you record your responses in any particular order or not, as we are not looking to this list for any purpose other than to estimate how clear your images are in general. If you prefer to close your eyes and relax in a comfortable position you may do so. Then imagine the following subjects or events with the best detail you can:

The President's face,	a 747 airplane,
Bambi,	the instrument panel of your car,
a guitar,	an ugly shoe,
face of your DVD,	the sound of breaking a twig,
the feeling of getting up in the morning,	the smell of earth in early rain,
the feeling of a large cool orange,	a dog running to fetch a ball,

the taste of a bell pepper,

feeling of cruising down the highway,

the feel of rain on your face,

loud noise in a party,

the smell of curry,

the sensation of being tired after a long hike,

the sensation of pain after a long hike,

the sensation of feeling sad,

a sound of opening a letter,

the feel of a warm cup of tea,

the feel of a computer mouse,

a loud motorcycle,

the smell of your favorite perfume,

the closing of a heavy door,

the sound of walking on dead leaves,

the sensation of feeling great.

If you really imagine these clearly, you should be able to describe the details of the subject or event with ease. For example, you should be able to describe where all the buttons of your DVD are. When you visualize breaking a twig, you should really feel the sound and the feel of the reaction of the twig breaking in your hand. The last event should really make you feel great. Otherwise, your imagination may be vague. Over the years, I have seen most students complain about their inability to visualize the smells of things (such as their favorite perfume or curry). Apparently, this is difficult for many people. Taste is not as hard, since we associate taste with particular food items, and we can quantify taste with adjectives and descriptors (such as sweet, sour, bland), whereas smells are not as easily described by descriptors.

Image Controllability[3] relates to how much control one may have on the flow or the sequence of a series of events or their timing. In this case, one has to be able to control how fast images change and different events progress. Controllability may in fact be more important than image clarity, as this may be a more important requirement for problem solving. Most solutions are not single events or individual processes but sequences, requiring multiple actions and motions or subcomponents. We will exercise with the following sequence of events to see how clear or vague your imagination may be. As before, it is better if someone else reads these to you while you imagine them in a convenient posture. Make notes whether the images are vivid and whether you can control the flow and the timing. The reader should read these with the same timing she or he expects you to imagine:

- The large doors of a movie theater opening and hundreds of people entering.
- A car coming toward you, passing you, and going behind you.
- A driver entering a car, putting on the seat belt, checking the mirror, starting the engine, putting it in gear, and slowly moving into traffic.
- A robot picking up a nut, moving toward a screw held in a jig, and slowly engaging the two and turning the nut a few turns.
- With a video camera in hand, standing on top of a hill and filming the panoramic view of the city below.
- A large airplane approaching the landing strip, slowly coming into contact with the tarmac, braking, and eventually coming to a complete stop at the terminal.
- A ball-shaped robot rolling toward a staircase, opening up and exposing two short legs, walking toward the stairs, growing the legs to 2 ft each, then climbing the stairs, collapsing into a sphere again, and rolling away.

Here, too, you should see all the details of each event if you are in clear control. For example, the camera should move slowly while you see the details of the city, tall buildings, and the river in the middle, while you feel the weight of the camera. The car should not pass you unless you are told so. It should continue coming toward you until you are asked to imagine it passing you. And can you describe the color of the robot that walked up the stairs? What was the shape of the feet? Did the body sway sideways as it walked? Was it noisy?

Image Synthesis relates to one's ability to synthesize images based on need, as in problem solving. As we have already seen many times, when you look for solutions by thinking about them, and when you visualize solutions in your mind, you are actually synthesizing images that do not yet exist. In this case, you need to create the situation based on need, without direction from others. You need to include details in the images, integrate motions when needed, come up with shapes, sizes, and other physical requirements, and also make sure that it works. Image synthesis is much more involved than visualizing preconfigured events. This requires both the ability to visualize clearly, with much control, but also the ability to create and synthesize what does not yet exist. As an exercise, do the following:

> Visualize solutions for possible ways to reduce or eliminate injuries in car accidents. As part of this, imagine yourself in a car, driving in some particular condition (sleepy, tired, speaking on the phone, in congested driving, on a long stretch of boring road. . .) when suddenly another vehicle comes toward you. Imagine what it feels like, what you may or may not be able to do, what may happen to you, and how you can prevent injuries. Imagine whether air bags (including side air bags, etc.) will be adequate or if you need other ones added. Imagine what it will be like if you were in a soft car and you would be hit by another soft car.
>
> Visualize that you are in a foreign country, where you do not speak the language, you cannot read the written word (imagine it is not the Latin alphabet), and you do not know anyone. What do you suggest must be done to make it possible for someone like you to be able to accomplish what you need, for example to get to a hotel, to find your way around the tourist areas, to order food, to buy souvenirs, or to pay for services. How will you communicate with the natives? How will you make a phone call? How will you thank someone who gives you directions?
>
> Visualize a solution for the page-turner for the disabled. We have already discussed many possible solutions and what may be required. Now try to visualize other possible solutions by imagining that you are the machine and you are to do the work. How would you, as a machine, accomplish the task?

Richard De Mille has used the phrase "directed fantasy" in his book "Put Your Mother on the Ceiling; Children's Imagination Games"[5] for a set of scenarios in which one directs other individuals to fantasize impossible situations. In one game, you are to imagine your mother on the ceiling doing different activities, thus the name of the book. You may either use the book or come up with your own scenarios. For example, what if you imagine that you are an engine? Try to feel the forces of combustion, the vacuum of intake stroke, the smell of gasoline as it is mixed in, or the heat of exhaust. The motions of an engine in high revolutions are mind boggling. See if you can visualize it, while the oil is splashed all over, and while the sensors control the action. What about visualizing playing a game such as hockey? Try to visualize all the details, the actions and the movements, the puck, the ice, the cheering spectators, everything. And what if your mouth was located on the side of your face? What would be the result? How would you eat? Would you see what you eat, could you find your mouth while your eyes are in their present locations? Imagine kissing your child (or someone else). Brush your teeth. Put on lipstick (or lip balm). Imagine a dentist bringing his drill close to your mouth. Do you see it? Is it better that you cannot see the instrument? On which side of your head would you prefer to have your mouth?

As you see, these games can go in many different directions with many possible scenarios. Invent your own.

A final word about imagination and exercising the fantasies mentioned above. There is a fundamental difference between drug-induced hallucinations or mind-altering practices and these exercises. The fundamental difference is that in those hallucinations, the control of the brain and its functions is lost to the drug, and as a result, the individual cannot return to reality until the effect of the drug is gone. In these exercises, the person can get back to real life instantly at any time. This is voluntarily imagining unrealistic situations in order to practice the art of visualization. It is not hallucination.

6.3 DRAWING

Drawing is used both as a mechanism to assist in visualization and bringing abstract thoughts into existence and for communicating thoughts with others. Imagine a problem-solving session where you have developed ideas in your mind. No one can guess or see what is on your mind unless it is described in words, in some form of drawing or sketch, or other physical form such as a model or an artifact. Words can be used in countless ways. When used properly, they are very effective, powerful, can arouse feelings, and they can be poetic and artistic or practical and realistic. However, they can be visualized and interpreted in many different ways by different people, as with reading story books, where every person imagines what is described by the author in his or her own way. But as we saw in Section 3.13, inability to communicate with words can become a big hindrance. Colleagues, students, physicians, artists, and spouses can regularly miscommunicate information, ideas, and their interpretations, causing disagreements, misunderstandings, and perhaps significant problems. Making physical models can also be very time consuming, expensive, or impossible to do when many ideas are present and making each one requires detailed design and much effort. The best choice is perhaps to draw or sketch the idea. A simple sketch, or at a higher level, a drawing, may solve many of these problems. Imagine someone calling a machine shop and ordering the manufacture of a part by describing it over the phone. The designer may have all the details figured out in his or her mind, and may to the best of his or her ability describe it to the technician. But still there is no guarantee, in fact not even a slight chance, that the other person will understand what is requested.

Additionally, to see the details involved, to understand the limitations of the solution, to see the relationship between different components, to solidify a thought, and to make a record of the ideas in order to remember them later, you may opt to draw those ideas. For example, consider the design of the page-turner. If you do not draw your ideas, you may completely forget about them later, you may not get credit for them, and you may not be able to compare different ideas during the selection phase. Alternately, consider the design of a power plant. In fact, in order to ensure that pipes and other physical components of the plant do not interfere with each other, they must be drawn and labeled. Otherwise, in the absence of physical models of the plant, which can be extraordinarily expensive, pipes and other components of the system may interfere with each other during construction. Similarly, unless an idea is sketched, you may

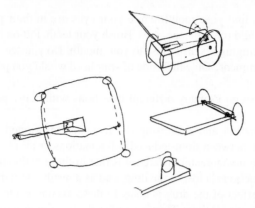

FIGURE 6.14 An example of drawing at the personal level.

have difficulty seeing the relationship between different parts, their dimensions, and interferences.

Drawings may be used at least at two different levels: personal, and interpersonal. A drawing used at the personal level may not follow any rules, standards, or notations. It is generally used to remember ideas and for personal communication and understanding, and therefore, it may be at any level of sophistication. One may just doodle a few lines, and that may be all that is necessary for one to remember all the details regarding the idea later. One may also draw complete pictorials, use labels and symbols, use shading, and take advantage of instruments and sophisticated CAD programs to create the drawing. So long as the drawing is used personally, there are no rules or requirements. In fact, the first sketch of an N1-M flying wing (a predecessor to today's stealth bombers) was sketched on the back of cocktail napkin during a lunch meeting. Looking at it, you may not be able to discern what it relates to, but obviously the inventor knew exactly what was on his mind. He just used the sketch to remember it. Figure 6.14 is a sample of a personal drawing (drawn by my junior high son during a discussion session for a mouse trap car). Only he might know exactly what they refer to. On the other hand, if he were to communicate these ideas with others at an interpersonal level, he would have to draw them very differently, enabling others to understand the details of the drawing. Figure 6.15 shows an example of a drawing at this level.

Drawings used for interpersonal communications may be divided into the following common types:

- Engineering drawings
  - Orthogonal drawings
  - Assembly drawings
  - Exploded views
  - Pictorials (axonometric projections)
    - Isometric drawings
    - Oblique drawings
- Realistic drawings
  - 1-point perspectives
  - 2-point perspectives
  - 3-point perspectives
- Sketches

We will discuss these techniques of drawing in the following sections.

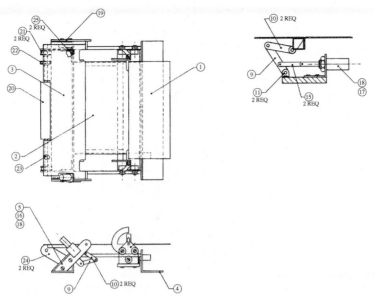

FIGURE 6.15 An example of a drawing at the interpersonal level. (Letter folder/stuffer courtesy of Gorospe and McGarraugh, 1986.)

6.4 ENGINEERING DRAWINGS

Engineering drawings[10] are used for many purposes, not only to create documents that can be used for a variety of engineering-related activities, but also to simply communicate with others in a clearly defined way. Due to the fact that we all learn how to read and interpret engineering drawings (almost) identically, and because we have developed clear methods and standards for dimensioning and specifying parts, there is a great chance that the part will be understood correctly. Engineering drawings are fundamentally unrealistic. It is impossible to look at a part and see it exactly as shown in an engineering drawing. This is due to the way projections are made in engineering drawings, where all projection lines are parallel. In reality, projection lines converge as the distance increases from the eye. Still, due to the superior way engineering drawings can communicate dimensions and relationships between different elements, they are by far the most commonly used method of design communication for engineering designers. Some designers opt to use realistic perspectives more often than design engineers.

Engineering drawing techniques and details are commonly taught at both the high school level and the college entry level, and by most accounts, the great majority of design students are familiar with it. As a result, we will only have a brief discussion about them here as related to their applications and differences with other methods. For more information about engineering drawings, please see other references.

Engineering drawings consist of orthogonal projections (views, or as called in architecture, elevations) and cutaways or sections, assembly drawings, and exploded views.

6.4.1 Orthogonal Projections

Orthogonal projections are the views of a part or system as seen on planes that are perpendicular to each other as if you placed the object in a cube. As shown in Figure 6.16, there can be up to six orthogonal views. These views show what would be seen if one would

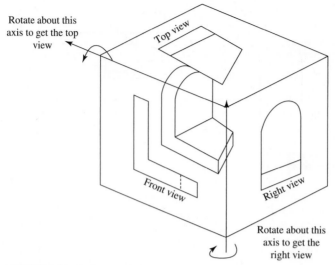

FIGURE 6.16 Orthogonal views.

look at the object with parallel, nondiverging, light. All hidden details that cannot be seen are shown by dotted lines. Figure 6.17 shows the six orthogonal views of a simple part (in architecture, the word elevation is used instead). These views include the front, top, bottom, left, right, and back views of the object or assembly. International and national standards govern the sizes, the margins, the dimensioning, and the symbols used for specifying the object and its characteristics.

One important point is the difference between the way drawings are arranged in the U.S. and in other countries. Most countries follow the ISO system. In this system, the front, top, and left views are drawn and arranged as shown in Figure 6.18, while in the English system used in the U.S., the front, top, and right views are drawn and arranged as in Figure 6.19. In reality, these are very similar and they convey similar information. Since the title block is generally on the lower right side (or all the way on the bottom) of

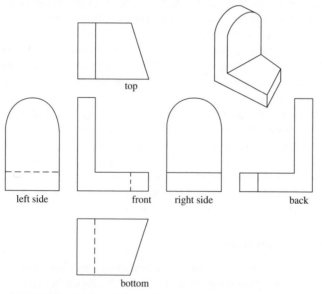

FIGURE 6.17 The 6 orthogonal views of a part.

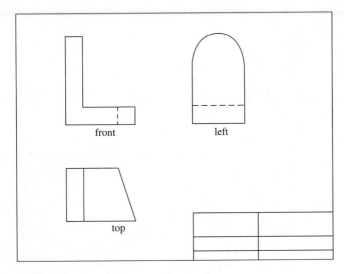

FIGURE 6.18 Arrangement of views according to the ISO standard.

the drawing, the ISO system allows for a better usage of the space available on the paper. Most engineers and designers have to work in a global environment where many others follow a set of rules that is different than the rules followed in the U.S. What is important for you is to realize the difference in this arrangement and be prepared to read and understand either system. Many foreign companies who work with large American companies do in fact follow the English method when they prepare their drawings. Still, it is important to be aware of the differences and be prepared to deal with them.

6.4.2 Assembly Drawings

Assembly drawings are used to show the relationship between the different elements of a machine when they are put together, assembled, or attached. The assembly drawing may be in a 3-dimensional pictorial form (as will be discussed later) or in orthogonal form. Usually no dimension is indicated on an assembly drawing, and hidden details are ignored. Otherwise, the drawing has the potential to become very involved and cluttered, and therefore, useless. In practice, a bill of materials table accompanies the assembly drawing with the number of all

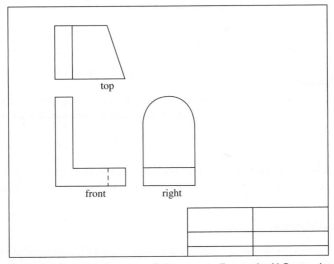

FIGURE 6.19 Arrangement of views according to the U.S. standard.

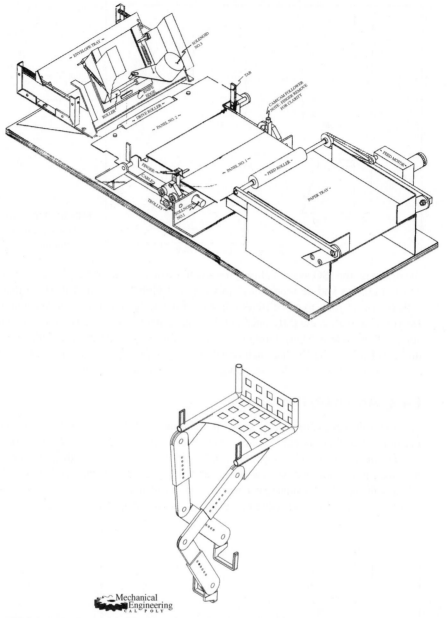

FIGURE 6.20 Examples of assembly drawing for a letter folder and stuffer and a backpacker (drawn by A. Gorospe and B. McGarraugh, 1986, and A. LaPlante, R. Paterson, K. Rowland, and K. Thor, 2007).

parts needed, materials, and a reference number. Each part is usually identified by a name, a number, or other unique characteristic. Figure 6.20 shows examples of assembly drawings.

6.4.3 Exploded Views

Exploded views are somewhat similar to assembly drawings, except that in exploded views, the different parts or subassemblies of a system are drawn in relation to each other as they would be

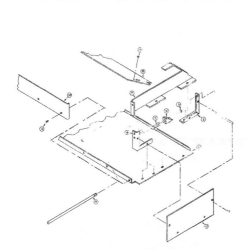

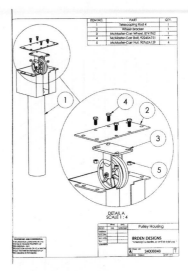

FIGURE 6.21 Examples of exploded view drawings. (Letter folder/stuffer, A. Gorospe and B. McGarraugh, 1986; pulley housing by J. Engel, D. Bonny, and C. Nikkel, 2007.)

assembled. These drawings are very useful as instructions for maintenance, repair, or assembly by others, but are also used for manufacturing. Exploded views can also help the designer in discovering potential problems in assembly or proper placing of parts relative to each other. Parts are usually identified by numbers, names, or other distinguishing characteristics. Figure 6.21 shows examples of exploded views.

6.4.4 Pictorials

Pictorials are three-dimensional (3-D) representations of objects. Because they are drawn in 3-D, pictorials can better represent certain details about the shape of the part and the relationship between its different parts and subassemblies. They are also easier to understand for individuals who lack adequate training to read orthogonal drawings. Thus, you may draw a pictorial for a child and the child will understand it without any prior specific knowledge. However, since the depth representation in pictorials is generally distorted, and since many elements such as circles and ellipses do not appear as circles or ellipses, drawing pictorials is generally more time consuming and sometimes more difficult than drawing orthogonal projections. Additionally, it is much more difficult to specify dimensions in a pictorial along all three axes than it is to do in three views of an engineering drawing.

Axonometric pictorials can be drawn according to many standards, including isometric, oblique, dimetric, and trimetric. Dimetric and trimetric pictorials are even more difficult to draw, and therefore, uncommon. Isometric and oblique pictorials are both very common and will be discussed next.

Isometric pictorials are based on the assumption that the x-, y-, and z-axes representing a coordinate frame are drawn with a 120° separation between them, as shown in Figure 6.22. Unlike realistic pictorials (perspectives), parallel lines do not converge and dimensions do not become smaller as depth increases. In reality, it is impossible to look at an object and see the three axes at these angles without the effect of depth on dimensions. However, drawing in isometric is much simpler than in perspective, and thus, it is used extensively in engineering and in design communication. The isometric grid of Figure 6.22 should be used as a tracing background to create your isometric drawings.

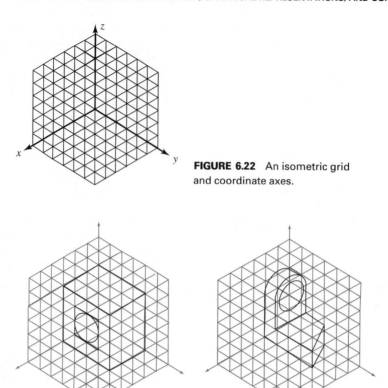

FIGURE 6.22 An isometric grid and coordinate axes.

FIGURE 6.23 The isometric representation of a simple object.

Figure 6.23 is an example of an isometric drawing. As you can see, most lines follow the three coordinate axes (if they are parallel to them). If a surface is not parallel to the three orthogonal planes formed by the three axes, the lines representing it will not be parallel to the axes either. Notice how the circles become ellipses in isometric. Since all three planes (x-y, y-z, and x-z) are at an angle, all circular lines become elliptical in the pictorial. Consequently, if there is a part with many circular elements (say an instrument panel with many knobs) which must be drawn as ellipses, drawing the part will be much more time consuming. Note how you may find the correct orientation of the ellipse and its size. If you assume that the circle is surrounded by a square, and first draw the square in isometric, you will be able to properly orient the ellipse and draw it correctly by placing the ellipse inside the parallelogram representing the square.

Dimetric and **Trimetric** pictorials are very similar to isometric, except that diametric angles are 110°, 110°, and 140° instead of 120°. Trimetric angles are 130°, 110°, and 120°.

Oblique pictorials are based on the assumption that two of the three coordinate axes are orthogonal. As a result, if you assume that the y-axis is horizontal and the z-axis is vertical, the x-axis will be off in between the two at some standard angle, as shown in Figure 6.24. In this figure, the y-axis is chosen to be at 45°. Other angles such as 30°, 60°, and 75° are common too. Obviously, like isometric pictorials, oblique drawings are unrealistic and impossible; it is not possible to look at an object and see the front of it straight on (with horizontal and vertical lines) and still see the depth of it (top and side planes) at

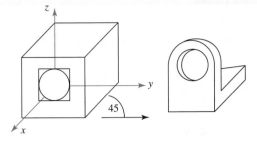

FIGURE 6.24 The oblique representation of a simple object.

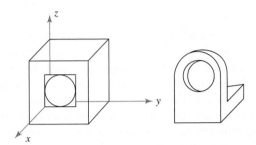

FIGURE 6.25 Oblique pictorials with scaled depth.

an angle. However, like isometric, oblique drawings are very common due to the fact that they are easier to draw than realistic pictorials (perspectives). In fact, oblique drawings are less realistic than isometric, but even easier to draw since circular shapes and curved lines that fall in the front plane retain their shapes. This is a great advantage when there are many curved surfaces with which we must deal. However, oblique pictorials do look distorted.

Another problem with oblique pictorials is the further distortion caused by the depth dimension. The length, width, and depth of the cube in Figure 6.24 are equal. But as you see, it looks too long. The same is true for the bracket shown in that figure. In order to make the pictorial more acceptable and less distorted, it is necessary to scale down the depth dimension by a factor related to the chosen angle. Common scale factors are the sine of the chosen angle, $\frac{1}{2}$ (called cabinet projection), 0.7, 0.86 or arbitrary numbers of $\frac{3}{4}$, $\frac{5}{8}$, and $\frac{3}{8}$. Figure 6.25 shows the same objects with a 0.7 scale factor.

To draw an oblique pictorial, first draw the coordinate frame that will represent it. Then draw an oblique box in which you can easily draw the object of interest. We will discuss the box method in more detail later.

6.5 REALISTIC DRAWINGS

Realistic pictorials or perspectives are just that: realistic. They are drawn as seen by the eyes, and they are to display the object as it appears in real form. This means that as distances increase, dimensions decrease, and therefore, lines converge into points called vanishing points (v.p.). In real life, this is exactly what we see. The rails get closer to each other as distance increases until they merge; telephone poles become smaller as they get farther away, until they vanish; objects get smaller as they move away. Realistic pictorials are drawn to represent the same characteristic (Figure 6.26).

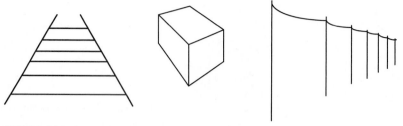

FIGURE 6.26 In perspectives, as distance increases, dimensions decrease.

The **Vanishing point** is the point to which all lines in one direction will converge. For example, in Figure 6.27, all the lines that go to the left will vanish to a point to the left, and all lines going to the right will vanish to a point to the right. There may be as many as four vanishing points in a drawing, as will be discussed shortly. The closer the vanishing points are to each other, the more distorted the object becomes. When vanishing points are close to each other, it means you are looking at the object close up, and therefore, the perspective becomes distorted. The farther away the vanishing points are from each other, the less distortion in the drawing. Obviously, there is a limit to this distance; as it goes to infinity, you are infinitely far away, and the drawing converges to an equivalent of an isometric drawing where the lines do not vanish anymore. A rule of thumb is that the vanishing points should be apart at least 4 times the radius of your cone of vision. The cone of vision is a cone where our vision is not distorted, and is about 60°. Figure 6.28 shows the effect of the location of vanishing points on a drawing. Notice how the lower right corner is elongated when the vanishing points are close to each other. Generally, the two construction lines that form the lower corner of the object should be more than 90° apart. Angles with lower values will cause severe distortion.

The **Horizon** is a line at eye level to which all lines vanish, and is the same line at the horizon that the observer sees (Figures 6.27 and 6.28). The horizon is always at eye level and parallel to the line between the eyes. Thus, if the observer moves up or down, or if the head is tilted sideways, the horizon will also move up or down or will tilt accordingly. A vertical line perpendicular to the horizon (or to the line connecting the eyes) will divide the horizon into two equal portions and is called Measuring Line (ML). All dimensions on this line will have true length. The measuring line and the horizon determine how a perspective is seen. For example, if an object is below your eye level (and thus, below your horizon), the top of the object will be seen and drawn, whereas if the

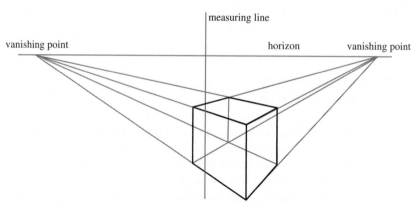

FIGURE 6.27 Vanishing point, the horizon, and the midline.

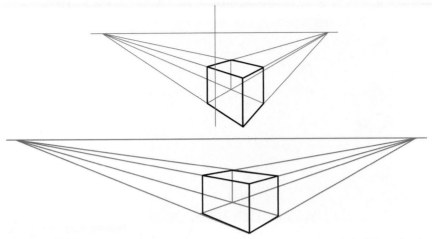

FIGURE 6.28 The effect of the location of vanishing points on the level of distortion in perspectives. As the vanishing points get closer, the drawing becomes more distorted.

object is above the horizon, the bottom of the object will be seen and drawn. Similarly, if the object is to the left of the measuring line, the right side of the object will be seen and vice versa. Perspectives are drawn based on 1, 2, or 3 vanishing points, representing the relative orientation of the object and the viewer, as discussed below.

6.5.1 Perspectives with One Vanishing Point

Perspectives with one vanishing point are drawn when the plane of the object is parallel to the eyes (or perpendicular to the midline of sight). For example, if a cube is in front of your eyes with the front plane parallel to your eyes, or if you stand in a room and look straight into the back wall, you will see all lines vanishing to one point on the horizon in front of you. Figure 6.29 shows how a 1-point perspective is drawn. Since the depth of different points on the object that are parallel to the eyes does not change, those points do not vanish. Others that are on other planes do vanish.

As you can see in Figure 6.30, the same is true in a real picture. All lines that are not parallel to the plane of eyes vanish to the same point. The location of the vanishing point determines what is seen. The vanishing point should be right in front of your eyes and at the eye level on the horizon. Thus, as you move up or down, or as you move sideways, the perspective changes accordingly.

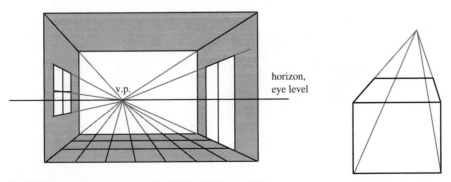

FIGURE 6.29 Perspectives with one vanishing point.

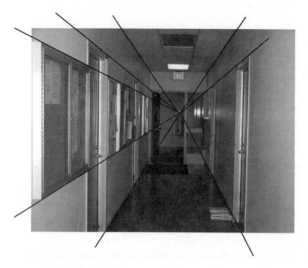

FIGURE 6.30 A real image with one vanishing point.

6.5.2   Perspectives with Two Vanishing Points

Perspectives with two vanishing points are drawn when the object is not parallel to the plane of eyes. This means that the features (planes) of the object make an angle relative to the eyes. In reality, there are always two vanishing points on the horizon. In the one-vanishing point drawing, one of the two becomes irrelevant. As soon as the plane of the object and the observer become nonparallel, both vanishing points become relevant and are used. As an example, imagine that you are standing in a room. If you are perpendicular to the back wall, as in Figure 6.29, only the vanishing point across from your eyes is relevant because the back wall and everything else in it do not vanish. However, at the same location in the room, if you turn a bit, your eyes will no longer be parallel to either of the walls to your right or left, and therefore, there will be two relevant vanishing points. If you continue to rotate until one or the other side-wall becomes parallel to your eyes, one of the vanishing points becomes irrelevant again. You will have a new vanishing point for the new back wall. Figures 6.31 and 6.32 show similar objects with two vanishing points, as well as a real image which follows the same principles.

6.5.3   Perspectives with Three Vanishing Points

Perspectives with three vanishing points are appropriate when vertical lines in the object are long, and therefore, vanish as well. This applies, for example, to a tall building, especially when you are close to it. In short objects, or when you are far away from an object,

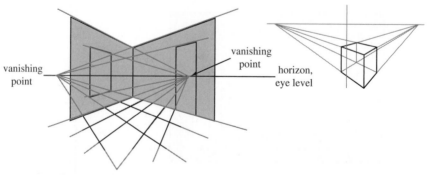

vanishing point

vanishing point

horizon, eye level

FIGURE 6.31 Perspectives with two vanishing points.

FIGURE 6.32 A real image with two vanishing points apparent in it.

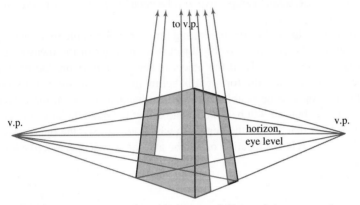

FIGURE 6.33 A perspective with three vanishing points.

vertical lines do not appear to vanish. Otherwise, vertical lines vanish to a point above the object (if the object is mostly above your eye level) or below the object (if the object is mostly below the horizon). Figure 6.33 is an example of a drawing with three vanishing points. Figure 6.34 is an example of a real image with the same effect. Perspectives with three vanishing points are more realistic and more accurate, and actually, with less

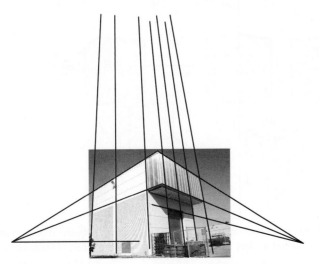

FIGURE 6.34 A real image with three vanishing points.

distortion. However, they are more difficult to draw, and unless the object is long, the third vanishing point is ignored (Note that if an object is very long and extends both above and below your eye level, there will be two vanishing points for vertical lines, one above, one below, rendering a 4-vanishing-point perspective).

6.6 PERSPECTIVES AND LOCATION OF OBJECTS

The way an object appears to our eyes is dependent on where it is, not only on distance, but also on whether it is above or below the horizon, and whether the object is to the left of our midvision line (measuring line) or to the right of it. Thus, when a perspective is drawn, you first have to decide about the relative location of the object and your eyes. And if the object moves, a new perspective will be needed to represent it in its new location and orientation.

If the object is to the left of the measuring line, we will be able to see the right side of the object and its related details. If the object is to the right of the measuring line, its left side will be visible. Similarly, if the object is above the horizon, the bottom of the object will be visible, but not the top. If it is below the horizon, we can no longer see the bottom, but the top will be visible. If the object happens to be on the horizon, then neither the top nor the bottom will be visible. The extent to which the object is away from either of the two lines determines the degree to which we can see the opposite side of it. Figure 6.35 shows the relative location of a cube and the way its perspective looks as its location varies.

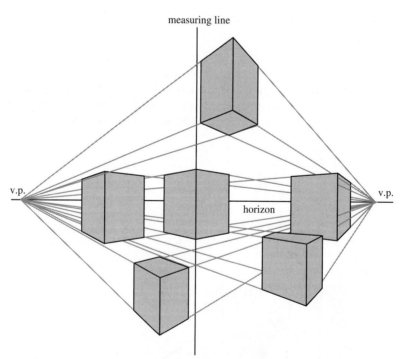

FIGURE 6.35 The relative location of an object and its perspective. (Objects drawn in this figure are not meant to be equal in size. They are drawn to show the effect of location on how they appear.)

6.7 DETERMINING DEPTH IN PERSPECTIVES: SCALE FACTORS

One of the fundamental issues in drawing perspectives is to come up with the correct scale factor for depth. This means that we need to establish a basic cube in perspective that will allow us to draw equal measures along the three axes. This will determine whether or not the object you draw looks too long and distorted.

Although any angle can be used to draw perspectives, we will discuss two methods that will allow us to establish a basic cube for the 45° grid and for the 30–60° grid.

The Basic Cube for a 45°grid will allow us to create a cube for drawing an object at 45° or as seen on the measuring line.[6] To do this, you may do the following (referring to Figure 6.36):

- Draw the horizon and the middle measuring line and choose your vanishing points. Relative to the size of your object, the vanishing points should not be placed too closely.
- On the measuring middle line, mark the location where the lower bottom corner of the cube will be (point A). Draw a line from point A to each vanishing point. The angle between the two lines should be larger than 90°, otherwise move point A on the middle line.
- Draw a horizontal line BC the size of your cube perpendicular to the measuring line.
- From B draw a true 45° line and mark a distance equal to BC on it (BD). Draw a line at D parallel to BC.
- Draw two vertical lines up from B and C to intersect this new line (E, F). Lines BE and CF are the edges of your cube. Construct the rest of the cube from these points.

This method will establish a true 45° cube that looks equal in all dimensions. All other cube sizes can be made from this basic cube, as will be discussed next.

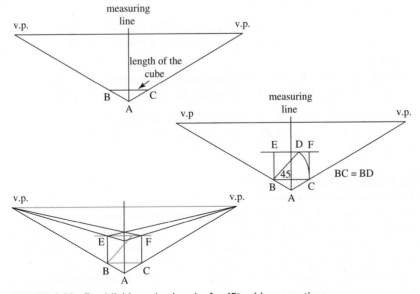

FIGURE 6.36 Establishing a basic cube for 45° grid perspectives.

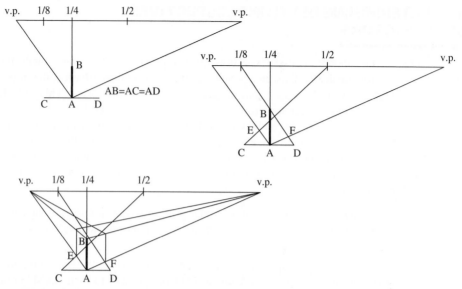

FIGURE 6.37 Establishing a basic 30–60° perspective cube.

The Basic Cube for a 30–60°grid will allow us to create a cube for drawing objects at 30–60°.[6] To do this, you may do the following (referring to Figure 6.37):

- Draw the horizon and choose your vanishing points. Relative to the size of your object, the vanishing points should not be placed too closely.

- On the horizon, divide the distance between the vanishing points in half. Then on one side divide the distance into $\frac{1}{4}$ and $\frac{1}{8}$. The measuring line will be on the $\frac{1}{4}$ mark.

- On the measuring line, mark the location where the lower bottom corner of the cube will be (point A). Draw a line from point A to each vanishing point. The angle between the two lines should be larger than 90°; otherwise, move point A.

- On the measuring line, choose the size of the cube AB. On a horizontal line at A, mark AC and AD equal to AB.

- Draw a line from D to the $\frac{1}{8}$ mark and from C to the $\frac{1}{2}$ mark on the horizon. Locate E and F where these lines intersect with the construction lines from A to the vanishing points.

- Draw lines from B to the vanishing points. Draw vertical lines from E and F up to these two lines. These vertical lines are the sidelines of the cube. Finish the cube.

This will result in a basic cube seen at 30–60° angle with the correct scaling in all three dimensions. For more information on grids and drawing perspectives, please see other references such as (4) and (6).

6.8 DRAWING PERSPECTIVES

Before you start to draw a perspective, you must decide where the object is relative to your midline of vision and the eye level (horizon). You must also decide how far the object is from you. This will determine where the vanishing points will be located. Then you must also decide whether the object is perpendicular to your eyes or at an angle, and whether the object is either too close or too tall. This will determine whether you should draw a 1-point, 2-point, or 3-point perspective.

First draw the horizon. The horizon is always a horizontal line, unless you tilt your head too. In that case, you must somehow show to the person who is looking at your drawing that your head is tilted! Otherwise, the horizon should be horizontal. Draw your measuring line too, as this will help you with the relative location of the drawing. Next, place the vanishing points, whether there are 1, 2, or 3 of them. Generally, as will be discussed later, drawing a box first is the best practice for creating drawings. Assume that whatever you want to draw is in a box. First determine the size of the box you need to draw. If it is a cube, use the methods already discussed to draw the basic box. If the box is not a cube, first draw a basic box and expand it to whatever size you need. We will discuss a simple method on how to expand the basic cube to other sizes in the next section. You can then continue drawing the rest of the object as you progress from one detail to another.

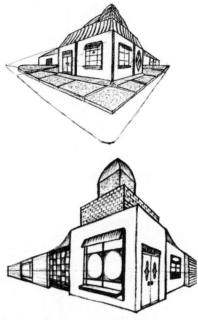

One general question most student novices have is how to find the correct relationship between different details of an object. The general answer to this question is that *whatever relationship exists in real life between the different elements of an object will exist in the perspective as well.* To understand this issue, let's assume that we want to draw the correct lines that will divide a cubic box into 27 equal portions. This requires that we draw lines that will divide each dimension into three equal portions. However, since in perspectives, as the depth increases the dimension decreases, we should expect that the first, second, and the third portions on each face will be different, not equal, sizes. Thus we need to find a way to divide a line into three portions that are not equal in length, but *appear to be equal to each other in perspective.* The critical relationship here is that if you look at the rectangle (or square) representing each face, the diagonal line will intersect the dividing horizontal and vertical lines at the same points (Figure 6.38e). The same must be true in the perspective. We then should be able to use this fact to construct the lines that will divide the box into 27 equal parts. The same relationship was used in Figures 6.29 and 6.31 to draw the floor tiles. Figure 6.38 shows the succession of operations that will accomplish this task as follows:

- First draw the box or the cube based on the methods discussed above.
- In Figure 6.38a, the vertical line in the front of the cube is parallel to our sight. Thus, if you divide it into three equal portions, it will still look correct. Draw construction lines between the vanishing point on the left and these portions.
- Next draw the diagonal line on the left face (Figure 6.38b) and draw vertical dividers at the intersections. This divides the left face into nine equal portions.
- Next draw construction lines between the vertical dividers and the vanishing point on the right, followed by another diagonal (Figure 6.38c). The intersection point between them will indicate where the construction lines must be drawn to the left vanishing point. This will divide the top face, with points for the right face.
- Repeat for the right face (Figure 6.38d).

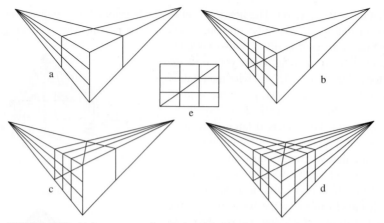

FIGURE 6.38 Using geometric relationships to draw correct perspectives.

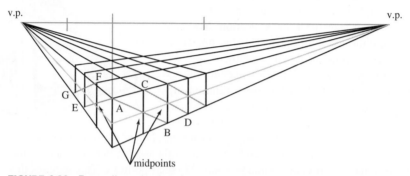

FIGURE 6.39 Expanding a basic cube to other sizes.

The same can be used to do the opposite, to enlarge a cube to other sizes or to create grids. If you draw the first basic cube, by using the diagonal relationships, you can build additional cubes around the basic cube and expand the size of your box, as in Figure 6.39. Note how the diagonal lines AB, CD, AE, and FG are used to build additional boxes next to the basic box with correct dimensions. Each diagonal is drawn from a corner to the middle of the opposite side. Other similar relationships are also possible.

Another example of the application of geometric relationships is how to draw a circle in perspective. Obviously, a circle will not be an ellipse when drawn in perspective, since the side that is farther away will shrink in size. To draw the circle, first draw a square that will contain the circle. Subsequently, sketch the circle in the square as you expect it to be in real life (Figure 6.40).

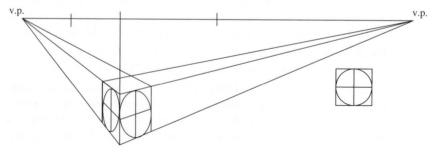

FIGURE 6.40 Drawing a circle in perspective.

6.9 SKETCHING

If you remember, in Section 3.16 we discussed inability to communicate ideas and thoughts with words. In that discussion, it was shown how difficult it was to convey to others the shape of a simple cube with a few cutouts on it using only descriptions. Alternately, through a simple sketch that may only take a few seconds, not only was the idea conveyed, but also the certainty that it was not misunderstood. This is why it is so important to be able to quickly sketch your ideas. Sketching does not require fancy drawings. It can be very simple. Obviously, the better you can sketch, the better the drawings. But your ability (or inability) to draw nicely should never be a roadblock to sketching.

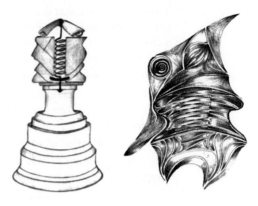

The following are a few very simple rules that will help you sketch more effectively. As you get better, you may try to advance your sketching techniques by following the rules and methods of drawing pictorials and perspectives. Figure 6.41 shows a few simple suggestions that will help you in this matter. For more information, please see references 3, 4, 6 and 7.

- To draw a straight line, you may either draw small lines and correct yourself as you go on or try to draw fast with one quick stroke. You may in fact draw a nice straight

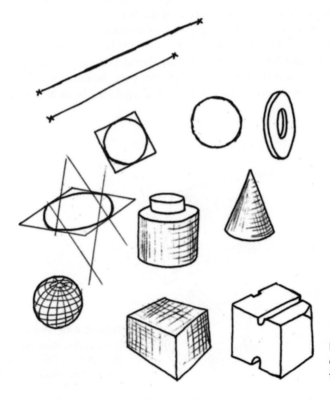

FIGURE 6.41 Simple examples of methods to use in sketching figures.

line this way. If you need to draw a line that reaches a specific point, look at the destination point while you draw. Your eye will direct your hand toward that point much more accurately than if you look at the pen and where you currently are, hoping to get to your destination by chance.

- To draw circles, either draw in small strokes and correct yourself as guided by your eyes and the sense of roundness it gives you, or try to draw a square first and match a circle inside it. The square allows you to see the relative shape of the four quarter-circles. The same can be used to draw ellipses by drawing a rectangle first and trying to place the four portions inside it.

- To draw an ellipse, you may also do the following. First, draw a parallelogram with equal sides. Draw four perpendicular lines at the midpoint of each side. The intersections of these four lines can be used as the centers of four arcs that will constitute the ellipse. Use a pair of compasses to draw the four arcs.

- Use simple shading to create a sense of three-dimensionality or round surfaces. Simple shading can add much to the beauty of a sketch.

6.10   PUT-IT-IN-THE-BOX TECHNIQUE

Everything you would like to sketch may be placed in a box first, no matter what it is. The best approach is to draw the box first, then continue with the details until you finish the drawing.

Placing the object in a box and starting with the box first allows you to establish the border of the object, the relative dimensions of the width, height, and depth, and if a perspective, the relationship between the object and the horizon, the measuring line, and the grid angle. This simple box will also help you see the relative dimensions of different parts and their location. As shown in Figure 6.42, a simple box was first drawn for the vice. Subsequently, smaller boxes required for smaller detail were drawn in order to establish the locations of each element. These elements were then drawn into the boxes. Without the guidance from the boxes, the final drawing would not be as closely drawn as in this picture.

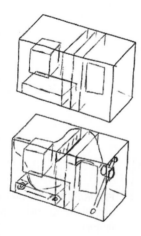

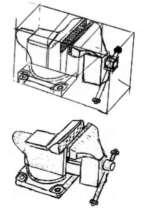

FIGURE 6.42   Place the object in a box and start with drawing the box first. This will enable you to draw much more easily.

You may wonder whether this technique can be applied to other shapes that are not geometric, say a person, an animal, a tree, or other nongeometric shape. The answer is a big **YES**. Anything you can imagine can be placed in a box first. Try this approach and see for yourself whether it helps you sketch objects and "things" more easily or not.

6.11   SKETCHING HUMAN FIGURES

This same method of "put it in a box" can also be adapted to sketching humans. In order to facilitate drawing a human, first try to draw simple boxes and cylinders that relate body proportions and orientations together, as shown in Figure 6.43. Drawing the rest becomes much easier, depending on how much detail you need to show. Figure 6.44 shows the approximate proportions of the human body which can be used in order to create the form in Figure 6.43. For more information on how to sketch human figures, please refer to other references such as 11.

FIGURE 6.43 Using the same methods of boxes and cylinders, a form of the human body can be created to assist in sketching it.

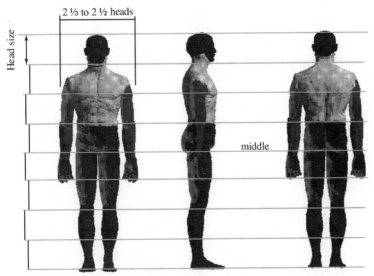

FIGURE 6.44 The approximate proportions for the human body. These proportions can be used in order to create the form of a human similar to Figure 6.43. (Figures created by NexGen Ergonomics ManneQuinPRO program.)

6.12 ANSWERS

Example 6.2

FIGURE 6.45 An object with E and B and G on its three faces.

Example 6.3

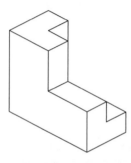

FIGURE 6.46 The object represented in Figure 6.5.

Example 6.4

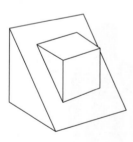 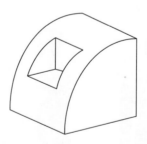

FIGURE 6.47 Two possible solutions for the object represented in Figure 6.6.

Example 6.5

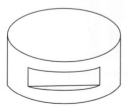 

FIGURE 6.48 The object represented by Figure 6.7.

REFERENCES

1. WOHLEBER, CURT, "The Work of the World," Invention and Technology, Winter 1992, pp. 44–52.
2. GAZZANIGA, MICHAEL S. "The Social Brain; Discovering the Networks of the Mind," Basic Books, New York, 1985.
3. McKIM, ROBERT, "Experiences in Visual Thinking," Brooks/Cole Publishing Company, Belmont, Calif., 1972.
4. HANKS, K. and L. BELLISTON, "Rapid Viz; A New Method for the Rapid Visualization of Ideas," William Kaufmann Inc., Los Altos, Calif., 1980.
5. DE MILLE, RICHARD, "Put Your Mother on the Ceiling; Children's Imagination Games," Ross-Erikson Publishers, Santa Barbara, 1981.
6. GERDS, DONALD, "Perspective; A Step by Step Guide for Mastering Perspective by Using the Grid System," 5th Edition, DAG Design, Santa Monica, 1989.
7. EDWARDS, BETTY, "Drawing on the Right Side of the Brain; A Course in Enhancing Creativity and Artistic Confidence," J. P. Tarcher, Inc., Los Angeles, 1979.
8. NIKU, SAEED, "Introduction to Robotics; Analysis, Systems, Applications," Prentice-Hall, New Jersey, 2001.
9. GIMMESTAD BAARTMANS, BEVERLEY and SHERYL A. SORBY, "Introduction to 3-D Spatial Visualization," Prentice-Hall, New Jersey, 1996.
10. EARLE, JAMES H., "Engineering Design Graphics," Addison-Wesley Publishing Co., Massachusetts, 1994.
11. FARRIS, EDMOND J., "Art Student's Anatomy," Dover Publishers, New York, 1944.

HOMEWORK

6.1 Change the order, letters, words, or other similar elements of a company name or trademark to make it fit better. Example: Kentucky Freed Chicken.

6.2 Development of a Commercial for a Phony Product. You are to consider a phony, nonexistent, ridiculous product or service that makes no sense, and make a 45–60-second commercial for it. The length of the commercial must be between 45 and 60 seconds. It must be recorded on a VHS tape (or a CD if your instructor allows) and must be handed in exactly at the beginning of the commercial. You or anyone else you choose may act in it. Commercials with scripts and acting are much preferred to announcements with written material only.

Check the final quality of your tape when you transfer video from tape to tape. Beware of wind noise if you record outdoors and of lighting if you record indoors. You may work in groups as assigned by your instructor, with one tape (one commercial) per group.

6.3 Which one of the following in Figure P.6.3 is the same as "a"?

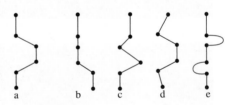

a   b   c   d   e

FIGURE P.6.3

6.4 Are the surfaces in Figure P.6.4. the same?

6.5 Which one(s) in Figure P.6.5 are the same objects as a?

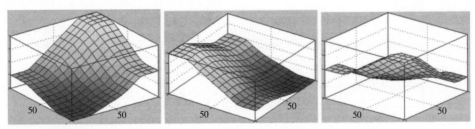

FIGURE P.6.4

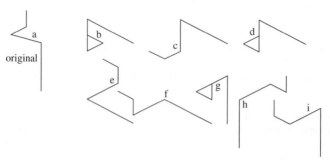

FIGURE P.6.5

6.6 Draw a pictorial perspective of one of the products in Figure P.6.6 in another angle. This means that you should visualize the product from another point of view and draw it accordingly. You may have to imagine what the details might look like in that view if they are not clear.

6.7 Draw a perspective pictorial of the object shown in Figure P.6.7 based on the given unit cube.

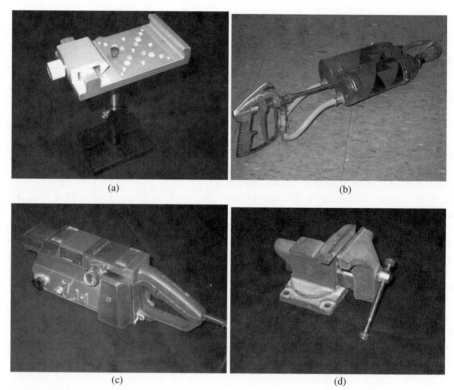

(a)

(b)

(c)

(d)

FIGURE P.6.6 Draw one of these products (a–d) in perspective as seen from another point of view.

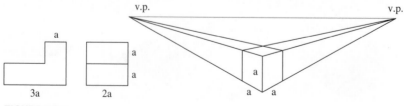

FIGURE P.6.7

6.8 Draw a $3l \times 2w \times 2h$ basic perspective cube in 45° and 30–60° grids.

6.9 Draw a figure similar to Figure P.6.9 which will be approximately valid whether upside down or not (turn the page to see the figure upside down).

FIGURE P.6.9

PART TWO

THE DESIGN ENVIRONMENT

CHAPTER 7

DESIGN CONSIDERATIONS, DECISIONS, AND CONSEQUENCES

A mathematician, a physicist, and an engineer were given a red rubber ball and were asked to measure its volume. The mathematician measured very accurately the diameter, wrote an elaborate triple integral equation, and evaluated it for the given data. The physicist put the ball in a container with water and measured the water displacement. And the engineer looked up the serial number and the model in his little red-rubber-ball catalog.[11]

You have freedom to choose. But you do not have freedom from the consequences of your choice.

As you may remember, I asked the following question in Chapter 1. If you have not already answered this question, please do so before you go on. Your answer will help you better relate to the following discussion:

> What is the last product you bought for your house or for someone else?
> Why did you pick that particular product?

7.1 INTRODUCTION

Design considerations are the concerns that we must address during the design phase of any system or product development or when a solution for a problem is sought. Each consideration requires a decision with its related consequences. It is best to address these concerns as early as possible, during the time when the solution is put together and major decisions are made, and certainly before a final design is achieved or manufacturing has begun. This is because during the design phase, it is much easier to change the design if needed or to change the details of a design if necessary with fewer negative and detrimental consequences. However, as a system or product is finalized, and especially if manufacturing is started, the cost of a change will be tremendously higher. The worst case is when a product is distributed and is in use, but a design flaw is found that warrants a change, leading to a recall of the product. In that case, the cost of recalling, correcting the problem, and instituting a design change becomes prohibitively expensive.

Examples abound in industry about the recall of products and how much they cost. The Chrysler Corporation spent $400,000,000 during the late 1990s just to recall its minivans in order to correct the back-door latch problem, in which the door would suddenly open and passengers would be thrown out if hit in the rear. In March 2005, Greco™ recalled 1.2 million toddler beds due to the possibility that toddler extremities or heads might get caught between the railings. Additionally, it was fined $4 million for not notifying the government about it. DELL and Apple computers recalled a total of 6.1 million laptops in August 2006 to change the SONY batteries that were used in the computer. These batteries would catch fire due to a short circuit. The Toyota Camry 1983 was recalled in mid-1980s in order to correct the problem of its timing belt. The engineers at Toyota had specified a timing belt made by two competitors. The expected life for a timing belt is about 60,000 miles before it deteriorates and must be inspected and changed. However, in many cases, the timing belt would disintegrate and break before 20,000 miles. Later, the engineers determined that the belts made by one of the two competitors were the culprit belts, while the ones made by the other company had a normal life. The company decided to recall all Camrys, inspect the timing belts, and change the ones from the first company. This involved removing the timing belt cover, inspecting the stamp on the belt to determine who the manufacturer was, replacing the cover if the belt was from one company, or replacing the belt with a new one and replacing the cover. The mechanics were allowed to charge the company for this inspection or the replacing of the product. Toyota also stocked the dealers with countless timing belts. If you consider the total cost of informing the owners, the cost of the belts, the cost of inspection, and the cost of replacing the unacceptable belts, you will see why this simple mistake was so costly. Fortunately, in this mistake, there were no injuries or property damage (other than broken timing belts). Otherwise, the cost would be significantly higher, both to the reputation of the company, as well as the financial health of the company. The life expectancy of the product and customer expectations are both among important design considerations with which the designer must be concerned. Correcting the mistake during the design phase would have been much easier, simpler, and less expensive. According to one estimate by a large company, if a change during the design phase cost $1, the same change would cost $15 during manufacturing phase and $92 if the product was already shipped. This does not include liability costs involving injuries or property damage or the negative effects on the reputation of the company.

In the following sections we will consider a number of design factors and issues that are deemed important. In fact, certain design considerations are so important that we will devote whole chapters to their discussion. These include human factors, safety, liability, aesthetics of design, economics of design, and intellectual property issues. Table 7.1 is a list of some of these factors.

It should be noted that this list is by no means complete; it is only a list of some of the more general and common concerns and their consequences. Every design may require many other considerations that are not even mentioned here. It is also possible that for certain problems, products, or systems, some of the above-mentioned concerns may not be applicable. It is important to remember that the designer has to recognize what must be considered and what other concerns must be added to the list for any given product or system. For example, resistance against roll-over is not a significant factor in the design of a toaster. The same is a crucial consideration for an automobile due to its footprint and the location of the center of mass (this has been a significant factor recently due to popularity of SUVs that have high center of mass, especially if they are narrow. The Ford Explorer was designed too narrow and had a high center of mass. To lower the center of mass, Ford recommended that the tire pressure be reduced to about 26 psi. At that pressure, due to additional friction between the tires and the road, especially in hot weather, the defective tires made by Firestone would explode. High center of mass, combined with the sudden explosion of a tire would roll the Explorer causing damage, injuries, and death).

Please also note that these factors are listed in a random order, not in order of importance. One factor may be more important for one product or system, while another factor may be more important for another. The level of importance of each factor for each product must be determined by the designer and the design team.

Please note that many of these design considerations are interrelated. For example, as you choose a particular material for your design, the choice will have a direct and significant effect on other considerations such as the methods of manufacture, weight, cost, the look and the feel of the product, the effects of environment on the product, and many more. Every decision has consequences. Imagine that you would make an automobile from stainless steel (as the De Lorean was). Stainless steel is heavier, is more difficult to machine and weld, and is much more expensive than steel. It requires more skilled labor to work with it too. However, it does not rust and does not need to be painted. On the other hand, since it would not be painted, all cars made from stainless steel would have the same color. Repairing the body after accidents would be more difficult and more expensive as well. The choice would also have an effect on the perception of the customers, the prestige of the car, its selling price, and its life. Thus, in this case, the choice of material affected many other design factors directly and indirectly. Similarly, quality, safety, and life expectancy of a product affect its price, the choice of materials, manufacturing methods, and other factors.

Table 7.1 provides a quick reference guide for some of these considerations in no particular order. These are used as a basis for discussion. For every design project, you must edit, add, subtract, or otherwise modify, the list for your own purpose:

TABLE 7.1 Physical, Structural, and Functional Considerations and Factors

Decisions/Consequences	Considered
Cost	_____
Size	_____
Weight	_____
Material selection	_____
Methods of fabrication	_____
Physical and structural standards	_____
Functional standards and expectations	_____
Performance	_____
Efficiency	_____
Reliability	_____
Company image and mission	_____
Quality	_____
Serviceability	_____
Styling, shape, aesthetics, packaging	_____
Safety	_____
Human factors	_____
Environmental effects (on or from the design)	_____
Disposability	_____
Sustainability	_____
Assembly/disassembly	_____
Life expectancy	_____
Ethical issues	_____
Patents and other intellectual property	_____
Legal matters	_____

In the following discussions about design considerations and decisions and their consequences, we will employ a number of example products and systems, including the page-turner, an automobile, a building, and a toaster. The automobile is a complicated system with many subassemblies, while a toaster is simple and straightforward. These are only used for comparison purposes.

7.2   COST

Unlike popular perception, the goal of the designer should not be to minimize the cost of a product, as it is not true that the least expensive product is the best choice. What matters is the relative relationship between cost and quality of the product or its features. As will be discussed in Chapter 11, the fundamental deciding factor is *product value* defined as

$$\text{Product value} = \frac{\text{What you get}}{\text{What you pay}} \tag{7.1}$$

In this relationship, when you get more for what you pay, the product is better. Obviously, reducing the cost of a product without reducing its features, quality, or utility, will add to the value of the product. However, if minimizing the cost were the only concern, you might end up with a product that costs little, but has little value as well. This happens when the designer eliminates necessary and valuable features or reduces the quality of the product only to reduce the cost. The final result is less cost, but also less value. On the other hand, if for the same quality and features the cost can be reduced by better design, better methods of manufacture, less required labor costs, etc., the value will increase and the product will be preferred by the consumer or the client. Then the goal of the designer should be to minimize the cost only in the context of value, and to come up with the minimum cost without sacrificing other requirements or specifications. An absolute determination to reduce cost will lead to unacceptable or unsafe products. I remember hearing a story from a friend about cost saving and its consequences. In that case, the company had developed an extensive checklist, printed on card-stock, that had to be checked off and certified before the system could be delivered for use. To save money, the company printed the next set on regular paper. However, individual sheet of paper could easily tear and be lost. In one case, one lost page held up delivery at the cost of $1 million per day due to delays until it could be reproduced. In another example, a cheaper gasket would leak an explosive gas into the next portion of a system, requiring that a purging device be added to get rid of the explosive gas. The additional cost of the purge system was orders of magnitude larger than the savings from the less expensive seal. However, since the two portions of the system were designed by separate groups, and because each team was trying to reduce the cost of their portion of the system, the problem caused by one decision and its consequences were not adequately addressed. Please also see Example 7.4 and the role of desired minimum cost in the design of the Ford Pinto.

As we discussed in Chapter 5, Quality Function Deployment (QFD) can be an effective asset in prioritizing this factor in our design. Based on both customer requirements and product characteristics, the role of cost in our design decisions can be evaluated and prioritized.

It should also be mentioned here that the final price of most products is a function of both the production costs as well as what market forces dictate. In most cases, the final cost of the product is set by marketing elements based on the perception about the product, whether it is a new product and protected by a patent or not, how badly it is needed, and advertising efforts, regardless of the production costs. For example, in many cases, even if a product could be sold for less, companies increase the price to what they think the market will accept in order to increase their profits. Customers, too, develop a perception about the quality of a product in relation to its cost; they might think that if a product is less expensive, it is probably cheaply made. As a result, the design engineer must bear in mind that the final price of the product is affected by many factors governing the design as well as its market. However, whenever the cost can be reduced without sacrificing or jeopardizing other factors, even if the final price in the market is not reduced accordingly, profit margins increase. We will discuss how the cost of a product is affected by other factors in much more detail in Chapter 11 when we discuss economics of design.

EXAMPLE 7.1

In 2008, gas prices increased significantly due to increased market pressures from new global markets (China, India), increased consumer consumption, crises in oil producing countries, and decreased production due to oil-well and refinery damages by natural disasters. Crude oil prices, which constitute about $\frac{1}{2}$ of the gas-pump prices, increased to over \$140, raising the average gas prices to over \$4 per gallon nationwide. Although the crude oil production continued to increase, and although refineries were brought back on line after each disaster, the prices did not go down significantly, providing for record profits for the oil companies. The market dictated the price of the gas at the pump regardless of profit margins, production levels, or even the crude oil prices. ∎

7.3 SIZE

Size consideration allows the designer to specify the approximate or desired size of the final product, whether a whole product, a subassembly, or a component. The general perception regarding size specification is to minimize it. Many assume that if everything becomes smaller it is better (generally because smaller size translates into less weight). However, the goal should be to find the correct and appropriate (optimum) size, not the minimum. For example, the final size of a family car is an important issue. Minimizing the size of a family car is not really appropriate. The goal should be to define and specify the optimum size for a family car. No doubt this specification will be different for an optimum two-person sports car, as it will be different for a large-size pickup truck. However, even when a specific range is considered, the designer may be able to make decisions about other considerations that positively affect the final size of the family car. I remember when we made the final choice in my family for a minivan, it was between two similar vans with similar price and specifications, but that one had a very small trunk space. For a minivan that is supposed to be used for family outings and camping, shopping, with strollers and other baby equipment, the small trunk of one choice was deemed unacceptable. This made the decision for us. Now consider a radio. It is actually possible to make a radio circuit that will be so small that it may be impossible to locate or handle. This size radio may have its own purposes in many different fields. But in the consumer market, it is impossible to sell such a miniature radio that cannot be seen or handled. As a result, the circuit may have to be placed in an artificially larger ''box.'' The same is true for majority of drugs. The size of the needed active ingredient of a drug is generally very small. But a pill that size is impossible to handle. Drug manufacturers add other fillers to the active ingredients in order to increase the size to what they think is the right size and shape for their drug. For our page-turner, we discussed the size of the books that the device would handle, which would ultimately dictate the size of the page-turner.

Similarly, the size considerations about most elements of a building are extremely important. One has to specify the desired number of rooms, offices, bathrooms, and other living- or working spaces and their sizes before a building can be designed.

In addition to utility and handling, the size of a product may also affect its price, its weight, its looks, its safety, and many other factors. It is very important that the design team or the individual designer consider the ramifications and the effects of the size of the product on the final outcome and choose an appropriate size for the product.

Similar to the cost of a product, the importance of the size of a product can be established and prioritized with QFD or similar methods.

EXAMPLE 7.2

The solid rocket boosters (SRB) for the space shuttle are very large and very long. To be able to transport them between the manufacturing plant (Morton Thiokol plant in Utah) and assembly plant (the Kennedy Space Center in Florida), they are made in cylindrical sections, each 12 feet in diameter, and when assembled, 150 ft long (Figure 7.1a). Therefore, an elaborate system of 177 pins, a

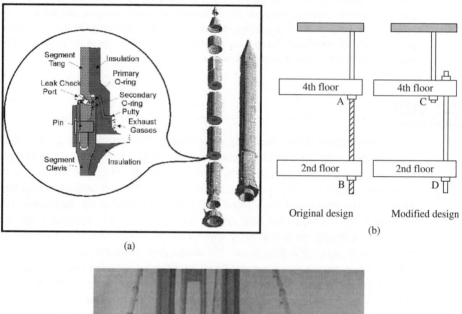

(a)

(b)

(c)

FIGURE 7.1 (a) Schematic representations of the joints in the solid rocket boosters for Challenger[12], (b) the Kansas City Hyatt Regency Hotel walkways, (c) the Tacoma Narrows Bridge (Prelinger Archives).

tang and clevis joint, putty, and multiple O-rings are used to seal the assembled joints. It was the leak through one of these O-rings due to cold-weather hardening that caused the explosion of the Challenger in 1986, killing all its crew members.   ∎

EXAMPLE 7.3

In 1981, the 4th floor walkway of the Kansas City Hyatt Hotel in Missouri collapsed onto the 2nd floor walkway, which collapsed onto the lobby floor where a large tea-party was held, killing 114 and injuring over 200 individuals. The reason was attributed to the way the walkways were designed and assembled. Originally, a long threaded bar was to be used to connect both the 2nd and the 4th floors to the ceiling beams (Figure 7.1b). In this design, each joint would be carrying the load of one floor, with the upper part of the bar carrying both loads (for which, it was supposedly designed). The company that was contracted to build the building deemed it impractical to thread the bar for two stories, as threading a bar for that long would be very time consuming and unnecessary. They changed the design to hang the 2nd floor from the 4th floor and hang the 4th floor from the ceiling, seemingly similar to the original design, but not realizing that in this case, the joint at the 4th floor connection was carrying twice the designed load. The builder claims to have obtained the approval of the designers, and the designers claim they never did so, although the changed plans were signed and approved by them. Joint C broke, collapsing the 4th floor.   ∎

EXAMPLE 7.4

In 1940, the brand new Tacoma Narrows Bridge in Puget Sound, Washington, collapsed due to wind-induced flutter. The fundamental reason for the violent movements was that the girders (deep I-beams used in the construction) reacted to the winds in the area by moving in torsion, and therefore, increasing the angle of attack between the I-beams and the wind, eventually increasing to failure. Therefore, the size of the components of the bridge resulted in violent movements, and ultimately, in its collapse.   ∎

In the preceding examples, due to size considerations, decisions were made that eventually severely affected the final outcome of the design, disastrously. In all cases, better designs would have worked safely, preventing disasters of historic proportions, even if the original sizes were kept.

It should be mentioned here that size also relates to the location of different elements relative to each other as well as the general dimension of each element. For example, the location of a radio in the dashboard of a car can severely affect the safety of the car. We will discuss these considerations in Chapters 8 (Human Factors and Safety) and 9 (Aesthetics).

7.4 WEIGHT

The most common response to the question of what the weight requirement for most products should be is to minimize it. The perception is that the lighter the weight, the better the design since less weight costs less, requires less energy (if it moves), is easier to handle, and requires less material resources; however, this perception is not correct.

A fundamental issue regarding the design of the foundation of a tall building against wind forces is to make it heavy enough to carry the wind-load on the building and not topple. Obviously, in this case it is not desired to make a lightweight foundation. The desired definition should be to make it as light as possible without the danger of toppling the building under those conditions. Now consider a car. If the goal for the weight of a car were to make it as light as possible without regard to other factors and their consequences, the designer might design the car out of titanium regardless of how much it would cost or how hard it would be to work with titanium. It is certainly lighter than steel or even aluminum (for the same strength). However, when other factors such as cost and manufacturability are considered, using titanium does not yield a satisfactory result, even if it is among the lightest materials available.

Obviously, the operating word for weight consideration is not to minimize it, but to optimize the weight. The weight of a product should be optimized in consideration of other factors that are affected by it. However, under the same conditions, it will be beneficial, even desirable, if the weight is reduced without endangering or compromising other factors. Thus the designer should strive to reduce the weight as much as possible without creating an unacceptable disadvantage in regard to other requirements.

EXAMPLE 7.5

The Ford Pinto became a disastrous product when it was found that during rear collisions the bolts extending outwardly from the differential would puncture the gas tank and the leaking gas would explode and catch fire. Ford had set a nonnegotiable goal of 2000 lb and $2,000 price for the car, and as a result, would not consider design changes to the gas tank location, addition of a rubber insert into the gas tank to prevent leaking, or any other remedy suggested by engineers. ∎

The weight of the product is obviously affected by the choice of materials used, the features required by the design, the loads to which the design is exposed, to the life expectancy of the product, quality of the product, and many other factors. One choice affects many others. The designer needs to set a particular acceptable range for the product and strive to achieve it without endangering other factors.

A common cliché is that anyone can build anything so long as there is enough material available. What this means is that given enough material anyone, even with no engineering knowledge, can eventually build, say a building, as large as desired. What an engineer does is to apply his or her knowledge of engineering to do it optimally and safely with the least possible amount of material. Therefore, when the weight of a product is an important factor, QFD may help in establishing its priority based on a weight factor.

Later, in Chapter 16, we will discuss some aspects of engineering design such as stress analysis and fatigue. These discussions will further clarify the role and importance of weight in the design of products and its consequences.

7.5 MATERIAL SELECTION

The choice of materials in the design can have significant consequences on the product, the way it is processed or manufactured, its quality, aesthetics, weight, size, and so on. The designer must consider the consequences of the choice on the product and on the other considerations. For example, let's say that the designer's choice for material in a product is plastics, except that the company has never made anything with plastics. The introduction of the expertise to design with plastics, the investment in the necessary equipment for manufacturing with plastics, the creation of necessary molds, and training the supporting staff to use the equipment may be very costly. However, this cost may be offset by the lower cost of manufacturing, especially if the volume of production is very high, and the fact that once the process is introduced into the company, plastics may be used in the future as well. In a case like this, choosing plastics may still be a wise decision. Otherwise, it may be better to continue using materials with which the work-force is familiar and owns the machinery to process the material. As mentioned above, the choice of material also affects the weight of the product, its aesthetics, its life span, its quality, and many others. The designer should consider all these effects in choosing the best material for the product or its components.

EXAMPLE 7.6

Examples of cases where different materials have been selected for the same purpose, but with different consequences abound in industry. For example, wood, metal, and many different types of plastic have been selected for automobile dashboards. Each material has its own characteristic weight, look, texture, life, environmental requirements and responses, and cost. Metals are more expensive and heavier, but last longer. They need to be painted, but due to their smooth finish, they can reflect light. Plastics are cheaper and lighter and can be molded much more easily and less expensively into many intricate shapes, but if the choice is not right, they shrink, crack due to long exposure to sunlight, and may have limited life. Most car bodies in recent decades were made of steel. However, more and more parts have been made from plastics to reduce the weight, affecting the quality and safety of the car, but also the way it is repaired (some cars have significant portions of the body made of composite materials which incur negligible damage in light impacts, but break apart in harder impacts beyond repair). The unibody of the Jaguar XJ8 series after 2005 is made of aluminum, rendering the car hundreds of pounds lighter, creating better acceleration for the same engine, but also affecting the cost of production and repairs. De Loreans were made out of stainless steel, not requiring any paint, but heavier, more expensive, and very difficult to repair. Soda con-tainers are made of aluminum, glass, and plastics, affecting their weight and production cost, the way they are stacked, the rate at which they may be cooled down, their recyclability, and even health claims (such as the link between aluminum and Alzheimer's disease and rat deposits on soda cans). Kitchen countertops can be made from a variety of materials too, including granite and mar-ble, artificial materials such as Corian, laminates like Formica, tile, solid wood, and stainless steel. Each material has its own advantages and disadvantages, including material cost, manufacturing and installation costs, life and durability, stiffness (e.g., granite is very durable, but also stiff, and glass items hitting it can easily break; whereas Formica is not stiff and is much more forgiving), maintenance, reaction to fire, appearance, porosity and probability of staining. ∎

Another aspect of this consideration is whether new materials that become available are appropriate for your application. New materials such as composites, ceramics, pho-tonic materials, biocompatible materials, polymers, and other newly developed or improved materials may provide an opportunity to redesign and improve existing prod-ucts or invent new products and applications.

Most of this discussion applies to building design as well, including the choice of traditional versus new materials and their consequences in methods of construction, weight, cost, and appearance. What do you think about this issue as related to the page-turner? Does it not apply the same way? Please see Chapter 10 for an in-depth discussion about common engineering material properties and selection techniques.

7.6 METHODS OF FABRICATION

Fabrication techniques and related required machinery are affected by the choice of mate-rials, the volume of production, and the required quality. In turn, they affect the quality of the product, production costs, and its final cost. Methods of manufacturing also affect the skill level of the required labor, the number of people that will be involved in manufactur-ing, and the required time that it takes to complete each operation.

Methods of manufacturing also have an effect on the environment and worker safety. Many techniques of fabrication require that the workers be protected against hazards and may require additional safety equipment. Some processes produce byproducts that are toxic, and they may endanger the lives of the people involved. For example, using lasers for welding requires safety equipment to safeguard the workers. Certain electronic manufacturing methods use solvents and chemicals as part of the manufacturing process, even if not as part of the material selected, that are hazardous and require special material handling equipment, storage, and disposal. Robotic manufacturing requires elaborate safety devices to prevent accidental injuries to workers and other equipment, especially when powerful and fast robots are used. Other manufacturing techniques may involve patent rights and other legal issues. All these factors and their consequences should be considered during the design to ensure that requirements will be addressed.

Another important consideration is the venue in which manufacturing will take place, especially with the advent of globalization. Many companies choose to either have their own manufacturing facilities in other countries or subcontract foreign manufacturers to build their products. Factors such as communication, quality of manufacture in other countries, cost of travel and shipping the manufactured goods, the differences in cultures, social and economical consequences of lost jobs and reduced manufacturing base of the country, as well as legal issues and patent rights must also be added to the considerations.

Please refer to Chapter 10 for common material processing methods, including Design for Manufacturability (DFM) and Design for Assembly (DFA). Please also see Chapter 12 about lean manufacturing and Chapter 11 for the relationship between manufacturing methods and economics of the design.

7.7   PHYSICAL AND STRUCTURAL STANDARDS

The U.S. standard railroad gauge (width between the two rails) is 4 ft, 8.5 in. That's an exceedingly odd number. Why was that gauge used? Because that's the way they built them in England, and English expatriates built the U.S. railroads. Why did the English build them like that? Legend has it that the first rail lines were built by the same people who built the pre-railroad tramways, and that's the gauge they used. But why did they use that gauge? Because the people who built the tramways used the same jigs and tools that they used for building wagons, which used that wheel spacing. Why did the wagons have that particular odd wheel spacing? Because if they tried to use any other spacing, the wagon wheels would break on some of the old, long-distance roads in England, because that's the spacing of the wheel ruts.

So who built those old rutted roads? The first long-distance roads in Europe (and England) were built by Imperial Rome for their legions. The roads have been used ever since. Roman war chariots first formed the initial ruts, which everyone else had to match for fear of destroying their wagon wheels. Since the chariots were made for (or by) Imperial Rome, they were all alike in the matter of wheel spacing.

The United States standard railroad gauge of 4 ft, 8.5 in. derives from the original specification for an Imperial Roman war chariot. Specifications and bureaucracies live forever. So the next time you are handed a specification and wonder what horse's behind came up with it, you may be exactly right, because the Imperial Roman war chariots were made just wide enough to accommodate the back ends of two war horses. And that is the answer to the original question. (From the Internet, Source unknown).

In response to a variety of needs, requirements, safety issues, and accidents, professional societies and governmental agencies have developed a large body of standards that govern many design decisions. These include codes and standards by professional engineering societies such as the American Society of Mechanical Engineers (ASME), American Society of Civil Engineers (ASCE), Institute of Electrical and Electronics Engineers (IEEE), American Society for Testing and Materials (ASTM, now ASTM-International), Society of Automotive Engineers (SAE), the American Institute of Architects (AIA), Underwriters Laboratories (UL), and a host of other entities. They also include governmental agencies such as the Environmental Protection Agency (EPA), the Nuclear Regulatory Commission (NRC), the Consumer Product Safety Commission (CPSC), the Occupational Safety and Health administration (OSHA), Military Specifications (Mil Specs) and many others. To this you must also add State and local government agencies (such as Cal-OSHA and local building codes) which have their own specifications, standards, and regulatory requirements. Designers, architects, and others involved in manufacturing, construction, and management of these activities must abide by these requirements and regulations in their design as well as subsequent activities.

During the 1800s and early 1900s, there were hundreds of cases in which boilers and other pressure vessels exploded, killing, injuring, and maiming thousands of bystanders and workers. Nowadays, you hardly ever hear about accidents like this related to pressure vessels and boilers. The reason is primarily due to the pressure-vessel code and standards that ASME has developed over the years that regulate the design of these vessels. Obviously, if your design is related to pressure vessels, you have no choice but to consult the code and design according to its standards. Otherwise, you will be liable for not following standard codes if anything happens.

EXAMPLE 7.7

Another issue related to the collapse of the Kansas City Hyatt Regency walkways was that apparently, the Kansas City Building Code had been violated by having safety factors below the required values. Additionally, the method that was used to connect hanger bars to the welded c-channels was determined to be inappropriate, causing failure. Both these occurred during the design of the building by ignoring the required codes. If the designers and engineers had followed common sense, as well as the related codes, the disaster may have never happened, even if the original design was questionable. ∎

Military specs (Mil-Specs) are a set of published specifications for materials, products, and services that the military uses for specifying the products and services it purchases. Although recently abandoned, the Mil-Specs for canned fruitcakes [MIL-F-1499F(1)] was many pages long. Then imagine what they may be for a fighter jet or a tank. However, if you consider that the fruitcake may need to be stored for long periods of time in high and low temperatures, damp and dry environments, or dusty locales, may need to be dropped off from air, must remain edible and nutritious under combat conditions, etc., then you will realize why the specs are so important and thorough. In reality, many products are designed to military specs, even if not used for the military use, both because these standards are common and available, and because the manufacturers hope that they will be used in the design of military equipment by others.

Consumer products are generally governed by Consumer Product Safety Commission (CPSC) regulations and standards. Household items such as ladders and lawn mowers and dryers, toys for all ages, baby products such as cribs and car seats and walkers, and other products used by consumers must be designed to follow the requirements of CPSC standards, whether manufactured in the United States or imported from abroad. For example, ladders must have a label attached to the top step indicating that it is extremely dangerous to step on the surface. Failure to attach a label to the ladder will result in liability for the designer and the manufacturer. It is also required by this code that all lawn mowers have a braking device on the engine that will stop the engine within 3 seconds if the user releases the handle on the lawn mower chassis. The code does not specify how to design the brake. It only states the specifications and requirements of it. If an architect designs a building, she or he will have to follow State and local building codes, which can be different from locality to locality. All steel and concrete-based structural designs must follow related codes developed by professional societies such as the American Society of Civil Engineers (ASCE).

EXAMPLE 7.8

The following is the CPSC Standard for power lawnmowers. If you purchase a walk-behind rotary lawnmower, the machine must meet the June 30, 1982 federal standard, and must be certified as complying with the regulation. Some of the safety features of the regulation include:

- *The blade brake control:* On all new mowers, an automatic brake stops the blade in three seconds when the operator releases his or her grip on the handle-mounted control bar. This feature prevents the rotary blade from operating unless the operator actuates the control. It also requires the operator to maintain continuous contact with the control to keep the blade operating, and stops the blade completely within three seconds when the operator releases the control. If the mower only has a manual start, the control must stop the blade without shutting down the engine, unless the manual starting control is located within 24 inches from the top of the handle, or the mower has a 360° foot shield. For user protection, the mower must also have a secondary control which must be activated before the mower can be operated.

- *Foot shield:* The area at the rear of the mower that might be reached by the foot when using the mower is subjected to a probing test using a foot-like probe. With the mower wheels on the ground, this area is probed to ensure that neither the foot probe nor any part of the mower (such as a trailing shield) will enter the path of the blade. Shields must either close automatically or prevent operation of the mower (when open) unless the grass catcher is present.

- *Labels:* New mowers must have a warning label near the discharge chute cautioning users to keep hands and feet away from the chute. The mower must also have a certification label with the inscription, ''Meets CPSC blade safety requirements.'' ∎

ISO Standards refer to the International Standards Organization (or International Organization for Standardization) rules and standards. Specifically, **ISO-9000** refers to a set of quality management standards that are periodically updated. These standards include both guidelines (about how to implement quality management) as well as requirements that must be met in order to be ISO-certified. ISO 9000:2005 relates to the quality management systems fundamentals and vocabulary. ISO 9001:2000 relates to requirements, and ISO 9004:2000 relates to guidelines for performance improvements. The purpose of these is to facilitate international trade by creating a single set of standards that can be recognized and used throughout the world. These standards apply to practically all industries, from manufacturing and electronics and legal services to forestry and printing and aerospace. Companies that incorporate these standards in their processes and quality management may be audited by an accreditation board, and if qualified, will be ISO 9000-certified. For more information, please see Reference (13).

Food and Drug Administration (FDA) regulates all foods, drugs, and medical devices. All new drugs must be tested and certified based on extensive clinical tests, and their safety, value, and side effects must be documented before they are allowed in the market, sometimes requiring years. The same is true for all medical devices that are used in the body, including stents, defibrillators, pace-makers, and artificial body parts. Food labeling and health claims must also follow an extensive set of regulatory rules. Please see FDA's website for detail information about these processes and requirements.

As will be discussed in Chapter 13, the designer will be held liable when required standards are not followed. Although there is no guarantee that no accidents occur when standards are followed, and although there is no way to completely prevent liability, following required codes enables the designer to design products that are safer and provides for more powerful defense if a liability suit is brought against the designer (or the company for which she or he works).

Except in particular situations, common engineering curricula do not prepare students for standards and codes. It is generally the responsibility of the designer to find out what codes and standards are applicable to the activities in which the designer is engaged. Thus, it is vital that when you engage in design activities you find out about the applicable standards and regulations, learn them, and make certain that they are followed.

Please also see Chapter 13 for more relevant information about standards and codes as they relate to liability.

7.8 FUNCTIONAL STANDARDS AND EXPECTATIONS

In addition to the required standards that govern materials, products, and services, there are many other standards that relate to functional expectations. Many of these standards are not even documented, are not part of any codes, and may not be accurately portrayed. In many cases, they change rapidly as time goes by, and in other cases, they are not very clear or defined. In fact, most are implied. But there are plenty of them around and the designer must be aware of them and take them into consideration, or risk the danger of falling behind competition or losing clients. Let me explain.

Customers have a certain set of expectations about what they should get in relation to the price they pay for a product. Although as discussed earlier, the ultimate measure is "value," even between different categories of similar products, there are certain minimum implied expectations that customers have. For example, not only is it the law, but also expected that a product offered for sale is safe. This is implied and expected. No one expects to get electrocuted by an electrical appliance. In addition, every product carries other expectations that may not be written, but are still expected. The timing belt of an automobile is expected to last a certain length of time. When it does not, no law is broken, but the company is still expected to fix the problem. When you buy a car in the $20,000 range your expectations are different than when you spend $40,000. These expectations are also implied, and if a car does not have those standards, no law is broken, but the product is condemned in the market (otherwise, why a Cadillac vs. Chevrolet and a Lexus vs. Toyota? Each pair is manufactured by the same company, but with different sets of expectations).

EXAMPLE 7.9

Before CD players came into widespread use, the standard media for playing music was a cassette player. In their heyday, you could buy a cassette player with radio at a variety store at about $20, a cassette deck from electronic stores in the $150–200 range, or one from a specialty store for more than $350–400. What was the difference between them? The $20 cassette player/radio set was not even stereo, the frequency response was around 7000 Hz, the quality of sound was low, and the life expectancy was short. For $20, no one expected to have anything more than the basic features. For $200, you would expect a nice stereo player, with frequency response in the 14 000–16 000 Hz range, some features, and perhaps a few years of life without problems. For a cassette player at $400, the expectation would be to have frequency response in excess of 18 000–19 000 Hz, many features, long life, and modern design. None of these specifications were written or expressed by any entity. It was the standard of the industry at the time. If one company could design and offer a deck for less than $100 that would be comparable to the $200 decks, it would change the dynamics of the market, would raise consumer expectations, and would drive the competition out of the market. Others would either have to catch up with the changed and updated standard expectation or they would lose. ■

The same is true for perhaps most other products. Automobiles, toasters, home appliances, travel services, entertainment, media, and even educational institutions have their own expected standards and need to compete with others in the same category. The page-turner is no exception; it is expected that it should be safe, turn the pages without damage, and be reasonable in cost. However, its features can vary based on the cost, its applications, and the user's abilities. Since there is a lack of competitive models in the market, the chosen features by the first major producer will set the standard for all future models. Thus, it is important for the designer to be aware of these changing expectations, standards, and requirements, and design to meet or exceed them.

7.9 PERFORMANCE

Performance is defined as how well the product performs its function. The better it performs its function, the higher the performance. This can be defined in concrete terms for certain products, but not for others. For example, a high performance car is one that has high acceleration (and thus, can achieve higher speeds in shorter times) and can pass a more rigorous slalom test (which relates to its handling and response). This means that a car with larger engine, lower weight, and better suspension will be considered high performance. However, this also means that a car with a smaller engine that uses less gasoline and is more environment friendly is not a high performance car. Nor is a car with heavy payload capability (like a truck) or a family car, leaving only select cars in this category. However, what most drivers want in a car is unrelated to this formal definition of performance. Therefore, for most customers, the formal definition of performance for cars may not be applicable. In fact, most drivers cannot test a car for its maximum peak performance during a short drive when they test a car, they cannot determine whether a car over-steers or under-steers (they probably do not even know what these mean), or whether it is prone to roll-over or not. Still, most people do consider the level at which the car of their choice performs its function as an important factor. If a car is safe, comfortable, and aesthetically pleasing, and if it requires less gas and less maintenance, it is performing well, even if not formally high performance. Thus the formal definition may be replaced by another.

On the other hand, many products do not have any official performance criteria that distinguish them from others. For instance, the performance of a toaster cannot be defined by terms other than whether it makes your toast the way you want it according to the settings you choose. If it does what you want, its performance is good (or high). At times, performance criteria may be contradictory as well. For example, the design criteria of a car may require higher weight for safety, but lighter weight for economy and higher acceleration, a more powerful engine for higher acceleration, but smaller one for economy and environment friendliness, small size for easy parking, but large to carry all your belongings, etc. The designer may need to compromise one requirement to achieve requirements for another.

During the space race to the moon, the primary concern for the designers of rockets and spacecraft systems was performance; they wanted to win the race to the moon for national pride, and consequently, cost was not a consideration. The ability to launch rockets into space enabled the industry to also launch satellites into orbit, almost without competition. However, by the 1980s and 90s, many other countries had developed the capability to launch satellites into orbit. To increase their competitiveness, American companies had to include cost as a major consideration into their designs, creating a whole new set of performance and reliability criteria (including the reusable space shuttle and rocket boosters, the issues related to the re-assembly of the solid rocket boosters that eventually led to the Challenger disaster).

Whether or not official criteria exist, the designer must consider the performance of the product and whether or not it actually performs as desired and as designed. When the product fails to perform its intended functions, it has failed. Ultimately, it is necessary that the designers determine what factors constitute performance criteria for the product, how to define these factors, and how to measure them. Then it will be possible to incorporate appropriate measures in the design to achieve the desired performance.

Personnel Performance: Why God would never receive tenure at any university:

- *He only had a couple of publications,*
- *They were in foreign languages,*
- *They had no references,*
- *They were not published in refereed journals,*
- *Some question who the real author was,*
- *He has not done much since He created the world,*
- *The scientific community cannot replicate his results,*

- *When a part of the experiment went awry, he tried to cover it up by drowning the subjects,*
- *He rarely came to class and just told students to ''read the book,''*
- *He never obtained permission from ethics boards to use human subjects,*
- *He expelled his first two students,*
- *He held office hours at mountaintops,*
- *Most students failed his ten simple requirements.*

—Source unknown

7.10 EFFICIENCY

Efficiency is defined as the ratio of what you get from a system to what you put into it. In engineering systems, when dealing with power, it is the ratio of power-out divided by power-in, or:

$$\varepsilon = \frac{\text{Power-out}}{\text{Power-in}} = \frac{\text{What you get out}}{\text{What you put in}} \tag{7.2}$$

For example, a power plant that generates electrical power requires a certain amount of input power (in the form of energy in the fuel) in order to generate output electrical energy. This ratio may be as high as the low 40 percents. The efficiency of an internal combustion engine may reach the high-20 percents to very low 30 percents. The rest of the input energy is wasted in the form of heat that must be dissipated from the system (sometimes requiring even more energy to run pumps and fans). As efficiency increases, less energy is required to do the same. Therefore, a high efficiency engine or power plant will generate more output power for the same input power. No system can have an efficiency of 100 percent, as this will defy fundamental thermodynamic laws, and there is always friction present, reducing efficiency. However, a better design can increase efficiency.

Defining efficiency for certain systems is easy. For others, it is not very clear how to define efficiency. For example, how could one define efficiency for a toaster? What is the output that is measured to calculate efficiency? Or how can one calculate efficiency for a page-turner or a building? However, you can consider the efficiency of the toaster in relation to its function. If you design the toaster in such a way as to require less energy to toast a piece of bread to the same level as another toaster, it will be more efficient. In this case, the power inputs required for the same function in two different designs can be compared in order to determine the relative efficiency of the two systems. This is why all toasters have a reflecting mirror-finished panel on either side to reflect the light and trap it inside. Otherwise, the light from the backside of the heating element would just warm the metal body and be dissipated. In a page-turner too, the less energy required to perform the same functions, the more efficient the system. Similarly, if a building is designed to require less energy for heating or cooling, the more efficient it will be. Additionally, if a building can be designed in such a way as to provide the same features and usable space in a smaller foot-print, the more efficient it will be, not only in heating and cooling costs, but also in terms of the cost of the building relative to its function. Consequently, it should be the goal of the designer to consider all possible ways that the efficiency of a system can be increased. This will help the product, the customer, the environment, and the resources that are available to us.

EXAMPLE 7.10

In 1983 Ford introduced a new family-class car (the Ford Tempo and Mercury Topaz) with 2300 cc, 98 hp engines. At about the same time, Toyota introduced the Camry family cars, with 2000 cc, 100 hp engine. The 1984 official EPA ratings for the 1985 models of these cars sold in California were 16 mpg for the Tempo and 24 mpg for the Camry.[17] Although these cars were of the same size, capacity, performance, and class, their efficiencies were vastly different. One could do much more for much less input gas energy. With the design of new hybrid cars, gas mileages of 50 and 60 mpg are now possible without any loss in performance or characteristics. ■

EXAMPLE 7.11

Although front-loading washing machines have been around for decades throughout the world, they have found a new favor in the United States. This is primarily due to the fact that, for the same washing capacity, they are much more efficient than top-loading machines. This increased efficiency is due to the way the articles are washed. Top loaders agitate the clothes in the drum in sideway motions, always in the water. Front loaders lift the clothes in the drum and continually drop them into the water, hitting it. This extra hitting action increases the scrubbing action of the machine and thus, its efficiency (conceivably, the same machine could also be used as a dryer, although they generally are not). Additionally, they require less water than top loaders. Unfortunately, these machines are unnecessarily more expensive. ■

7.11 Reliability

Reliability is the ratio of how many times a system performs as intended over how many times it is used, or

$$R = \frac{\text{How many times worked satisfactorily}}{\text{How many times attempted}} \tag{7.3}$$

For example, if you use a light switch 1,000 times and it works properly (or satis-factorily) 978 times, its reliability is 978/1000 or 97.8 percent or simply 0.978.

If a system is made up of only one part (or component), the reliability of the system is the same as the reliability of the part or the component. However, if there is more than one part in a system, the total reliability of the system is the product of the reliabilities of all its components or parts, or

$$R = r_1 \cdot r_2 \cdot r_3 \, \cdots \tag{7.4}$$

where $r_1, r_2, \ldots$ are the individual reliabilities of each part or component. If a system has three parts, with reliabilities of 80 percent, 90 percent, and 85 percent, the total reliability of the system will be $0.8 \times 0.9 \times 0.85 = 0.612$ or about 61 percent. Since the reliability of any system is always less than 1 (even if very close to 1), as the number of parts or components increases, the reliability decreases. For example, imagine that a system would have 1,000 parts, components, or subsystems (how many parts or components do you think the space shuttle or a car may have?). If we assume that the reliability of each and every component or part is 99 percent, the total reliability of the system will be

$$R = (0.99)^{1000} = 0.000043 = 0.0043\% \tag{7.5}$$

This means that with component reliabilities of 99 percent, the system will work properly 43 times out of 1,000,000 times attempted, or once out of every 23,000 times. Would you like to buy such a system? Would the space shuttle be expected to even fly at any given time?

Increasing the reliability of each component or part to 0.999 (or 99.9 percent) will increase the total reliability of the system to 0.37 or 37 percent, and reliabilities of 0.9999 will yield a total reliability of 90 percent.

Still assuming that a system has 1,000 parts, what would be an acceptable reliability for the system? Is 90 percent adequate? Would you like your car, your toaster or page-turner, or the space shuttle to have this level of reliability? Consider the following case. By the time of this writing, there have been five space shuttles built and launched 115 times since the beginning of the program in 1981. Assuming that all orbiters have been launched equal times, this yields an average of less than 23 times per orbiter. So far, unfortunately, two orbiters have been lost due to malfunction. This yields a reliability of about 95 percent. However, these numbers are not accurate because since Challenger exploded in 1986, there had been only four orbiters left to launch. Columbia was on its 28th mission, yielding 96 percent reliability. Different orbiters have flown a different number of times as well. As you see, even at this level of reliability, the result can be disastrous.

As the expected level of reliability increases, the required individual reliabilities of components increase as well. According to NASA, each space shuttle is assembled from 2.5 million parts (including 230 miles of wire, 1060 valves, and 1440 circuit breakers). To achieve a reliability of 99.9 percent with 2.5 million parts, the required reliability of indi-vidual parts should be about 0.999,999,999,5. In other words, we should expect that each individual part should fail less than five times in 10 billion times. This means that each part becomes exceedingly expensive, increasing the overall cost of the system.

Reliability can be improved through redundancy of components in the system, assuming that if one system fails, another will work. This is one of the reasons why the space shuttle's heat resistance system consists of three components (heat-resistant tiles, blankets, and the RCC panels). However, each redundant component or subassembly will increase the cost, the weight, and the total number of components.

EXAMPLE 7.12

The late-1970s Fiat 128 had a tray of 16 relays under the hood for controlling different functions of the engine and the car, ranging from the fuel pump and the radiator to the door locks and the air conditioning. If the relays had been made sufficiently air tight to prevent moisture from entering the relay compartments, the additional control provided by these relays would have added to the overall control of the car and its overall performance. However, early morning moisture could render any one of these relays nonfunctional at any given day in winter, reducing the overall reliability of the car. In that case, fewer relays would mean better reliability. ■

It is also important to consider the danger each failure creates and its severity or consequences. For example, if the engine of a car fails, it may leave the occupants stranded. Engine failure in an airplane may kill several hundred people. A failed tire on a bicycle may be a minor issue. On an unstable SUV it may cause rollover, injury, and death. Failure of a bolt in a washing machine may ruin the articles and the machine and create a nuisance. Failure of the bolts in the 4th floor walkway of the Hyatt Regency hotel in Kansas City killed 114 people and injured over 200. In May 1979, a McDonnell-Douglas DC-10 airplane lost an engine when the pylon failed during a critical point during takeoff, causing it to stall asymmetrically and roll. The plane crashed into an open field right outside of Chicago O'Hare's airport, killing 273 (including two persons on the ground, injuring an additional two, and destroying a hangar, several cars, and a mobile home). The damage to the pylon had been caused by faulty maintenance routines.[1] In July 2006 the bolts of the ceiling panels in the Big Dig highway tunnel connecting city of Boston to the Logan Airport failed due to wrong epoxy glue, dropping the panels onto a car and killing a woman instantly and injuring her husband.[14]

One way to increase the reliability of a system is to increase the quality and reliability of each component or subsystem. Of course, this increase in quality and reliability increases the cost of the component or system (space- and aviation-grade parts are prohibitively expensive for the same reason). Another way is to include redundancies in the system, also increasing the cost and space requirements. The human body has very few redundancies; perhaps the lungs and kidneys can be considered somewhat redundant, as we can live with even one of them. However, we do need the full capacity of both for normal life. The human body is equipped with many sensors, a powerful controller, and the power of healing to increase its reliability. The alternative, of course, is to simplify the design to require fewer parts. As the total required number of parts decreases (decreasing the total cost too), either the reliability of the system increases if individual reliabilities of components remain the same or individual reliabilities can be reduced (further decreasing cost) for the same overall reliability of the system. This is why simplicity in design and the use of fewer parts in a system are so important. With more parts, there are more chances that something may go wrong. Always strive to design with simplicity in mind; it pays.

Please see Chapter 12 for discussion on quality, Taguchi Methods, and Failure Modes and Effects Analysis (FMEA). Please also see Chapter 13 for additional discussion on product liability.

7.12   COMPANY IMAGE AND MISSION

An important consideration for the designers is the perceived Company image. As part of their business, and to be able to compete with their competitors, companies plan for, and through their years of market activities, develop a certain image that defines them and their products. The image they put forward may be that of a pioneer, an innovative company, a leader, a low- or high-tech company, high quality product manufacturer, etc. In all these cases, the products they make and sell has to match their image; otherwise their perception in the market may be tarnished or changed. In the 1980s, a mid-level Japanese electronics manufacturer that had a particular (mid-level) quality image offered a top of the line cassette deck, granted at a high price, that changed its market image for a long time afterwards. Similarly, until its crisis years of the late 1970s and early 1980s, Chrysler Corporation had a tarnished image which only changed after their restructuring. Therefore, it is vital that the designer also consider the image of the Company and the consequences of deviating from it.

7.13   QUALITY

Quality is one of the prime measures of the worth of a product or system. As was mentioned in Equation (7.1) the actual value of a product or system is measured by what you get versus what you pay. An important measure of what you get is the quality of the product. Thus, the higher the quality of the product or system, the more you get out of the product or system (and possibly, for a longer period of time too), and consequently, the higher the value. According to Genichi Taguchi, a major international leader and educator on this subject, quality is the inverse of the loss that a product causes to society after being shipped, rather than losses caused by its intrinsic functions. What this really means is that if the total loss to society caused by a product is considered, then the higher the quality, the lower the loss to society. For example, if you buy a product that is defective and you must spend time to return it either for replacement, adjustment, or repair, or if you decide you do not want to keep the defective product at all, the time you spend is wasted, thus a loss to society. Moreover, when you find a new product defective, you lose your trust in the company and its products and that is even more loss to the society, because the manufacturer, a member of the society, has lost. If a product does not perform as promised or expected, if it does not last long enough and has to be replaced with a new product not because its life is over or that you need a different model, but because it is not a good product, then society has lost value. Considering the total loss to society and not just the loss to a company, your view of quality in design and in manufacturing will change. Figure 7.2 is a simple depiction of this concept. As the range of acceptable tolerances is increased the cost of production decreases, but the cost to society increases. The best value is where the total cost is the minimum.

Since WWII, theories and practices developed by W. Edwards Deming and Genichi Taguchi have had a profound effect on the concept of quality in design and manufacturing. Concepts such as robust design, Quality Function Deployment, and Lean Production have advanced the quality of many products in the global market, while at the same time, many other products have lost their quality due to economic pressures of saving money and lower costs.

Taguchi Methods [1,2] are a set of methods that are used to specify and design parts with better quality. Some of the methods such as Total Cost to Society are philosophical in nature; others such as the Design of Experiments are very mathematical with roots in statistics.

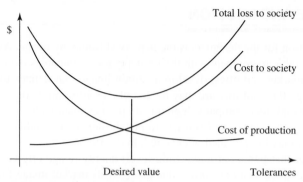

FIGURE 7.2 Total cost to society.

Robust Design is another concept that affects quality. The major premise in robust design is to do whatever possible to reduce the adverse effects of variations in input to a system on the output of the system. If the response (or behavior) of a product under varying and different conditions remain unaffected by changes or variations in the input, the product will continue to behave predictably. As an example, consider the cooling system of an engine. Obviously it will be unacceptable if the engine were to heat up during hot weather and to run cold when the weather is cold. But if the engine runs at a predictable and stable temperature regardless of the outside temperature, the output of the system is unaffected by the input (temperature). Chapter 12 presents additional information on how to design for robustness.

Please refer to Chapter 12 for an in-depth discussion on the subject of quality, including the concepts proposed by Deming, the Taguchi methods, robust design, tolerances, lean production, and examples of how to use these concepts.

7.14 SERVICEABILITY

Have you ever seen the labels on the back of some electronic devices stating "no serviceable parts inside"? This indicates that even if a small part of the system stops functioning correctly, you either have to replace the device or you will need the services of a trained technician to replace it. Certainly it cannot be expected that a layperson would be able to fix an electronic device. But what if a simple light goes out? Should you not be able to replace it?

Extending this to other products and systems, one would expect that some, at least simple, maintenance and repair jobs should be performed by the users and owners. This will certainly reduce the cost of operating and maintaining the product, will enhance the understanding of the product and its functions, and will aid the owner in emergency situations. On the other hand, due to the increased complexity of products and systems, fewer and fewer repairs and maintenance tasks can be performed by the owners and users. No one expects that complicated repairs or maintenance tasks be made so simple that everyone would be able to do it (although it sounds tempting); but simple and straightforward tasks should be.

Unfortunately, with the high cost of repairing products, except in the case of more expensive items such as automobiles, owners prefer to throw away the product and buy a new one than to have the defective product repaired. Although this sounds advantageous to the company who sells more products (and certainly some companies count on this and even plan this into their products), the total effect of this on resources, the economy, the environment, and the total cost to society must also be considered. When a product, which otherwise could be repaired and used for much longer, is thrown away, the environment is negatively affected, the owner may lose trust in the company, more resources are used to create yet more products, or more money is spent on repair and maintenance of simple items.

Many individual owners do expect that they should be able to perform simple maintenance and repair on their products. For example, many owners do routine oil changes on their car. If the designers do not consider this expectation in their design, they will create dissatisfaction in their customers. There are countless examples of these expectations. They must be considered when a product is designed. For a 1970s model of VW Rabbit you would need a specialty wrench to remove the distributor cap. Changing the spark plugs of the Fiat 128 was so difficult that many mechanics would not even accept the job. Replacing the spark plugs of a Toyota minivan also requires a special wrench, more expensive than the cost of having it done at a dealer. A broken wire in an electric iron where it enters the product cannot be repaired or fixed. Replacing a muffler gasket costs $77 in labor and $0.40 in material. Ask anyone who is inclined to do these tasks and they will have similar stories. This indicates that many manufacturers and designers ignore this requirement and render their products difficult to service and repair.

EXAMPLE 7.13

The oil filter of Toyota Previa is located under the minivan (Figure 7.3). In addition to the difficulty of reaching it unless the car is raised on a rack or placed over a pit, the filter sits right above a portion of the chassis. As a result, when the filter is unscrewed, inevitably some oil drips into the chassis railing. Even if the oil is wiped, some will continue to drip from the chassis for a while afterwards, dirtying the driveway or your garage. The remedy to this simple oversight is to place a simple piece of sheet metal over the railing to divert the oil above the chassis railing and into the collection pan.

Oil filter

Chassis

FIGURE 7.3 Oil filter located above the chassis railing. ■

EXAMPLE 7.14

Proper timing of internal combustion engines (valves and ignition) is accomplished by either timing belts or timing chains. Most engines are designed with a timing belt, which has a limited life (about 60,000 miles) and must be replaced periodically. However, if the belt breaks, depending on the design of the engine, there may be little or significant damage to the engine. If the engine is noninterference, it will have little or no damage because the valve stems and the piston will not run into each other when the belt breaks. Thus, all the owner needs to do is to change the belt. If the engine is not a noninterference type, there may be significant damage to the engine and its components. The proper design of the engine during the design process will result in significant cost savings later if the timing belt fails. ■

7.15 STYLING, SHAPE, AESTHETICS, AND PACKAGING

Styling and aesthetics relate to the shape of the product and the way it appears to the customer. The shape of the product (if displayed) or its packaging (if presented in its packaging) is the first thing that attracts the customer, even before price, quality, features, or brand name are seen or tested or considered by the customer. If you do not like the appearance of a product, no matter what the features are, what the brand name is, or what quality the product has, you will not be attracted to the product. Still, most designers, especially engineering designers, ignore the significance of it.

The aesthetics of a product are affected by many fundamental elements, including the relationship between the form and the function of the product, the combination of lines, colors, textures, and composition of the different parts of the product, the balance between the harmony and variety of the different parts of the product, proportions of the different dimensions, and its styling. But it is also affected by engineering elements such as the materials used, the sounds produced by the product, vibrations, and many others. Additionally, the safety of the product may be severely affected by the decisions that are made solely based on the aesthetics of the product. For example, if a sharp bend or curvature is made into a load-carrying part, due to stress concentrations caused by these, the part may fail.

Because of the importance of aesthetic elements on products, we will devote Chapter 9 to this subject. Please refer to that chapter for more information and discussion.

7.16 SAFETY

Safety of the product or system is an essential requirement. According to the law, any product or system offered for sale is presumed to be safe. This is called implied warranty, meaning that it does not have to be explicitly expressed by the manufacturer that the product is safe. It is automatically assumed. If there are any safety issues related to your product or system, you are required to address the concern and find a remedy. This requirement must be weighed against two other requirements, that introduction of safety devices to remedy a hazard should not render the product useless or change its utility to the point of rendering it useless, and that the price of the product should not increase due to safety devices to the point that the product will be noncompetitive. Otherwise, adding warnings will not replace needed safety equipment. For example, a knife is meant to cut. Eliminating this danger will make the knife useless. However, there is no reason why a product such as a table or stereo or other similar products should cut your finger. The function of a sander is to be abrasive. Removing this function from it will render the sander useless. Still, a belt sander must be designed with a good grip, the control switches must be placed in appropriate places for easy and immediate access in emergencies, and it must have adequate safety guards. Thus, the designer must evaluate the utility and the danger associated with the design and create a safe but useful product or system (in 2007, 811,000 Skil® brand circular saws were recalled due to a malfunctioning safety lock-out).

Of course, the severity of the consequences of the hazard, the probable frequency of it, and the seriousness of the consequences are very important matters. If the probability of something happening is very low, the need to remedy it and the level at which it must be remedied might be different than if the probability is high. Similarly, if the consequence of the hazard is a cut finger the needed remedy might be very different than if the hazard results in death or serious injury. For example, as was mentioned earlier, a blown bicycle tire is a hazard, but not the same hazard as a blown tire on an unstable automobile that may result in a rollover and death.

All rotating parts of a machine require a cover. All nip points require attention and safety devices to prevent accidents. The Consumer Product Safety Commission's literature has regulations for safety requirements for many consumer products. Other standards such as ASME Pressure Vessel Code, ASCE building code, City and County codes, and many others have established minimum required safety factors for many products and buildings. As the designer, you must consult with these codes and take their requirements and suggestions into account while a product or system is designed.

EXAMPLE 7.15

The Challenger space shuttle exploded 73 seconds after it was launched in January 1986 primarily because the o-rings used in the solid booster rockets had hardened due to very cold temperatures, allowing a blow-by and melting of the attaching strut that resulted in the disintegration of the spacecraft. There are hundreds of reports and position papers published on this disaster, blaming the management, government officials, engineers, and administrators. However, one interesting point related to the safety was the philosophy with which safety was approached. NASA had always maintained the policy that it was required to prove a launch was safe. Due to the particular circumstances related to this launch, NASA management changed their philosophy, requiring proof that a launch was not safe. To prove that a launch is safe or to prove that it is not safe is fundamentally different. Although engineers at Morton Thiokol (the manufacturer of the external solid booster rockets) had incomplete, but adequate, data to feel that the launch was not safe (and expressed this in their refusal to authorize launch), they could not prove that it was not safe, since adequate data were not available for the low temperatures of that day. The consequence was that Challenger and its seven astronauts perished. ∎

For more information and a discussion about responsibility of a designer and the liabilities associated with products and systems, please refer to Chapter 13. Please also see Chapter 8 for more information and discussion about human factors, safety, and associated consequences.

7.17   HUMAN FACTORS

Human Factors relate to the application of anthropometric data and principles as well as other human-related information to the design of products and systems in order to make the product or system safer, more comfortable, easier to work with, and more appealing. These include human capabilities and limitations, physical dimensions, psychological expectations, fatigue, learning curves, human–machine interfacing, and many more. Application of human factors to the design of products has been an important issue for decades, but it is becoming even more important today. With liability suits against manufacturers abounding, with more competition among products, and with higher level of product awareness and expectations among consumers, human factors play a more influencing role in the design. Due to the importance of human factors, we will devote a complete chapter to this subject. Please refer to Chapter 8 for more information and discussion about this subject.

7.18   ENVIRONMENTAL EFFECTS (ON THE DESIGN AND CAUSED BY THE DESIGN)

If you have studied engineering in the past, you may be familiar with a concept called *entropy*. Entropy is a measure of order that is present in a system. The more order a system has, the less its entropy. Nature always tries to reduce order in a system, thus increasing entropy. Humans increase order in systems and reduce the entropy. So, nature is in a constant battle with systems we humans build, trying to increase the entropy. As an example, have you ever seen a house naturally built with a particular order as we build them? Obviously not. We design and build a house by creating a particular order in the materials we use. However, even when we create a particular order by arranging parts in a certain order, eventually calling it a building, nature tries to destroy the order by any means it can, from the forces of nature to termites to oxidation to earthquakes. Similarly, have you ever seen a car built by nature? It does not occur naturally because it is against the principle of entropy. So, even if we create order by arranging all the parts of the car together and making a product that is capable of doing something useful, nature tries to destroy it. I am also certain that you have never experienced a situation in which oxygen and nitrogen, all mixed together in air, separate in a room, all the oxygen in one side, all the nitrogen in another corner, leaving you scrambling to locate the oxygen. That is because even if we were to forcefully separate the oxygen and nitrogen in a room, and thus, giving it some sort of order, nature would destroy the order by mixing it again. Leave a sugar cube or some salt in a glass of water overnight and see what entropy does to that order. The same is true for an ice cube in hot water. It will not last, as the artificial order will disappear by the exchange of heat between the two to reduce the order. Even mountains are washed off and boulders are broken by expanding freezing water and earthquakes. All living systems, including plants, animals, and humans, fall under the same category. We are all progressing toward increased entropy (less order). Although discussing the details of entropy, even as fascinating as it is, is beyond the scope of this book, let it suffice to say that entropy governs the life of all systems and tries to reduce the order through a number of mechanisms that are at its disposal. This includes oxidation, effects of ultraviolet (UV) rays, forces of nature (such as the expansion of volume of water as it freezes), and the biological clock in our DNA.

This shows the reality of nature. Thus, whenever we design and create a system by creating order, we need to consider the effect of nature (environment) on it. Metals that oxidize easily must be protected by paint or other means, which only increases their life incrementally. Plastics and similar materials fade, decompose, or deteriorate due to weather and sunlight. Even the paint used to protect other products (whether a car, a house, cabinets, a deck, or anything else) is itself subject to entropy and can fade, deteriorate, or chip (due to the effect of moisture, freezing, expansion, or UV). It is the designer's responsibility to consider the effects of environment on products, in any form, and to provide for an appropriate and adequate defense against nature for the expected life of the product. Nothing will last forever, but we can protect the product for a longer, if limited, period of time.

The opposite of this is also true. Many products have an effect on the environment as the environment has an effect on the product, usually called pollution. However, pollution can have many forms, including solid waste pollution, air pollution, water pollution, auditory (sound) pollution, and even what is called visual pollution and social pollution.

In general, a product or system can have a negative effect on the environment during the life of the product, from cradle to grave (or cradle to cradle if it is recycled), as follows:

- *Birth of the product, when it is manufactured, built, or installed.* Many manufacturing processes create solid, water, or air pollution. This includes leftovers, parts that are below standard and are discarded, scrap materials, toxic material that is generated during manufacturing, chemicals, solvents, lubricants, and many others.

When the designer chooses a process that creates less waste, less scrap material, and requires fewer processes, chemicals, and other nonessential or indirect materials, not only the cost of production, but also the environment will be positively affected.

Figure 7.4 shows a drain spout for rain located right above a light fixture. Obviously, the architect did not consider the location of these two relative to each other, and the consequences of this placement. To remedy the problem, a piece of flexible rubber hose was added to the drain to divert the water from the light fixture. Does this not look familiar?

FIGURE 7.4   The placement of a rain drain above a light fixture was a mistake that had to be solved by the addition of a rubber hose to the drain. Should it have been placed there?

EXAMPLE 7.16

Nylon-6 is a durable, soft, and strong artificial fiber that is used to make many products such as socks, curtains, and many types of cloth. During the manufacture of Nylon-6, many chemicals (some toxic) are produced as byproducts that must be disposed of. Additionally, during the process of extruding the fibers, due to a variety of reasons including adjustments, large globs of nylon material are wasted which cannot be reused for extrusion. This material is either thrown away, or at times, it may be used for making other products such as air filters or padding.   ■

- *Life of the product, when the product is used by the owners.* This includes many different types of pollution, including operational byproducts such as air, water, or other waste pollution, but also indirect pollution as a result of deteriorating materials, defects, lack of adjustments, and damage. A car's engine may pollute the air all its life if not designed or manufactured properly. The Dodge Ram 2004 pickup truck is rated at about 9 mpg in the city and 12 on the highway. Hybrids get 50–60 mpg in the city. But also because it is possible that the engine may go out of adjustment without warning and pollute even further without anyone noticing until it is checked for other purposes. Many wood products are treated with chemicals for different purposes. These chemicals leak into the air as time goes by, creating many ill effects, from unpleasant odors to sickness. Lead-containing paint was used in many products for decades, making thousands of babies sick when they chewed on their cribs and toys.

• *Death of the product, when the product is no longer used and is discarded.* Many products are built to last a long time. However, if they do not work properly, or are not needed, they are discarded. These discards usually end up in a dumpsite or landfill, where, covered by layers upon layers of other material, with no sunlight and no moisture, they can last hundreds of years. According to one report,[2] trash bags may last 30–100 years, magazines 4–10 years, disposable diapers 20–80 years, plastic bottles and Styrofoam 100+ years, cardboard 3–12 years, and tin cans 20–50 years. In this case, the volume of solid waste keeps increasing. On the other hand, when they do deteriorate, these products may generate yet more toxic pollutants or other unwanted materials that harm the environment. With the widespread popularity of one-time use products (such as plastic ware, food packaging, papers and cardboards, cameras, cleaning items, juice boxes, water bottles, and soda cans), this has become even a more serious problem in recent years. A new law, taking effect in California in 2006, makes it illegal to throw away electronic waste (e-Waste), including batteries, cellular phones, fluorescent lights, printers, VCRs, answering machines, computers, monitors, and other household products containing hazardous materials into the garbage. These articles must be taken to collection centers that are equipped to handle the potentially hazardous material contained in them.[3]

• *Recycling of the product, both during collection, separation, and reuse.* Recycling plants generate their own pollutants and byproducts as well. However, even reuse of many recyclable materials creates further pollutants. For example, many products have painted parts. Recycling these parts may require removal of the paint (which requires solvents) or burning off the paint (which creates many toxic fumes and may affect the material of interest). Additionally, many parts of machines are built from different materials that are integrated into one assembly (e.g., an electric motor) that cannot be recycled together, making it necessary to disassemble or separate the parts. As a result, it becomes imperative to think ahead about the question of disassembly of the product for reuse. We will discuss this later.

The more common pollutions such as solid waste, water pollution, air pollution, and global warming are generally well understood by most designers. However, another important issue regarding effects of the product on the environment is the wider definition of environment, not just the nature, but also the environment in which we live. As was mentioned earlier, we refer to this as auditory (sound) pollution, visual pollution, and social pollution.

Please also see Section 7.18 on a discussion about disposability, as well as Section 7.20 on a discussion about designing for disassembly.

Auditory pollution refers to the effects that a product may have on the people around it through sounds and noise. Many products create excessive sounds, sometimes deliberately, other times because they are not designed properly. In some cases, such as in jet engines and cars, quieter products are either required by industry standards or by governmental agencies, and as a result, much is done to make these products quieter. In other cases, products are not regulated, and as such, they may produce too much noise. Many industrial machines are in this category. The result is high noise pollution in factories, requiring that the workers wear protective earmuffs or other noise reduction products, or else be exposed to too much noise and run the risk of hearing damage or loss. If the machinery used in factories was designed to be quieter, the risk would be eliminated and the excess cost as well as the discomfort of wearing protective products would be eliminated. On the other hand, many products are designed to create excessive noise, even if the noise is detrimental to hearing. This includes exorbitantly powerful sound systems used in concerts and in cars. The fact that we can produce such high output sound systems is one thing, using it inappropriately without regard to the damage it causes is another. Infant toy rattles have been recalled for the danger they pose to babies due to the risk of breakage and small parts that can be swallowed, but not because they generate as high as 95 dB noise near the infant's ears. We all have seen cars with powerful sound systems that can be heard from hundreds of feet in busy streets, loud muffler noises, and construction noises. In some restaurants, loud music is used to entice customers to come in for the excitement of the atmosphere, and in others they are used to drive the customers out in order to replace them with new ones at the tables after the initial excitement is replaced by frustration. Lawn mowers, fans (such as bathroom fans), motorcycles, leaf blowers, and many other products are the same. More information about the effects of loud noises on human hearing is presented in Chapter 8.

Social pollution includes the effects that a product may have on the society, its expectations, or its behavior. After many years of popular acceptance of cell phones by all people around the world, it is still strange to see a person, say in a bus or on the street, talking, apparently, to no one. In the past, if someone would claim telepathic abilities, he would be considered insane. Many ended up in mental institutions for talking to invisible others. On the one hand, this has changed the expectations and the behavior of the users, and on the other hand, has created a certain level of acceptance about a behavior that used to be considered a mental problem until a few years ago. Still, both cell phone and other devices such as personal audio systems, do isolate the user from their environment. The individual using these products usually loses contact with the people around him or her, including while driving; she or he misses clues about the people and situations such as emergencies, and lacks attentiveness to what is said to them. All these, although minor in general, do have a social impact that is artificial, primarily caused by the presence of the product.

Visual pollution is the effect of products on the collective beauty of the environment as we see them. You probably have seen the chaotic display of signs and billboards on commercial streets. Most people dislike this array of protruding and sometimes obnoxious signs with no harmony or uniformity in them. Although commercially valuable and perhaps necessary, the signs do not look good. However, there is no reason why they cannot be beautiful and pleasing, blending with the environment without negatively affecting it. Visual pollution, unlike other forms of pollution, is more subjective and cannot be measured or proven one way or another (Figure 7.5). Every person may have a different opinion about the standards of beauty and ugliness. What should concern the designer is whether there is anything that must be done to minimize the negative visual effects of the product on its environment. One example of this is when, during the remodeling of a building in downtown San Luis Obispo, an artificial façade made up of a painted cloth was hung in front of the building, covering the construction work behind it in order to retain the beauty of the downtown area. Similarly, boarded sidewalks during construction have been painted to reduce the negative effect of several tens of 4 by 8 boards on the street. Many other examples are presented in Chapter 9 when we discuss aesthetics of design.

(a)   (b)

FIGURE 7.5 Examples of temporary visual pollution due to a construction project. However, this could be made much nicer by adding colors or pictures to the wood panels.

7.19 DISPOSABILITY

As part of the cradle-to-grave cycle of life of a product, and sometimes as required by law, it is extremely important that the designer also consider what will happen to a product at the end of its life, both because of regional and international laws and regulations that govern the disposal of many products, as well as the effect of the product on the environment and the limited ability of consumers to discard or recycle many items. For one thing, many products such as computers and computer monitors may no longer be disposed of by discarding them into the trash, and thrift stores or dumpsites no longer accept them. There is a large glut of these used and obsolete products around the world with no apparent solution. These products are toxic to the environment if improperly disposed of, and are not designed to be easily recyclable. No one needs them anywhere either. What is the user supposed to do?

In September 2003, California enacted its e-Waste Collection and Recycling Act. This act requires manufacturers to phase out toxic and dangerous substances in their equipment by 2007. As mentioned above, the e-Waste ban, instituted by the California Department of Toxic Substances Control went into effect in February 2006. This ban makes it illegal to toss used batteries, cellular phones, fluorescent lights, and other household hazardous material into the garbage[3] and requires users to take such material to local centers that are specially equipped to handle these potentially toxic agents. This law also requires communities to establish some kind of household waste collection system for these products. The ban includes products that contain mercury or heavy metals, including all household batteries, mercury thermometers, printers, VCRs, answering machines, radios, and computer and television monitors. e-Waste often contains valuable materials such as copper, gold, and zinc that can be recycled. Other heavy metals such as lead, mercury, and cadmium can pose a threat to public health and the environment. Considering that a color monitor may contain as much as 4–5 lb of lead as well as other toxic substances, and a computer can contain mercury, cadmium, and hexavalent chromium, it is no wonder that their disposal is an environmental issue.[4] Starting July 1, 2006, any retailer that sells cell phones will also be required to collect used cell phones at no charge, and most retailers that sell rechargeable batteries will also collect used ones at no charge. According to the Environmental Protection Agency, Americans have about 500 million obsolete, broken, or otherwise unused cell phones, and about 130 million more are added each year. Less than 2 percent are recycled.

There are also many international requirements for toxic material disposal. According to the Reuters news agency, in December 2001, Dutch authorities seized more than 1.3 million Playstations[TM] made by Sony Corporation because they contained too much cadmium in the console's cables. [4] You must make sure your products abide by international laws and regulations. A program called PLM, developed by Agile Software Corporation is supposed to include a module for product governance and compliance to keep track of environmental regulations and policies that affect an engineering company's products.

As mentioned earlier, one policy that has been considered is to force each manufacturer (or retailer) to take back what they make (or sell) when the useful life of the product is over. Although this is economically difficult to enforce, it may actually be the ultimate solution to force the designer and the manufacturer of products to address the issue of disposability during the design phase.

Additionally, more and more products are designed and made to be used once, including packaging of food items, plastic ware, paper products, cleaning tissues, and cameras. Just to see the extent of the effect of this issue, compare an apple with a carton of apple juice. Consider the effects of eating a fresh apple, its nutritional value, its fiber, how it cleans and exercises your teeth as you bite into it, what is disposed of after the apple is eaten, and its effect on the environment, in contrast with the use of the juice, the process it goes through for preparation (extraction, pasteurization, concentration, transportation, reconstitution, packaging, etc.), the nutritional value of the juice, comparative cost, total caloric input, and finally the disposal of the carton, especially that the carton is made to last a very long time in the presence of moisture. Now extend the same idea to many other disposable products and the appropriateness of their design or application.

Figure 7.6 shows a simple, disposable, paper cup that is made in the shape of a simple pocket. Even though the use of paper cups in some situations may be inevitable, a simple innovative design can reduce the detrimental side effects of the disposable item on the environment by reducing the magnitude of the material and its nature. Here, the amount of the material used for this paper pocket is much less than a regular paper cup, it requires much simpler manufacturing, and is at least friendlier than Styrofoam or waxed paper. It also requires much less space.

FIGURE 7.6 A simpler pocket-paper cup can be friendlier to the environment because it requires less paper, simpler material, and simpler manufacturing. It also requires less space.

7.20 SUSTAINABILITY

Sustainability, although an old concept that has always been a design issue, has nowadays become a hot topic and is discussed in many different circles. Sustainability is no longer just an issue related to the design of products and systems, but also to any entity (a household, a university, a company, a city, a country, or even the world). In 2006, it is estimated that the ''sustainable lifestyle'' is a $230 billion industry in the United States alone. This includes hybrid vehicles, ecotravel, organic food, solar and wind power, sustainable buildings, and other similar products and systems.

In a nutshell, sustainability is the answer to the question of whether the system under consideration can be sustained (indefinitely). Can the needs of today be met without compromising the future? For example, can the energy resources available to the world markets be sustained (go on) as they are, and if so, at what cost, and for how long? If not, what needs to be changed to make it sustainable? Can a household be sustained with the income and expenditures involved, and if so, how? If not, what has to be changed in order to make it sustainable? Can the production of computers and their disposal be sustained the way it is now? What has to be changed in order to make the system sustainable?

Sustainable systems use the least amount of resources, but also return what is taken to its source. If a new tree were planted for every tree cut down, the forests would possibly remain sustainable. If everyone used fluorescent light bulbs, the existing power plants could be sustained for much longer without the need for new resources. If most houses used solar energy for heating (as well as producing electricity), much less energy would be needed to sustain the status quo. If most (or hopefully all) parts of a product were recycled, the system might remain sustainable. If a system would use less and fewer resources, the system might be sustained much longer. Each one of these remedies could contribute to the sustainability of the whole system. In the concept of ''cradle to cradle,'' the assumption is that most of the parts of any product would be recycled. The European Union has a mandate regarding End of Life of Vehicles (ELV) that requires manufacturers to take back the vehicles they manufacture. Due to this requirement, more and more auto manufacturers are designing their cars with recyclability as a goal. This may involve millions of cars per year, thus sustaining the system a bit longer. And if the scraps from one manufacturing process were used as raw material for another, fewer new resources would be needed to manufacture goods.

EXAMPLE 7.17

Nylon-6 fibers are used in the manufacture of hosiery, curtains, and a variety of cloths. During the process of manufacturing the fibers, much waste is produced when the extruding machines malfunction, or when the fibers are correctly made but are not wrapped around a spool. Two ideas were suggested for the reuse of the waste as raw material for new products. In one, a tray would be placed under the machine where the wasted fibers would be dumped to form 1/2-inch thick panels that could be used as lightweight ceiling panels. In the other, the fine fibers would be collected and used to make air filters for heating systems. In both cases, the material that was wasted would become raw material for another product. ■

EXAMPLE 7.18

In many areas of the world, especially in arid areas, fresh (versus ocean) water is scarce, and therefore precious. Many countries have fought many a battle to preserve their water resources. However, much fresh water is wasted all the time without any attempt for its collection and reuse (such

FIGURE 7.7   City of San Luis Obispo's sign for recycled irrigation water.

as in household reverse-osmosis filters, where 4–5 times as much water as filtered is wasted). Similarly, a huge amount of drinking water is used for watering shrubbery, greenery, and irrigation.

Now consider the following. San Luis Obispo County's average per annum rainfall is about 23 inches of rain, which is not too bad, except that most of it falls between December and March, without a drop during the rest of the year. A typical house of 1500 ft^2 area has at least the same amount of roof. If the rain from the gutters were diverted into a drain pipe and collected in an underground tank with a pump, one household could conceivably collect as much as 3000 ft^3 (about 22 000 gallons) of fresh water. Assuming that one would want to water plants and shrubbery for the remainder of the year (240 days), this would yield over 90 gallons per day. There is no question that installing an underground tank will cost money. It will also be necessary that a pump be installed over the tank for pumping the water. However, this would make the household completely sustainable for irrigation and gardening. In fact, the City of San Luis Obispo now irrigates large portions of the city's shrubbery with recycled grey water from sewage treatment plants (see Figure 7.7).   ■

One of the main concerns about sustainability is population growth. As reported in 2006, every 7 seconds a baby is born, every 13 seconds someone dies, and every 31 seconds a new immigrant enters the United States. This means that every 2820 seconds, the U.S. population increases by 227, equal to 8,500 per day, or 3.1 million per year. This is over 1 percent net increase. As the world's population grows, more resources are needed to sustain the population. Additionally, as the standard of life increases, even more resources will be needed to sustain the increase in expectations (a big part of increase in gasoline consumption in the United States comes from larger and more powerful engines in SUVs, cars, and trucks). These resources include basic and fundamental necessities such as food, shelter, transportation, and basic health requirements, but also other necessities that everyone expects, including education, health care, representation, respect, entertainment, vacation, travel, and retirement. The more advanced countries have so far been able to better serve these needs, perhaps by somewhat sacrificing certain elements of the future. Other countries and smaller societies are not able to provide for this increase in needs and expectations. Many countries have run into food shortages that are exacerbated by increased populations. However, will it be sustainable to have food donations from other countries pour into these communities, and if so, for how long? When health care improves, mortality rates decrease and populations increase. However, the additional population requires more housing, more education, more food, and more resources, negatively affecting the environment. How long will this be sustainable?

There are certain resources around us that are not fully tapped or used. For example, wind energy, wave energy, solar energy, hydrogen cells, and nuclear energy are some examples of energy sources that are not yet fully utilized. In fact, nuclear energy was much more popular in the past, but due to environmental concerns, no new nuclear plants have been built in the United States for years. However, one has to consider the alternative, the fossil fuels, and the consequences of relying on these resources alone, and ascertain whether or not these alternatives are sustainable. In 2006, as part of the Sustainability Campaign, Cal Poly allowed SunEdison Company to install 1,000 solar panels, totaling 14,000 ft^2, on top of the flat roof of Engineering-West building. Although most of the power generated is sold by the company, it also provides for 25 percent of the power needed by the building, saving over $5,000 per year. But more importantly, a roof area that laid useless is now used to produce renewable energy.[15]

Sustainability must be assessed against the needs of the consumers as well. In certain cases, the immediate or long-term needs of the consumers may be more important than immediate or long-term sustainability. However, in all these cases, the fundamental issues that create the unsustainable conditions must also be addressed. For example, the immediate needs for providing food for a country's population in many areas of the world may be more important than saving the forests. However, deforestation cannot be sustained indefinitely. Thus, the underlying cause might be overpopulation and what causes it rather than deforestation itself. Similarly, during a war, the immediate needs of an army may require certain sacrifices that are not sustainable indefinitely. However, the leaders of the governments involved should consider the root cause of the war and try to address those problems. The same can be extended to family concerns, products, and other systems.

7.21 ASSEMBLY/DISASSEMBLY

Methods of assembly are heavily intertwined with the way a product is designed, the choice of the materials used, and the methods of manufacturing selected. Most products and systems are in fact collections of other subassemblies and systems that must be assembled into an integrated unit. Many manufacturers such as Toyota make few parts themselves; they are system integrators more than parts manufacturers. Other manufacturers and subcontractors make all the systems and subassemblies. Toyota just assembles those parts and subsystems into automobiles. Similar to the methods of manufacturing, the choice and the method of assembly, the way each component is designed, and the total number of assembly operations that is needed will determine the cost of assembly. To this you must add the effect of the cost of labor at different skill levels, the number of individuals needed to complete the assembly, and the time that it takes to finish the job. In industry, this is called Designing for Assembly or DFA. Please refer to Chapter 10 for discussion about DFA and how this is accomplished using DFA programs. Please also see Chapter 11 for the relationship between the economics of design and how it is affected by the cost of assembly.

The other side of this issue is designing for disassembly (DFD). Disassembly of a product relates to the need for maintenance and repair, but also to the issue of recycling and disposal of the product, which we have already discussed above. If you think about natural products and systems such as trees and domestic animals, we have found techniques and products to enable us to use perhaps 100 percent of most natural products in different forms and completely recycle the remaining with little waste. Trees, in addition to their use for shade, for their CO_2/O_2 exchange, their fruits, leaves, sap, their beauty, and countless other attributes while alive, provide countless other benefits to us when they are dead. Their disposal means that we use the wood, the wood shavings (in the form of other products such as veneer, plywood, particle boards, etc.) the bark, and the roots. An animal such as a cow is the same. It is used in countless ways while alive, and still, we use all the body parts in many other ways when dead, including the meat, the skin, the hair, the horns, etc. In fact, we should attempt to design our manufactured products to be as close to 100 percent recyclable as possible. This means that while a product is designed, the designer should attempt to foresee the required motions and actions that are needed to enable others to disassemble the product for repair or disposal as easily as possible. There are many techniques and software programs[5,6] available that allow a designer to include safeguards for easy disassembly. For example, whenever possible, non-compatible products should be attached in such a way to enable a disassembler to easily separate them for recycling into appropriate bins. This may be as simple as using tabs that can easily be broken off, to the application of materials that are all similar and do not need disassembly. If a product can be disassembled easily, it will be much more feasible to expect that recyclers will sort the materials used in the product and recycle them.

Obviously, there is a difference between disassembly for repairing and disassembly for recycling. After repairing, the product must be reassembled, and in this process, parts should not be damaged. In disassembly for recycling, parts may be broken or forced out without concern. However, the design methodology for both is similar and aids the designer in both improving serviceability of a product as well as its recyclability. BMW had designed one of its 1990s models such that it could be disassembled in a few minutes. The down side of this attempt (and the designer should also be aware of these social issues) was that thieves could also disassemble the car in record times.

It is always easier to disassemble than to assemble, even with living organisms, including humans. One can easily disassemble a human body or a tree by operating, cutting, chopping, or separating the parts. Assembling them is much more difficult, and in the case of living things, perhaps yet impossible. However, the fact that it is easier to disassemble should not mean that designing for it is trivial. Unless safeguards for proper disassembly are planned into the product during the design, it may simply be ignored when the product's life is over.

Please refer to the list of References 5–10 for more information and resources on this subject.

If you have never dissected a machine or product before, I suggest that you find (or buy) a product that is discarded and dissect it. This "hands-on" activity will teach you many different things. For one, you will learn how the machine works, how its different parts are designed to work together, how different subassemblies are manufactured and put together, etc. Two, you will learn how easy or difficult it might be to disassemble the machine, and whether or not you can put it back together without much difficulty. Third, you will learn that if the machine is relatively complicated, it will be necessary to somehow keep track of the order of assembly of the parts and their orientations. For example, if a machine has many different types and sizes of screws, you will quickly learn that you somehow have to mark these screws; otherwise you will forget which screw belongs to what part or location. Ultimately, a good practice is to also sketch the parts and describe their functions.

7.22 LIFE EXPECTANCY

Life expectancy relates to the planned and expected life of a product and is affected not only by usage, maintenance, and the way the product is treated by the users, but also by the quality of its design and manufacture and the type and quality of its components. It is also affected by the type of loads that the product is subjected to. For example, the life of a component can be significantly different depending on whether the loads are static or dynamic (fatigue loading). Life expectancy is also affected by the type of material selected and the way it is protected against wear and deterioration. For certain products, the life expectancy is built into the product, and in such cases, it is expected that the product will start to deteriorate and become useless during the planned life. These components are designed to have a finite life. Some are inherently short lived, and yet others may deteriorate due to natural environmental factors and general wear and aging.

Before we discuss the issues that are affected by life expectancy, let's consider one important engineering subject that relates to the life expectancy of many products. This will clarify the meaning of planned life expectancy. We will have a more in-depth discussion of this subject in Chapter 16, where we will study engineering analysis of fatigue loading. However, the following is a short introduction to help us understand this subject better.

Fatigue Loading and Planned Life Expectancy One general characteristic of a machine versus a structure is moving parts; machines have moving parts, structures do not. A general characteristic of moving parts is varying loads and fatigue phenomenon. Loads, either forces or torques (or moments) are very important subjects in the mechanical design of most products. Every product that has substantial loads compared to the size of the elements that carry the loads must be designed properly to carry the loads. Otherwise, these loads have the potential to break the components and destroy the product. For example, suppose that there is a large load on a shaft connected to a motor. If the shaft is not strong enough to carry the load, it may bend, deform, or break. Even if the load is small, if the diameter of the shaft is also very small, the load may still damage (bend, deform, or break) the shaft. If the load is generally constant, as in the loads in a static structure, what is used to design the part is the strength of the material (e.g., the yield strength S_y). As long as the stress in the part, multiplied by a safety factor, is less than the yield strength, the part will be able to carry the load. Then, in a very simplified case:

$$S_y \geq k\sigma \tag{7.6}$$

where S_y is the yield strength, k is the safety factor desired, and σ is the stress in the element. So, the design engineer will compare the stresses caused by the loads with the strength of the materials used in the product in order to come up with the size of the parts capable of carrying the loads. However, in many cases such as the wings of an airplane or on a rotating shaft, the loads are not constant, but vary over time. If you imagine the airplane on the ground, the load on the wings is due to gravity, and therefore, is downward. As the plane becomes airborne, the load is due to the higher air pressure on the bottom surface of the wing, and therefore, is upwards. Every time the plane lands or takes off, the load changes. In the case of a rotating shaft too (Figure 7.8), imagine that the shaft is pushed down due to a downward load

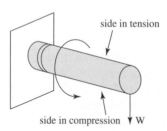

side in tension

side in compression W

FIGURE 7.8 Fatigue loads on a rotating shaft.

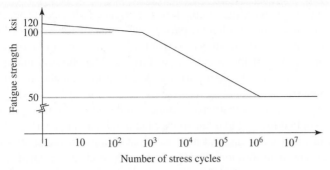

FIGURE 7.9 Fatigue strength for UNS G41300 steel, normalized. (Data from NACA Technical Note 3866, December 1966.)

at any given instant. This means that the material on the bottom of the shaft will be in compression and the material at the top will be in tension. As soon as the shaft rotates 180°, these points on the shaft will be on opposite sides and the stresses on them will switch from tension to compression and compression to tension. As the shaft continues to rotate, the load continually changes from tension to zero, to compression, to zero, and so on. This is called *fatigue loading*. In these situations, the calculation of the required size of the part is not based on the yield strength of the material, but on what is called fatigue strength (or endurance limit). *Fatigue strength* is a function of the strength of the material and the number of cycles the load changes. For steel, the fatigue strength decreases as the number of repeated loads increases, until about 10^6 cycles, at which point it remains constant (Figure 7.9). This means that for steel there will be a need to increase the size of the part as the number of repeated load cycles increases until this limit (endurance limit). Otherwise, the part will have a finite life. It is possible, for example, to design a part to last a finite time, say 100,000 cycles. For other materials, including aluminum, there is no endurance limit; the larger the number of load cycles, the lower the fatigue strength (this means that parts made from materials such as aluminum will eventually fail due to fatigue, whereas parts made from steel can be designed to have "infinite" fatigue life). This also means that when the designer designs a product, the life expectancy of the product should be used to estimate the number of cycles of load that will be applied to the product. Based on the type of material and the fatigue strength, the designer determines the size of the parts. Obviously, if the product lasts beyond this life expectancy, chances that the parts of the product will fail increase accordingly. This is one of the ways life expectancy is planned for a product by the designer.

Figure 7.10 shows a bicycle pedal crank that was broken due to fatigue. The part is made from cast and stamped aluminum. Similar to a shaft, as the pedal was pressed

FIGURE 7.10 An example of a part broken due to fatigue loading.

during the down- and up-strokes, the crank stem experienced alternating tension and compression. Exactly at the point where a part of the name of the product (mighty) was pressed into the stem, due to stress concentration, a crack developed which resulted in premature failing of the crank due to fatigue. The chevron marks can be seen on the cross section of the broken part too. We will discuss the actual fatigue mechanism in Chapter 16.

There are other similar parts that have a finite life dependent on the type of load, number of cycles of applied loads, and other similar factors, including bearings, belts and chain drives, materials that wear out, and materials that change characteristics through hardening, softening, and deterioration. The total life of a product is related to the life expectancy of these components and the way they are used in the design.

The life expectancy has a strong relationship with the economy of the product, as will be discussed in Chapter 11. These issues include resources and their availability, obsolescence, product improvements, and repeat (future) business. For example, a toy needs to last relatively a short time, since most children do not play with their toys for more than a short time, but still needs to be of an adequate quality to be safe. Computers also need to last a relatively short time, since they quickly become obsolete as the state of the art advances. Thus, a computer with a life expectancy of 15 years will be completely useless in much shorter time. However, when discarded, due to its higher quality and because it is designed to last a long time, it will also last longer in a dumpsite. A car that is designed to last 30 years may also be excessive, since many advances in automobile technology render the car less useful as time goes by, reducing its utility. For example, many cars of the 1960s did not have a seat belt or a headrest, let alone air bags, since they were not required. Not only would they be dangerous today, they might even be illegal. Many cars did not utilize any air pollution control sensors or devices either, as these were not required at the time. Therefore, a limited life expectancy would force these cars out of the market and would result in new and better cars.

On the other hand, as every new product is built, it requires new resources, including new materials and energy for conversion and manufacturing, and it affects the environment. The more new products we make, the more resources we need, unless we recycle the old materials. Furthermore, if all products would last a long time, and if they would maintain their utility throughout their life, few people would buy new products, severely limiting the vitality of the economy, job market, and livelihoods.

Considering all these, the designer should decide what the expected life of the product or system should be and design the product or system accordingly by planning the life of the components and the quality of the product.

7.23 ETHICAL ISSUES

The National Society of Professional Engineers (NSPE) has developed a Code of Ethics for Engineers, shown on pages 248-249, that governs the ethical behavior of engineers. Similar codes of ethics exist for other professions, including but not limited to, physicians. Designers, whether in engineering, industrial design, architecture, or other professions, should read, understand, and regularly consult the Code whenever necessary, in order to ensure that their decisions are ethical.

One of the fundamental ethical issues for engineers is public safety. Engineers are *to hold paramount the safety, health, and well-being of the public*, a Fundamental Canon.

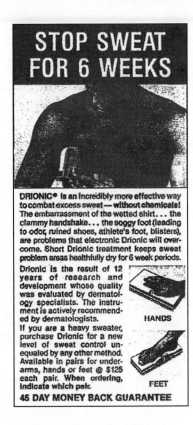

FIGURE 7.11 An advertisement.

As we have already discussed, many ill-conceived decisions in industry have resulted in disasters, from the Challenger and Columbia disasters to the Union Carbide Bhopal plant in India, and to the Ford Pinto and Explorer problems. However, sometimes, even minor cases create an ethical dilemma. For example, Figure 7.11 shows an advertisement for a product that stops sweat for 6 weeks by effectively paralyzing sweat glands. Should a designer create a device to prevent sweating, a natural phenomenon, for unsuspecting consumers for profit?

Obviously, there are many other ethical concerns as well, including conflict of interest, transparency, competence, and unfair competition, which the designer must consider. The National Society of Professional Engineers (NSPE) periodically publishes recent cases for which their Board of Ethical Review has had to make judgments about ethical issues involving member engineers.[16]

No doubt there are many different points of view, values, and standards applied to any situation that poses an ethical dilemma. Great philosophers, throughout history, have discussed ethics, have developed theories, and have argued their points of view as to what people should or should not do. Obviously, this book is not an appropriate outlet for the philosophical arguments and theories on ethics. However, the professional codes of ethics relate to the basic responsibilities of the members of the profession, and in general, are not disputed by anyone. Therefore, it is imperative that designers follow their professional codes of ethics in order to ensure that their decisions are supported by their peers. These codes can serve as a guiding light for all engineers to decide whether or not the decisions they make are ethical.

Code of Ethics for Engineers

Preamble

Engineering is an important and learned profession. As members of this profession, engineers are expected to exhibit the highest standards of honesty and integrity. Engineering has a direct and vital impact on the quality of life for all people. Accordingly, the services provided by engineers require honesty, impartiality, fairness, and equity, and must be dedicated to the protection of the public health, safety, and welfare. Engineers must perform under a standard of professional behavior that requires adherence to the highest principles of ethical conduct.

I. Fundamental Canons

Engineers, in the fulfillment of their professional duties, shall:

1. Hold paramount the safety, health, and welfare of the public.
2. Perform services only in areas of their competence.
3. Issue public statements only in an objective and truthful manner.
4. Act for each employer or client as faithful agents or trustees.
5. Avoid deceptive acts.
6. Conduct themselves honorably, responsibly, ethically, and lawfully so as to enhance the honor, reputation, and usefulness of the profession.

II. Rules of Practice

1. Engineers shall hold paramount the safety, health, and welfare of the public.
   a. If engineers' judgment is overruled under circumstances that endanger life or property, they shall notify their employer or client and such other authority as may be appropriate.
   b. Engineers shall approve only those engineering documents that are in conformity with applicable standards.
   c. Engineers shall not reveal facts, data, or information without the prior consent of the client or employer except as authorized or required by law or this Code.
   d. Engineers shall not permit the use of their name or associate in business ventures with any person or firm that they believe is engaged in fraudulent or dishonest enterprise.
   e. Engineers shall not aid or abet the unlawful practice of engineering by a person or firm.
   f. Engineers having knowledge of any alleged violation of this Code shall report thereon to appropriate professional bodies and, when relevant, also to public authorities, and cooperate with the proper authorities in furnishing such information or assistance as may be required.
2. Engineers shall perform services only in the areas of their competence.
   a. Engineers shall undertake assignments only when qualified by education or experience in the specific technical fields involved.
   b. Engineers shall not affix their signatures to any plans or documents dealing with subject matter in which they lack competence, nor to any plan or document not prepared under their direction and control.
   c. Engineers may accept assignments and assume responsibility for coordination of an entire project and sign and seal the engineering documents for the entire project, provided that each technical segment is signed and sealed only by the qualified engineers who prepared the segment.
3. Engineers shall issue public statements only in an objective and truthful manner.
   a. Engineers shall be objective and truthful in professional reports, statements, or testimony. They shall include all relevant and pertinent information in such reports, statements, or testimony, which should bear the date indicating when it was current.
   b. Engineers may express publicly technical opinions that are founded upon knowledge of the facts and competence in the subject matter.
   c. Engineers shall issue no statements, criticisms, or arguments on technical matters that are inspired or paid for by interested parties, unless they have prefaced their comments by explicitly identifying the interested parties on whose behalf they are speaking, and by revealing the existence of any interest the engineers may have in the matters.
4. Engineers shall act for each employer or client as faithful agents or trustees.
   a. Engineers shall disclose all known or potential conflicts of interest that could influence or appear to influence their judgment or the quality of their services.
   b. Engineers shall not accept compensation, financial or otherwise, from more than one party for services on the same project, or for services

pertaining to the same project, unless the circumstances are fully disclosed and agreed to by all interested parties.
   c. Engineers shall not solicit or accept financial or other valuable consideration, directly or indirectly, from outside agents in connection with the work for which they are responsible.
   d. Engineers in public service as members, advisors, or employees of a governmental or quasi-governmental body or department shall not participate in decisions with respect to services solicited or provided by them or their organizations in private or public engineering practice.
   e. Engineers shall not solicit or accept a contract from a governmental body on which a principal or officer of their organization serves as a member.
5. Engineers shall avoid deceptive acts.
   a. Engineers shall not falsify their qualifications or permit misrepresentation of their or their associates' qualifications. They shall not misrepresent or exaggerate their responsibility in or for the subject matter of prior assignments. Brochures or other presentations incident to the solicitation of employment shall not misrepresent pertinent facts concerning employers, employees, associates, joint venturers, or past accomplishments.
   b. Engineers shall not offer, give, solicit, or receive, either directly or indirectly, any contribution to influence the award of a contract by public authority, or which may be reasonably construed by the public as having the effect or intent of influencing the awarding of a contract. They shall not offer any gift or other valuable consideration in order to secure work. They shall not pay a commission, percentage, or brokerage fee in order to secure work, except to a bona fide employee or bona fide established commercial or marketing agencies retained by them.

III. Professional Obligations

1. Engineers shall be guided in all their relations by the highest standards of honesty and integrity.
   a. Engineers shall acknowledge their errors and shall not distort or alter the facts.
   b. Engineers shall advise their clients or employers when they believe a project will not be successful.
   c. Engineers shall not accept outside employment to the detriment of their regular work or interest. Before accepting any outside engineering employment, they will notify their employers.
   d. Engineers shall not attempt to attract an engineer from another employer by false or misleading pretenses.
   e. Engineers shall not promote their own interest at the expense of the dignity and integrity of the profession.
2. Engineers shall at all times strive to serve the public interest.
   a. Engineers shall seek opportunities to participate in civic affairs; career guidance for youths; and work for the advancement of the safety, health, and well-being of their community.
   b. Engineers shall not complete, sign, or seal plans and/or specifications that are not in conformity with applicable engineering standards. If the client or employer insists on such unprofessional conduct, they shall notify the proper authorities and withdraw from further service on the project.
   c. Engineers shall endeavor to extend public knowledge and appreciation of engineering and its achievements.
3. Engineers shall avoid all conduct or practice that deceives the public.
   a. Engineers shall avoid the use of statements containing a material misrepresentation of fact or omitting a material fact.
   b. Consistent with the foregoing, engineers may advertise for recruitment of personnel.
   c. Consistent with the foregoing, engineers may prepare articles for the lay or technical press, but such articles shall not imply credit to the author for work performed by others.
4. Engineers shall not disclose, without consent, confidential information concerning the business affairs or technical processes of any present or former client or employer, or public body on which they serve.
   a. Engineers shall not, without the consent of all interested parties, promote or arrange for new employment or practice in connection with a specific project for which the engineer has gained particular and specialized knowledge.

b. Engineers shall not, without the consent of all interested parties, participate in or represent an adversary interest in connection with a specific project or proceeding in which the engineer has gained particular specialized knowledge on behalf of a former client or employer.

5. Engineers shall not be influenced in their professional duties by conflicting interests.

a. Engineers shall not accept financial or other considerations, including free engineering designs, from material or equipment suppliers for specifying their product.

b. Engineers shall not accept commissions or allowances, directly or indirectly, from contractors or other parties dealing with clients or employers of the engineer in connection with work for which the engineer is responsible.

6. Engineers shall not attempt to obtain employment or advancement or professional engagements by untruthfully criticizing other engineers, or by other improper or questionable methods.

a. Engineers shall not request, propose, or accept a commission on a contingent basis under circumstances in which their judgment may be compromised.

b. Engineers in salaried positions shall accept part-time engineering work only to the extent consistent with policies of the employer and in accordance with ethical considerations.

c. Engineers shall not, without consent, use equipment, supplies, laboratory, or office facilities of an employer to carry on outside private practice.

7. Engineers shall not attempt to injure, maliciously or falsely, directly or indirectly, the professional reputation, prospects, practice, or employment of other engineers. Engineers who believe others are guilty of unethical or illegal practice shall present such information to the proper authority for action.

a. Engineers in private practice shall not review the work of another engineer for the same client, except with the knowledge of such engineer, or unless the connection of such engineer with the work has been terminated.

b. Engineers in governmental, industrial, or educational employ are entitled to review and evaluate the work of other engineers when so required by their employment duties.

c. Engineers in sales or industrial employ are entitled to make engineering comparisons of represented products with products of other suppliers.

8. Engineers shall accept personal responsibility for their professional activities, provided, however, that engineers may seek indemnification for services arising out of their practice for other than gross negligence, where the engineer's interests cannot otherwise be protected.

a. Engineers shall conform with state registration laws in the practice of engineering.

b. Engineers shall not use association with a nonengineer, a corporation, or partnership as a "cloak" for unethical acts.

9. Engineers shall give credit for engineering work to those to whom credit is due, and will recognize the proprietary interests of others.

a. Engineers shall, whenever possible, name the person or persons who may be individually responsible for designs, inventions, writings, or other accomplishments.

b. Engineers using designs supplied by a client recognize that the designs remain the property of the client and may not be duplicated by the engineer for others without express permission.

c. Engineers, before undertaking work for others in connection with which the engineer may make improvements, plans, designs, inventions, or other records that may justify copyrights or patents, should enter into a positive agreement regarding ownership.

d. Engineers' designs, data, records, and notes referring exclusively to an employer's work are the employer's property. The employer should indemnify the engineer for use of the information for any purpose other than the original purpose.

e. Engineers shall continue their professional development throughout their careers and should keep current in their specialty fields by engaging in professional practice, participating in continuing education courses, reading in the technical literature, and attending professional meetings and seminars.

As Revised January 2003

"By order of the United States District Court for the District of Columbia, former Section 11(c) of the NSPE Code of Ethics prohibiting competitive bidding, and all policy statements, opinions, rulings or other guidelines interpreting its scope, have been rescinded as unlawfully interfering with the legal right of engineers, protected under the antitrust laws, to provide price information to prospective clients; accordingly, nothing contained in the NSPE Code of Ethics, policy statements, opinions, rulings or other guidelines prohibits the submission of price quotations or competitive bids for engineering services at any time or in any amount."

Statement by NSPE Executive Committee

In order to correct misunderstandings which have been indicated in some instances since the issuance of the Supreme Court decision and the entry of the Final Judgment, it is noted that in its decision of April 25, 1978, the Supreme Court of the United States declared: "The Sherman Act does not require competitive bidding."

It is further noted that as made clear in the Supreme Court decision:

1. Engineers and firms may individually refuse to bid for engineering services.
2. Clients are not required to seek bids for engineering services.
3. Federal, state, and local laws governing procedures to procure engineering services are not affected, and remain in full force and effect.
4. State societies and local chapters are free to actively and aggressively seek legislation for professional selection and negotiation procedures by public agencies.
5. State registration board rules of professional conduct, including rules prohibiting competitive bidding for engineering services, are not affected and remain in full force and effect. State registration boards with authority to adopt rules of professional conduct may adopt rules governing procedures to obtain engineering services.
6. As noted by the Supreme Court, "nothing in the judgment prevents NSPE and its members from attempting to influence governmental action . . ."

Note:

In regard to the question of application of the Code to corporations vis-a-vis real persons, business form or type should not negate nor influence conformance of individuals to the Code. The Code deals with professional services, which services must be performed by real persons. Real persons in turn establish and implement policies within business structures. The Code is clearly written to apply to the Engineer, and it is incumbent on members of NSPE to endeavor to live up to its provisions. This applies to all pertinent sections of the Code.

National Society of Professional Engineers®

1420 King Street
Alexandria, Virginia 22314-2794
703/684-2800 • Fax:703/836-4875
www.nspe.org
Publication date as revised: January 2003 • Publication #1102

THE CITY OF NEW ORLEANS AND HURRICANE KATRINA

In August 2005, Hurricane Katrina made landfall on the Gulf Coast, destroying lives, leveling homes and leaving thousands of survivors homeless. At least 1,836 people lost their lives in Hurricane Katrina and in the subsequent floods, making it the deadliest U.S. hurricane since the 1928 Okeechobee Hurricane. The storm is estimated to have been responsible for $81.2 billion in damage, making it the costliest natural disaster in U.S. history. According to a report aired on Friday, May 19, 2006 on National Public Radio, in 1965, when the Army Corps was trying to figure out how to protect the city against a big storm, the initial plan was to build flood-surge gates at the eastern edge of Lake Pontchartrain, to keep storm surge out of the lake. Otherwise, the entire levee along the southern edge of the lake would have to be made higher and the canal walls would have to be raised. The floodgates would have blocked water at two deep channels, called Chef Menteur and the Rigolets. If a storm came, the gates could be lowered to block the two passes. However, environmentalists had concerns about the plan, worrying that the gates would make it easier for developers to drain wetlands and build houses and reduce the flow of water into the lake.

In 1977, a judge ruled that the Army Corps had not completely evaluated the impact of the floodgates on the environment, even though the gates would only close during a storm. There was also a second plan put forward by the Army Corps in 1990 which considered putting floodgates (called ''butterfly gates'') right at the end of the canal, where it meets the lake. This time the opposition came from the city officials. So in the end, the Army Corps decided to leave the canal open to the lake. Instead, the canal walls were made higher.

It was not the failure of engineers and engineering that caused New Orleans to be flooded. It was the shortsightedness of nonengineers who did not understand the realities.

Should the United States have stopped using nuclear power generation due to opposition by others too? Is the environment in a better shape? This too has to be analyzed further, and proper decisions must be made to save the planet.

7.24 PATENTS AND OTHER INTELLECTUAL PROPERTY

Patents are an extremely important issue. It is very easy to knowingly or unknowingly infringe on someone else's patent and get into legal trouble. Thus, it is important for the designer to be aware of patent laws and patents that are related to their subjects as well as their own rights and responsibilities.

Patents are issued by the government in order to exclude others from the right to use the patented subject. There are three types of patents: design, utility, and plant patents. Design patents last 14 years and relate to the shape and physical specifications of subjects, not the way they work or their application. For example, the body panels of a car may be design-patented in order to exclude after-market manufacturers or other auto manufacturers from copying the shape of the car or producing competitive parts. Plant patents are issued for plants only and are unrelated to this subject. However, patents, when unqualified, mean utility patents. A utility patent is issued for 20 years from the date of first application and gives the patent owner the right to exclude others from using, selling, producing, or importing the product. Patent law requires that the subject of the patent be:

- A process, a machine, a manufacture, a composite of matter, or new uses of the above or improvements of the above
- Yours
- New and useful

- Novel

- Nonobvious

Patents are issued separately in each country. There is no such thing as an international patent. Thus, when a patent is issued in the United States, there is no protection in any other country unless the patent owner also applies and secures a patent in each country.

We will discuss this issue in much more detail in Chapter 14. However, as it relates to this discussion, it should be pointed out that a designer must consider at least two issues regarding patents and design. On one hand, it is very important to ensure that the ideas the designer is considering are not patented. Regardless of whether or not the designer has come up with an original idea and whether the idea is completely his or hers or not, she or he should make certain that the idea is not already patented. Examples abound of individuals with an idea, believing that the idea is original, and therefore, should be unique. But you should be aware that there is a very good chance that someone else may have thought of the same idea, and therefore, a good chance that the idea may already be patented. In one case, I found 30 patents similar to an original idea I had about a computer-based book. A simple search revealed that what I thought was an original idea had already been patented by another 30 individuals with many variations and different utilities. Thus, to ensure that there are no infringements, it is vital that the designer do a patent search to make certain that the idea is not patented. On the other hand, the designer must also protect his or her intellectual property. Thus, it is important that proper documentation be kept throughout the design process, with proper witnesses and acceptable signatures, and to ensure that the idea is kept confidential until protection by law is obtained. Otherwise, the commercial value of the design may simply be lost to others who may quickly copy it. Please refer to Chapter 14 for more information, and always consult a professional patent attorney regarding protection of your ideas and infringement of other people's ideas.

7.25   LEGAL MATTERS

As in any other professional or business activity, design activities also require that the designer look into the legal issues that affect the profession. These include required permits from regulatory agencies (FDA, CPSC, UL), issues related to patent rights (as discussed above), liability issues, advertising requirements and standards, local, county, state and federal requirements, and many more. These legal issues may be simple or very intricate and require the services of professionals who are familiar with the law and know what needs to be done. Designers who work for larger entities, such as companies, universities, or research institutions, may have access to legal departments supported by these entities who can assist the designer and guide him or her in what needs to be done. In cases where the designer is working for smaller entities or as a consultant, it may be necessary to seek professional services of attorneys to ensure that all required laws and regulations are followed. This provides an opportunity for the designer to better defend his or her actions and decisions if problems arise. Please also see Chapter 15 regarding entrepreneurship and starting new businesses.

In this chapter, we discussed a collection of different factors and issues and their consequences that must be considered by the designer and addressed during the design phase, when it is the easiest and the least expensive to make changes if necessary, and to ensure that the design is appropriate and will work properly. In the next chapters we will discuss some of the more important issues in more depth.

REFERENCES

1. "National Transportation Safety Board Report on the 1979 Chicago Crash," Report NTSB-AAR-79-17, Washington, DC, December 1979.

2. *Lubbock Avalanche Journal*, March 6, 1990.

3. "New law Makes it Illegal to Throw Out Electronic Waste," Cal Poly Mustang Daily, Reported by Associated Press, February 10, 2006.

4. THILMANY, JEAN, "Up to Snuff," *Mechanical Engineering*, March 2005, pp. 28–30.

5. LOWE, ALLEN, S. B. NIKU, "Methodology for Design for Disassembly," ASME Publication # DE-VOL. 81, March 1995, pp. 47–53.

6. BOOTHROYD, G., "Assembly Automation and Product Design," Marcel Dekker, New York, 1992.

7. FORCUCCI, TOMPKINS, "Automotive interiors: Design For Recyclability," Designing for Recyclability and Reuse of Automotive Plastics, SAE Special Publication 867, February 91.

8. KAHMEYER, LEICHT, "Dismantling Facilitated," translated from Kunststoffe German Plastics, Vol. 12, 1991, pp. 30–32.

9. SIMON, "Design For Dismantling," *Professional Engineering*, November 1991, pp. 20–22.

10. ADLER, SHWAGER, "Software makes DMFA Child's Play," *Machine Design*, April 1992.

11. "A Prairie Home Companion Pretty Good Joke Book," new 3rd Edition, Highbridge Company, Minnesota 2003.

12. http://www.me.utexas.edu/~uer/challenger/chall2.html.

13. http://www.praxiom.com/iso-intro.htm.

14. WALD, MATTHEW, "Late Design Change is Cited in Collapse of Tunnel Ceiling," New York Times, November 2, 2006, p. A17.

15. KRANE, JOSHUA, "Renewable Energy Shines on Cal Poly," *Mustang Daily*, October 3, 2006, pp. 1–2.

16. "Opinions of the Board of Ethical Review," National Society of Professional Engineers, Vol. VIII, Publication Number 1106-G, 1999.

17. "EPA Fuel Ratings for 1985 Vehicles Sold in California," Los Angeles Times, Part IV, September 24, 1984.

HOMEWORK

7.1 Consider a common product such as a chair or sofa. List all different materials that have been used to make it. For each one, list the characteristics of the material as it relates to the chair or sofa, its advantages and disadvantages, and associated consequences such as cost, durability, appearance, life, comfort, and maintenance. Also list which material might be advantageous for which application (garden chairs, outdoors/indoors, automobiles, home use, classroom use, etc.).

7.2 Select a common household product with which you work every day. Write a set of performance criteria for the product which you feel is necessary. Also determine how these criteria can be measured and evaluated.

7.3 Define efficiency for a common product you or your family commonly uses. This includes household appliances, television and stereo sets, computers, and toys. Determine how to define efficiency for this product. Determine ways that this factor may be measured and evaluated.

7.4 Assume that a product is made of 100 parts. Further, assume that all parts have similar reliabilities. Calculate the minimum reliability that each part should have to achieve a product reliability of (a) 90 percent, (b) 99 percent.

7.5 A product is made up of eight subassemblies. The reliability of five of these subassemblies has been measured as 0.99, 0.93, 0.95, 0.97, and 0.99. Determine the range of acceptable reliability of the remaining three assemblies in order to render a product reliability of 80 percent.

7.6 Consider a product that is available at different levels of sophistication, quality, and functions. For example, a digital camera can be purchased at a range of about $40 to hundreds of dollars. Compare the specifications of three products made by different manufacturers and list their functions, apparent quality, and specifications. Which one renders the highest value for you?

7.7 List the problems you have encountered when attempting to repair or service a product of your choice (such as a car, VCR, or the like). If you have never attempted to repair or service a product, secure a discarded product from a thrift shop and try to repair it.

7.8 Investigate the extent of safety devices used in a washing machine. List these devices and how they work. Assess whether in your opinion any other device must be added to the machine.

7.9 Choose a common household product. Investigate what measures the engineers have taken to protect the product from environmental effects and list them. Assess whether these measures are adequate and working. If not, how would you improve them?

7.10 Investigate how a consumer product of your choice affects the environment around it. Include visual, social, auditory, and environment-related effects.

7.11 The garment industry produces huge amounts of scrap cloth. Find at least three uses for these scrap as raw material for other products. As an extension of this exercise, devise a plan for collecting, sorting, and reusing the scrap cloth in your products.

7.12 Certain areas of India receive only a few inches of rain per year. There is not enough water for even daily activities such as bathing, cooking, and washing, let alone luxuries. However, every night, the humidity in the air condenses and a layer of pure dew forms on roofs. Devise a system to collect this water and use it for personal uses during the day.

7.13 The sewage water collected in water treatment plants must be treated and cleaned before it is released. The grey water is not potable, but is clean enough for gardening and agriculture use. In most cases, the water is instead pumped into the ground, or is released into larger bodies of water (oceans, lakes). Estimate the cost of investing in a system to distribute this water into a city through a separate piping system for gardening and amateur agriculture use. Also estimate how much money will be saved by not producing and using clean drinking water for watering gardens and amateur agriculture. At what level would this system be feasible and useful? Will this system enable a city to be more sustainable?

7.14 A machine component is subjected to a calculated stress of 15 000 psi. It is desired to have a safety factor of at least 2.5 for the component. Select a material that would be appropriate for this application (you will need to consult manufacturer's catalogs or a machine design reference book for material properties).

7.15 A machine component is subjected to fatigue loading which results in estimated stress of 75 000 psi. Using Figure 7.9, estimate the life of the component. If a safety factor of 1.3 is desired, how would this life expectancy be affected? By how much should the stress be reduced to approach infinite life?

7.16 Dissect a discarded machine. Compile a complete list of the parts and components used in the machine and their function. Describe the function of the product and how it is accomplished through the subassemblies or parts. Assess how easy or difficult it was to dissect and reassemble the machine.

7.17 Using Internet search engines, find the military specs for a hammer. Does such a thing exist?

7.18 Perform a patent search through the United States Patent and Trademark Office's website (http://www.uspto.gov) for a product or an idea. Change your keywords to see how it affects the search results.

CHAPTER 8

HUMAN FACTORS IN DESIGN

8.1 INTRODUCTION

Many products and systems are created specifically for humans. Many others are operated and controlled by humans, and still, other products and systems affect the environment in which humans live. Alternately, humans have many physical, perceptual, psychological, and biological capabilities and limitations that can have a significant effect on the human–machine system. For these reasons, we must consider the relationship between humans and machines and their interface with each other as a design issue with which we must deal during the design phase. Human factors, also known as ergonomics, is the study of the measurements, capabilities, and limitations of the human body and how the humans and machines interact with each other.

The military has been collecting information about human measurements, their performance, and their responses since the mid-1930s. However, the human factors engineering as a subject of study came into focus during WWII when the positive effects of incorporating the data into the design of military equipment became evident. The designers of military equipment noticed that whenever machines were designed with consideration of the limitations and capabilities of the operators of those machines, those operators tended to work better and safer, they were more comfortable in those machines, and they could operate them for a longer period of time without tiring. The collected information has been tabulated into anthropometric tables that are now an integral part of human factors data. Later, the anthropometric data were expanded significantly to include nonmilitary populations, other races, women, children, and citizens of other countries. The human factors science also expanded to include many other aspects of human nature, including the physiological aspect of human senses, human psychology, learning capabilities, and human responses to stimuli. Nowadays, this information can be found in human factors books, reference books, and computer databases, and it is used in the design of most products, machines, and factory operations. Human factors has changed the expectations of the users of machines, has made many products much safer and better, and as a result of competition, has forced most major products to have a higher standard of quality and performance.

In the proceeding sections, we will discuss the background information about human factors, its advantages and restrictions, and how to use the available information in the design of products and machines. We will also discuss the physical and psychological capabilities, limitations, and preferences of human.

8.2 HUMAN–MACHINE INTERFACE

Consider a human–machine system such as a driver and a car. The driver and the car interface with each other through two mediums; the workspace that contains both of them, and the information that is exchanged between the driver and the car.

Workspace is the union of the spaces that constitute the user and the machine, in this case, the car and the driver. The driver must fit in the car, be able to reach its controls, be able to adequately see outside of the car, easily see the controls and the gauges, and feel comfortable. This in fact is one of the first issues encountered by the driver. Can the driver reach the steering wheel or the pedals? Is there enough adjustment in the seat to make the driver feel comfortable without straining the back, the legs, and the arms? What is the visibility range of the driver? Can the driver see the front and the back of the car in order to maneuver and park the car without accidents? Can the driver control the automobile's audio system without jeopardizing his and the passengers' safety? These questions relate to the physical dimensions of the user and the machine and their interaction with each other.

Information Exchange is the flow of information that the machine provides to the user and the control decisions from the user that flow back to the machine. The user must receive, understand, and correctly interpret the information in order to decide what control action to take. The control action must then be related to the machine, which will respond to the control command. In our example, the car provides information to the driver both actively, through instruments, and passively, through other mediums. For example, the car provides information to the user about the speed of the car, the temperature of the engine, the level of remaining gas in the tank, and the total mileage traveled. It also provides additional information to the driver through passive means such as the sound of the engine, the vibration of the steering wheel, and the vertical motion of the car as a result of driving over a bump. It is extremely important to know whether or not the information provided by the machine to the user is adequate, accurate, timely, and useful. Unnecessary information may be as bad as inaccurate or inadequate information. Similarly, information that causes reduced or compromised concentration while driving can result in accidents. The user uses this information to make decisions about the machine. For example, if the engine temperature gauge shows that the engine is overheating, the driver may slow down or turn off the air conditioning to reduce the load on the engine and to allow it to cool down. Similarly, if the sound from the engine is not right, indicating that it is rotating too fast or the load is too high, the driver may gear down or slow down. In this case, it is necessary for the machine to receive the control signal correctly, and to respond to the signal adequately and in a timely manner.

Additionally, there are other factors that play an important role in this information exchange, including rate of learning, level of comprehension, physical fatigue, mental fatigue, boredom, and psychological reactions. For example, if there are not enough challenges involved in driving to prevent the driver from getting bored, she or he may get drowsy and fall asleep while driving.

In the next sections we will discuss these issues as they relate to the design of a product or system as well as the interface between a machine and the user. To do this, we will first discuss the anthropometric data that can be used in the design of the workspace. Later, we will analyze the issues related to the design of a workspace as well as the exchange of information and control.

8.3 ANTHROPOMETRIC DATA

Anthropometric data is the collection of the measured physical dimensions, capabilities, and limitations of the human body, including the strength of the human musculoskeletal system under different conditions and configurations. The data are usually tabulated to make it easy to find the range of each dimension or strength, with standard deviations and major percentiles of each set of data. Please see Tables 8.1 and 8.2. As shown, major dimensions or strengths are referred to by description. In most tables, a range is given,

TABLE 8.1a   Anthropometric Table of Body Measurements[1] (_N_ = 4000)

Measurement	Range	Mean	Standard Deviation	Percentiles				
				1st	5th	50th	95th	99th
Weights								
1. Weight (lb) .	104–265	163.66	20.86	123.1	132.5	161.9	200.8	215.9
Body Lengths								
2. Body Stature	59.45–77.56	69.11	2.44	63.5	65.2	69.1	73.1	74.9
3. Nasal root height	56.30–73.23	64.95	2.39	59.4	61	65	68.9	70.7
4. Eye height	56.30–73.23	64.69	2.38	59.2	60.8	64.7	68.6	70.3
5. Acromial height	54.72–74.41	63.92	2.39	58.4	60	64	67.8	69.6
6. Cervical height	50.39–66.39	59.08	2.31	53.7	55.3	59.2	62.9	64.6
7. Shoulder height	47.24–64.17	56.5	2.28	51.2	52.8	56.6	60.2	61.9
8. Suprasternale height	48.03–63.78	56.28	2.19	51.3	52.7	56.3	59.9	61.5
9. Nipple height	42.13–57.09	50.41	2.08	45.6	47	50.4	53.9	55.3
10. Substernale height	41.34–55.51	48.71	2.02	44	45.6	48.7	52.1	53.5
11. Elbow height	36.61–49.21	43.5	1.77	39.5	40.6	43.5	46.4	47.7
12. Waist height	34.65–48.82	42.02	1.81	37.7	39.1	42.1	45	46.4
13. Penale height	27.95–41.34	34.52	1.75	30.6	31.6	34.5	37.4	38.7
14. Wrist height	27.56–39.76	33.52	1.54	30.1	31	33.6	36.1	37.1
15. Crotch height (inseam)	26.77–38.19	32.83	1.73	29.3	30.4	32.8	35.7	37
16. Gluteal furrow height	25.20–37.01	31.57	1.62	27.9	29	31.6	34.3	35.5
17. Knuckle height	24.80–35.04	30.04	1.45	26.7	27.7	30	32.4	33.5
18. Kneecap height	15.75–23.23	20.22	1.03	17.9	18.4	20.2	21.9	22.7

All dimensions are in inches.

These data were adapted from (24), citing (1).
Human figurines produced by Mannequin Pro program by Nexgen.

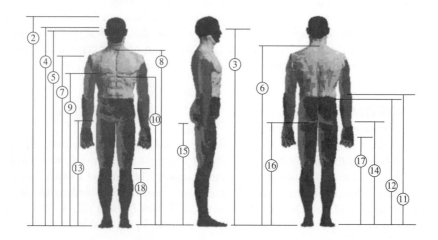

TABLE 8.1b Anthropometric Table of Body Measurements

Measurement	Range	Mean	Standard Deviation	Percentiles				
				1st	5th	50th	95th	99th
19. Sitting height	29.92–40.16	35.94	1.29	32.9	33.8	36	38	38.9
20. Eye	26.38–36.61	31.47	1.27	28.5	29.4	31.5	33.5	34.4
21. Shoulder	18.90–27.17	23.26	1.14	20.6	21.3	23.3	25.1	25.8
22. Waist height, sitting	6.30–12.99	9.24	0.76	7.4	7.9	9.3	10.4	10.9
23. Elbow rest height, sitting	4.33–12.99	9.12	1.04	6.6	7.4	9.1	10.8	11.5
24. Thigh clearance height	3.94–7.09	5.61	0.52	4.5	4.8	5.6	6.5	
25. Knee height, sitting	17.32–24.80	21.67	0.99	19.5	20.1	21.7	23.3	24
26. Popliteal height, sitting	14.17–19.29	16.97	0.77	15.3	15.7	17	18.2	18.8
27. Buttock-knee length	18.50–27.56	23.62	1.06	21.2	21.9	23.6	25.4	26.2
28. Buttock-leg length	35.43–50.00	42.7	2.04	38.2	39.4	42.7	46.1	47.7
29. Shoulder-elbow length	11.42–18.11	14.32	0.69	12.8	13.2	14.3	15.4	15.9
30. Forearm-hand length	15.35–22.05	18.86	0.81	17	17.6	18.9	20.2	20.7
31. Span	58.27–82.28	70.8	2.94	63.9	65.9	70.8	75.6	77.6
32. Arm reach from wall	27.56–39.76	34.59	1.65	30.9	31.9	34.6	37.3	38.6
33. Maximum reach from wall	31.10–46.06	38.59	1.9	34.1	35.4	38.6	41.7	43.2
34. Functional reach	26.77–40.55	32.33	1.63	28.8	29.7	32.3	35	36.4

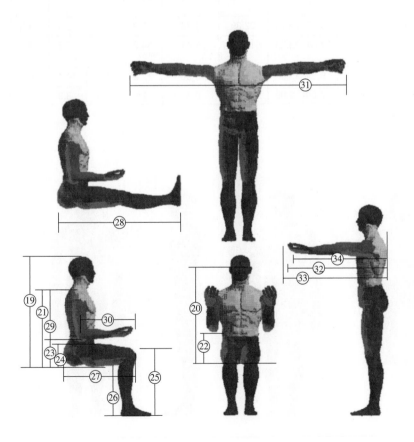

TABLE 8.1c Anthropometric Table of Body Measurements

Measurement	Range	Mean	Standard Deviation	Percentiles				
				1st	5th	50th	95th	99th
35. Elbow-to-elbow breadth	11.42–23.62	17.28	1.42	14.5	15.2	17.2	19.8	20.9
36. Hip breadth, sitting	11.42–18.11	13.97	0.87	12.2	12.7	13.9	15.4	16.2
37. Knee-to-knee breadth	6.30–10.24	7.93	0.52	7	7.2	7.9	8.8	9.4
38. Biacromial diameter	12.60–18.50	15.75	0.74	14	14.6	15.8	16.9	17.4
39. Shoulder breadth	14.57–22.83	17.88	0.91	15.9	16.5	17.9	19.4	20.1
40. Chest breadth	9.45–15.35	12.03	0.8	10.4	10.8	12	13.4	14.1
41. Waist breadth	7.87–15.35	10.66	0.94	8.9	9.4	10.6	12.3	13.3
42. Hip breadth	8.27–15.75	13.17	0.73	11.3	12.1	13.2	14.4	15.2
43. Chest depth	6.69–12.99	9.06	0.75	7.6	8	9	10.4	11.1
44. Waist depth	5.51–11.81	7.94	0.88	6.3	6.7	7.9	9.5	10.3
45. Buttock depth	6.30–11.81	8.81	0.82	7.2	7.6	8.8	10.2	10.9
46. Neck circumference	10.24–19.29	14.96	0.74	13.3	13.8	14.9	16.2	16.8
47. Shoulder circumference	35.43–56.69	45.25	2.43	40.2	41.6	45.1	49.4	51.5
48. Chest circumference	31.10–49.61	38.8	2.45	33.7	35.1	38.7	43.2	44.8
49. Waist circumference	24.41–47.24	32.04	3.02	26.5	27.8	31.7	37.5	40.1
50. Buttock circumference	29.92–46.85	37.78	2.29	33	34.3	37.7	41.8	43.5
51. Thigh circumference	14.57–28.74	22.39	1.74	18.3	19.6	22.4	25.3	26.4
52. Lower thigh circumference	11.81–23.23	17.33	1.41	14.2	15.1	17.3	19.6	20.9
53. Calf circumference	9.84–18.5	14.4	0.96	12.2	12.9	14.4	16	16.7
54. Ankle circumference	7.09–12.99	8.93	0.57	7.8	8.1	8.9	9.8	10.5

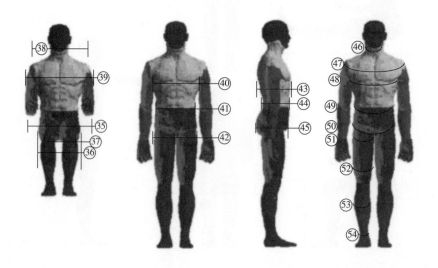

TABLE 8.2a   Maximal Static Hand Forces, Exerted on a Vertical Handgrip by Seated Males[2] **(N = 55)**

	Left hand						Right hand				
Direction of Force	Elbow Angle (deg)	Standard Deviation	Mean	Percentiles		Direction of Force	Elbow Angle (deg)	Standard Deviation	Mean	Percentiles	
				5th	95th					5th	95th
Push Horizontal	60	31	80	22	164	Push Horizontal	60	38	92	34	150
	90	35	83	22	172		90	33	87	36	154
	120	42	99	26	180		120	43		36	172
	150	48	111	30	192		150	45	123	42	194
	180	47	126	42	196		180	49	138	50	210
Pull Horizontal	60	23	64	26	110	Pull Horizontal	60	22	63	24	74
	90	28	80	32	122		90	30	88	37	135
	120	34	94	34	152		120	31	104	42	154
	150	37	112	42	168		150	36	122	56	189
	180	37	117	50	172		180	37	121	52	171
To the Right Horizontal	60	17	32	12	62	To the Right Horizontal	60	19	52	20	87
	90	19	33	10	72		90	23	50	18	97
	120	18	30	10	68		120	26	53	22	100
	150	20	29	8	66		150	25	54	20	104
	180	20	30	8	64		180	26	50	20	104
To the Left Horizontal	60	21	50	17	83	To the Left Horizontal	60	20	42	17	82
	90	22	48	16	87		90	18	37	16	68
	120	21	45	20	89		120	17	31	15	62
	150	27	47	15	113		150	18	33	15	64
	180	22	43	13	92		180	24	35	14	62
Up Vertical	60	18	44	15	82	Up Vertical	60	18	49	20	82
	90	22	52	17	100		90	22	56	20	106
	120	25	54	17	102		120	24	60	24	124
	150	27	52	15	110		150	28	66	18	118
	180	23	41	9	83		180	22	43	14	88
Down Vertical	60	18	46	18	76	Down Vertical	60	21	51	20	89
	90	20	49	21	92		90	20	54	26	88
	120	23	51	21	102		120	23	58	26	98
	150	16	41	18	74		150	18	47	20	80
	180	15	35	13	72		180	18	41	17	82

These data were adapted from (7) citing (2)

TABLE 8.2b Maximal Static Hand Forces, Exerted on a Horizontal Handgrip by Seated Males[3]
($N = 30$)

	Left hand						Right hand				
Direction of Force	Elbow Angle (deg)	Standard Deviation	Mean	Percentiles		Direction of Force	Elbow Angle (deg)	Standard Deviation	Mean	Percentiles	
				5th	95th					5th	95th
Push Horizontal	60	35	86	33	138	Push Horizontal	60	36	94	40	156
	90	28	60	27	93		90	24	65	25	100
	120	17	43	17	71		120	15	46	23	70
	150	18	37	15	69		150	18	40	18	66
	180	13	32	12	59		180	12	32	17	59
Pull Horizontal	60	18	39	20	64	Pull Horizontal	60	16	37	13	50
	90	18	37	17	65		90	13	32	14	54
	120	14	30	12	56		120	10	26	13	43
	150	13	32	15	52		150	10	29	12	48
	180	15	34	16	61		180	12	28	11	48
To the Right Horizontal	60	15	42	20	66	To the Right Horizontal	60	19	41	19	72
	90	12	38	17	60		90	15	31	12	64
	120	8	34	17	53		120	13	26	9	53
	150	11	31	17	54		150	11	21	9	39
	180	8	28	15	41		180	7	19	10	34
To the Left Horizontal	60	15	36	18	51	To the Left Horizontal	60	18	48	16	73
	90	11	27	11	54		90	15	39	16	59
	120	10	22	10	39		120	11	34	15	47
	150	16	23	9	53		150	7	32	18	45
	180	13	20	10	49		180	13	31	16	57
Up Vertical	60	22	57	22	100	Up Vertical	60	20	49	23	79
	90	24	77	37	123		90	29	69	28	112
	120	30	91	45	145		120	30	91	41	138
	150	32	100	58	159		150	38	99	43	165
	180	11	101	47	171		180	35	95	35	156
Down Vertical	60	35	74	18	139	Down Vertical	60	35	81	23	158
	90	34	75	23	136		90	35	83	22	142
	120	40	75	29	148		120	35	92	37	161
	150	29	79	39	136		150	34	90	40	154
	180	31	76	34	138		180	31	87	41	143

These data were adapted from (7), citing (3).

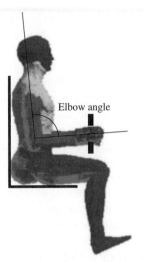

Elbow angle

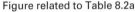

Figure related to Table 8.2a

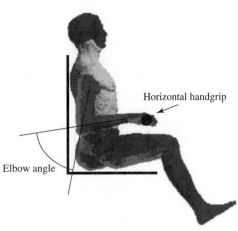

Horizontal handgrip

Elbow angle

Figure related to Table 8.2b

followed by standard deviation, 1st percentile, 5th percentile, 50th percentile, 95th percentile, and 99th percentile.

A percentile is the percent of people who are below the given number. Therefore, for the 5th percentile, 5 percent of the people measured have dimensions or strengths below the given number (called a quantile). Similarly, for the 99th percentile, 99 percent of people have dimensions or strengths below the given level. The 50th percentiles give the value at which 50 percent of measured individuals have dimensions or strength below the value or above it. Thus, the 50th percentile is the median of the measured values. However, please remember that the median, in general, is not the same as the average (which sometimes is given in a separate column). For perfect bell-shaped distributions, the average and median (as well as the mode) are practically the same. But in general, for data sets that are not perfectly bell-shaped, the average, median, and mode are different.

Imagine that you are designing a product that will be used by humans (can you think of many products that are not used by humans?). Let's assume that in this design, you need to place a control knob such that a person can reach it comfortably and safely. To determine the actual dimension for this placement, you may measure your own reach. In that case, it is obvious that if you are the only user of this product, you will comfortably reach the control knob, and you will safely operate the machine. But what about others? Will other users be able to reach the knob comfortably and safely? What if you measure the reach of someone else as well? Then you will have a range, or average, of two measurements, which perhaps is a better indicator of what the actual designed reach should be. In that case, anyone who happens to match either of the two dimensions, or falls somewhere in between, will comfortably and safely operate the machine. What if you measure a few more people? Obviously, the more measurements you acquire, the more accurately you will decide about the appropriate dimension that should be used for this purpose. Ultimately, if you were to measure a large population of possible users, the range would be such that most people would be able to use your design safely and comfortably. The data in anthropometric tables do exactly that. It is the collection of measurements from a very large group of people, representing many different races, backgrounds, ages, and capabilities, all tabulated together. This is why it is important to integrate these data into any design that is used by humans. In the following sections we will discuss both shortcomings and deficiencies of the tables, as well as how they can be used.

Let's consider the second row of Table 8.1a, called stature. As can be seen in the related figure, stature is the measure of the height of the person measured as the person stands upright. The second column of the table demonstrates the range of the measured values, in this case, 59.45–77.56 inches. The third column is the mean (or average) of the measured values, at 69.11 inches. Obviously, the mean value is not the midvalue of the range, but the average of the measured values. The standard deviation of the distribution is given as well. The next five columns are the percentiles. Only 1 percent of the measured values fall below the 1st percentile value of 63.5 inches. Similarly, only 5 percent of the measured values are less than the 5th percentile value of 65.2 inches. The 50th percentile is the median of the measured values and indicates the value at which equal number of subject are below and above the indicated value of 69.1 inches. Although the range of all measured values ($77.56 - 59.45 = 18.11$) is much bigger than the difference between the 99th and the 1st percentiles ($74.9 - 63.5 = 11.4$), the additional difference in the values ($18.11 - 11.4 = 6.71$ which is 37 percent of the total range) is only 2 percent of the measured values, a relatively small number of subjects. In fact, the mean $\pm 3\sigma$ ($69.11 \pm 3 \times 2.44$ which equals 61.79 and 76.43) correspond to 99.73 percent of the subjects. This means that only 0.27 percent of the subjects fall between 59.45 and 61.79 or 76.43 and 77.56 inches. Assuming that the total number of measured individuals is 4,000, this corresponds to about 11 individuals. Therefore, it is obvious that limiting the selected range of measurements given in these tables to specific percentiles, say 1st and 99th, or 5th and 95th, will correspond to few individuals falling outside of the selected range.

Tables 8.2a and 8.2b are similar to the above, but they contain information about maximal static forces. As shown in 8.2a, the table is divided into two portions, one for the left hand, another for the right. This is because the forces developed by the two hands are different. Of course, this is a function of whether the individual is left-handed or right-handed, and since there is no information about how many of the measured individuals are either left- or right-handed, it is unclear what the actual forces developed by a right- or left-handed population might be.

Additionally, as you notice, the forces developed in push versus pull motions in all directions are also significantly different. This is because opposing motions are generated by different muscle groups with different capabilities. Each section also includes information about the forces related to the relative angles between the upper and lower arms. This is due to two different facts; one that as the angle changes, the projected moment arm between the upper and lower arms varies as well; two, because muscles develop their maximum force at their midpoint. This means that as a muscle is extended or contracted away from the midpoint, its force generating capability reduces.

Clearly, in order to use the data in this type of anthropometric table, it must closely match the particular application. Otherwise, the data may not be appropriate. For more anthropometric data please refer to References (4) through (14).

EXAMPLE 8.1

Imagine that we are in the process of designing the seat of a car. There are a variety of measurements that must be used to design the seat properly. However, to simplify the problem for this example, consider only the range of adjustment needed for the seat and the back of the seat. If we assume that the tables we are using have been created by measuring a large cross section of individuals and that they represent the general population adequately, we may then choose a certain minimum and maximum range of individuals that we would like to accommodate. Assuming a 98 percent range, we may use Table 8.1 to determine the range of dimensions given in the 1st and 99th percentiles for knee height, buttock-knee length, forearm-hand length, arm reach, and functional reach. These ranges will allow us to design the seat with an adequate range of adjustments to accommodate most people within the ranges we have chosen (we will discuss common industrial ranges later).

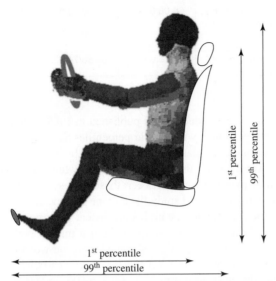

1st percentile
99th percentile

FIGURE 8.1   A schematic of a driver and the seat of a car.   ■

As you may have noticed the provided tables do not have adequate information for what is needed for this purpose. For example, there is no information in Table 8.1 about the total length of the arm with a small bend at the elbow, necessary for driving a car. There is also no information about the reach of the leg necessary in the driving position, as shown in Figure 8.1. As a result, we needed to compromise on the dimensions chosen. In other words, it is possible that there may not be enough information available in the tables about what is desired, or the information provided might not be appropriate for what is needed. Later, we will discuss these and other inadequacies of the anthropometric table in more detail. We will also discuss the difficulties with the application of these data and how to proceed with the design using alternative information.

8.4   THE CIVILIAN AMERICAN AND EUROPEAN SURFACE ANTHROPOMETRY RESOURCE (CAESAR) PROJECT

Equally sampled groups of people based on gender, ethnicity, and age from the United States, the Netherlands, and Italy were measured in order to update anthropometric data. This project, called CAESAR, was published in 2002. It contains information about more than 100 different body measurements with 1st, 2nd, 3rd, 5th, 10th, 20th, 25th, 50th, 75th, 80th, 90th, 95th, 97th, 98th, and 99th percentiles and associated quantiles (their magnitudes), for both men and women, in both metric and English units.[11] These data are presented in individual tables and therefore, cannot directly be accumulated in a simple table. The information, when needed, can easily be obtained from these tables.

TABLE 8.3 Comparison of Data for Stature from Three Different Sources for Women and Men

| Source | Percentiles | | | | |
	1st (women/men)	5th (women/men)	50th (women/men)	95th (women/men)	99th (women/men)
Hertzberg et al., 1955	63.5	65.2	69.1	73.1	74.9
U.S. Army personnel, 1988	58.39/63.10	60.15/64.84	64.06/69.09	68.40/73.48	70.09/75.14
CAESAR 2002	58.69/63.69	60.08/64.97	64.15/69.25	69.58/74.83	71.85/77.36

Additionally, a similar study was conducted and published in 1989 with an anthropometric survey about U.S. Army personnel with similar percentiles and quantiles.[12]

What is of interest to us is not the raw data; rather, we should be interested in seeing the differences between the newer data and the data in Table 8.1 which are from 1955. Table 8.3 shows a comparison of measurements between these data sources for stature. Neither the CAESAR nor U.S. Army personnel study list a range or a mean value. Therefore, only the percentiles and their associated quantiles are listed. Hertzberg's data are mostly military men. The U.S. Army personnel study is from a mix of both military women and men. CAESAR's data are from equally distributed men and women from the general populations. As you notice, the quantiles for men are very close to each other. For example, for the 50th percentile, the quantiles for men are 69.1, 69.09, and 69.25. Assuming that the population in the first study could have had a few women in it, the differences become negligible, although an increasing trend is evident in most measurements. The measurements for women at the lower percentiles are very similar too, but at the higher end, the differences are larger. This can be attributed to the fact that there are more women in the Netherlands, included in the CAESAR study, at the higher percentiles. Table 8.4 is similarly fashioned for the sitting height. The same conclusions may be drawn regarding these data.

Therefore, for most applications, even the old data are still valid and definitely more accurate than single measurements by the designer. However, whenever possible, newer, more accurate data should be used in the design of products and systems. Please refer to these studies for detailed information about anthropometric measurements.

TABLE 8.4 Comparison of Data for Sitting Height from Three Different Sources for Women and Men

| Source | Percentiles | | | | |
	1st (women/men)	5th (women/men)	50th (women/men)	95th (women/men)	99th (women/men)
Hertzberg et al., 1955	32.9	33.8	36	38	38.9
U.S. Army personnel, 1988	30.50/32.59	31.31/33.64	33.52/35.99	35.84/38.26	36.74/39.03
CAESAR 2002	30.83/32.57	31.86/33.70	33.91/36.18	36.41/38.78	37.50/40.08

8.5 BASIC APPLICATIONS OF THE ANTHROPOMETRIC DATA

The information embedded in the anthropometric data can be used in a variety of different ways and for different purposes, including designs based on minimum dimensions, maximum dimensions, average dimensions, and for specific ranges of dimensions.

Imagine that you are deciding on the placement of a water fountain, the door knob, or the location of an elevator control panel. These need to be reached by children, adults, men, women, and short or tall individuals. If you place them too high, or even at an average height, although easier for taller individuals, shorter people and children will not be able to reach them.

Obviously, if we have the luxury of multiple fountains, we may choose to place them at different heights for convenience. But if there is only one fountain available (or since there is usually only one control panel in an elevator and one door knob on a door), it is logical to place the fountain, the knob, or the panel at a minimum height to enable the largest possible population to reach it. In this case, you may need to look for the lower percentiles of heights in the anthropometric tables, perhaps 1st or 5th percentiles.

Now imagine that you are deciding about the size of an entrance to a vehicle, a room, an airplane, or a similar product. Obviously, it will make sense to make the entrance larger to accommodate taller individuals, perhaps at the 95th or 99th percentile. In this case, individuals who are not as tall can still go through the doorway without hindrance, but taller individuals will also be able to go through without hitting their heads. So, here, the preference is for a design based on maximum heights.

Next, suppose that we are deciding about the height of a product such as a table that must be used by a large variety of users without any adjustments. The table must be designed to be accessible and useful for all heights, genders, and races, whether short or tall, child or adult. In this case, using the minimum or maximum dimensions will exclude too many users in the population. And although, as will be discussed in more detail later, averages are very deceiving, we may need to design the table based on an average height (50th percentile versus mean, whichever is available). In this case, more individuals toward the middle range will be more comfortable, and fewer on the extremes will be equally uncomfortable.

Lastly, imagine a product that can be adjusted, for example the seat of a car. In this case, since we have the option of providing adjustments for variations in height, we should design the seat for a predetermined range, say 1st to 99th percentiles, or 5th to 95th percentiles. In general, if specific gender-based data are available, the preferred range of measurements is 10th percentile for women to 95th percentile for men. If specific gender-based data are not available, 5th to 95th percentiles are also used. In either case, there are obviously a number of individuals that will not be accommodated by these ranges. However, the benefit of extending the range must be weighed against the increased cost of additional adjustments. In a car, for instance, it is possible to increase the range of seat adjustment without too much cost. However, as the range increases, the available leg room for the passengers in the back seat is decreased, the cost of the seat is increased, and at times, the length of the car may need to be compromised to accommodate the increased range. The design engineer must then make a judgment as to what range is the best choice.

It should also be mentioned here that there are other variations to the above-mentioned choices. As an example, consider the design of a push–pull handle as shown in Figure 8.2. The inside diameter of the circular hole must be designed to accommodate a

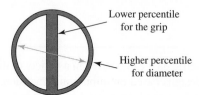

FIGURE 8.2 The design of a gripper like this may require both the lower and the higher percentile data.

large 95-percentile hand, while the gripper must be designed to accommodate a small 5-percentile hand. Thus, the different dimensions of the same part may require different data limits. As a designer, you must always consider what accommodations are needed and what limits are appropriate for the function.

8.6 SOFTWARE-BASED APPLICATION OF THE ANTHROPOMETRIC DATA

As you may have noticed, application of anthropometric data for different ranges, different genders, and different races is a difficult and time-consuming task, especially when multiple measurements must be used for a design. The combination of multiple pieces of information poses a significant challenge in using the data correctly and efficiently.

To overcome this problem, a clear-plastic template called the Anthropometric Data Application Mannikin (ADAM) was introduced in the 1970s, with front, top, and side views of $\frac{1}{4}$ size 5th percentile and 95th percentile templates. The templates had articulate joints that could be moved in order to orientate the body section to any desired configuration. These could be used to design a workspace or a product relatively quickly and easily without the need to refer to the anthropometric tables and without having to read and transfer individual data. A similar set of anthropometric charts, suitable for mounting, was published by Henry Dreyfuss [8] that could be used to create mannequins for design.

Today, software-based anthropometric data can be used for this purpose with much better efficiency and utility. Examples of this type of products are the NexGen Ergonomics, Inc. ManneQuin™ and ManneQuinPRO™ programs.[5] These human modeling systems allow the user to create anthropometric models of humans, humanoids, skeletons, robots, or stick figures between 2.5 and 97.5 percentiles, for heavy, average, and thin body types, and for a variety of races from different countries. The user has a choice of front, top, and side views, in many different poses. Additionally, each individual body section is a separate module that can be moved and reoriented. Therefore, it is easily possible to create custom models of humans that can be directly used in the program for analysis and design, or saved and integrated into other computer aided design (CAD) programs for further use. Other programs may similarly be used for the same purpose of facilitating the application of anthropometric data in the design of products, human–machine workspaces and human–machine interfaces. Figure 8.3 shows a variety of mannequins created by this program.

It should be mentioned here that this and other similar programs have additional features and capabilities that are not mentioned or discussed here. Please see the manufacturer's catalogs for more information.

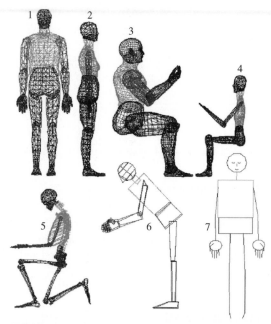

FIGURE 8.3 Mannequins created by NexGen Ergonomics ManneQuinPRO program (actual figures are in color).

1. U.S. adult male, 50 percent, average body type, human figure.
2. U.S. adult female, 50 percent, average body type, human figure.
3. Northern Europe adult male, 97.5 percent, heavy body type, humanoid.
4. Indian adult female, 2.5 percent, thin body type, human figure.
5. Like 1, but skeleton.
6. Like 1, but robot figure.
7. Like 1, but stick figure.

8.7 INSIDE-OUT VERSUS OUTSIDE-IN DESIGN

This concept relates to the way the interface between the human and the machine is approached. If the human requirements of a system or product are first integrated into the design, followed by the product requirements, the approach is inside-out design. If the product requirements are first developed followed by the human requirements, the approach will be outside-in design. As an example, consider a cockpit or a car or similar workspace. As shown in Figure 8.4, it is possible to first satisfy the requirements of the human operator or the user for comfort, safety, and efficiency, such as field of vision, ranges of motion, reach, and so on. The rest of the system is then designed and integrated with the human requirements. Alternately, it is possible to first design the workspace (e.g., the car exterior), and subsequently, try to accommodate the user or the operator into the remaining space. In most cases, the outside-in design will not satisfy all requirements set by the design. Consequently, the design may not be comfortable, the line of sight may be blocked or be inadequate, and there may be safety hazards present.

It should also be mentioned that there is a risk that inside-out design approach may yield an overall design that is not satisfactory in terms of the aesthetics, size, or other requirements. Thus, it is important to strike an appropriate balance between the human factor requirements and other design requirements. As an example, a commuter airplane that travels short distances requires small fuselage diameter for light weight, low fuel

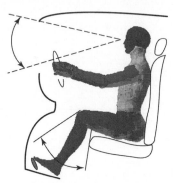

FIGURE 8.4 Inside-out versus outside-in design approach.

consumption, and short takeoff and landing distances. On the other hand, small fuselage diameter restricts passenger comfort. Therefore, there has to be a compromise dimension that yields the best result for both competing requirements. However, good and thoughtful design can improve this situation.

8.8 DEFICIENCIES AND SHORTCOMINGS OF ANTHROPOMETRIC DATA

Anthropometric data, whether raw, in tables, or in computer programs, has a number of deficiencies and shortcomings. These include the data collection methods, the measured populations, data tabulation, and data application. In the following sections we will discuss a number of these deficiencies and shortcomings, and in certain cases, how to improve the information.

Subjects Measured One of the first considerations about the relevancy of the data in the anthropometric tables is the population. As an example, Table 8.1 is the data collected from 4,000 male USAF personnel. Obviously this population is a selected population and does not represent the general population at large. Table 8.2a is from 55 people and 8.2b from 30, all military personnel. In this case, not only is the population a selected one, it is also very small, and therefore, not a very good measure of the variations in the general public. If the data are only for use by the military, the data may be a better representation of the population and may be used with better confidence. However, if the data are to be used for the general public, they may not represent the actual variations of the subjects accurately. Clearly, data from even a few subjects (such as in Table 8.2b with a population of 30 individuals) are still better than the data collected by the designer from even fewer subjects. But it does not mean that the lower and higher percentiles or the medians represent the general population with a high level of confidence. Therefore, it is possible that in reality, the equipment designed based on these data may not fit a large percentage of individuals.

Data from more recent studies such as CAESAR are more representative. The population is larger and more varied, but it only includes subjects from select countries. In the global markets of today, the designer must design for a world market that includes a large variety of people with varied characteristics. The anthropometric data may not necessarily represent all these varieties.

It is important to note that in most cases, most of the anthropometric data have been collected by the military. Military personnel are a select group of individuals. In many

TABLE 8.5 Comparison of Fighter and Bomber Pilots

	Fighter Pilots	Bomber Pilots
Number of subjects	210	1184
Age	27	29
Height	68.8 in	69.4 in
Weight	159 lb	166 lb

cases, if recruits are too short or small, too tall or big, too heavy or too light, or with any physical deficiencies, they will be dismissed or not admitted into the military. Personnel in certain branches of the military are selected even more stringently to fit particular anthropometric ranges, because the military does not have enough variety of uniforms and equipment sizes to match them, or the equipment is narrowly designed based on other specifications. As an example, one study[7] found that the average height and weight of 210 fighter pilots were 68.8 inches and 159 lb, whereas the same data for 1,184 bomber pilots were 69.4 and 166, respectively (see Table 8.5). This indicates that there is a difference even between the pilots in the same branch of the military. The reason is probably that since a fighter jet is to be more agile, and therefore designed to be smaller and lighter, the cockpit is also smaller. As a result, unlike a bomber airplane that can be larger and heavier with a larger cockpit, the pilots chosen for the fighters are generally somewhat smaller to fit the smaller cockpits. This is a requirement set by the design specifications of the product, affecting the type of personnel who operate it. This example indicates that the anthropometric data from these individuals alone will not be sufficient, or that they can be somewhat misleading.

One way to overcome this deficiency is to increase the range of the data found in these tables, if limited or selected individuals have been measured, to ensure that more individuals will comfortably fit the design. However, these data still provide a starting point for estimating a useful range for our design purposes.

Gender Bias Figure 8.5 is a sample comparison of a 50th percentile, U.S., average body type, adult, male and female.[5, 11] As shown, there is a significant difference between the average heights and weights of males and females in this population. Some tables present the data specifically for a particular gender. This type of specific information has applications in the design of equipment and wardrobe for the specific genders. However,

	Males	Females
Height	69.25 in	64.15 in
Weight	171.3 lb	135 lb

FIGURE 8.5 A sample 50 percent, U.S., adult male and female human (generated by NexGen Ergonomics ManneQuinPRO v.7).

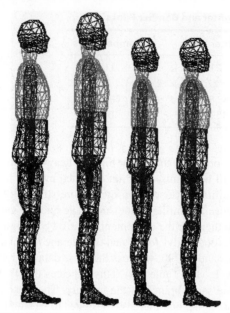

	American	Northern European	Japanese	South Chinese
Height	69.1 in	71.3 in	66.1 in	65.4 in
Weight	171.3 lb	166 lb	132.7 lb	132.1 lb

FIGURE 8.6 Comparison of a 50 percent, male, average body type, adult, Americans, Northern Europeans, Japanese, and South Chinese (generated by NexGen Ergonomics ManneQuinPRO v.7).

if the information presented is for the general population, it is very important that the measured subjects include similar numbers of males and females. Otherwise, as is clearly indicated, because the averages of the genders differ so much, the lack of enough subjects from one of the genders may severely distort the data. Unfortunately, much of the data collected are from military personnel, mostly in the past decades, when females were far outnumbered by males in all levels and branches of the military.

Obviously, this is also true for the representation of different ages and children in the data, or lack thereof. Military personnel are mostly young, healthy, adults over the age of 18. The data do not contain any information about the younger or older adults. CAESAR data were collected from men and women between the ages of 18 and 65.

Race Bias   Figure 8.6 is a sample comparison of the 50th percentile, male, average body type, adult, Americans, Northern Europeans, Japanese, and South Chinese. As is shown, there is a significant difference between the four different groups. Obviously, if a product or machine is to be used by a population that is mixed and includes a significant number of individuals from many different races, then the measured population should also include a similar representation of those individuals.

Table 8.6 shows the average dimensions for 50th percentile, adult, male, average body type, U.S. Army and South Chinese (data from NexGen Ergonomics ManniQuin-PRO v.7). As you notice, the difference between the statures of the subjects is 95 mm (3.75 in). However, the difference between the sitting heights (a measure of the upper body height) is small compared to the difference of buttock-heel lengths (a measure of legs). This indicates that the South Chinese have a relatively longer upper body, but the legs may be comparatively shorter.

TABLE 8.6   Comparison of Body Measurements of Races

	U.S. Army	South Chinese	Difference
Stature	1755 mm	1660 mm	95 mm
Sitting Height	851 mm	840 mm	11 mm
Buttock-Heel Length	1080 mm	1010 mm	70 mm

If a product or machine is designed to be predominantly marketed in a particular single-race society, then it should be designed for that population. As a result, for example, if a car is designed for body proportions found in South China, as long as it is predominantly marketed in that region, it might serve its purpose. However, if it is sold in the United States, on average, the product may be too small for the market, or its adjustments may not be adequate for the U.S. population.

It is important to note that most data presented in anthropometric tables are from a relatively uniform pool of individuals and do not necessarily represent a wide range of different ethnic backgrounds. CAESAR's data are from the United States, the Netherlands, and Italy.

Date of Measurements   Table 8.7 shows the result of the comparison between the soldiers of WWI and WWII (7). As is shown, the soldiers of the WWII era were heavier and taller than the soldiers of WWI, indicating that on average, in about 30 years, male Americans had grown in both weight and height. We also hear numerous reports about the average weight increase in the U.S. population, as we eat more and (perhaps) better, more nutritious foods (or at least fortified foods). Although in general, people are more aware of their weight, and although more people exercise routinely, and although more and more people are on a variety of diets, the average weight keeps increasing. As a result, if the anthropometric data were collected in the previous decades, They may be significantly off compared to the actual present-day data. In certain situations, the data may be updated on a regular basis. But in general, the data are old, and the designer must be careful about these measurements.

In another study,[8] it was found that the average growth rate of the U.S. population for both males and females is about 3 inches per century. Therefore, the average height of Americans has grown from 67 inches in 1900 to about 70 inches in 2000 for males and from 62 inches in 1900 to 65 inches in 2000 for females. During the same time period, average weights have grown from 140 lb to 165 lb for males and from 127 lb to 145 lb for females. However, according to the U.S. government data (as reported by the National Public Radio in October 2004), since the 1960s, the average weight of American adults has increased by 25 lb (from 166 lb to 191 lb). The average height as also increased.

TABLE 8.7   Comparison of WWI and WWII Soldiers

	WW I, 1919	WW II, 1946
Number	97,000	25,000
Age	23	25
Height	7.7 in	68.4 in
Weight	142 lb	155 lb

TABLE 8.8 Average Heights for Males and Females as Related to Their Age Groups

Age	Average Height: Male	Average Height: Female
20–24	68.7	64.0
25–29	68.7	63.7
30–34	68.5	63.6
35–39	68.4	63.4
40–49	68.0	63.2
50–59	67.3	62.8
60–69	66.8	62.2
70–79	66.5	61.8
80–89	66.1	N/A

The data in Table 8.8 attest to the same fact. As shown, the younger the population, the higher the average height for both males and females. This shows that due to better nutrition and more plentiful diets, the average height and weight of the population has grown steadily during the reported 70 years.

In 2006, the Coast Guard required that the operators of ferries change their average weight estimates of passengers from 140 lb to 185 lb per person. This was in reaction to the drowning of a ferry in Maryland a year prior that was attributed to being overweight.

What Is Measured and How Another important factor in the application of anthropometric data is the lack of knowledge about the exact way the subjects are measured, or what is measured. If you look at Table 8.1b, you will notice that the last three rows are arm reach from wall, maximum reach from wall, and functional reach. This shows that, in fact, it is important to recognize what is actually measured, and how. Imagine a subject being measured while sitting. If the individual sits relaxed or upright, there will be a significant difference between the measured values for sitting height. Similarly, if a person stands relaxed or erect, the stature will be different. The anthropometric data should indicate how the data were collected, and what was exactly measured.

Averages and Associated Concepts Table 8.9 shows the average values of body portion lengths for 50 percent, U.S. (military), male adults.

As you notice, the sum of averages of body portion lengths is very similar to the average of the body length, which is expected because the sum of averages should be the same as the average of the sums for normal distributions (although using the actual data reveals that these values do not add up to the same averages). However, this should not lead the designer to assume that someone with average height will have average body portion lengths.

TABLE 8.9 Averages of Body Portion Lengths for 50 percent, U.S. (military), Male Adults

	Head	Neck	Chest	Abdomen	Thigh	Lower leg	Foot	Sum of averages	Average height
Length (in)	7.3	3.3	15.5	6.6	16.7	16.6	3.2	69.2	69.1

Comparing the actual individual data from one population with the averages of the same population, the following was observed:

Out of 4,063 men,	1,055 had approximately average stature	(25.9%)
Of the same 1,055 men,	302 had approximately average chest circumference	(7.4%)
Of the same 302 men,	143 had approximately average sleeve length	(3.5%)
Of the same 143 men,	73 had approximately average crotch length	(1.8%)
Of the same 73 men,	28 had approximately average torso circumference	(0.69%)
Of the same 28 men,	12 had approximately average hip circumference	(0.29%)
Of the same 12 men,	6 had approximately average neck circumference	(0.14%)
Of the same 6 men,	3 had approximately average waist circumference	(0.07%)
Of the same 3 men,	2 had approximately average thigh circumference	(0.04%)
Of the same 2 men,	0 had approximately average crotch length	(0.0%)

As is clear from these data, there is hardly anyone who is average. It is incorrect to assume that if someone's height is average, his other body dimensions are also average. It is equally incorrect to assume that if someone is in the nth percentile range, the other body dimensions for the individual will also be at the same nth percentile. For the same reason, designing machines and products with the same assumption would be ill advised. For example, clothing designed for average or any other percentile would practically not match anyone completely.

To see this in a simpler way, imagine three blocks of wood, with height, width, and breadth of 1,2,3 and 2,1,3 and 3,1,2, respectively. The average of these three blocks will be 2,2,2. However, only one of these dimensions matches the average value.

8.9   WHAT CAN BE DONE

Clearly, we need to have reliable anthropometric data to appropriately design machines and products, but as we have seen, the available data are not perfect. Nevertheless, we are not individually equipped to collect these data on our own and to find the right measurements for all sorts of populations either. We must then rely on these studies and on available data, even if incomplete and somewhat inadequate, and try to minimize the inadequacies.

Damon, et al.[9] suggest that the anthropometric data be evaluated under the following two scenarios and adjustments or modifications be made based on the results. First, if the equipment is already prototyped and a mock up is available for evaluation, that a small population of test subjects, preferably 10 which include extreme lowest and highest percentile representatives, use the equipment for evaluation. It is probably adequate to use the weight and height of the subjects as an indication of their percentiles. The reported results by the subjects should then be used to compensate or adjust the design for best fit. Of course, the frequency, severity, and importance of the problems must be considered for correction. For example, even if 20 percent of the subjects have difficulty seeing a critical gauge compared to 100 percent having difficulty reaching a knob that is not critical, it is still more crucial to correct the gauge problem. Second, if the design of the machine or product is new, the designer should seek the most reliable anthropometric data from any available source, and use the data from the earliest stages of the design. In doing so, the designer must consider the dynamic nature of machines and equipment, the conditions in which the product or equipment will be used, and physical and psychological stresses that the operator may be subjected to.

Whitestone and Robinette[10] discuss a similar approach for the design of head-mounted displays. In their approach, they have integrated the testing of a prototype with 3-dimensional scanning of the head with and without the helmet to test for fit. Whether this approach can be used for other products needs further work.

I suggest that you also check out the new data, or if you are using the data on a regular basis, to join listings that publish and disseminate new information. You may also check http://www.hec.afrl.af.mil/HECP/cardindex.shtml, the Defense Technical Information Center (DTIC), or other similar websites for more information. For additional information on how to use human factors data, please see References (11) through (15).

8.10 LEFT-HANDED VERSUS RIGHT-HANDED USERS

Although not accurate, it is estimated that 10–15 percent of people are left-handed. In one study[16], it was claimed that the left-handed population has a life expectancy of up to 9 years shorter than the general population. The speculation was that this was due to the additional stress left-handed individuals experience, as well as due to the increased risk of accidents caused by products that are designed specifically for right-handed users. This claim has been seriously rejected by other researchers who have not found the same evidence, or have shown that the sample population was not appropriate. However, there is no doubt that many products are designed for the right-handed user. Examples range from fitted scissors and travel coffee mugs to power tools and from classroom desks to industrial machinery (Figure 8.7). If you are left-handed, you may have already experienced the problem. If not, you may try to do so as an experiment to realize how hard it is at times to operate these machines or products, from using a pair of scissors with your left hand to trying to write on a classroom desk with your left hand. Many power tools have a switch with a safety lock-out button that must be pushed in before the switch can be pressed (Figure 8.7). In most cases, the lock-out button is designed and located to be pushed in by the right thumb. However, a left-handed user will either have difficulty pushing this button, or by grabbing the power tool with the left hand, he will inadvertently push in the lock-out, rendering it useless. In this case, the switch can be accidentally pressed, creating a hazard. It is important that the designer includes left-handed comfort and safety concerns in the design requirements.

 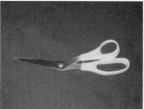 

FIGURE 8.7 Examples of products specifically designed for the right-hand user. A power tool may be a hazardous product for the left-handed user if the safety lock-out for the switch is not properly designed or appropriately located. Scissors can be very uncomfortable if molded for the right hand, and a travel coffee mug may be awkward if the opening is not reversible.

8.11   SENSORY INFORMATION EXCHANGE

As was discussed briefly earlier, a human–machine interface may generally provide information to the user in many different forms. Some information is passive, others are active. Passive information comes from sources within the system without any instrumentation. For example, as you drive your car, the sound from the engine provides information about how the engine is working, whether it is running too fast or too slow, whether it is overloaded or not, and whether it is running smoothly or not. However, you may not be able to tell how hot the engine is unless an instrument provides this additional information to you. This is an active source, designed specifically to provide the information to the user.

Although many different senses may be used to receive the information, vision and hearing are the most common. Visual displays and auditory signals provide the most common instrumentation in human–machine interfaces.

Important considerations in the design of sensory information exchange devices include:

- Is the information needed?
- What type of information is needed?
- When is the information needed?
- How much information is needed?
- Is the information comprehensible?

Is the Information Needed?   The designer must determine whether or not the information is actually needed. If not, the additional information provided may either distract the user or overwhelm his capacity to process all the information. For example, although it is possible to provide information about a lawn mower's engine speed, its value to the user is questionable. Many audio devices also display information about the content of the sound that in most cases is useless or cannot be comprehended. Except for aesthetic and marketing reasons, the information exchange is unnecessary.

What Type of Information Is Needed?   Analog speedometers are very common in many cars. The analog speedometer provides an appropriate type of signal for both velocity and acceleration. As the automobile accelerates or decelerates, the needle moves up and down accordingly, providing easy to follow and continuous information. Additionally, due to the continuous nature of an analog signal, as the speed of the car changes slightly while driving, the needle moves slowly and slightly, indicating the change. Drivers do not need to know the *exact* velocity of the car anyway, for example, whether it is 59, 60, or 61 mph. They need to know that the speed is about 60 mph. However, a digital speedometer, while more accurately displaying the speed, can be difficult to read as it changes continuously. Due to its discrete nature, as the speed changes slightly while driving, the displayed numbers will change too. Additionally, as the car accelerates or decelerates, the numbers change rapidly as well, making it difficult to read the information accurately. Clearly, the digital speedometer provides too much information that is not needed, in a form that is hard to follow. On the other hand, a pressure gauge with a digital read-out may be much better suited to many applications that require accurate measurements than a gauge with an analog read-out. Thus, the same method of conveying similar information may be appropriate for one application, but not another.

When Is the Information Needed? Engine temperature is an important piece of information for drivers in order to ensure that the engine is in good working condition and that it will not be damaged. Some car manufacturers provide information about engine temperature in the form of a red light in the instrument panel that turns on when the engine is hot. Although this instrument does provide temperature information about the engine, it is not at the right time. By the time the red light goes on, it is perhaps too late to save the engine. Obviously, it is more useful to provide this information in a timely manner, such that the driver will react in a timely manner to slow down, turn off the air conditioning, or check the engine for other problems if the temperature keeps rising. Therefore, a continuous temperature dial is much preferred to a red light and it does provide much additional but timely information as well. Similarly, imagine that a red light would turn on when the temperature of the core in a nuclear power plant reaches critical values. Obviously, this information would be useless at that point.

How Much Information Is Needed? In certain car models, it is possible to retrieve statistical data about the car, such as average gas mileage for a given period of time, total miles driven, and average consumption while driving. This information can generally be useful for learning how to drive more efficiently and how to take care of the car. However, it is seriously doubtful that this information should be available to the driver while driving the car at highway speeds, repeatedly pressing a button and reading the displayed data that causes what is referred to as ''secondary task distraction.''

Is the Information Comprehensible? In many cases, the information displayed may not be easily understood. For example, many instrument panels use icons, symbols, or abbreviations that are incomprehensible, especially when the user is not familiar with them. Imagine a driver in a rental car seeing an icon in the instrument panel blinking while driving. If the icon or the abbreviation is not clear, what can the driver do? How can the driver even find access to the meaning of the symbol if it is unclear? Therefore, it is very important that the information be in a form that can be understood without need for training or memorization.

Can you think of other displays that provide inadequate, improper, untimely, or unnecessary information?

8.12 HUMAN–MACHINE INTERFACE

Human–machine interfaces are the primary medium for interaction between machine and the people who use or operate them. The design of an interface is very crucial in whether the information is properly transferred to the user or not. A poor design will hinder the understanding of the information or may even cause hazardous situations. Therefore it is very important that the designer consider the human related factors in the design of interfaces. We have already discussed many points about the interaction between the user and the machine. All those points are valid here too. Additionally, you should also ensure that there is adequate redundancy in critical information transfer when human inadequacies may create hazardous situations. For example, if the information provided is color coded, a color blind user will not be able to perceive the information correctly or adequately. In this case, it may be necessary to also provide auditory information in order to ensure that the user or the operator will perceive the information adequately. As will be discussed

FIGURE 8.8 Part of the instrument panel of a 2004 Chevy Express cargo van and the location and orientation of the gauges.

later, psychological, physiological, and physical limitations of the user must always be considered when interfaces are designed.

Figure 8.8 shows the gauges in an instrument panel of a cargo van. As you notice, the gauges are situated on two sides of the steering wheel. However, also notice how the gauges are arranged such that at normal, each needle is in a different direction (except for the fuel gauge which has no normal state).

Why is this a poor design? First a story. In the 1960s there was a dramatic increase in the rate of accidents caused by a new-year model of an already established car. The car was basically the same as the previous years, except for some modifications in the interior, including the instrument panel. After extensive analysis it was determined that the cause was the new location of the radio in the instrument panel. The radio had been moved further away, and consequently, it was necessary for the driver to turn his or her head away from the front in order to see the knobs and to change stations or control the volume. Considering that the reaction time of humans to emergency situations is perhaps 1/10th of a second, the simple diversion of attention from the road was deemed as the reason why the new model car had such an increased rate of accidents. Now compare the gauges of Figure 8.8 to Figure 8.9. Unlike the one above, in Figure 8.9 the gauges are arranged such that all needles are at 12 o'clock position when normal. The driver only needs to quickly glance at the panel and see if everything is fine. It would take much longer to do the same with the arrangement in Figure 8.8, especially with two on one side, two on the other side of the steering wheel, perhaps contributing to hazardous situations.

Figure 8.10 shows the speedometer of a popular sports car. Can you quickly guess the speed of the car at this instant? Now imagine others who may not be as good as you in math, trying to guess the present speed of their car. This can easily be improved by changing the divisions between the major lines.

FIGURE 8.9 An alternative gauge arrangement for the instrument panel of the Chevy cargo van of Figure 8.8.

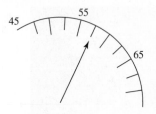

FIGURE 8.10 The schematic drawing of a speedometer of a popular sports car. What is the speed of the car as shown?

EXAMPLE 8.2

Figure 8.11 shows the schematic drawing of the oil pressure gauge of a 1971 U.S.-made sports car. The desired running oil pressure was 60 psi. Where would the needle be in this gauge at running pressure? How would you improve it?

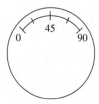

FIGURE 8.11 If the running oil pressure for this gauge is 60 psi, where will the needle be? ∎

EXAMPLE 8.3

Airplane instrument panels are crowded by countless gauges, knobs, and dials. Among these is an airspeed gauge, which indicates the speed of the airplane in air. The air speed is measured from the ram effect of air into a small, hollow tube called a pitot tube that is located outside of the airplane on the wing or fuselage. As the speed changes, the pressure in the pitot tube changes as well. This pressure difference is used to measure the air speed. Unfortunately, the pitot tube can freeze in very low temperatures, and therefore, may not show the air speed. To remedy this, a heater, enveloping the tube, is used to melt the ice. This is accomplished when the pilot (or the copilot) throws a switch located somewhere on the instrument panel. In Boeing 707s, the heater switch is on a panel overhead next to other similar switches.

In one particular case during a night flight the pilot observed that the air speed indicated by the gauge was falling rapidly and became convinced that the airplane was stalling. He lowered the airplane's nose trying to gain speed, but none was shown. Fortunately, the engineer officer noticed that the heater switch was off, but the one next to it was on. Apparently, the wrong switch had been turned on previously. He threw the correct one, at which time the gauge showed very high speeds. In this case, the mistake was caught in time, preventing a disaster involving hundreds of passengers. Later, the FAA studied the accident and made a recommendation for correction, requiring the addition of rubber padding on this switch to distinguish it from the rest. A worthless piece of rubber padding solved the problem, perhaps preventing other similar accidents. ∎

In certain cases, it is actually very difficult to determine the expected or preferred relationships between inputs and outputs of an interface. For example, imagine that a handle is used to control the up and down motions of an overhead crane (Figure 8.12). If the

FIGURE 8.12 Correspondence between motions of a control handle and the response from the machine cannot always be determined easily.

handle is also oriented so as to allow the operator to move it up and down to control the motions, it is clear that the expectation would be that the upward motion of the handle should control the upward motion of the crane. But there is no clear preference or expectation if the handle moves forward and back or sideways. Some associate the forward and backward motions of the handle to the motions of the control stick of an airplane and expect that pulling the handle back should correspond to upward motion of the crane. Others prefer that the forward motion of the handle (increasing distance from the operator) correspond to moving up (also increased distance from the operator). The worst is sideways motions, where there is no clear expectation, let alone a preference, that movement to the left or right should correspond to the up or down motions of the crane.

EXAMPLE 8.4

Study the location and arrangement of the knobs of your stove or gas burner. If the arrangement is similar to Figure 8.13a, do you have difficulty remembering which knob relates to what burner? Would a visitor to your house have difficulty? What would be your own preference? The fact is that there is no consensus about the best arrangement or what most people prefer. Obviously there are many different possibilities to improve this. Arrangements shown in Figure 8.13b and 8.13c are two possibilities, where the geometry of the locations of either the knobs or the burners is used to relate each knob to each burner. Can you think of any other arrangement?

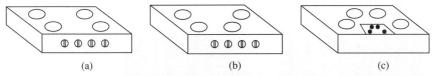

(a)   (b)   (c)

FIGURE 8.13   The placement of the burners and the knobs in a stove create a challenging problem for designers unless care is taken to somehow relate their positions to each other.   ■

As was mentioned earlier, the interface between a machine and the user must be easily discernable. Otherwise, many different problems may develop, from frustration and anger to hazardous situations. Have you not heard from many individuals about how difficult it is for them to program their VCRs? That is only due to the fact that the interface is ill-designed. In fact, since many products have numerous functions that require input from the user, designers have integrated multiple functions into one knob or button in order to decrease the total number of buttons in the interface. Consequently, one knob or button many have multiple functions that must be accessed in a particular sequence. This requires that the users remember what buttons to press in what sequences in order to accomplish one thing, sometimes with unrealistic symbols and abbreviations.

EXAMPLE 8.5

Figure 8.14 shows the interface of a car audio system. First, compare the size of the buttons with the size of the display. Also notice the amount of information displayed, including the loudness indicator bars on the two sides. On the small button on left the letter A is printed with the word Audio next to it. The next button is F with the word Func. Other abbreviations include Ent, DSP, and Disp.

Now imagine that you are driving and want to pause the CD player. What will you do? The instruction manual asks you to repeatedly press Function (Func button) until Pause appears in the display. Then press the up-arrow key to turn it on. Press the down-arrow key to turn it off. You must remember these instructions and do them while driving and looking at the display. Now guess how you change the CD track that is currently playing. And guess what RDM stands for? Yes, Random. And what was Ent? Entertainment. But what does this relate to?

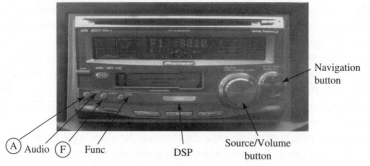

Navigation button

Source/Volume button

(A) Audio (F) Func DSP

FIGURE 8.14 A car audio interface. ∎

EXAMPLE 8.6

Interfaces between humans and machines are not just about information exchange. Figure 8.15 shows the door knob of the metal double-doors in the Engineering building on Cal Poly campus. The knobs are parallelogram shaped, smooth, rounded, and less than 3 inches long. To open the door, one must turn this knob and pull (not push) the heavy metal door simultaneously, which is extremely difficult when the other hand cannot be used (say if you are carrying books).

FIGURE 8.15 The design of this door knob and that the door must be pulled, not pushed, make it very difficult to open the door, especially if you are carrying things in one hand. ∎

8.13 PHYSICAL AND PSYCHOLOGICAL CHARACTERISTICS OF HUMANS

In this section we will discuss the physical capabilities and limitations of humans, including auditory characteristics, visual characteristics, acceleration capabilities, and temperature and humidity tolerances. We will also discuss psychological characteristics of humans, including learning curves, fatigue, boredom, attention span, mental challenges, and preferences.

Table 8.10 is a quick reference guide to human comfort zones and tolerance limits for a number of different factors.[8]

TABLE 8.10   Human Comfort Zones and Tolerance Limits [8]

Factor	Tolerance Zone Comfort–Tolerance	Discomfort and/or Physiological Harm Zone
Humidity	30%–70%	<10% or >90%
Temperature	65°–75°F	< 30° or > 100°F
Atmospheric pressure	10–20 psia	<8 psia
Acceleration	0–1 g	>1 g
Electrical current at 60 Hz	0–1 ma	>10 ma
Ventilation	13–20 ft^3/min	<5 or >50 ft^3/min
Oxygen	15–60%	>60%
Carbon dioxide	0–1700 ppm	40,000 ppm
Carbon monoxide	0–100 ppm	3000 ppm
Shock waves	0–2.5 psig	>7 psig
Mechanical vibration	0–1 cps	10 cps
	0–.005 in.	0.05 in.
Noise	0–85 dB	>94 dB
Light	20 – 100 ft-c	10,000 ft-c
Nuclear radiation	0–0.2 rem/yr	15 rem/yr

8.14   VISUAL CHARACTERISTICS OF HUMANS

Acuity   Conventionally, normal acuity is defined as seeing a 6-minute arc at the distance of 20 ft. An individual with normal eyesight is described as having a 20/20 vision, indicating that the individual can see the 6-minute arc at the 20-ft distance. All other measurements are described relative to the 20-ft numerator. Thus, a 20/15 (>1) acuity means that the individual can see at 20 ft what someone with normal eye sight can see at 15 ft, thus better than normal. Similarly, a 20/25 (<1) acuity indicates that the individual can see at 20 ft what someone with normal vision can see at 25 ft. The normal visual field is 50°–55° above horizon and 70°–80° below horizon. Visual acuity can be reduced by 50 percent at 7 g. Acuity is affected by age (older individuals become far-sighted), astigmatism, and near- or far-sightedness.

While designing products, including control panels and human–machine interfaces, the designer may have a tendency to assume that users have normal acuity and can see without difficulty. This assumption is not valid, as most people will not know whether they have normal vision or not until they are tested. Since change in acuity is gradual, even people who wear corrective glasses (or lenses) may not know that their acuity has changed until they are tested again. To ensure that the user will not miss important data or instructions, and for the sake of safety, it is important that the designer designs the product such that a person with less than normal acuity will still be able to read the data and instructions. This includes systems such as automobile dashboards, control panels, computer printouts, and gauges.

Color Detection   The cones of the retina specialize in seeing colors of red, green, and blue (RGB). With deficiencies in any cones, the corresponding colors may not be seen, or the color perception may be deficient. Thus, depending on the health of their cones,

TABLE 8.11 Ten Spectral Wavelengths and Corresponding Colors That Can Be Identified Accurately with Little Training

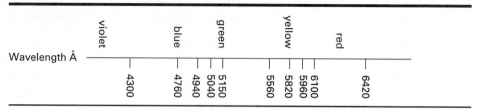

different people may see the same colors differently. About 0.4 percent of women and 8 percent of males have some color deficiency, making them color-deficient. About 0.003 percent of males are truly color-blind. At least 25 percent of people see colors somewhat differently from the majority. Color detection may also be affected by age. The fluid within the eye becomes yellow due to age, and as a result, greens become more yellow.

If you have ever tried to pick colors for painting a room or other objects, you have probably noticed that we can detect hundreds of different colors. However, unless compared side by side, it is very difficult to distinguish between similar colors. Therefore, only a distinct few colors should be considered for use in human–machine interfaces to identify different situations. For example, the color red is usually used to indicate danger or stop, while green is used for safe or go. However, the designer should not use similar colors (say shades of red) for conveying either different information or the intensity of a situation (such as level of danger). It is estimated that 97.5 percent of people can distinguish eight standard colors, 95 percent can distinguish 11 standard colors, and 90 percent can distinguish 15 colors. Table 8.11 shows ten spectral wavelengths and corresponding colors that can be identified accurately with little training.[7] The 2006 Ford Mustang's instrument panel light can be changed to about 120 different colors.

As was mentioned earlier, about 8 percent of males and 0.4 percent of females have some level of color deficiency, including color blindness. As a result, when exposed to different wavelengths, the individual may not see the colors as expected. This can create safety concerns if the operator misses perception of the information that is transferred by color or by its intensity. Therefore, the designer should always consider redundant means of presenting crucial information. For example, one may provide an auditory signal as well as visual signal when the temperature of a device reaches dangerous levels, so that a person with color deficiencies and a person with inadequate hearing can both receive the crucial information.

Green and blue colors are impossible to distinguish from afar, as are yellow and orange. Red and green (or blue) are easiest to recognize, followed by white. Yellow is the most luminous (most visible), orange and red-orange hold maximum attention value, while blue is hazy and indistinct. The most legible color combinations are black on white, black on yellow, (dark) green on white, red on white, white on blue.

Figure 8.16 shows the ranges for peripheral color detection. Notice that depending on the color, the range of peripheral color detection varies between 150° for white and 60° for green. Limit of color discrimination is 30° above and 40° below line of sight.

Human color perception is also affected (temporarily) by previous exposure to bright or strong colors. As an example, if you stare at bright (neon-color) Martian green for a few minutes, your eyes will lose the ability to detect green for a few minutes. As a result, you will see everything with a shade of purple for a while. In design too, the designer must consider the effect of previous exposure to the present sensitivity of the eyes.

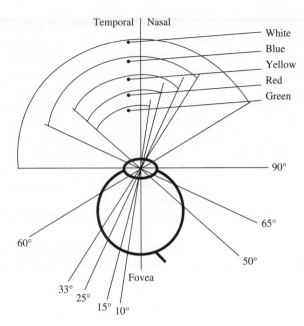

FIGURE 8.16 Peripheral color detection ranges for humans.

Illumination   Illuminance is a measure of how much light reaches a surface. Illuminance is measured in lux (lx in SI) or in foot-candle (ft-c in English units), where

$$1\, \text{lumen/m}^2 = 1\, \text{lux} = 0.093\, \text{ft-c} \qquad (8.1)$$

Different light sources provide different illumination levels. Different tasks, depending on the size of the object, detail of the task, and the type of work being performed, require different illumination levels too. While designing tasks, workstations, or human–machine interfaces, the designer should consult appropriate tables that indicate the required illumination levels for the task or the interface, and pick an appropriate number of light sources to provide the required illumination. The information about light sources is available from manufacturers. Table 8.12 is a short list of some tasks and associated required illuminance levels.

Contrast   Contrast is a measure of the difference between two illumination (brightness) levels that are next to each other. Contrast may be expressed by

$$C = \frac{B_1 - B_2}{B_1} \qquad (8.2)$$

TABLE 8.12   Recommended Levels of Illuminance for Different Tasks

Type of Activity	Ranges of Illuminance	
	lx	ft-c
Public spaces with dark surroundings	20–50	2–5
Workspaces with occasional visual tasks	100–200	10–20
Visual tasks with large contrast or large sizes	200–500	20–50
Visual tasks with medium contrast or small sizes	500–1000	50–100
Visual tasks with low contrast or very small sizes	1000–2000	100–200
Prolonged visual tasks with low contrast or very small sizes	2000–5000	200–500
Very prolonged or detailed visual tasks	5000–10000	500–1000

where B_1 is the brighter and B_2 is the darker brightness levels of the two adjacent areas or objects. As the difference between the brightness levels increases, it becomes easier to distinguish between the objects. However, the pupils of the eyes will eventually react to the brightness of the brighter source by closing (to let less light into the eye and to protect it). As a result, increasing contrast may prevent the eye from seeing the less bright object altogether. The opposite can also occur, where in low brightness levels (such as in nighttime driving) the pupils open to let more light into the eye for better recognition. In this case, if a sudden bright object is encountered, the large contrast will saturate the eye and will prevent it from seeing correctly or comfortably. Examples include bright hi-beam light from incoming traffic on two-way roads especially with more intense lights used in newer cars, and the brightness level of the instrument panels compared to the surrounding light which may make it difficult to see either the instruments or the road. At least 30 percent contrast is needed for good recognition in industrial situations.

In addition to the brightness levels, contrast can also be created by appreciable differences in colors. The better the color contrast, the easier it will be to distinguish between the objects.

EXAMPLE 8.7

The abbreviated descriptions of three of the pushbuttons of the car audio system shown in Figure 8.14 are printed in white letters on frosted plastic body, while the rest of the buttons are marked in black letters on a dark grey body. The contrast on all the buttons is extremely low and cannot be read easily. It would have been much better if the white letters were used on the dark grey buttons and the black letters were printed on the frosted plastic buttons. ∎

Character Recognition   The Human brain has the ability to recognize a great number of different fonts although they may look very different. This is true because the brain recognizes the basic building blocks of the characters (and numbers or shape) and eliminates the variations. However, certain fonts are easier to recognize under different conditions of lighting, color, and size than others. The designer of a product, system, or interface must consider the appeal of the font as it relates to the aesthetics of the product, but also the ability of the individual to easily recognize the letters under the specific conditions of use. As fonts become more elaborate, recognition becomes more difficult, creating either a nuisance or a hazard.

8.15   AUDITORY CHARACTERISTICS OF HUMANS

Sound results from changes in the air pressure as the air vibrates. Healthy, young, human ears can hear frequencies as low as 20 Hz and as high as 20 000 Hz. Sound intensity is measured by comparing the sound pressure to a standard level. The standard level P_0, estimated to be the lowest threshold of hearing, is the experimental value of the lowest sound pressure that a healthy young ear can hear.[15] This has been measured to be 0.00002 N/m^2 or 0.0002 dynes/cm^2 ($1\,N = 1 \times 10^5$ dynes). The theoretical largest air pressure that a healthy ear can bear is about 20 000 dynes/cm^2, or 2000 N/m^2. Since the ratio of the largest to smallest audible sound pressures is in the order of 1×10^8, a logarithmic scale called decibel (dB) is used to express the sound pressure intensity. A decibel, defined below, is the unit of sound pressure level (SPL):

$$\text{SPL, dB} = (10)\log_{10}\left(\frac{P}{P_0}\right)^2 = (20)\log_{10}\left(\frac{P}{P_0}\right) \tag{8.3}$$

where

SPL, dB = Sound pressure level, decibel

P = Sound pressure level of the noise in N/m^2 or dynes/cm^2

P_0 = Reference sound pressure level, 0.00002 N/m^2 or 0.0002 dynes/cm^2

Substituting the minimum and maximum sound pressure levels into Equation (8.3) yields

$$\text{SPL, dB} = (20)\log_{10}\left(\frac{0.0002}{0.0002}\right) = 0 \text{ dB}$$

$$\text{SPL, dB} = (20)\log_{10}\left(\frac{20,000}{0.0002}\right) = 160 \text{ dB} \tag{8.4}$$

Thus, the theoretical upper and lower thresholds of hearing are approximately 0 and 160 dB, although as we will see, 160 dB is beyond the threshold of pain.

Successive doubling of any sound pressure level P corresponds to a 6-dB increase in SPL,dB as shown below. For example,

$$\text{SPL, dB} = (20)\log_{10}\left(\frac{P=2}{0.0002}\right) = 80$$

$$\text{SPL, dB} = (20)\log_{10}\left(\frac{4}{0.0002}\right) = 86 \tag{8.5}$$

$$\text{SPL, dB} = (20)\log_{10}\left(\frac{8}{0.0002}\right) = 92$$

This means that as the sound pressure doubles, the logarithmic scale changes by 6 units. Thus, if the sound level doubles, it will be 6 dB higher. It also means that if there are two sources of noise present in a location, say each with 80-dB sound pressure level, the summation of the two sources together will generate a total of 86-dB noise, not 160-dB. Note that every 20 dB corresponds to a sound pressure level 10 times as large. It is estimated that a 3-dB change in a sound pressure level will be hardly detectable by human ears.

Another measure of sound intensity is the sound power level, which is defined as

$$\text{PWL, dB} = (10)\log_{10}\left(\frac{W}{W_0}\right) \tag{8.6}$$

where

PWL,dB = Power watt level, decibel

W = Acoustic power of the noise, watts

W_0 = Reference power level, 1×10^{-12} watt

Since $(10)\log_{10}\left(\frac{1}{10^{-12}}\right)$ is equal to 120, Equation (8.6) can be written as

$$\text{PWL, dB} = (10)\log W + 120 \tag{8.7}$$

Substituting the following noise power levels into Equation (8.7) we will get

$$\begin{aligned}
\text{PWL, dB} &= (10)\log(1) + 120 = 120 \\
\text{PWL, dB} &= (10)\log(2) + 120 = 123 \\
\text{PWL, dB} &= (10)\log(4) + 120 = 126 \\
\text{PWL, dB} &= (10)\log(8) + 120 = 129
\end{aligned} \tag{8.8}$$

As you see, for every doubling of the noise power, the PWL decibel increases by 3 dB. This means that every 3 dB corresponds to a doubling of the noise power.

Noise Level and Damage to Hearing Noise is one of the most pervasive occupational hazards found in a wide range of industries. Noise causes nervousness, irritability, and fatigue. Continuous noise can also cause permanent damage to hearing, even at lower levels. Lack of any noise, if continued, can also cause irritability and damage. Noise has been used as a weapon, creating temporary as well as permanent damage to tympanic membrane (bombs that create very loud noises to temporarily or permanently deafen the enemy soldiers). Some eating establishments have used music at elevated levels to excite their newly arrived customers as well as to force their current customers to leave, creating more space.

Low frequencies make high frequencies hard to hear, but high frequencies have little effect on low frequencies. Noise sounds louder in the presence of vibration, and vibrations are more annoying in the presence of noise.

It is estimated that a child can hear up to 20 000 Hz. This drops to about 15 000 Hz at age 30 and 13 000 Hz at age 50, with about 2 to 3.5 percent per year drop afterwards. Men lose more in the higher frequencies, while women lose more in the lower frequency and loudness capability. Men are also more likely to lose hearing (60% vs. 40%). Whites are twice as likely as blacks to have hearing loss.

Table 8.13 shows the dB levels for a number of different environments and systems. These are usually measured with a sound-level meter (dB meter, Figure 8.17). Obviously, the sound levels vary with different brands of the same products or systems. Additionally, the sound intensity measured is greatly affected by distance. In most cases, these are not

TABLE 8.13   Sound Pressure Levels (dB) for Different Environments and Systems

Sound level	dB
Threshold of pain	140
Ramjet, turbo jet	120–170
Drop hammer	100–130
Rivet gun	100–130
Thunder overhead	110–120
Circular saw	100–115
Woodworking shop, rock band	110
Band saw on sheet metal	105
Electric drill, belt sander	95
Subway train	90–95
Heavy trucks at 20 ft	92
City bus	90
Average factory	80–90
Light truck, 20 ft	80
Conversation at dinner table with children	70–80
Light traffic, 20 ft	70–80
Bathroom fan, 3 ft	65
Conversational speech	60
Quiet residence or office, turning newspaper	30
Radio studio, recital halls	30
Recording studio	25
Whisper at 5 ft	20
Sound picture studio, rustle of leaves	10
Threshold of hearing (young ears)	0

FIGURE 8.17 A typical sound-level meter (dB meter).

clearly indicated. So, the dB values listed in this table should be taken as an average estimated value and not an accurate representation of any particular machine.

Table 8.14 shows the maximum exposure times to any noise level as required by the Occupational Safety and Health Administration (OSHA). As you see, an individual may be exposed to continuous noise at the 80-dB level for up to 32 hours at a time. This time limit is reduced by a factor of 2 for every 5-dB increase in the noise level. Therefore, at 90 dB, the maximum will be 8 hours, or one shift in a factory. If an individual is required

TABLE 8.14   Maximum Time of Exposure for Different Noise Levels

A-Weighted Sound Level (dB) L	Duration (hr) T	A-Weighted Sound Level (dB) L	Duration (hr) T
80	32	106	0.87
81	27.9	107	0.76
82	24.3	108	0.66
83	21.1	109	0.57
84	18.4	110	0.5
85	16	111	0.44
86	13.9	112	0.38
87	12.1	113	0.33
88	10.6	114	0.29
89	9.2	115	0.25
90	8	116	0.22
91	7.0	117	0.19
92	6.1	118	0.16
93	5.3	119	0.14
94	4.6	120	0.125
95	4	121	0.11
96	3.5	122	0.095
97	3.0	123	0.082
98	2.6	124	0.072
99	2.3	125	0.063
100	2	126	0.054
101	1.7	127	0.047
102	1.5	128	0.041
103	1.3	129	0.036
104	1.1	130	0.031
105	1		

to stay in that environment any longer, either the noise level must be reduced or the employer must provide hearing protection and the individual must wear the hearing protection. Otherwise, as a result of prolonged exposure to high-level noise, hearing loss may result. It should be mentioned that since most people are exposed to additional noise in their life, including at home, during transportation, and other activities, these numbers are actually high, and they do cause permanent hearing damage. Employers are required to test their employees at standard time intervals to establish possible levels of hearing loss. If noise levels are high, OSHA requires hearing protection be provided to the employees. However, the same is true for individuals in places other than their work which is covered by OSHA. Loud noise from sport activities, driving, traffic noise, music, office noise, and many other background noises also contribute to permanent hearing loss.

In Table 8.14 the reference duration, T, is computed by

$$T = \frac{8}{2^{(L-90)/5}} \qquad (8.9)$$

where L is the measured A-weighted sound level. However, in many cases, the noise is not continuous and cannot be subjected to the above table. In these cases, a device called an audiodosimeter is used to measure the "dose" of the noise, which is converted to an 8-hour time-weighted average sound level (TWA). Table 8.15 lists the conversion factors for converting the reading of a dosimeter into TWA. For example, a dose of 91 percent over an 8-hour day results in a TWA of 89.3 dB, and, a dose of 50 percent corresponds to a TWA of 85 dB. For more information about noise exposure requirements, please see www.osha.gov.

Occupational hearing loss can either be noise-induced or due to acoustic trauma. Noise-induced hearing loss develops slowly over a long period of time (several years) as the result of exposure to continuous or intermittent loud noise and is always sensory-neural, affecting hair cells in the inner ear. Since most noise exposures are symmetric, the hearing loss is typically similar in both ears. Occupational acoustic trauma is a sudden change in hearing as a result of a single exposure to a sudden burst of sound, such as an explosive blast.

The rate of hearing loss due to aging increases over time. However, the rate of noise induced hearing loss is usually greatest during the first 10–15 years of exposure and it decreases as the hearing threshold increases and the individual is less sensitive to noise. Similarly, the risk to children is greater than the risk to older individuals because younger children have a lower threshold of hearing. The risk of hearing loss also increases with chronic exposures above 85 dB for 8-hour, time-weighted average (TWA). Continuous noise causes more damage than disrupted noise that allows the ear to have a rest period. Still, very high-level noises, even if short, can cause significant damage to hearing as well. The loss of hearing due to noise is not reversible.

Since people listen to their personal audio devices such as iPods at high volumes, there is an increased chance of hearing loss compared to devices that use common speakers. Users of these systems must be careful about the volume of the sound, especially if they listen to it often.

Physiological and Psychological Effects of Noise   Noise can cause stress, arousal, and attitude changes. It can also have a negative effect on performance and productivity of individuals. As an example, it has been found that if a worker is subjected to random, intermittent, high frequency loud noise, the individual's productivity will be negatively affected and his tolerance for complicated problems will be reduced even when the source of noise is removed. It is thought that one reason for this behavior is that the individual feels a lack of control over the environment in which he works, and as a result,

TABLE 8.15   Conversion Factors from Dose to an 8-Hour Time-Weighted Average Sound Level

Dose or Percent Noise Exposure	TWA	Dose or Percent Noise Exposure	TWA	Dose or Percent Noise Exposure	TWA
10	73.4	350	99.0	690	103.9
20	78.4	360	99.2	700	104.0
30	81.3	370	99.4	710	104.1
40	83.4	380	99.6	720	104.2
50	85.0	390	99.8	730	104.3
60	86.3	400	100.0	740	104.4
70	87.4	410	100.2	750	104.5
80	88.4	420	100.4	760	104.6
90	89.2	430	100.5	770	104.7
100	90.0	440	100.7	780	104.8
110	90.7	450	100.8	790	104.9
120	91.3	460	101.0	800	105.0
130	91.9	470	101.2	810	105.1
140	92.4	480	101.3	820	105.2
150	92.9	490	101.5	830	105.3
160	93.4	500	101.6	840	105.4
170	93.8	510	101.8	850	105.4
180	94.2	520	101.9	860	105.5
190	94.6	530	102.0	870	105.6
200	95.0	540	102.2	880	105.7
210	95.4	550	102.3	890	105.8
220	95.7	560	102.4	900	105.8
230	96.0	570	102.6	910	105.9
240	96.3	580	102.7	920	106.0
250	96.6	590	102.8	930	106.1
260	96.9	600	102.9	940	106.2
270	97.2	610	103.0	950	106.2
280	97.4	620	103.2	960	106.3
290	97.7	630	103.3	970	106.4
300	97.9	640	103.4	980	106.5
310	98.2	650	103.5	990	106.5
320	98.4	660	103.6	999	106.6
330	98.6	670	103.7		
340	98.8	680	103.8		

loses his tolerance of dealing with complicated situations. We are also familiar with the stress that a sudden, loud noise can induce in someone, including fear, elevated heart rate, and even heart attack. On the other hand, the noise of a waterfall or a stream of water, although continuous and sometimes loud, can have a soothing effect that can increase awareness of the person who enjoys the associated environment. Many individuals have also learned to enjoy music while they work, even if the music to one person may be noise to another. Additionally, consumers attribute specific characteristics to a product based on the product's sound. For example, a motorcycle projects a certain image to many consumers depending on the sound of the engine. Harley Davidson motorcycles generate a particular sound that is unique to them, coming from the 45° engines. This particular sound projects power, so much that a few years ago, the company (unsuccessfully) tried to patent the noise the engines make. The same may be true for other products such as vacuum cleaners, lawn mowers, etc. For these products, customers may attribute low noise to low power. Thus the designer must be careful about these psychological effects of noise as well.

Design Considerations   An important design issue to consider is the type and level of noise that products and systems produce. If a product or system is noisy, it will add to the overall level of noise in the environment. If the product or the system is used in a factory or other workplace, the added noise will increase the overall noise level, requiring more hearing protection, irritating the workers, and decreasing overall productivity of the employees. If it is used in places other than places of work, the added noise may be annoying, disruptive, or dangerous. Examples include noisy toys, loud power tools, and traffic noise.

Many products we design create noise, some wanted, some unwanted. Guns are among the sources of loudest sounds. Music is another source of everyday noise, as is traffic noise. Countless research studies have shown an increase in the rate of hearing loss among the "baby-boomers," including former President Clinton. Musical artists such as Pete Townshend and Ted Nugent have acquired substantial hearing loss and are now advocating quieter concerts.[18] Farmers, craftspeople, machine operators, and professional drivers are more likely to have hearing loss. Administrators, professionals, and individuals in service and sales are less likely to have hearing loss.

On the other hand, we must also be careful about reducing the level of noise. Nowadays, automobile manufacturers try to acoustically isolate a car from outside noise, thereby reducing the interior noise. However, many drivers listen to music (or their cellphones) while driving, and if present, children can also be very noisy. As a result, the driver may no longer hear exterior noise that carries vital information, including engine noise, the siren from emergency vehicles, and the like. This is why the rate of accidents between emergency vehicles and other cars has risen to such high levels, so that the drivers of these vehicles have to stop at all crossroads to ensure that there is no traffic (in 2002, in California alone, there were 15 deaths and 1,498 injury accidents involving emergency vehicles, as reported by the California Department of Transportation (CalTrans). Although not all of these accidents are related to this problem, there is no doubt that this is a significant contributory effect).

On the other hand, there is much that can be done to reduce unwanted noise. As was discussed earlier, high frequency noise attenuates much faster than low frequency noise. In one case, the low frequency noise from a large fan upon a hill was too disturbing to the residents of the area. The low-count fan blades were replaced by higher-count blades. The resulting higher frequency noise could not travel as far, eliminating the problem. This is the common complaint of the individuals who live near wind generators. The low frequency noise of the blades can be heard from long distances. Increasing the frequency would reduce this problem.

8.16   TEMPERATURE AND HUMIDITY

One of the mechanisms that the human body uses to dissipate the generated heat in the body is sweat. The resulting humidity over the skin absorbs heat from the skin to evaporate, thus dissipating body heat. The faster the sweat can evaporate, the cooler the skin

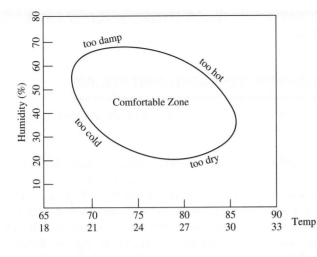

FIGURE 8.18 Approximate temperature and humidity for comfortable surroundings.

(and the body) will be. However, the rate of evaporation into the ambient air is greatly affected by the relative humidity of the ambient air. As the relative humidity increases, the rate of evaporation decreases. As a result, in humid air, the body sweat will not evaporate as readily, reducing the rate of heat loss from the body and causing it to feel warmer. Obviously, the amount of clothing, the vapor permeability of the fabrics involved, and the surrounding wind speed have a significant effect on this as well.

This has opposite effects on the body in winter and summer. Humid air in winter creates a feeling that the air is warmer than it actually is, making it more comfortable. In summer, humid air feels warmer than it actually is, making it more uncomfortable. Dry air has the opposite effect. Figure 8.18 shows this relationship.

One way to characterize the relationship between temperature and humidity is Wet Bulb Globe Temperature (WBGT), developed in the 1950s for the U.S. Marine Corps, but now part of ISO 7243. WBGT is calculated by a three-temperature device that measures the following:

- T_g, the black globe temperature, measured by a thermometer inside a black 6-in globe, representing the effects of radiation and wind
- T_{nwb}, the natural wet-bulb temperature, measured by a thermometer with its bulb covered with a wet cotton wick, representing the effects of humidity, wind, and radiation
- T_a, the air temperature in the shade, measured by a thermometer shaded from radiation, representing the air temperature

The WBGT can be calculated by

$$\text{WGBT} = 0.7T_{nwb} + 0.2T_g + 0.1T_a \tag{8.10}$$

If a Wet Bulb Globe device is not available, simplifying equations that only require the air temperature and relative humidity can be used instead.[19]

The American Society of Heating, Refrigeration, and Air Conditioning Engineers (ASHRAE) has published Standard 55-1992 for acceptable ranges of operative temperature and humidity for people in summer and winter clothing during light activities. If you design work or living spaces, or if you participate in the design of machines and systems that generate significant heat or humidity, you must consider the effect this heat and humidity will have on the occupants. Since environmental discomfort has significant effect on the performance of the users, you should consider ways to make the environment comfortable for them.

8.17   ACCELERATION CAPABILITIES AND LIMITATIONS

Humans may experience four different types of acceleration: continuous acceleration, impact, vibrations, and subgravity.

Continuous Accelerations   Under favorable physical conditioning, humans may learn to withstand as high as 9-g accelerations (g, or acceleration of gravity, is 32.2ft/sec^2 or 9.81 m/sec^2), although under special circumstances and for a short time, this may be even larger. This is an important issue in certain circumstances such as high-speed airplanes, space travel, and racing. One of the consequences of exposure to high continuous accelerations is that body fluids and parts can be displaced in the opposite direction of the acceleration, creating hazards. For example, it was noticed that many pilots who were accelerating at high rates to gain rapid velocities or were maneuvering in fast airplanes, were temporarily blinded due to the lack of oxygen-rich blood reaching their heads as it was pushed down to their legs. Later, (anti) g-suits were designed to restrict the flow of blood to the legs under high-g conditions by restricting the large blood vessels in the legs and lower abdomen, preventing blackout (through blowing more air into the lower part of the suit). Another important issue is the reaction of the body in deceleration, which throws the body in the opposite direction. For example, during rapid braking, the body of passengers in an automobile (and other vehicles such as in rides) will be thrown forward. The designer needs to consider the hazards related to these circumstances and design safety devices or harnesses to prevent injuries or damage to property. In October 2003, a Staten Island ferry ran into its pier at the St. George Terminal in New York. The rapid deceleration caused by this accident and the lack of any harnesses on the ferry threw the passengers into the pier and around the ferry, killing at least 10 individuals and injuring at least 60. If the passengers were still sitting in their seats and were strapped by seat belts, the injuries and deaths would be far lass, perhaps negligible.

Impact   Impact is a sudden acceleration (or deceleration). If the impact is shorter than 0.1 seconds, human body reacts to it as a rigid body. This means that the body parts do not have enough time to move significantly during the impact. Therefore, the body moves as a unit (rigid) body. If the length of the impact is longer than 0.1 seconds, the body parts and body fluids (such as blood) can move relative to the body, causing blackouts, dislocations at the joints, and other injuries. Automobile and sport accidents are caused by this type of impact. To reduce injuries, air bags are used in automobiles to prevent the throwing of the body against the windshield. Unfortunately, in numerous cases, the extremely rapid deployment of the airbags in certain cars has caused large enough impact loads to also injure the drivers and the passengers.

Vibrations   Vibration is the result of changing accelerations. The system's response, due to its inertia, is vibration. In other words, the acceleration times the mass of the system is a force. As the acceleration changes, so does the force. The changing force will cause the system to vibrate. While the response to a constant or sustained acceleration is also continuous or sustained, the response to a constantly varying acceleration is vibration. Vibration in a system may or may not be desirable. In fact in many systems the designer designs the system to vibrate—e.g., vibrating cell phones, electric shavers, and speakers. In other systems, vibration may not be desirable, but may be unavoidable—e.g., engines.

NATURAL FREQUENCY

Imagine that you have a simple mechanical system consisting a mass m and a spring with stiffness k as shown in Figure 8.19. If we stretch the spring by pulling on the mass, and then release it, the mass will start oscillating. Although all natural systems have some level of damping (or internal resistance to movement), assuming that the damping is low, the mass will oscillate for a long time without any external energy. The rate of oscillation (number of complete oscillations per second) is called *natural frequency* (ω_n), and is expressed by

$$\omega_n = \sqrt{\frac{k}{m}} \quad \text{Hz} \qquad (8.11)$$

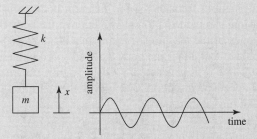

FIGURE 8.19 A simple mass-spring system.

The larger the mass, the smaller the natural frequency, and the larger the stiffness, the larger the natural frequency. Thus, a massive system oscillates slowly, while a stiff system with low mass oscillates very fast.

Now suppose that we desire to make the mass oscillate at a rate other than the natural frequency. In that case, we will have to exert an external force unto the mass at the desired frequency that will force the mass to oscillate at a frequency other than its natural frequency (Figure 8.20). This difference between the two conditions is very important. As you notice, for oscillating at the natural frequency, the system requires no external energy. In fact, theoretically, the system may oscillate at infinite amplitude with little external effort. An example of this phenomenon is the violent vibrations caused by small imbalances in a tire.

If there is any imbalance in a tire due to imperfections in the material or mounting, as the tire rotates, it generates a rotating external force. As long as the tire is rotating at rates not close to its natural frequency, it will require large forces to oscillate. However, if the speed of the car is such that the tire rotates close to its natural frequency, even with the small force of a small imbalance (perhaps a few grams), it will oscillate violently, even shaking the whole car.

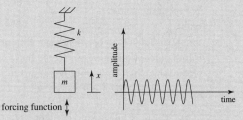

FIGURE 8.20 A forcing function is required to oscillate a mass at a frequency other than its natural value.

Similarly, the human body and its different parts have natural frequencies, at which rate, they oscillate easily. We naturally take advantage of this fact by moving these body parts at these rates with little effort. For example, while walking, we tend to move our arms and legs back and forth at about the natural frequency rate of the arms and the legs. This requires very little energy, and as a result, we can walk and move our arms and legs for a very long time without tiring. However, as soon as we walk faster than the normal rate, or if we carry a weight in our hand and thus change its natural frequency, we will tire quickly (this is exactly why people are told to carry a weight and walk fast to increase their heart rate and exercise). The same is true for breathing, our heart rate, and producing sounds with our vocal chords.

However, for the same reason, if different body parts are subjected to external oscillating forces that are close to their natural frequency, they will shake violently, creating a nuisance or eventually a hazard. Imagine a machine that produces a varying force that is close to the natural frequency of an arm while the person is in contact with it. As a result of this forcing function, the arm may oscillate violently to the point of physical damage.

TABLE 8.16 Natural Frequencies for Organs and Body Parts

Body Part	Natural Frequency (cps)
Head and shoulders (traverse)	2–3
Head with respect to body	20–30
Eyeball	60–90
Hand	30–40
Whole body, standing	5–12
Pelvis	6–9
Whole body, sitting	4–6
Body seated on cushion	2–3
Whole body, spine	3–4
Skull	300–400
Inner ear fluids (can cause nausea with prolonged exposure)	30
Depth perception, greatly affected by	25–40 and 60–90 cps
Noise which can be felt but not heard.	1–30 cps.

Table 8.16 shows the approximate natural frequencies for different body parts.[8] When you design machines and systems that can induce a forcing function to the body or its parts, you must ensure that the frequency of the machine or the system is not close to the natural frequencies of any body parts and that either the body is not exposed to these frequencies for any appreciable length of time or it is isolated from the source. Can you calculate the minimum acceptable idle speed of an engine that will not be harmful to human body? Do you know of any car that runs at that speed?

Subgravity Although subgravity is a relatively new environment for humans, more and more people are exposed to it. Subgravity can occur in space, but also in flying machines and underwater. Short exposures to subgravity are inconsequential; there is no harm. However, long exposures to subgravity can harm the bone density and muscle strength of individuals. Astronauts who return to earth after spending long periods of time in space may suffer from muscle pain and bone fractures, and therefore, must be rehabilitated to regain their strength. Since humans are not normally trained to function in subgravity, they must be trained for it. The strength of human muscles is adapted for Earth's gravity. When in an environment with lower gravity, the same muscle strengths will create large movements with which the individual must deal. Thus, training is required to learn how to control one's movements in subgravity.

8.18 PSYCHOLOGICAL FACTORS

Psychological factors include learning curves, boredom, mental fatigue, attention span, fear, and other stressors. These factors can play a significant role in the way a product is perceived or understood, and consequently, the way it is used. Psychological factors can also create significant hazards if not considered and dealt with. In the following sections we will consider a few of these factors with examples. For more information you must refer to human factors references that deal with this issue in more depth.

Stressors are factors that create a stressful situation. They include danger, fear, anxiety, competition, crowding, and information overload. Some stressors can be positive, at least to a certain degree, others are negative. For example, to the point where an individual feels adequate, competitive situations can increase the output level of a person and make the individual more productive and effective. However, if the level of competition increases beyond the capability of the individual, he may feel inadequate and get frustrated, his productivity level may drop, he may give up or create a dangerous situation by overreacting to the situation. Fear, for example, may cause a person to be more careful, and consequently, safer. However, fear may also negatively affect a person's ability to function properly. In Chapter 3, we considered mental barriers that fear and other psychological factors can create, negatively affecting the individual's creative ability. Similarly, many individuals find it difficult to perform well during a test, while they do much better when this particular stressor is not present.

Earlier we also discussed human–machine interfaces that provide too much information, many of which may be unnecessary or untimely. All these are also examples of information overload. When too much information is provided quickly, the brain may overlook most of it, or if it does process this information, other information may be overlooked, creating a hazard. For example, while driving, if too much detailed information is provided to the driver by the instruments crowding the instrument panel, he may overlook the information provided by other means such as the noise from the engine, an emergency vehicle's siren, a child playing on the street, etc. and get into a hazardous situation.

In early 1980s, an elderly couple was driving in an upper-scale foreign-made car that had received a safety award for its good design as it related to safety devices, including its seat belt system.[20] The unfortunate couple got into a head-on accident which killed the woman who was sitting in the passenger seat, while the man who was driving did not sustain any significant injuries. Later investigation revealed that the woman had been thrown into the front windshield and dashboard and had died as a result of the sustained injuries. It was also revealed that she had worn the seat belt correctly, and that "the seat belt had functioned properly as designed." The puzzling issue was that although the automobile in question had received safety awards for its seat belt design, and that the passenger was wearing the seat belt correctly, she had been thrown into the windshield. When asked, most respondents express one of the following possibilities:

- She had not been wearing the waist part of the belt, but she was.
- She was too short and the seat belt was too high and she slipped from under the belt, but she was not.
- The seat belt had malfunctioned and did not lock during the accident, but as reported, it had functioned properly as designed.
- The seat belt had torn, but it had not.

In reality, what had happened was related to the psychological reaction of a person in danger. As a common reaction to the realization that a car is coming toward them head-on, they may have tried to brace themselves by grabbing something close to them. The husband, driving, had grabbed the wheel. The wife, sitting in the passenger side, had grabbed the door handle on her right and either grabbed or raised the seat belt emergency-release lever on her left side, unaware that she had disengaged the seat belt tensioner. The mechanism, correctly, was designed to disengage the seat belt in emergencies. However, it was not designed for inadvertent disengagement. So, although it functioned properly as designed, the design was flawed for not considering a common human reaction to danger. Later, the seat belt was redesigned with a feature similar to the safety lock-out button of power tools to prevent accidental disengagement.

Humans react differently to different situations, especially when it comes to fear or dangerous situations. Some react positively, by taking action to prevent danger or to

TABLE 8.17 Summary of the Percentages of Deaths in Car Accidents with Different-Age Drivers and Different Types of Cars

Highest	Rate	Drivers under Age 30	Male Drivers	Lowest	Rate	Drivers under Age 30	Male Drivers
Chevy Corvette	5.2	30%	85%	Cadillac DeVille	1.1	8%	61%
Chevy Camaro	4.9	47%	79%	Olds Cutlass Ciera	1.1	9%	54%
Ford Mustang	4.4	50%	75%	Toyota Cressida	1.1	20%	66%
Nissan 300ZX	4.2	41%	79%	VW Jetta	1.1	33%	61%
Chevy Chevette	4.1	21%	46%	Lincoln Town Car	0.8	6%	6%
Honda Civic CRX	3.9	60%	64%	Ford Taurus Station wagon	0.7	8%	62%
Pontiac Fiero	3.6	48%	68%	Volvo 740/760	0.6	7%	62%

contain its consequences, some react negatively by withdrawing and helplessness. A good designer will consider these situations and will create appropriate safeguards to prevent dangerous situations, and eliminate, contain, or reduce the consequences of the inevitable dangerous situations that may relate to the product or system. Please also see Section 8.19 on hazard recognition and elimination, as well as Chapter 13 on product liability.

Another aspect of the psychological effects of human factors is a product's perception and its effects on the consumer. Table 8.17 is the summary of the percentages of deaths in car accidents with different-age drivers and different types of cars.[22] As is shown, certain cars with a fast, sporty image also have the largest percentages of deaths, while others that are advertised as safe cars have the lowest percentages. The data also show the difference between younger drivers with faster cars and older drivers with family cars, and the perception these vehicles project. In fact, other than the perception, many of these cars have similar power ratings, maximum speed capabilities, and power, but the way they are driven is different.

Learning Curve, Boredom, Mental Fatigue, and Attention Span Humans can learn from experience. Consequently, as they perform the same or a similar function more often, they become more experienced and more productive (e.g., working with a new radio in your car, a new machine tool, or flying an airplane). However, the learning follows an asymptotic curve similar to Figure 8.21; as time increases, learning lessens. On the other hand, as the operation becomes easier to perform, it becomes less challenging and more boring. A repetitive, monotonous operation causes lack of interest and care too. Consequently, the individual may not perform as well, or ultimately may create hazardous situations. Similarly, a function that is too easy and does not require much thinking can be boring and tiring, while a function that is too

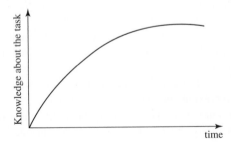

time **FIGURE 8.21** A typical learning curve.

difficult will create mental fatigue and carelessness. Therefore, it is imperative that the designer always considers the correct balance between how difficult an operation is, how much learning is necessary, whether there is enough challenge to keep the mind interested without fatigue, and whether or not there are enough variations in the task to keep it interesting. Otherwise, as is true with physical and physiological factors, the psychological effect will create their own unacceptable and hazardous situations.

8.19   HUMAN FACTORS AND SAFETY

Remember Murphy's Law: *If Anything Can Go Wrong, It Will.*

As has already been discussed and demonstrated, human factors play a significant role in the safety of a product or system and their associated liability. The following sections are provided as a simple guide for the designer to check for possible safety issues that must be considered to ensure that adequate safeguards are integrated into the product or system.

8.19.1   Hazard Recognition

Hazards must be recognized during the early stages of design. This recognition [17, 21] and provisions for safety features to eliminate the hazards is of utmost importance. You should look for all types of possible hazards and their possible frequency or severity, including the following:

- *Kinematic hazards:* Includes rotating machinery, moving elements such as linkages and chain drives, and nip points (points where a moving element comes close to a nonmoving element). Body parts, such as fingers, can get caught in nip points and easily be injured.

- *Energy hazards:* Includes mechanical energy such as potential and kinetic energies (springs, flywheels, rotating elements), electrical energy such as batteries, capacitors, and AC power sources, chemical energy such as fuels and explosives, and microwave energy.

- *Electrical hazards:* Includes electrical shocks, electrical shorts, electromagnetic fields (EMF), and electrical interference between different devices (e.g., effects of electromagnetic waves on pacemakers).

- *Chemical hazards:* Includes toxins, chemical reactions between different materials, oils, vapors, interference between different materials, oxidation, and other chemical interference.

- *Material hazards:* Includes reactions among different materials, toxic materials, hazardous material, and material-handling issues.

- *Environmental hazards:* Includes noise and noise levels, lighting, ultraviolet (UVA and UVB) rays, bacteria, viruses, fungi and yeast, oxidation and fading of materials and colors. It also includes the effects of the product or system on the environment, including air pollution, water pollution, solid waste, product's end-of-life issues, and recycling issues.

- *Human factors hazards:* Includes factors related to the physical, physiological, and psychological limitations and capabilities of humans as they deal with the product or system, as discussed in this chapter.

- *Misuse and abuse:* Includes the unexpected but predictable uses of the product, unintended uses and abuses of the product or system, and possible mistakes that the users may make. It also includes untrained users, children using the system with or without permission of adults, and misunderstandings created by the lack of knowledge of the language or differences in cultures.

8.19.2 Hazard Elimination

The designer may consider the following methods among others to find the most appropriate solution for the elimination or containment of the hazard:

- *Human factors considerations* The hazard may be eliminated by inclusion or integration of human factors into the design. In many cases, the addition or integration of human factors principles into the design of a system or product adds very little to the cost of the system, but can significantly improve its comfort and safety. This also helps in improving the image of the product and its commercial success.

- *Design change* If the hazard cannot be adequately contained or controlled, a design change may be necessary. The change may be relatively simple and inexpensive if it is made at the early stages of the design, but very expensive when the product is released. Nevertheless, if a hazard exists in a product or system, it may be necessary to recall all released products for correction while the design of the product is also changed.

- *Safety devices* At times, it may be adequate to add safety devices to a product. However, it is imperative to make sure that the safety device is adequate, and that it cannot be easily overridden. For example, the tip of a chain saw is among the most effective areas of the saw, but also the most dangerous to use. [20] If the chain saw is caught in a knot or a nail or other similar tough material while cutting at the tip, it can kick back and hit the user. Manufacturers have at times added a safety guard to the tip of chain saws to prevent the use of their tip for cutting. However, many users remove the guard and use the tip nevertheless. A design change, adding an embedded groove into the body of the chain saw at the tip, will prevent the use of the tip for cutting, but will also prevent the user from removing the safety feature. In this case, just adding a cover may not be adequate, while a design change may be.

- *Guards, switches, etc.* It is unlawful to leave rotating machinery without guards, even if the machine was designed and manufactured in the past. All rotating machinery must be safeguarded by covers. In addition, there should be adequate safety switches in products and systems to ensure that unsafe situations are controlled. It is the responsibility of the designer to decide whether switches and guards are adequate or not. For example, in a top-loading washing machine, the addition of a safety switch to the door will reduce the danger of small children opening the lid and getting their hands in the rotating drum during operation. However, adding a brake to the drum to stop it immediately if the door is opened will further add to the safety of the product. Better yet, the machine will be safer if a sensor (or latch) is added to prevent the door from opening until the drum stops completely. Similarly, in a microwave oven, there must be a safety switch that prevents the opening of the oven's door while it is turned on.

  Unfortunately, in the first generation of microwaves, there was no safety switch, and many users would open the oven door while the microwave was on. Later, a simple switch was added to turn off the machine when the door was opened. However, enterprising users would override the switch by pressing it in order to leave the oven on with the door open. Later designs have an internally installed switch which is triggered by a hook-type extension on the oven door, making it much more difficult to override the safety switch.

- *Warnings and instructions* It may also be possible to adequately safeguard a hazard by warnings. However, it must be noted here that if a safety problem exists that can be solved by other means, including any of the above, simply warning a user against the hazard will not be adequate and the courts do not favorably consider this simple solution in a liability suit.

8.20 OCCUPATIONAL INJURIES

Worker may be exposed to many situations during work that can cause injuries. These include back injuries, repetitive stress injuries, carpal tunnel syndrome, and injuries due to heavy lifting. These are in addition to other possible occupational hazards that we have already discussed, including auditory injuries, toxins, explosions, and others.

Carpal Tunnel Syndrome (CTS) Coming from the Greek word *karpos*, or wrist, carpal tunnel is the passageway, composed of bones and ligaments, through which a major nerve system of the forearm passes into the hand. Due to wear and tear of repeated movements, the lubricating membrane of the tendons in the tunnel thickens and the nerves get pressed against the hard bone (called nerve entrapment), causing pain. CTS usually affects the dominant hand, and as an occupational injury, is brought on by repetitive motions, including from repetitive motions at work, in sports, in typing, and others. Thousands of people are diagnosed with it each year. As far as a designer is concerned, much attention must be paid to the design of machinery and equipment as well as to the motions needed to perform occupational tasks (such as assembly motions, manufacturing motions, handling of parts) that require repetitive motions, as these motions can cause CTS, a condition that if left untreated can cause permanent damage.

To avoid CTS at home and at work,

- keep wrists straight.
- lift objects with whole hand, or both hands.
- make workstation comfortable. At a keyboard, make sure that the fingers are lower than the wrists. Do not rest the heel of hands on the keyboard.
- take frequent breaks.
- type with soft touch not pounding on the keys. If hands ache while doing something, such as sports, ease up or stop.

Repetitive Stress Injury (RSI), (Repetitive Strain Injury and Cumulative Trauma Disorder) RSI is caused by repeatedly putting stress on joints. Repetitive motions are generally considered to be more than five times per minute. Especially in men and women over age 30, this repetitive stressing of the joints can lead to tendonitis (the inflammation of the tendons) or bursitis (the inflammation of the bursae, the small sacs filled with fluid that act as cushions between tendons and bones). Repetitive stress injury may occur due to repetitive actions in workplace or in sports (such as tennis elbow and golfer's elbow). Typing and using a computer mouse are among the most common causes of RSI.

Low Back Pain (LBP) Low back pain is attributed to manual lifting activities, and despite efforts at educating the workers and improving work conditions, continues to be a major problem in industry. Generally, low back pain episodes are sporadic, but usually happen in the same region. This is an indication of injury to the bones and disks in the area. Repeated injuries eventually lead to chronic low back pain.

Both personal and workspace-related characteristics contribute to low back pain. Personal characteristics include age, sex, muscle strength, medical history, fatigue, genetic factors, and others. Work-related factors include lifting heavy objects, moving and lifting bulky objects, lifting object from the floor, frequent lifting, bending, slips and falls, vibration, and trauma. Low back injuries are affected by both the moment and the compressive stresses on the back. As the distance of the load from the spine increases, the moment on the back increases as well. The compressive stress is related to the weight of the load. As the weight increases, so does the compressive stress on the low back.

The load on the back can be calculated both by applying proper equations [23] or through programs such as the NexGen Ergonomics, Inc. ManneQuinPRO[TM] programs [5] discussed earlier.

The National Institute for Occupational Safety and Health (NIOSH) and the Department of Labor have recognized this problem and have developed guidelines for workers' safety. One of these guidelines is called Lifting Index (LI) which is defines as

$$LI = \frac{\text{Load weight}}{\text{Recommended weight limit}} = \frac{L}{\text{RWL}} \qquad (8.12)$$

where L is the weight of the object to be lifted. RWL can be calculated from

$$RWL = LC \times HM \times VM \times DM \times AM \times FM \times CM \qquad (8.13)$$

These factors are shown in Table 8.18, where

- H is the distance of the hands holding the load from the midpoint between the ankles;
- V is the vertical distance of the hands above the floor;
- D is the absolute difference between the vertical heights of the points of origin and destination of the load; and
- A is the angle between the midplane (mid-saggital plane) and the worker's body at the origin or destination of the load.

Low back pain may also be caused by improper sitting posture, improper gait, athletic activities, or shoes. In fact, countless innovative designs and products have been

TABLE 8.18 Factors Used in the Calculation of the Lifting Index

		Metric	U.S. Customary Units
LC	Load constant	23 kg	51 lb
HM	Horizontal multiplier	25/H	10/H
VM	Vertical multiplier	$1 - (0.003|V - 75|)$	$1 - (0.0075|V - 30|)$
DM	Distance multiplier	$0.82 + (4.5/D)$	$0.82 + (1.8/D)$
AM	Asymmetric multiplier	$1 - (0.0032A)$	$1 - (0.0032A)$
FM	Frequency multiplier	See note below	See note below
CM	Coupling multiplier	0.9-1.00	0.9-1.00

The Frequency Multiplier varies between 0 and 1 depending on how often a load is lifted (up to 15 times per minute) and work duration (<1 hour, >1 but <2, >2 but <8).

designed to help people with their back pain as a result of driving or sitting behind a desk. When the spine is exposed to long period of sitting without adequate movement, especially when the posture is not right, back pain may develop.

EXAMPLE 8.8

Figure 8.22 shows four individuals walking across a street. As you notice, the three individuals in the front are wearing shoes with short heels. As shown, their legs are normally stretched as they walk, and their posture is normal. The last individual is wearing high heel shoes, and as shown, her posture is not straight and the legs are bent. This can contribute to back problems and bad gait. Many other items of clothing, though fashionable at times, can similarly contribute to health problems with their respective effects.

FIGURE 8.22   The effect of high heels on gait and posture. Other items of clothing can have ■
similar effects on the human body. (Reprinted by permission from Mustang Daily.)

Rapid Upper Limb Assessment (RULA) is a program which allows for a quick assessment of upper limb stresses. The conditions under which the upper limbs operate are entered. A score is reported back, indicating whether or not any corrective actions must be taken. [25]

Permissible Exposure Limits (PELs) define the limits of exposure for approximately 500 chemical and other substances. These are contained in the U.S. Department of Labor's OSHA standard 29 CFR 1910.1000. For details about these limits please see Reference 26.

8.21   HUMAN FACTORS AND MAINTAINABILITY

The following list is from the MIL-STD-1472 military standards for maintainability:

- Avoid the use of special tools. Not only are special tools expensive and not always available, in emergency situations, they may not be accessible when needed, creating a hazardous situations. Their application should only be considered in extreme situations where it is deliberately intended that unauthorized individuals not be able to operate a machine or interfere with a process. For example, fire hydrants require a five-sided wrench, preventing unauthorized individuals from opening the valves.

- Make wear-out items replaceable as modules. This allows the repair or maintenance of the product cheaper, easier, and faster.

- Design adjustments to be mutually independent. This means that each adjustment must be independent of other adjustments, allowing the personnel to set the product or machine easily and quickly. Otherwise, as one variable is adjusted, others may change accordingly, including others who are already adjusted and set.
- Provide failure indications to make troubleshooting fast, simple, and easy.
- Check for proper accessibility to parts.
- Design for goof-proof re-assembly by using noninterchangeable fittings, distinctive patterns, color codes, etc.
- Use knobs for adjustments that are made frequently.
- Minimize the number of fastener types. This helps prevent mistakes while providing for quick assembly and re-assembly.
- Provide captive fasteners if dropping or loss of the item may result in damage to equipment or danger to personnel. As an example, the inadvertent dropping of a bolt into the oxygen tank of a Titan-1 rocket during modification work could have caused an explosion of the rocket destroying the launch pad and possibly injury or death of personnel if the impeller of the tank hit the bolt causing a spark. [15] This was only prevented because the quality assurance supervisor counted the final number of remaining parts and noticed that one was missing, reported back, and insisted that the rocket be rechecked.
- The equipment must be rendered inoperative when maintenance workers have to get any portion of their bodies inside machines with moving parts.

REFERENCES

1. HERTZBERG, H. T. E., G. S. DANIELS, and E. CHURCHILL, "Anthropometry of Flying Personnel-1950," WADC Technical Report 52-321, USAF, Wright Air Development Center, Wright-Patterson AFB, Ohio, September 1954.
2. HUNSICKER, P. A., "Arm Strength at Selected Degrees of Flexion," Report No. WADC-TR-54-548, 1955, Wright Air Development Center, WPAFB, Ohio.
3. HUNSICKER, P. A., "A Study of Muscle Force and fatigue," Report No. WADC-TR-57-586, 1957, Wright Air Development Center, WPAFB, Ohio.
4. PHILLIPS, C. A., "Human Factors Engineering," John Wiley and Sons, New York, 2000.
5. www.nexgenergo.com and www.humancad.com.
6. MORGAN, CLIFFORD, J. S. COOK, A. CHAPANIS, and M. W. LUND, Editors, "Human Engineering Guide to Equipment Design," Sponsored by Joint Army-Navy-Air Force Steering Committee, U.S. Printing Office, Washington D.C., 1963.
7. VAN COTT, HAROLD P. and R. KINKADE, Editors, "Human Engineering Guide to Equipment Design," Revised Edition, U.S. Government Printing Office, Washington, D.C., 1972.
8. DREYFUSS, HENRY, "The measure of Man; Human Factors in Design," Second Edition, Whitney Publications, Inc., New York, 1967.

9. http://www.hec.afrl.af.mil/HECP/cardindex.shtml.
10. WHITESTONE, J. J. and K. M. ROBINETTE, "Fitting to Maximize Performance of HMD Systems," from Head-Mounted Displays: Designing for the User, James Melzer and Kirk Moffitt, Editors, McGraw Hill, New York.
11. HARRISON, CATHERINE R. and KATHLEEN M. ROBINETTE, "CAESAR: Summary Statistics for the Adult Population (Ages 18–65) of the United States of America," AFRL-HE-WP-TR-2002-0170, Society of Automotive Engineers (SAE), Warrendale, PA June 2002.
12. GORDON, CLAIRE C., et al., "1988 Anthropometric Survey of U.S. Army Personnel: Methods and Summary Statistics," United States Army Natick, Research, development and Engineering Center, Massachusetts, 1989.
13. TILLMAN, PEGGY and BARRY TILLMAN, "Human Factors Essentials; An Ergonomic Guide for Designers, Engineers, Scientists, and Managers," McGraw Hill, New York, 1991.
14. PHILLIPS, C. A., "Human Factors Engineering," John Wiley and Sons, New York, 2000.
15. KONZ, S. and S. JOHNSON, "Work Design; Industrial Ergonomics," Holcomb Hathaway Publishers, Scottsdale, Arizona, 2000.
16. "University of California at Berkeley Wellness Letter," February 1992, p. 2.

17. CHILES, J. R., "Systems Give Warning before They Break up. The Trick Is to Listen—and Learn," *Mechanical Engineering* New York, March 2004, pp. 36–38.

18. SUTER, ALICE H., "Noise Sources and Effects—A New Look," *Sound and Vibrations,* Bay Village, Ohio, January 1992, pp. 18–32.

19. http://www.bom.gov.au/info/thermal_stress/.

20. TALBOT, THOMAS F., "Chain Saw Safety Features," Paper # 86-WA/DE-16. Presented at the Winter Annual Meeting of the American Society of Mechanical Engineering, Anaheim, CA, December 7–12. 1986.

21. GIBSON-HARRIS, SHEREE, "Expecting the Unexpected: Hazard Recognition in Machine Design," Paper # 86-WA/DE-11. Presented at the Winter Annual Meeting of the American Society of Mechanical Engineering, Anaheim, California, December 7–12. 1986.

22. "Question of Survival," *U.S. News and World Report,* January 8, 1990.

23. CHAFFIN, DON B., G. B. J. ANDERSSON and B. MARTIN, "Occupational Biomechanics," third Edition, John Wiley and Sons, New York, 1999.

24. BEAKLEY, GEORGE C. and ERNEST G. CHILTON, "Design, Serving the Needs of Man," McMillan Publishing Co., New York, 1974.

25. http://www.rula.co.uk/index.html.

26. http://www.osha.gov/SLTC/pel/recognition.html.

HOMEWORK

8.1 Using a program such as NexGen Ergonomics ManneQuinPRO or anthropometric tables, decide the internal diameter of the fuselage of a commuter airplane based on the following assumptions:

- The cross section of the fuselage will be a circle.

- There will be three persons seated in each row, two to one side, and one on the opposite side, with a walkway in between.

- The comfort level of the passengers has to be weighed against the drag, extra weight, and higher operating cost of a bigger diameter fuselage.

- A commuter airplane flies short distances between cities. An example may be a flight between San Luis Obispo and Los Angeles or between New York and Boston (say up to 1.5 hours long).

Make a copy of the set-up, including the human model and print the result indicating the dimension of the cross section of the fuselage.

8.2 Using a program such as NexGen Ergonomics ManneQuinPRO or anthropometric tables, design a seat for a commuter airplane and determine its height, width, and depth. As you can probably imagine, the level of comfort of a passenger in an airplane must be weighed against the additional cost of operating the airplane. Decide the internal diameter of the fuselage of a commuter airplane based on the following assumptions:

- The cross section of the fuselage may be an ellipse, with the ratio of major to minor axes no more than 1.1.

- There will be three persons seated in each row, two to one side, and one on the opposite side, with a walkway in between.

- The comfort level of the passengers has to be weighed against the drag, extra weight, and higher operating cost of a bigger diameter fuselage.

- A commuter airplane flies short distances between cities. An example may be a flight between San Luis Obispo and Los Angeles or between New York and Boston (say up to 1.5 hours long).

Make a copy of the set-up, including the human model, the dimensions of the seat you designed, the distances between the seats, and the fuselage cross section. Indicate what percentiles you used for each dimension. Print the results.

8.3 Study a few different faucets and the direction the handles are turned for on or off. Do all faucets behave the same way? Can you predict what way a faucet must be turned for on or off? What is your preference? Ask at least five other individuals about their preference, and compare with your preference.

8.4 Study common logos used by major institutions such as American Society of Mechanical Engineers (ASME) or American Society for Engineering Education (ASEE). Would you arrange the letters similarly? What would be your preference?

8.5 Design a few signs for an airport. The signs must be easily understood by others from other countries as well as young children and older individuals.

8.6 Study and criticize five signs used in an airport or similar throughway. How would you improve these signs?

8.7 Study the design of the interface of a video tape recorder and the way it must be programmed by a user. Is it adequate or appropriate? How would you improve it?

8.8 Analyze the gauge-cluster of a 1980s GM bus, shown in Figure P.8.8.

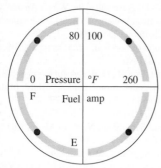

FIGURE P8.8 The gauge-cluster of a 1980s GM bus.

8.9 How can you improve the arrangement of the burners and the knobs in a six- or eight-burner stove (Figure P.8.9)?

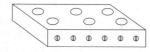

FIGURE P8.9

8.10 Analyze the shower set-up of Figure P 8.10, and suggest how it could be improved.

FIGURE P8.10

8.11 Which one is better, to have a heater in a room to warm it, or to place a kettle of water on the same heater? Why?

8.12 Analyze your car radio's interface. Report on any shortcoming you notice. Make suggestions about how to improve it.

CHAPTER **9**

AESTHETICS OF DESIGN

Beauty is only skin deep; ugliness goes deep to the bone.
—Unknown source

9.1 INTRODUCTION

One of the most important deciding factors for selecting a product, especially for consumer products, is the aesthetics of the product. Aside from brand name loyalties or word of mouth, the first thing that attracts a customer to a product is the way it looks, even though other factors such as price and specifications may play a vital role in final decision making. You can probably remember many occasions too when you were either attracted or repulsed by the shape of a product and consequently, either at least considered the product or completely ignored it. Just go to a local dealer of automobiles, audio store, or even a clothing store and you will find products that you will dislike as well as products you will like, all based on the way they look, regardless of their quality, price, or specifications.

Aesthetics of a product or system relate to the way the product or system looks, its colors, shape, lines, textures, proportions of different elements, how the elements relate to each other, and the way they develop a perception about the qualities of the product. Just depending on elements of aesthetics such as color and line or texture combinations, a product may look stable or unstable, strong or weak, modern or old, and elegant or cheap. For example, a customer's perception of a camera may be very different depending on whether the camera is gray or yellow (perhaps with animal figures on it). Although otherwise the two cameras may be exactly the same, one projects a sense of serious product, high quality, made for adults; the other childish or youth oriented with lower quality, made for fun.

Aesthetic elements can have two distinct qualities to them, emotional, and intellectual or logical. Emotional qualities of aesthetics relate to their ability to arouse a sense of emotional satisfaction and enjoyment (or dissatisfaction and repulsion). Emotional qualities are generally subjective and can be perceived and interpreted differently by different people. For example, you may see one person like and adore one product, while another hates it and cannot stand it. Neither one can tell you why. They just know they either like or dislike the item. In many cases, you may be able to trace this to past experiences, and to some extent, to the educational background of the person, including exposure and propaganda. You may also be able to describe the reasons behind these emotions. Still, the emotional qualities are subjective and vary from one person to another. Many companies spend great resources in market analysis and research to find out what people like and what they do not like in order to be able to design their products for success.

Intellectual and logical qualities of the aesthetic elements are more related to the way the brain understands the logical relationship between different elements and the

305

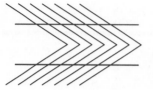

FIGURE 9.1 Although the two lines are parallel, they do not appear so.

environment. For example, there is a strong relationship between the shape of a part and the way it behaves mechanically, between its sharp corners and safety, and between its grooves with the way the part behaves in fatigue loading. Even if the observer is not familiar with engineering concepts such as fatigue loading, stress concentration, or safety requirements, she or he may be able to recognize the relationship or the effect each one has on the consequences, and therefore, there will be a feeling of dissatisfaction or dislike for the product. As another example, consider the way lines are perceived in different backgrounds. Two parallel lines in a crosshatched background look nonparallel. This is not just a feeling, it is seen as such, and it is perceived by the brain as being wrong. If such a thing happens on a product without a good reason, the individual may feel that the quality of the product is less than expected since lines that should be parallel are not (Figure 9.1). Camouflage patterns are the same. We will discuss these qualities in more detail as we discuss elements of aesthetics.

Even design of foods can make a big difference in its value, price, appeal, and pleasure. The combination of colors, arrangement, and design, including the size of each element, play a distinct role in its appeal, and thus, its value. Many cookbooks and magazine recipes are designed to be works of art in their own right. They look great, tasty, healthy, and appealing (Figure 9.2).

According to the psychologist Abraham Maslow, human needs form a spectrum, from basic needs such as hunger to complex needs such as prestige and satisfaction with ones job. Hunger takes precedence over satisfaction with job; even if you are not completely satisfied with your job you may continue to do it in order to eat, although you may eventually try to get another that satisfies you. So are our preferences for aesthetic satisfaction. We prefer to see our aesthetic preferences applied to our daily needs. Thus, we look for objects and environments that satisfy our aesthetic preferences and needs. Some of these preferences are inborn; some are acquired through experience, exposure, or training.

Table 9.1 is a very simplified list of different schools of thought in Art History. The table shows how arts affected everyday lives of the people through products, architecture, posters, and performing arts.

(a)   (b)

FIGURE 9.2 Beautiful foods are attractive too. To see color photos, go to www.wiley.com/college/niku

TABLE 9.1   A Short History of Arts

Industrial Revolution	1760–1840	
Victorian Era	1820s–1900	Ornate design to show wealth, include Gothic architecture, colored printing Words secondary to images, spaces filled
Arts and Crafts	1850s–1900	Crafts preferred to mass production, reaction to industrialization Less ornamental, more object oriented, organic decorations
Art Nouveau	1890–1910	Floral, curvilinear motifs, international practices, simplified forms Text and images combined, free-flowing forms, also applied to architecture
Cubism	1911–	Figures into geometric frames, bold colors, flat planes
DADA		Affected by Futurism, opposition to WWI, anti-art, negative, anti-elements-of-design
Surrealism	1934	Faith in man and spirit, intuition and feelings, opposition to DADA
Russian Suprematism	1914–1932	Elemental abstraction, rejected political and social roles, seeked artistic roles, perceptual effects of colors
Constructivism	1922	Opposed to artistry, seeked ideology, opposite of suprematism Natural materials and textures, architectural elements, asymmetrical balance, use of halftones
De Stijl	1917	Equilibrium and harmony in art, influenced by Cubism, use of primary colors only Art should be part of everyday life, part of everything
Bauhaus	1919–1933	To deal with life at an aesthetic level, simple geometric shapes, influenced shape of buildings in the U.S. and Europe, emphasis on colors and shapes Ultimate goal: functional aesthetics
Art Deco	1920–1930s	Unlike Bauhaus, this was about decorative arts for a society dominated by wealth. Notion of speed, indicating modern society. Women were an important element. Yellow and orange favored
Swiss Design	1950s–	Objective clarity, simple shapes, little change, sometimes boring Asymmetrical organization of design elements, mathematical proportions, strong grid behind the design
American Design	1950s–1970s	Pragmatic, experimentation in techniques, originality of concepts Utilitarian and aesthetically pleasing, logical, ethical design
Post-Modernism	1960s–	Started with applications in architecture, but expanded to other fields, anti-Bauhaus and Modernism, antiorganization Pastel colors common
Digital Age	1990s–	Advent of digital design, computers

9.2   AESTHETIC ELEMENTS OF DESIGN

Aesthetics (or esthetics) of design are the combination of a number of different elements that collectively create the shape and the feeling that the particular shape presents. These include, but are not limited to, form and function, line, color, harmony, variety, style, balance, proportion, and composition. These elements work together in creating the particular perception that they project to the customer, and are affected by each other. In the following sections we will discuss these elements with some examples.

It should be mentioned here that many engineering designers shy away from this subject, feeling that this is just an art-related issue, and due to lack of formal training, they will never be able to either understand it or effectively apply it. Although this may be true for some individuals, many designers are assigned or are expected to work on the aesthetics of a product as part of their jobs. Therefore, it is important to know about aesthetics for two reasons. One, because it is in fact very possible that an engineering designer may end up working on the aesthetics of a product, and two, because in most situations, teams of engineers, designers, industrial designers and even artists may work together on a product. Even some fundamental understanding of aesthetics can be a great asset to engineering designers as much as any designer. Additionally, there is a strong relationship between creative design and aesthetics of a design. As a creative designer, whether an engineer or not, one needs to have a basic understanding of aesthetics in order to be able to both contribute to and understand the aesthetic development of a product.

9.3   FORM AND FUNCTION RELATIONSHIP

Before reading this section, please first look at Figure 9.3. If this object were attached to a wall, perhaps 16 inches from the ground, sticking out about 4 inches, what do you think it might be?

Form is the collection of the shapes or arrangement of the different parts of an object. For example, the form of an airplane is the collection of the shapes of the fuselage, the wings, the rudder, and the engines in a particular arrangement. The form of a toaster is the collection of the shapes of the body, the knobs, and the slots. In this section, we will discuss the form of an object and its relationship to the function it performs. Of course, the form in itself is not an independent entity. It is composed of the other aesthetic elements such as lines, textures, colors, proportions, etc.

The form of a product has been likened to the interface between the inputs and outputs of two systems, one technical, another biological. The product in question is the technical system while the biological system is the human user. For example, the inputs to a stereo (which are outputs from the user) are the signals, the volume level, the tone, and other ''commands'' by the user. The outputs of the stereo (which are inputs to the user) are the sound, its level, and its qualities, but also the appearance of the stereo. The components of the stereo constitute part of its form (the shape), and its function constitutes another aspect of the form. This form is the only means through which the biological system engages the technical system. This is why form and its elements are such an important aspect of the aesthetics of the product.

Figure 9.4 shows some of the elements that constitute and affect a form according to the Gestalt psychology, including rhythm, symmetry, geometry, number of elements, cohesiveness, and relative location of elements.

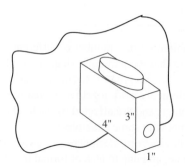

FIGURE 9.3   If you were to see this object attached to a wall, what do you think it might be?

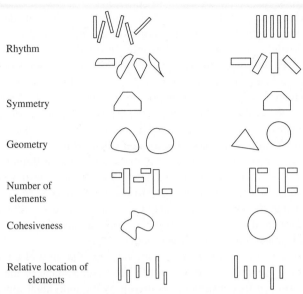

FIGURE 9.4 Form is affected by these elements and their relationship within the composition.

Rhythm is the repetition of elements, either by their shape, position, color, or other qualities. Similar buttons and knobs on a stereo panel comprise an example. Symmetry may relate to the whole form or to the elements of the form. Symmetrical shapes are sometimes deliberately avoided in order to create specific effects. For example, most tables are symmetrical, and yet a nonsymmetrical table may appear more modern, and to some, more desirable. Geometry relates to the shapes of individual elements and the effect they have on the form. Similarly, the form is affected by the number of elements and how they are arranged. In many cases, the number of elements is deliberately reduced by combining functions, by making elements look the same, or by hiding the elements from view. The sleekness of many modern cars is attributed to this reduction in the number of elements as headlights are covered by the same grill pattern or hidden when not in use, handles and bumpers are painted the same color as the body, and curves are reduced to blend different parts of the body together. Cohesiveness is defined as the ratio of the length of the element (the perimeter) to its surface. A circle has the lowest value. Additional curvatures create larger values. And finally, the location of elements creates order or lack thereof. For example, the buttons on a panel are arranged in straight lines or a square pattern in order to create a feeling of order. Buttons that are scattered around the panel will lack the same feeling. Similarly, if letters of text are not written in a straight line, they project a sense of disorder. This was used in a poster about confused children to transmit the feeling of confusion to the observer.

In the first half of the nineteenth century, Horatio Greenough, a famous sculptor, published a book entitled ''Form and Function,'' in which, he put forward ideas about form and function and their relationship. In the 1860s, Fredrick Law Olmsted and other landscape architects popularized the concept of detached houses planted inside grassy areas like flowers (incidentally, compact lawn mowers came out at about the same time). These ideas became the basis of the early twentieth century architectural functionalism and the popular idea that ''Form Follows Function,'' borrowed from Louis Sullivan, an American architect of early twentieth century.

Greenough discussed the idea of inside-out design versus outside-in design in architecture.[1] According to this concept, the overall design of an object should be the sum of its requirements from within, and not the opposite. For example, consider the design of a house in the fashion of the ''Little House on the Prairie'' log cabin, where a rectangular shaped structure with straight walls all around would be divided internally into the required rooms, staircase, bathroom(s), kitchen, etc. In this case, the design is outside-in. The outside is given and set; the inside is matched into it. Conversely, consider an inside-out design concept, in which, based on necessities and desires (how many bedrooms, how many bathrooms, where is the best place for the kitchen to provide the best traffic, what is the best view and for what purpose, . . .) the different elements of the house would be placed together first. The form of the house would then be the collection of these elements. This, unlike the given rectangular design, provides a completely different form for each set of requirements. The same ideas can be extended to mechanical design and product design, in which, instead of using predetermined forms in which to place the components, one would expect that the form of the object or product would be influenced by what the product is supposed to do. Thus, unnecessary ornamentations, facades, and unrelated spaces have no place in the final form of the product.

In the *function-directed* or *function-dictated* form concept, the form of a product is supposed to be driven by its function. In other words, the unified shape of the product would be the collective form of its constituents as driven by the individual functions of each part. As an example, consider the sewing machines of Figure 9.5. As you see, a somewhat antique sewing machine of the 1950s and the brand new sewing machine of the 2000s look practically the same, except for the differences in colors, materials, and modern lines. In fact, if you were to look at the first sewing machines developed in the 1850s by Singer, you would see that it practically looked the same. Why? Because the collection of the necessary shapes of its parts dictated the form of the sewing machine. The sewing machine, as invented originally and as it still functions, requires stitches that are made by looping one thread and passing another thread inside the loop, performed from under and above the cloth. As such, the material had to be accessed from under and over the cloth. Additionally, since the cloth was held in place by the left hand, and since the design of the mechanism required a rotary motion, the right hand was used along with a handle to provide the input motion (never mind the few of us who are left-handed). There had to be enough distance between these two parts to equal the distance between the two arms, but also to provide space for the cloth. The spool had to be placed somewhere too. All of these together created the original shape or form of the sewing machine, which has not significantly changed since then. In fact, even though the design was later improved to accommodate operation with a foot pedal and crank (treadle), the other parts of the sewing machine remained the same, even if unnecessary. And later, even when an

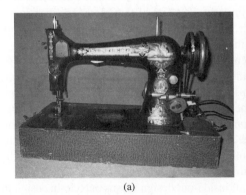
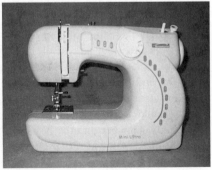

(a) (b)

FIGURE 9.5 Sewing machines, as in the past and present.

(a)   (b)   (c)

FIGURE 9.6 Vacuum cleaners and coffee-makers, and even an old washing machine, like most other products, are designed based on the function-dictated form concept.

electric motor was added to rotate the mechanism, the basic shape remained the same. All because the basic shape followed the function of the product, which made sense and was practical and useful. A few years ago, I saw a small, handheld sewing machine in the shape of a date-stamping machine used in offices, whereby a handle is pushed up and down to sew. It was extremely inexpensive ($6 at the time), and could be used for small repair jobs. Since its sewing function was different, the machine looked completely differently from the classic sewing machines of Figure 9.5.

The same can be said for majority of other devices and products we use everyday. For example, the first vacuum cleaners (called suction sweepers, invented by Murray Spangler in 1908 and later taken over by Hoover) were essentially the same form as today, although the styling has changed much and you will see many different features in each model. The early suction sweepers needed to contact the surface (floor) while moving on a wheel, which resulted in the front suction area. The motor, running a vacuum pump, was placed right behind it. A handle was attached to the machine for handling it, with a bag, hung from the handle to collect the dust and dirt. The same concept is still valid, although all these elements could have been placed in a big rectangular box, with all details hidden inside. And what about an early washing machine with its integrated water-squeeze or a coffee-maker (Figure 9.6)? Is an airplane not designed based on the same concept? In fact, it is. The fuselage, the wings, the rudder, and the engines are put together based on their function, dictating the final form of the airplane. There are no extra, conventional, unnecessary, elements in the design that do not belong. Is it possible then to design an airplane in a shape other than the conventional shape? Of course it is. There are hundreds of variations in the basic design, many of them patently novel. But still they all follow the same basic logic embedded in the function-dictated form concept.

An important issue related to this concept is the fact that when function dictates the form, it is much easier for the consumer to understand what the product is about, what it is supposed to do, how it works, and how to handle and use it. This is very important both for the manufacturer and for the customer. For the manufacturer, it is useful in exposing customers to the product without much effort. The customer will guess what the product is by the virtue of looking at it, will understand it, will be able to operate it with fewer instructions, and will follow the function of the product in more detail. For the customer, too, this is important in understanding the product and how to work with it, but also in using it more productively and safely, and even more importantly, knowing what to do in

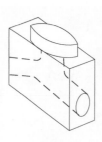

FIGURE 9.7 A faucet and its functional design.

emergencies. When the customer better understands the function of the product, she or he can react more decisively in emergency situations when the product must be stopped, moved, changed, or manipulated.

By now you have probably guessed what the object shown in Figure 9.3 was. Clearly, it was a faucet. Or could you not tell? Obviously, you could easily make a faucet with a rectangular cubic body that functions exactly the same as a regular faucet. In fact, you can actually buy variations of this design at stores. However, unless deliberately designed so, the disadvantage of this design is that it can be completely unclear to others as to what it is and how to use it, even in an emergency. Taking away the façade and only keeping the functional portions of the design (Figure 9.7) will yield the common form of faucets that we all understand. There are hundreds of variations to the design of faucets, some more beautiful and clever, some very typical, some easy to use, some very difficult to figure out. Still, most faucets have a functional form.

Form and function are also related in the opposite way, namely form-dictated function. For example, imagine a peanut-butter maker. This machine is intended for making fresh peanut butter by grinding peanuts and collecting the peanut butter in a container, as schematically shown in Figure 9.8. Further, imagine that you are trying to wash off the peanut butter in the bottom of the container. A sharp corner at the bottom may lead to difficulty in cleaning the container, whereas a rounded bottom renders it very easy. Although this minor point may have no other effect on the product, this simple element of the form can affect its function.

Furthermore, in critically loaded machine parts, the shape of the part can significantly affect the performance and life, and therefore, the function. This is related to what is called stress concentration (discussed in more detail in Chapter 16).

Imagine a bar with a cross section of A, loaded axially with a force F. The average normal stress in the part will be (Figure 9.9a):

$$\sigma = \frac{F}{A} \tag{9.1}$$

FIGURE 9.8 Schematic of a peanut butter maker.

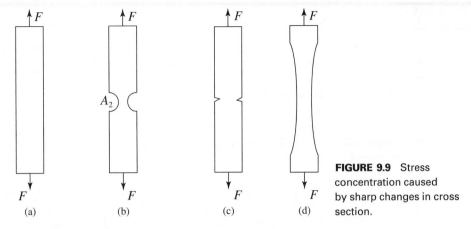

FIGURE 9.9 Stress concentration caused by sharp changes in cross section.

(a)   (b)   (c)   (d)

If there is a reduction in the area, then stress will increase proportionally to the reduced area. The average stress is (Figure 9.9b):

$$\sigma_{ave} = \frac{F}{A_2} \tag{9.2}$$

However, if the reduction in the cross section of the bar is sharp, the maximum stress can be significantly larger than the average stress. The increase depends both on how sharp the change is as well as whether the material is ductile (it has the capacity to yield, deform, and redistribute the higher stresses more evenly) or brittle (it does not yield or deform). In that case, the maximum stress is multiplied by a stress concentration factor K that can be as high as 5 or more:

$$\sigma_{max} = K \frac{F}{A_2} \tag{9.3}$$

To test the effect of a sharp change in the area in the load-carrying capacity of a part, pull a rubber band and notice how much load it takes before it breaks. Next, cut a small notch in the rubber band and repeat (Figure 9.9c). You will see how easily the rubber band breaks when you pull it, even though it is so ductile. Repeat the same, but this time cut a smooth, long notch (Figure 9.9d). Although the area is the same as the notched band, it carries almost the same load as the original rubber band. Another example for a brittle material is glass. As soon as you create a scratch on the surface with a diamond tool, it will easily break at the scratch.

You will need to consult a machine design book for stress concentration values. However, what is important in this discussion is that if the designer is not careful in the way the shape of an object or part is designed, and inadvertently creates sharp changes, notches, creases, or grooves in the part, the part may be subjected to very high stresses that will damage or ruin the product. This is even more important if the load is varying, such as in fatigue loads. The 1972 Ford Capri's shifter had a deep groove that caused it to fail in fatigue. Eventually, the company had to recall the car in order to change the shifter. This was only due to its shape.

It should be mentioned that function-dictated form is only one of many concepts. Clearly, there are many other possibilities, from the older school of outside-in design to modern designs to many other theories developed over the years. In fact, there are many products that do not follow this concept anyway. So the designer should be aware of the power of this concept, especially in the American consumer product market, but also remember that innovation and breaking rules is the name of the game.

9.4 LINES

Lines can be defined as a mathematical entity with characteristics such as length, direction, and shape (straight, curved, skew lines, etc.), or as an aesthetic component with potent emotional value.

Aesthetic lines are created intentionally as a geometric entity by drawing a line for emphasis, attention, direction, or similar purposes (such as racing stripes on a car's exterior), or are the result of variations in colors, textures, surfaces, contrasts, and materials. For example, when two surfaces come together they create a line, as do two different colors or materials or textures. Whether intentionally drawn or the result of other variations, lines can be very powerful emotional elements.

Figure 9.10 shows a few simple lines. Each one of these lines may induce a particular feeling that is subjective, but nevertheless, real. See if you can identify the type of emotion these lines arouse in you. Perhaps, as in many others, you may feel a sense of calm or sadness in 1, aimlessness or broken heartedness in 2, playfulness or growth in 3, anger in 4, and of course, happiness, sadness, or anger in 5, 6, and 7.

Figure 9.11 shows two simple examples of the above-mentioned effects. Airliners are most often decorated with straight-line paintings along the window line. Almost certainly you have never seen an airplane painted with wiggly lines, since this imparts a feeling of turbulent flight and is unacceptable. Similarly, the smile on the nose is painted to impart friendliness, whereas fire would never be used in an airliner. But what do you see on a fighter jet? Obviously these are just lines and have absolutely no effect on the way the airplane flies, behaves, or reacts to disturbances. But the perception that these lines create can be extremely different.

Now imagine that you have designed a product, say a toaster, with a crease in the front body panel to strengthen it, and with two knobs, one for time, one for darkness (Figure 9.12). If not careful, you may create a happy or sad feeling in the product that can be a subliminal message to a customer. If you do not care for the feeling that the smiling toaster projects it is due to another effect. This particular shape feels as if it can rock, and therefore, it looks less stable than the sad toaster. We will discuss this effect later. This is why the designer should be careful about the effects of lines in the design.

Lines can be deceptive, either intentionally or unintentionally, and can create illusions. For example, as shown in Figure 9.13, a rectangle with horizontal lines may appear to be shorter and wider than the same rectangle filled with vertical lines. This is used in the garment industry to induce a feeling of being bigger (more muscular) or thinner and taller. Similarly, camouflage patterns are designed to deceive the viewer by reducing the ability to see the perimeter of the object. The patterns are not random lines and colors, but are researched to create the largest effect.

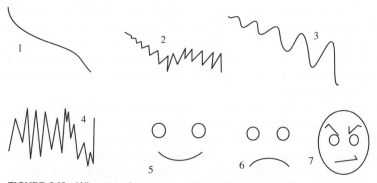

FIGURE 9.10 What emotions can you detect in these lines?

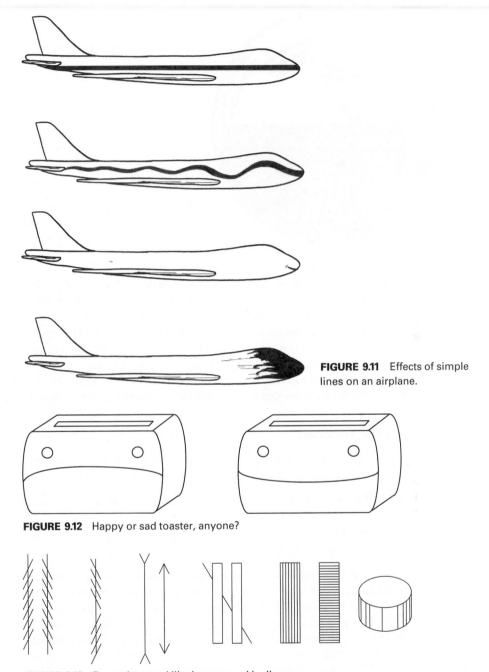

FIGURE 9.11 Effects of simple lines on an airplane.

FIGURE 9.12 Happy or sad toaster, anyone?

FIGURE 9.13 Deceptions and illusions caused by lines.

9.5 COLOR

Color is a potent aesthetic element, capable of imparting a variety of emotions. Like lines, color too has a logical or intellectual quality, e.g., its wavelength, heat generation or energy capability, diffraction qualities, and many more. But it also has powerful emotional qualities which we will discuss shortly.

FIGURE 9.14 When rotated, the Mason's wheel will appear colorful even though there are no colors in the original figure.

Colors do not exist per se, except as an effect on the cells of our retina. It is only the effect of the electromagnetic waves within a narrow range (0.4–0.8 micrometers) on our eyes that we perceive as color. Within this narrow range, our eyes can detect perhaps millions of different tints and color variations. Luckiesh[2] suggests a Mason wheel, shown in Figure 9.14 as a test of this. If you copy this figure, cut it and mount it on a rotating wheel (I suggest pasting it on a cardboard and attaching it to the eraser of a common pencil) and rotate it; you will notice a number of different colors appearing on the wheel. This shows that colors can be seen even though they originally do not exist in the figure.

9.5.1 Color Theories

There are two basic color theories that deal with its source and combinations (this does not include electromagnetic wave theory which is the physics of colors):

- **Munsell's** theory is based on lights, stating that lights are additive, and that when all colors are added, they create white light.
- **Ostwald's** theory is based on pigments, stating that they are subtractive, and that when all pigments are added, they create black.

As you see, these relate to two different concepts. For example, sunlight contains all different colors, from red to violet, and when passed through a prism, it separates to the colors of the rainbow. If those colors are added together again, they will create white light. This is what is used in creating color images on the television screen or a monitor. There are in fact three sets of red, green, and blue pixels on a television screen that create three separate images in red, green, and blue, each individually scanned and reproduced. When superimposed, they create all other colors on the monitor (we will talk about other possibilities shortly). On the other hand, pigments (such as paints) are subtractive. They absorb most of the light projected onto them. The colors they do not absorb are reflected, thus the color we see in the material. If all the light is absorbed, the object will appear to be black (lack of color or light). Thus, if white light is projected onto an object that appears red, it means that all other color lights are absorbed except the red, which is reflected and we see it. Thus, if the same object is seen in a light that has no red, no light will be reflected, and thus, the object will appear to be black. This is why the same color will appear to be different if seen in the sunlight, florescent light, or incandescent light.

9.5.2   Primary Colors and Their Derivatives

There are three primary colors for lights and three primary colors for pigments. All other colors can be created by changing the proportions of these primary colors. The three primary colors for lights are red, green, and blue (Figure 9.15). The three primary colors for pigments are red, blue, and yellow. Primary colors cannot be produced from other colors.

For lights, red and green create yellow, green and blue create cyan, and blue and red create magenta. Red, blue, and green together create white light. In pigments, red and yellow produce orange, yellow and blue produce green, and blue and red produce violet. Orange, violet, and green are called secondary colors. Similarly, combining primary and secondary colors at different proportions produce intermediate colors. Mixing of three primaries at different proportions creates tertiary colors.

An alternative to the use of the three primary colors of red, green, and blue is to use the three secondary colors of cyan, magenta, and yellow. If printed, combining cyan, magenta, and yellow creates grays not black. Thus it is also necessary to add black ink to be able to print color pictures. A system using these colors is called CMYK, for cyan, magenta, yellow, and black. Combining these colors in correct proportions can create all other colors, both in print and in lights. Figure 9.16 shows a picture and its CMYK constituents. Notice the intensity of the colors for each combination.

In artistic applications, there are three primary color characterizations: Pure colors (hue), white, and black. There are also four secondary characterizations:

- **Tint** is a color plus white.
- **Shade** is a color plus black.
- **Gray** is white plus black.
- **Tone** is a color plus gray.

Chromatic and Achromatic Colors   Chromatic colors are colors, as we know them, defined by their hue, such as yellow, red, green, and blue. Achromatic colors are the variations of gray shades between black and white. Grays are obtained by combining the three primary pigments. Grays vary from white to black.

9.5.3   Color Characteristics

Colors are specified by three qualities:

(a) **Hue**, which is the name of the color such as red, green, blue, brown.

(b) **Saturation** (intensity, brilliance, purity, strength) which is the color's brightness or dullness. The best way to change saturation without changing the value of the color (see below for value) is by adding gray to the color. As the gray intensity increases, the saturation decreases. The maximum saturation level of a color is equal to the

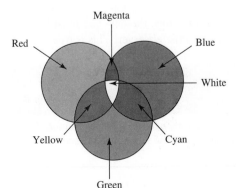

FIGURE 9.15 Primary and secondary colors for lights. To see color photo, go to www.wiley.com/college/niku

The original picture

Cyan   Magenta   Yellow   Black

FIGURE 9.16   An image and its CMYK constituents. To see color photos, go to www.wiley.com/college/niku

intensity of the color as white light leaves a prism. Neon colors are high in intensity (low in grey) and therefore are very bright.

(c) Value, which is quality of darkness or brightness or amount of light reflected from the color. For achromatic colors, black, white, and all grays have zero saturation, but different values. Similarly, for any given hue, as more white is added to the color, its value decreases. Pale colors (pink) are low in value. Vivid colors (red) are high in value.

Both chromatic and achromatic colors are affected by their background, and thus, the way they appear to the observer changes as the background changes. Figure 9.17 shows a set of nine circles, all at the same gray level. However, the same gray levels, when viewed in different backgrounds, appear to be different. This shows the importance of the background to the perception of intensity or value of colors.

The same is true for chromatic colors. For example, as shown in Figure 9.18, the same orange color appears to be very different in a background that is yellow or red. The same can be expected for all other colors. Thus, the designer must be careful about color combinations and the way they appear to the observer. This is true even in clothing

FIGURE 9.17   The same gray level appears differently in different backgrounds. To see that this is true, cover the backgrounds with a piece of paper to see that the circles are all at the same grayness level.

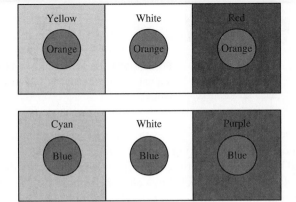

FIGURE 9.18 The same color appears to be different in different backgrounds. To see color photos, go to www. wiley.com/college/niku

design. The same shirt, worn with different jackets, will appear differently, as would a tie with different color shirts. We will discuss other effects of color combinations shortly.

9.5.4 Visual Effects of Colors

Colors have many psychological and visual effects, both positive and negative. For example, certain color combinations are preferred by men and others by women. Different colors can be seen at different distances, and depending on the situation in which they are used, these effects may be very important. As an example, orange and yellow are very outgoing colors and are noticed more than other colors, say blue or brown. Orange and red-orange hold max attention, while blue is hazy and indistinct. As a result, many designers use these colors in order to attract attention (such as in window decorations) or to divert attention (such as in hiding obtrusive features). On the other hand, red and green are visible from a large distance, while blue and violet are not. Thus, if a traffic-related sign is made in blue, it may not be seen from a distance, creating a hazard. Similarly, it is impossible to distinguish yellow and orange from afar, as is green and blue. The most legible color combinations are black on white, black on yellow, green on white, red on white, and white on blue. Table 9.2 is a sample of some of these effects.[4]

Light pastel colors such as pink, light blue or green, yellow, cream, and white, are serene and comforting. Thus, they are used in painting living rooms, classrooms, factories and offices, and other living or working quarters. Purple and violet are royal, splendid, and elegant, while reds are initially exciting and lively, but frustrating if too much. Some restaurants use these colors to create excitement in their customers as they enter their

TABLE 9.2 Sample of Some Visual Effects of Colors

Color Preferences	For men: Blue, red, violet, green, orange, yellow
	For children: Orange, red, blue, green, violet, yellow
	For women: Red, blue, violet, green, orange, yellow
Color Visibility	To men: Black, red, green (all on white background)
	To women: Red, green, black (all on white background)
Maximum Range of Visibility	Red 3.5 miles
	Green 3 miles
	White 2.5 miles
	Yellow 1.5 miles
	Blue, violet 0.75 miles

FIGURE 9.19 Darker colors and black appear heavier than lighter colors and white.

establishments, but also to get rid of them soon after they are served in order to create space for new customers. Too much blue can bring laziness, but green will not.

It should be mentioned here that many of the above effects are different for different people. Not everyone reacts to colors or is affected by colors the same way. The cultural value of colors is also different. In Western cultures, red is a sign of danger, left, and communism, as well as love, while yellow is cowardly. In Asian cultures, yellow and red are signs of life and happiness.

Lower frequency colors such as yellow, orange, and red are warmer, primarily because as we move toward infrared, the heat energy of the color increases. In the opposite extreme of the color spectrum, as we get close to violet and ultraviolet, colors have little heat energy, but more chemical energy. Therefore, they are cooler, but have more penetrating capacity. Similarly, since black and other dark color pigments absorb most of the light projected onto them, reflecting little, they get warmer. Thus, products with darker paints can also get warmer (such as an automobile), affecting their feel and behavior in the summer or winter.

Colors have an associated weight or heaviness as well. Darker colors feel heavier while lighter colors feel lighter too. Black and darker gray colors also appear to be heavier than lighter grays and white. For instance, as seen in Figure 9.19, the same product on the right feels top-heavy, and perhaps unstable, compared to the one on the left which appears bottom heavy, and thus, more stable.

Additionally, since black and dark colors absorb most of the light energy that is projected onto them, an object that is black or dark appears to be smaller as well. This is why most evening gowns are in these colors, not only because they are elegant, but also because the person wearing them appears thinner and taller.

Table 9.3 is a list of psychological effects of colors and their associations. As mentioned earlier, these associations are affected by culture and background.

TABLE 9.3   Colors and Their Psychological Associations and Effects

Color	Mood Created	Sample Association
red	stimulating; positive	fire, danger, blood, stop, hot, left
yellow	cheerful, warm, outgoing	sun, caution
orange	like red	Heat
yellow green	neutral	
green	restful, cool, serene	nature, vegetation, go
blue	opposite of red, cool, subduing	sky, water
purple	solemnity	royalty, mourning
white	neutral to positive effects	snow, marriage, purity, (death in China)
gray	neutral, contemplation	Overcast
black	neutral, sorrow	death, darkness

Chapter 8 contains additional information on colors and their effects. Please refer to Section 8.14 for more information.

9.6 HARMONY, UNITY

Every design is composed of a collection of shapes, lines, colors, or other elements. Harmony is the way the visual elements in a composition relate to each other, achieved through the repetition of similar but not identical forms.

Unity may take many different forms. For example, in a painting, the use of large brush strokes throughout the painting may establish a sort of harmony, although the painting still has different colors, shapes, etc. Monet's drawings have this characteristic. Harmony in a building may be achieved by the use of similar windows and doors, all of similar composition and material, although not the same size or proportions. The exterior of a building can benefit from the repeating shape of rectangular bricks, even though the repetition may be broken by other elements such as doors and windows and angles, or different materials integrated into the building. And finally, the design of a car may have the same repeating element throughout the design that gives it unity. The Ford Taurus of the late 1990s had repeating oval-shaped elements such as windows, lights, front grill, rear trunk panel, and door handles.

The lack of a unifying element can look chaotic. The campus of California Polytechnic State University in San Luis Obispo is populated with buildings that have no relationship to each other. One huge building is made of concrete, while the next one is brick, the next one stucco, and so on. Every building seems to have a different material, tone of color, different roof, and different character. You can find a modern-looking building next to a classical and old structure, next to another that has no common features with the first two. In contrast, the core area of Stanford University is all ornamented stone buildings with similar roofing and similar exteriors and repeating characters, although the different buildings do look different from each other.

Similarly, imagine that you buy the best audio and video components you can find and try to use them together. You may end up with wood grain color, matte silver, black paint, or polished chrome next to each other, one with large round knobs, one with small pushbuttons, one with red indicator lights, the others with LEDs in blue and green and orange, all with different widths. There may also be a jungle of wires in the back of the system. The combined system looks ugly and chaotic because it lacks unity or harmony. Instead, imagine that all the components, sometimes designed and sold as a package by some manufacturers, would have the same face plates, similar knobs and lights, same color finish, and same width. Although each component may look different based on its function, they collectively have similar elements that relate them to each other.

The same is true with reports, newsprints, books, and posters. When too many different sizes and fonts are used together, they lack unity. As mentioned earlier, in the design of a poster about confused high school children, many different fonts were intentionally used together in order to emphasize the main point of the poster.[5]

On the other hand, as we will see next, it is important that the designer provides enough variations in the application of the unifying element; otherwise, the design may become boring and dull.

9.7 VARIETY

Variety is the opposite of unity. Too many similar repetitions of the same element will create a monotonous and boring design. For example, the over-application of the same elliptical shape in the design of the late 1990s Ford Taurus can take away from its beauty.

A building with millions of red bricks everywhere can be uninteresting, and a stereo system with all exact shapes repeated throughout will be bland. Thus, the designer must strike a balance between enough repetition of similar elements to unify the design, but simultaneously create enough variations to make it interesting. One of the reasons that flagstones are used is that they provide variation in shapes and tone, but also create unity through material and repeating colors and shades (Figure 9.20).

In addition to economic factors and the quality of workmanship, track houses are generally less desirable because they are not varied enough. However, if an adequate number of variations in the design are provided among any group of track houses, the similarity of the materials and colors and shapes in the design can in fact make the block more appealing than an area where every house is completely different in shape, material, color, roofs, etc. Common, popular cars have the same problem. Although they are popular for a variety of reasons, owners try to "customize" their car by adding other features to the car that make it unique.

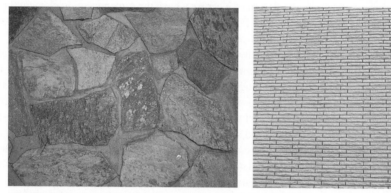

FIGURE 9.20 Flagstones provide unity in material and colors and shades as well as variety through varied shapes and tones. Overuse of bricks can be dull and bland.

9.8 BALANCE

Balance is the feeling of equality of elements in weight, attention, color, attraction, etc. For example, in Figure 9.21 the small square is darker and thus appears to be relatively heavier than the bigger but lighter square. These two can balance each other if properly painted, sized, and positioned. Otherwise, two equal shapes with different colors may project a feeling of imbalance. The same is true for other characteristics of different elements. A design in which one dimension is excessively larger than another may also create a feeling of imbalance. You may see this in the design of some cars as well. For example, the 1980s Subaru Sports Coupe was designed with an excessively long front

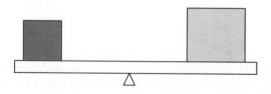

FIGURE 9.21 Balance is a combination of size, weight, and location.

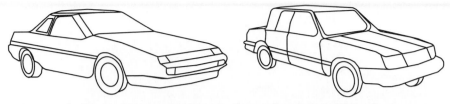

FIGURE 9.22 Schematic representations of a Subaru Coupe and a Dodge K-car.

engine compartment and trunk and with slanted lines throughout. The Dodge K-cars were long in the front, short in the back, with slanted lines in the front and vertical lines in the back. The trunk of both cars were shorter than the front engine compartments. At least to this observer, they both looked extremely ugly, both for their proportions as well as the use of lines that were exaggerated. In both cases, the relative dimensions did not relate properly to the remaining dimensions and the direction of lines in one area clashed with the lines in another area. Balance was lost (Figure 9.22).

9.9 PROPORTION

Proportion is the relationship between sizes, numbers, positions, angles, space, colors, and scales of the elements in the design.

A common ratio used in many structural and natural designs, usually referred to as the "golden ratio," is 1 to 0.618, which is equal to 1.618. What is interesting about this ratio is that the ratio of all successive *Fibonacci numbers* yields this ratio of 0.618 (after the first 5). Fibonacci numbers can be found by adding any two previous Fibonacci numbers, starting with 1. The second Fibonacci number is also 1, the third is $1 + 1 = 2$, the fourth is $1 + 2 = 3$, next is $2 + 3 = 5$, $3 + 5 = 8$, $5 + 8 = 13$, $8 + 13 = 21, \ldots$ Dividing the Fibonacci numbers, we will get

$$1/2 = 0.5, \quad 2/3 = 0.667, \quad 3/5 = 0.600, \quad 5/8 = 0.625, \quad 8/13 = 0.615,$$
$$13/21 = 0.619, \quad 21/34 = 0.618, \quad 34/55 = 0.618, \quad\quad\quad 55/89 = 0.619$$

As you see, the ratio quickly converges to the Golden ratio. You may try this for hundreds of numbers, and they all hold true. It should be mentioned that there are infinite different Fibonacci series, and in fact you can start anywhere. For example, with a starting number of 20, multiply by 1.618 to get 32, and then add $20 + 32 = 52$, the ratio of 32/52 is also 0.615, the next number is $52/84 = 0.619$.

The Golden ratio has been found to be a significant ratio in many respects. For one thing, it is a comfortable and pleasing ratio to humans, and thus, has been incorporated into the design of many products and structures, including 3×5 cards, 5×8 cards, 8.5×14 legal paper size, and 210×297 mm paper size (the standard ISO-A4 paper used in most countries). But more significantly, this ratio has been used extensively in the design of classic structures for the width to height ratio of buildings, windows, and doors (see Figure 9.23). Many body proportions are also close to this ratio (although not quite exactly the same).

As shown in Figure 9.24, drawing an arc at a radius of the distance from the midpoint to an apex of a square will yield a Golden ratio ($\sqrt{0.5^2 + 1^2} = 1.118$ and $1.118 - 0.5 = 0.618$). If you draw a rectangle with Golden ratio (#1) and then add successive squares to the result, they all remain at the same Golden ratio. Drawing successive arcs inside the squares will yield a spiral.

FIGURE 9.23 Golden proportions used in classic buildings (Fontana di Trevi, Rome).

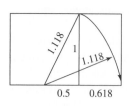

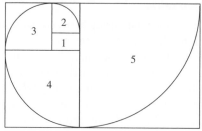

FIGURE 9.24 Golden ratio and a spiral.

Interestingly enough, many natural spirals follow the same pattern, including plants and leaves, ram's horn, shells, and pinecones. The number of seeds in sunflower rows, three sections in bananas, cucumbers, and bell peppers, five fingers, five appendages in starfish, five seed pods in apples and pomegranates, eight in oranges, and many others are additional examples of the Fibonacci numbers in nature.[6] This is used as evidence that the Golden proportions are a natural occurrence and that when the same proportions are used in our designs, they feel natural as well. You may verify this by measuring the dimensions of common products, from mirrors and napkin dispensers and common-size books to tables and wastebaskets (these ratios are not exactly 0.618, but are generally close). In many cases, the choice of dimensions may not even be intentional, but they just appear to be the right proportions.

Obviously, this does not mean that all proportions should follow the Golden ratio, and fortunately, they do not. Otherwise, it would be a boring world to have everything follow this proportion. However, the designer should always be careful about dimensional proportions.

9.10 COMPOSITION

Composition relates to the way all the different aesthetic elements of a design are integrated, and includes the size, shape, and location of elements, their color, their number, and their importance. For example, as was mentioned earlier, a dark color seems heavier

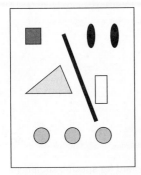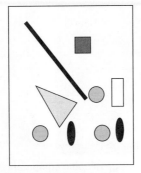

FIGURE 9.25 Compositions can be changed to create balance between different shapes, colors, intensities, and locations.

FIGURE 9.26 This water color painting is composed to show simultaneous balance and stability. Does it convey the same feeling to you?

than a light color. So, if different colors are used in a design, the elements with these different colors will induce a different weight, as well as emphasis, on the design. This can be used when an emphasis is needed or when the designer wants to attract attention to a particular characteristic, for example, to create a perception of bottom-heaviness in a product to emphasize stability. Otherwise, there may be a need to place these elements in such a way as to create a balanced appearance. Figure 9.25 shows two compositions, one with more balanced integration, one with less. Figure 9.26 tries to demonstrate simultaneous balance and stability by distributing the elements in relation to their colors, and yet, a relatively bottom-heavy arrangement. Does it convey the same feeling to you? The designer needs to use the different elements of the design in a way to create the balance required for satisfaction by customers.

9.11 STYLE

Style, or content, is the essence of design, and relates to what the design is supposed to convey to the observer. In product design, it relates to the way the product appears to the customer. As you watch a product, it may relate a certain characteristic to you that tells of its quality, workmanship, and function, but also about the type of product it is. For example, a product may be designed as modern, as art deco, as old-fashioned, as home-style, western, or many others. The PT-Cruiser of Daimler-Chrysler was designed as a look-alike of the cars of the past, as was the BMW Roadster. A Ferrari and a Camry are designed to convey sporty elegance or family-oriented images. The difference between different refrigerators is not in their compressors, as most of them are made by a couple of companies. The difference is in the product's styling, the way everything is put

FIGURE 9.27 Examples of design of audio equipment designed for specific audiences (JVC™ and Nostalgic 4-in-1 audio system).

together and integrated, where the doors are, what the shelves are made of, how many different adjustments are available, and how many different minor functions it performs. The same is true for video and audio equipment. Practically all major video components are the same, made by the same company or two who supply them to other manufacturers. The difference is mostly in their styling and the different gadgets that go along with the video components.

The very popular Ford Mustang was designed based on the same chassis, engine, and transmission of the Ford Falcon, which was a failure. Most internal components of the two were the same; just the cab was redesigned and named Mustang. But the different styling of the Mustang was so popular that even after nearly 40 years and many design changes over the decades, the car is still popular and the old Mustangs are still in high demand by fans. Many internal components of the German Ford Capri of the 1970s were the same as the Ford Pinto. The body style was very different (as was the gas tank!). The Pinto was a disaster; the Capri was very popular. This shows the power of styling in a product. It can make or break a product. It can make the product a success, or it can make it a failure. Next time you are in an audio/video store, look at the styling of different products and what the style tells you. Look for the difference between a slim-looking clean design and a product designed apparently for the young (at heart?), populated with large speaker grilles and countless huge knobs and buttons in different colors, compared to a nostalgic design of another that attempts to bring back memories of the past (Figure 9.27). In fact, there is no reason that products such as these are any better or worse than any other design. The actual quality and worth of the product is related to the way it is designed and manufactured. However, the styling of the product can convey a completely different feeling to the customer.

On the other hand, it is still important to remember that although styling, especially at the onset of the relationship between a product and a customer, is extremely important, it is equally important to have a quality product that will be satisfactory to the customer. In the 1960s, there was an advertisement by Volkswagen of America, Inc., in which six 1949 cars were displayed next to each other, including a VW Bug, a Packard, a Studebaker, a Tucker, and a Desoto. The bottom of the ad read ''Where Are They Now?,'' indicating that all those mentioned, except for the VW, were long gone. The connotation was that the American companies were paying attention only to the styling of the cars, making them larger and bigger, with more elaborate radiator grills and more chromium-laden exterior parts and not many improvements in the mechanical components of the car. In fact, to a large extent, this was true as the American manufacturers were changing models every year, but using the same chassis, engine, and transmission. VW was the opposite, where the body style was modified in a minor way, but apparently, the mechanical components were improved. As a result, the advertisement claimed, we are still

(a)   (b)

FIGURE 9.28 Advertisement posters from the late 1950s and early 1960s.

around; they are not. However, the lack of variation and change in styling eventually caught up with VW as well and it was taken off the market, except in Brazil. This shows how the designer must strike a balance between quality and styling, improving both the function and the appearance of the product. They are both important, and they must both be addressed. Figure 9.28 shows two advertisements from the late 1950s and early 1960s.

The phenomenon of annual style change or *rapid obsolescence* is used in many industries, including the automobile, camera equipment, electronics, computer, and garment industries. As the style changes, the older versions become obsolete and lose their value. However, if the turn-around time for a major, substantive design change is longer than the frequency of styling changes, it is obvious that only minor, superficial improvements are possible, sometimes even detrimental. For example, Harley Earl, a designer with the Oldsmobile division of General Motors, inspired by the wrap-around design of the Lockheed's P-38 fighter plane's windscreen, incorporated a similar design into the 1954 Olds, capturing a significant lead in the market. However, it was soon discovered that the curvatures in the windshield could create a serious hazard due to distortions at the curves as well as glare and double vision from its sculptured surfaces.[1] More testing with these innovations and improving the windshields before releasing the product into the market could have prevented the hazards.

Styling can also be important in comfort, performance, and the price of a product. Highly (and beautifully) stylized products can be priced higher if customers associate more sentimental value to them. The styling of a product such as a chair can also affect its comfort and utility, as well as the way it is manufactured and used. Many internationally famous designers have involved their name and expertise in the design of everyday consumer products, as evidenced by the brand names and signatures these products carry. At times, the styling and the brand name behind the product make a huge difference in the way the product is perceived by the customer. And yet, there is no reason why industrial designers, as well as engineering designers, should not be able to design equally stunning and beautiful products that are valuable and satisfying to the customer.

9.12 ART AND TECHNOLOGY; A RECIPROCAL RELATIONSHIP

So far, we have discussed the influence of arts on technology. However, a reciprocating relationship exists between technology and art and how technology affects arts and at times, influences it. For example, nondestructive evaluation techniques have been introduced to unravel hidden images under current art, detect forgeries, and devise strategies for preservation.[7,8] Some of these techniques have even found their way into the movies. What was impossible before can now be done with the aid of technology.

Additionally, new technologies have found their way into the modern aesthetic aspects of products and systems. New paints and glosses, powder coating techniques, sculptured laminated wood, mass produced paintings, and digital arts are but a few simple examples of how technology is used in the aesthetic design of our everyday products and systems. Artists are becoming experts in metal forming, vapor deposition, and many others. Pictures of deep space, taken by Hubble telescope, are sold as posters. Even digital pictures of scientific works are finding their way into the world of arts.[9] There is no doubt that as new technologies emerge, more and more will find their way into the arts and aesthetic aspects of products.

In this chapter we discussed some basic and fundamental ideas about aesthetics of design. For more information about the artistic aspects of design aesthetics, please refer to other references such as (3).

REFERENCES

1. SPARKE, PENNY, "An Introduction to Design and Culture in the Twentieth Century," Harper and Row, Publishers, New York, 1986.
2. LUCKIESH, M., "Visual Illusions: Their Causes, Characteristics, and Applications," Dover Publications, Inc., New York, 1965.
3. OCVIRK, O., R. STINSON, P. WIGG, R. BONE, and D. CAYTON, "Art Fundamentals: Theory and Practice," McGraw Hill, New York, 1998.
4. HARRISBERGER, LEE, "Engineersmanship; a Philosophy of Design," Brooks/Cole Publishing Co., Belmont, California, 1966.
5. RACKOW, LEO, "Postercraft," Sterling Publishing Co., New York, 1972.
6. HAMMEL GARLAND, TRUDI, "Fibonacci Numbers in Nature," poster, Dale Seymour Publishers, Palo Alto, 1988.
7. CORTES-COMERER, NHORA, "The Second Unveiling," *Mechanical Engineering*, September 1987, pp. 36–42.
8. CORTES-COMERER, NHORA, "Out From the Shadows," *Mechanical Engineering*, September 1987, pp. 45–50
9. BAER, TONY, "Art and Technology: Reciprocal Inspiration," *Mechanical Engineering*, October 1987, pp. 60–63.
10. WOODHAM, JONATHAN M., "Twentieth-Century Design," Oxford University Press, New York, 1997.

HOMEWORK

9.1 Visit your local museum and look at the displays of paintings, furniture, and other articles. Look for variations in these articles from different times and see if you can describe differences between them. Analyze the use of colors, the composition, and other aesthetic elements.

9.2 If you have a chance, visit a contemporary or modern museum and look at their displays of modern art. See if you can describe the differences between classical art and modern art. Which one is your favorite? Why?

9.3 Consider two tables or chairs seen in a showroom, a museum, or a catalog. Compare the aesthetic elements of the two and list their differences and how they affect the way these two appear to you. Which one is your favorite and why?

9.4 Visit a local coffee shop and look at the coffee makers they have on display. Compare their designs, colors, features, as well as aesthetic elements of the two. Describe how these elements affect the appearance of the coffee-makers.

9.5 Consider two cars, one your favorite, and another you do not like. Compare the aesthetic elements of the two and list their differences and how they affect the way these two appear to you and why.

9.6 Different people prefer different atmospheres for their living quarters, some like it modern, some country style, some "Martha Stewart" home style, etc. What is your favorite and why? Can you verbalize the effects of this environment on yourself?

9.7 Compare two successive generations of cell phones and discuss how their aesthetics have changed. Describe the effects of these changes on the utility and appeal of the cell phones. Can you guess what the next generation might look like?

9.8 Select one product whose aesthetic you do not like (not its function). Redesign the product into a shape, form, and composition that you prefer. Then describe what these changes were and how they improved the design.

9.9 Draw your favorite futuristic concept car. Describe why you think it is nice and why you like it.

9.10 Design a new table, both the top surface and the legs. Describe why it is a good design, both in terms of its utility and its aesthetics.

9.11 Pick one of the pictures in Figure P.9.11 and analyze it for its composition, the use of colors, and other aesthetics elements.

(a)

(b)

FIGURE P.9.11   To see color photos, go to www.wiley.com/college/niku

CHAPTER *10*

MATERIAL PROPERTIES, SELECTION, AND PROCESSING

10.1 INTRODUCTION

There is an exceedingly large array of material choices available to the designer, from natural materials such as wood to man-made ones such as plastics, and from common materials like cement and steel to exotic superalloys and biomaterials. You may, in fact, get advanced degrees on the science and engineering of materials, in production, applications, and research. Obviously, in this book we will not be able to cover all the knowledge you will need about materials to be an effective designer. However, we will discuss a number of issues that relate to the relationship between design issues and material properties, selection, and manufacturing. The suggested references at the end of this chapter should also help you with further readings about this subject.

As was discussed in Chapter 7, there is a direct relationship between the choice of materials and the consequences of this choice. These range from cost, weight, methods of manufacture, and aesthetics, to life expectancy, quality, and others. The designer should learn about material properties and manufacturing methods in order to be able to select proper materials and proper methods of manufacturing a product or system.

In this chapter we will discuss some general material properties and standards, followed by methods of material selection, and finally, some common methods of material processing. Without a doubt, hands-on experience with material processing (such as shop courses and building your own projects) learning about material properties, alloys, plastics, and modern material, as well as learning about standard practices are essential tools in successful design. We will discuss some of these within this chapter at an introductory level, but expect that the designer will learn more within appropriate courses and through experience.

10.2 MATERIAL PROPERTIES

In Chapter 16 we will discuss some material properties that relate to engineering materials such as steel in the context of stress analysis and fatigue strength. These common properties and material behavior are not unique to steel; most materials have similar behavior and can be defined by the same properties. Of course this does not mean that every material has the same property, but that different materials can be defined by how they behave under similar conditions. The following is a summary of some common material properties that can be used to classify materials for selection:

Proportional Strength, Elastic Strength, and Ultimate Strength As will be discussed in Chapter 16, when a specimen of a material is loaded with a force or moment, its deformation can be studied in order to classify its strength. Most materials in nature will deform (elongate, shorten, bend, or rotate) when subjected to loads (tensile,

330

TABLE 10.1   Common Properties for Steel Alloys

	E (psi)	S_y(kpsi)	S_u(kpsi)	Brinell Hardness
Carbon Steel	$27-30 \times 10^6$	27–110	40–273	86–388
Alloy Steel	$27-30 \times 10^6$	53–260	110–273	150–627
Stainless Steel	$27-30 \times 10^6$	30–80	75–120	137–595

compressive, bending, or torsional). A general plot of stress versus strain may be used to specify proportional strength, elastic strength, yield strength, ultimate strength as well as fracture strength. Please see the discussion in Section 16.4.

Table 10-1 shows the range of values for modulus of elasticity, yield strength, tensile strength, and Brinell hardness for different steels. Table 10.2 is a summary of yield and ultimate strengths as well as Brinell hardness for some common steels and other metals.

Table 10.3 is the list of modulus of elasticity (E) and modulus of rigidity (G) for a number of different materials.

Hardness   This property is defined as resistance to local penetration, scratching, wear, and machining, but is also a measure of yield and true ultimate strength, work hardening, and modulus of elasticity. One common way of measuring it, called Brinell Hardness

TABLE 10.2   Common Properties for Metals

UNS Number	Type	Yield Strength (kpsi)	Tensile Strength (kpsi)	Brinell Hardness
Cast iron		8–40	18–60	100–300
Wrought iron		25–35	45–55	100
Structural steel		30–40	50–65	120
G10100	CD	44	53	105
	HR	26	47	95
G10150	CD	47	56	111
	HR	27	50	101
G10180	CD	54	64	126
G10180	HR	32	58	116
G10200		30	55	
G10350	CD	67	80	163
G10350	HR	39	72	143
G10400	CD	71	85	170
G10400	HR	42	76	149
G10500	CD	84	100	197
G10500	HR	49	90	179
Stainless steel	18-S	30–35	85–95	145–160
Aluminum, pure	rolled	5–21	13–24	23–44
Brass		8–80	40–120	50–170

CD: Cold drawn; HR: Hot rolled

TABLE 10.3   Moduli of Elasticity and Rigidity for Different Materials

Material	Modulus of Elasticity		Modulus of Rigidity	
	Mpsi	GPa	Mpsi	GPa
Aluminum alloys	10.3	71.0	3.8	26.2
Brass	15.4	106.0	5.8	40.1
Carbon steel	30.0	207.0	11.5	79.3
Cast iron, grey	14.5	100.0	6.0	41.4
Copper	17.2	119.0	6.5	44.7
Douglas Fir	1.6	11.0	0.6	4.1
Glass	6.7	46.2	2.7	18.6
Inconel	31.0	214.0	11.0	75.8
Stainless steel	27.6	190.0	10.6	73.1

Number (BHN), is to force a hardened sphere under a known load into the surface of a material and measuring the diameter of the indentation left behind. The BHN is then calculated by dividing the load by the area of the indentation. Table 10.2 lists the BHN values for some common materials.

Toughness and Fracture Toughness are defined as resistance to crack propagation. To measure them, a sample of the material with a deliberate crack in it is loaded in tension until the crack grows. The toughness and fracture toughness are calculated from these values. You should consult a handbook for values of toughness and fracture toughness if your application includes design calculations with these values.

Machinability refers to the ease at which the material can be machined, and influences the cost, both in time as well as the type of machines that will be needed for manufacturing. Machinability is measured experimentally as the relative cutting speed for a fixed tool life under prescribed cutting standards. It is influenced by thermal conductivity, hardness, and reduction in area at fracture.

Creep is defined as permanent deformation of materials under constant load at higher temperatures, even if the load is below the yield strength. This can happen to metals as well as other materials. Metals, especially at high temperatures, but also at room temperature are subject to creep. So, it is important to consider long-term effects of creep in buildings, bridges, structures, and machine. For other materials, creep may be much more significant. For example, wood deforms relatively easily under constant load. This is a common problem in wooden beams, bookshelves, and many other products. Leather is the same, although at times this can be used to our advantage while forming the material.

Thermal properties of materials include their rate of expansion as temperature increases, rate of thermal conductivity, as well as the effects of temperature on other material properties. For example, most steels become very brittle at low (near freezing) temperatures, and are thus subject to fracture (in 1943, the S.S. Schenectady broke in half as it was being launched off the coast of Portland, Oregon).

Electrical and magnetic properties include electrical conductivity, dielectric constant, and electromagnetic properties such as their magnetic response. Ferrous materials respond to magnetic fields. However, some stainless steels (containing nickel) are not magnetic. Aluminum, copper, and other nonferrous materials are not magnetic.

10.3 TYPES OF MATERIALS

Ashby[1] postulates that at the beginning of the design process, when the product or system is conceptually designed, there may be thousands of candidate materials available. As limits on mechanical, thermal, electrical, and other technical requirements are applied to the range of possible available materials, the total will be reduced to 10–50 candidates. Further modeling, experimentation, and prototyping will reduce the choices to 5–10. And finally, application of working prototypes, final design requirements, and engineering analysis will further reduce the choices to 1–2, from which, the designer picks the best candidate.

So what are the thousands of choices before us? Consider a ''seating product,'' perhaps a chair or stool. Think about all the possible materials that can be used to make the chair, natural materials, man-made materials, all sorts of woods and wood products, metals and alloys, cloth, leather, endless arrays of stones and other similar material, countless types of plastics, resins, composites, and the like, as well as strings, ropes, and cables. To this, you should also add the thousands more combinations of different materials for different parts. But when you apply the requirements such as strength, mechanical properties needed in manufacturing, cost, aesthetics, comfort, and utility, the choices are reduced significantly. These are further reduced as you start to model prototypes and when you learn whether or not what you want can be accomplished with the particular material at hand. And finally, you will have a couple of choices from which you pick the best one. Is this not true for an automobile too? A coffee-maker? A pencil?

In the following sections we will discuss material properties for some common and useful materials. There are countless other materials on the market. You must consult other references for material properties.

10.4 METALS, ALLOYS

The following are some general characteristics and compositions, as well as a numbering system for metals and alloys.

10.4.1 The Numbering System

The Unified Numbering System for Metals and Alloys (UNS) provides a means for specifying metal alloys as well as for correlating many internationally used metal and alloy numbering systems currently administered by different societies, trade associations, and countries. The UNS has established 18 designations for metals and alloys (Table 10.4).

As an example, common low-carbon construction steel can be specified as UNS G10100 HR or CD for hot-rolled or cold-drawn; cold drawing increases the strength of the steel. (Please see Chapter 16).

10.4.2 Steel and Its Alloys

Steel is one of the most common engineering materials. It is strong, relatively inexpensive, and easy to machine. However, there are endless variations to its characteristics when additives are added to pure iron. These include carbon, phosphorous, silicon, sulfur, and other minerals, but also other metals that form alloys with iron, including chromium, molybdenum, magnesium, nickel, and others.

- **Carbon steel** is simply iron and carbon, and accounts for 90 percent of steel production. Higher carbon content increases the strength of steel, but makes it less

TABLE 10.4 UNS Designations for Metals and Alloys

- (A) Aluminum and aluminum alloys
- (C) Copper and copper alloys
- (D) Specified mechanical properties steels (D00001-D99999)
- (E) Rare earths and similar metals and alloys
- (F) Cast irons
- (G) AISI and SAE carbon and alloy steels

G10050-10950	Carbon steel
G11080-11510	Resulfurized carbon steel
G12110-12150	Rephosphorized-resulfurized carbon steel
G13300-13450	Manganese alloy steel
G15130-15900	Carbon steel
G23, 25	Nickel alloy steel
G31, 33	Nickel-chromium steel
G40120-G48200	Molybdenum alloy steel, Cr-Mo alloy steel, Ni-Cr-Mo alloy steel
G81150-G88220	Ni-Cr-Mo alloy steel
G50150-G52986	Chromium alloy steel, Cr-B alloy steel
G61180-G61500	Chromium-vanadium alloy steel
G92540-G98500	Chromium, silicone, manganese, nickel, and sulfur alloy steel

- (H) AISI and SAE H-series steels
- (J) Cast steels
- (K) Miscellaneous steels and ferrous alloys
- (L) Low melting metals and alloys
- (M) Miscellaneous nonferrous metals and alloys
- (N) Nickel and nickel alloys
- (P) Precious metals and alloys
- (R) Reactive and refractory metals and alloys
- (S) Heat and corrosion resistant steels, stainless steel, and iron-based superalloys
- (T) Tool steels, wrought and cast
- (W) Welding filler metals
- (Z) Zinc and zinc alloys

ductile as well. Pure iron can be temporarily magnetized when in a magnetic field; carbon steel can be magnetized permanently.

- **HSLA steel alloy** (high strength, low alloy) has small additions of other elements, typically 1.5 percent manganese, to provide additional strength for a modest price increase.

- **Low alloy steel** may have other elements such as molybdenum, manganese, chromium, or nickel in amounts of up to 10 percent by weight. This improves the ability of the thick sections of the alloy for hardening.

- **Stainless steels** and **surgical stainless steels** contain a minimum of 10 percent chromium, often combined with nickel, to resist corrosion (rust). Stainless steels can be grouped into three categories:

  Austenitic: Typically with 18 percent chromium and 8 percent nickel (known as 18-8). This stainless is nonmagnetic. It is typically used for cutlery, utensils, sinks and food machinery, and appliances.

    Ferritic: Typically contains little nickel, with 12 or 17 percent chromium. This state is magnetic.

    Mertensitic: Typically contains 12 percent chromium and no nickel. It is magnetic too.

    It should be mentioned here that stainless steels are not completely immune to rusting. Many species rust when exposed to natural elements. This is a function of their chromium content and quality. Additionally, if stainless steel is polished or worked with steel brushes, small particles of the brush material can embed in the stainless steel and subsequently rust. Inexpensive barbeques are an example of this.

- **Hadfield steel** (after Sir Robert Hadfield), also known as Manganese steel, contains 12–14 percent manganese. The skin of this alloy can be hardened to superior strengths which resist wearing. This alloy is used for tank tracks, bulldozer blade edges, and cutting blades on the jaws of life.

Internet sites such as http://www.matweb.com/search/SearchUNS.asp allow you to find information on any metal, including yield strength, hardness, tensile strength, and moduli of elasticity and rigidity, but also electrical and thermal characteristics.

10.5 WOODS

Wood, one of the most common natural products used in industry and crafts, is versatile, warm, beautiful, renewable, inexpensive, and strong. Although not as tough as steel or concrete, it is nevertheless extremely useful. In addition to its cost savings and ease of application, because it is more flexible than concrete and bricks and mortar, wood is extensively utilized in earthquake-prone areas like California for building structures. This is a good example of the application of special characteristics of a material to our needs. A structure that can move with the earthquake motions will be less prone to destruction than one that is very rigid; wood provides this compliance, and therefore, is very useful.

Different woods have vastly different strengths and other characteristics. Some woods are strong but also difficult to work with, some splinter, some have longer life when exposed to the environment, some have shorter grains, some longer, and some easily break and therefore are only good for burning. Eucalyptus trees burn very hot in order to help their seed-pods burst and spread their seeds. This natural characteristic makes the wood useful for one application, namely burning, but not useful for others, since it also cracks easily. Table 10.5 is a list of some common woods and their characteristics.

Man-made wood products such as particle boards, plywoods, and veneers have varying characteristics depending on they way they are made, the material used, the cores, thicknesses, etc. For more detail information on these you will have to refer to manufacturers' information.

Even today, after many millennia of using woods for making products and structures, there are new applications for woods as well as new processes that allow us to make new products with them. Different types of plywood, veneers, particle boards, melamine boards, and "engineered" woods are relatively new products. They have been around for perhaps a few decades compared to thousands of years for wood planks and pieces. But you can still see newer wood products and applications in the market. Bamboos have been used for endless applications in the areas where they grow, but their application as engineered wood-flooring planks and cutting boards is relatively new.

As a natural material with its variety of grains, colors, and patterns, wood is an important element in many products. It adds to the beauty of the product, to its aesthetic

TABLE 10.5 Common Woods and Their Characteristics

Tree Species	Average Oven Dry Specific Gravity	Modulus of Elasticity Static, Bending (E) ×10⁶ (psi)	Tensile Strength Perpendicular to Grain (psi)	Compression Parallel to Grain, Max Crushing Strength (psi)	Compression Perpendicular to Grain (psi)	Shear Parallel to Grain, Max Shear Strength (psi)
Hardwoods						
Ash, White	0.64	1.77	940	7,410	1,160	1,950
Basswood	0.40	1.46	350	4,730	370	990
Beech	0.67	1.72	1010	7,300	1,010	2,010
Birch, Yellow	0.66	2.01	920	8,170	970	1,880
Cherry, Black	0.53	1.49	560	7,110	690	1,700
Elm, American	0.55	1.34	660	5,520	690	1,510
Hickory, Shagbark	0.77	2.16		9,210	1,760	2,430
Magnolia, Cucumbertree	0.48	1.82		6,310	570	1,340
Mahogany	0.51	1.50	750		1,100	1,230
Maple, Sugar	0.68	1.83		7,830	1,470	2,330
Oak, Northern Red	0.66	1.82	800	6,760	1,010	1,780
Oak, Pin	0.63	1.73		6,820	1,020	2,080
Oak, White	0.71	1.78	800	7,440	1,070	2,000
Poplar, yellow	0.45	1.58	540	5,540	500	1,190
Sycamore, American	0.49	1.42		5,380	700	1,470
Walnut, Black	0.55	1.68	690	7,580	1,010	1,370
Willow, Black	0.39	1.01		4,100	430	1,250

Softwoods						
Cedar, Alaska	0.44	1.42		6,310	620	1,130
Cedar, Atlantic White	0.32	0.93		4,700	410	800
Cedar, Western Red	0.34	1.11	220	4,560	460	990
Douglas-fir, Coast	0.51	1.95	340	7,230	800	1,130
Hemlock, Eastern	0.43	1.20		5,410	650	1,060
Hemlock, Western	0.44	1.63	340	7,200	550	1,290
Larch, western	0.59	1.87	430	7,620	930	1,360
Pine, Eastern white	0.37	1.24	310	4,800	440	900
Pine, Ponderosa	0.42	1.29	420	5,320	580	1,130
Pine, Red	0.46	1.63	460	6,070	600	1,210
Pine, Shortleaf	0.54	1.75	470	7,270	820	1,390
Redwood	0.42	1.34	240	6,150	700	940
Spruce, Black	0.42	1.61		5,960	550	1,230
Spruce, Sitka	0.42	1.57	370	5,610	580	1,150
Spruce, White	0.45	1.43	360	5,180	430	970

quality, and to its warmth. A coffee pot manufacturer (Chemex™) owes its popularity to the fact that the product is made of glass, with a wooden neck-handle and a leather string. Simulated wood-grain panels used in automobiles, stereos, and cabinets are also evidence of the popularity of wood, although artificial.

10.6 PLASTICS, POLYMERS, AND SYNTHETIC MATERIALS

There are thousands of polymers, each with a different characteristic, too many to even mention here. The vast variety of polymers and plastics available allow us to choose a relevant product for almost any need, whether by strength, long life, low density, price, or any combination of countless other characteristics and requirements. It also makes it that much more difficult to sort through all available choices. One way to try to understand plastics is to look at their general categories which differentiate them from each other. Most polymers are made of the simplest of atoms; hydrogen, carbon, oxygen, chlorine, and at times, nitrogen and fluoride. However, these atoms can be linked together in many different ways that give us an endless variety. A most important characteristic of polymers is that their molecules can chain up together to create continuous, long strands that can be formed, machined, cast, or worked into almost any shape and form (Figure 10.1). Leather, wood, wool, and natural rubber are nature's polymers; synthetic polymers are chemists' contributions to the world of materials. The following is a summary of man-made materials that are called polymers and plastics.

10.6.1 Thermoplastics

Thermoplastics soften and eventually melt when heated, but harden again when cooled. This allows us to mold them into shape, to join parts together, and also to recycle them by re-forming them if necessary. It also allows us to change their physical characteristics by controlling their temperature if needed. Most thermoplastic polymers accept coloring pigments and fillers and fire-retarding agents too.

Thermoplastics include the following. Table 10.6 shows the general characteristics of crystalline and amorphous resins.

Acetal
Acrylonitrile-butadiene-styrene (ABS)
Cellulose
Ionomers
Nylons
Polyamide (PI, Nylon)
Polycarbonate (PC)
Polyetheretherkeyton (PEEK)
Polyethylene (PE)
Polymethylmethaacrylate (PMMA)
Polypropylene (PP)
Polystyrene (PS)
Polytetrafluoroethylene (PTFE or Teflon)
Polyvinylechloride (PVC)
polyethylene terephthalate (PET), a form of polyester

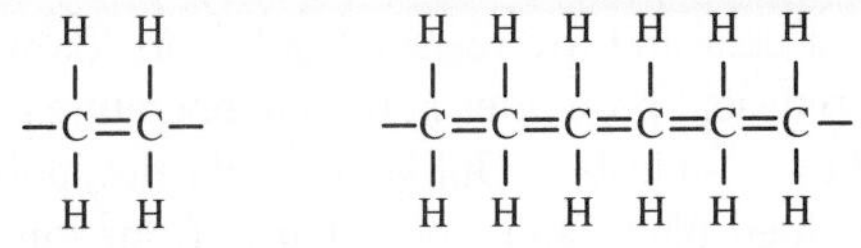

```
  H   H              H   H   H   H   H   H
  |   |              |   |   |   |   |   |
 —C = C—           —C = C = C = C = C = C—
  |   |              |   |   |   |   |   |
  H   H              H   H   H   H   H   H
```

Ethylene monomer   Polyethylene polymer

FIGURE 10.1 A monomer and a polymer. The polymer is a long chain of monomers linked together which can be formed, machined, cast, or worked into shape.

Acetal resins have high tensile strength, stiffness, resilience, fatigue endurance, and moderate toughness. Among thermoplastics, acetals have high resistance to organic solvents, excellent dimensional stability, a low coefficient of friction, and outstanding abrasion resistance. Acetals absorb low amounts of moisture and are resistant to a wide range of solvents, but are vulnerable to strong acids and oxidizing agents. Delrin is in this family. Several grades of acetal are approved by the FDA for food contact applications. Typical applications include bushings, buckles, valves, and gears.

ABS is typically used for parts with important aesthetic requirements. It is used for bicycle helmets, telephones, computer housing and keyboards, and the interior of automobiles.

Acrylic thermoplastics are strong, stable, and weather resistant. The transparency, gloss, and dimensional shape of acrylics are virtually unaffected by years of exposure to the elements, salt spray, corrosive atmospheres, or light from fluorescent lamps. Clear acrylics are normally formulated to stop the ultraviolet radiations below a 370 nm wave length. The light weight and toughness of clear acrylic qualify it as a replacement for glass for many jobs (Plexiglas^{TM} and Lucite^{TM}). Typical applications include lighting panels and skylights, frame glasses, instrument panels, and watch crystals.

Nylons are characterized by their toughness, low coefficient of friction, and good abrasion resistance. Using nylon reduces lubrication requirements, eliminates galling and corrosion problems, and improves wear resistance and sound dampening characteristics. Nylon 6/6 (extruded), the most widely used of the nylon plastics, is available in a number of formulations for molding and extrusion. Nylon 6 (cast) is the second most widely used of the nylons. Its properties are similar to those of 6/6, but it absorbs moisture more rapidly and its melting point is lower. Nylon-6 is used to make fibers for many different types of cloth and socks.

TABLE 10.6 Typical Characteristics of Crystalline and Amorphous Resins

	Crystalline	Amorphous
Resin Family	Acetal	Acrylic
	Nylon	ABS
	Polyethylene	Polycarbonate
	Polypropylene	Polystyrene
		PVC
Typical Properties	Opaque	Clear
	Higher shrinkage	Lower shrinkage
	Lower impact strength	Higher impact strength
	Higher chemical resistance	Lower chemical resistance
	Higher flow	Lower flow
	Higher strength	Lower strength
	Quick solidification	Slower solidification

Polyimides are some of the most heat-and fire-resistant polymers known and have excellent retention to mechanical and physical properties at high temperatures. Polyimides are formulated as both thermosets and thermoplastics. Unlike most plastics, polyimides are available as laminates and shapes, molded parts, and stock shapes from some materials producers and as resins from others. Polyimides can also be extruded or injection molded.

Polyetheretherketone (PEEK) resins have desirable engineering properties, such as toughness, strength, rigidness, and creep resistance. They are resistant to organic solvents, to dynamic fatigue, and retain their ductility on short-term heat aging. PEEK has also been used to make artificial hip joints.

Polyethylene (PE), a polyolefin, is the largest volume thermoplastic polymer used today. It is characterized by its toughness, excellent chemical resistance and electrical properties, low coefficient of friction, near-zero moisture absorption, light weight, and ease of processing and is available in a wide variety of grades and formulations, including LDPE, MDPE, and HDPE. Low-Density PE (LDPE) is characterized by good toughness and flexibility, relatively low heat resistance, low-temperature impact resistance, and clarity in film form. Typical applications include containers, squeeze bottles, covers, and Tupperware^{TM}. High-Density PE (HDPE) resins' rigidity and tensile strength are considerably higher than those of the low and medium density materials. Typical applications include toys, milk bottles, buckets, bottles, and bottle caps.

Polypropylene (PP) thermoplastics offer a good balance of thermal, chemical and electrical properties with moderate strength. They possess a good strength to weight ratio and are extremely stable. Polypropylene can be processed via extrusion, injection molding, blow molding, compression molding and rotational molding. Typical applications include molded-in hinges, bottle caps, appliance parts, and industrial containers.

Polytetrafluoroethylene (PTFE) was the first fluorocarbon and is the most chemically resistant plastic known; only a few chemicals react with it. Mechanical properties are low compared to other engineering plastics, but its properties remain at a useful level over a large temperature range. PTFE properties include excellent thermal and electrical insulation and a low coefficient of friction. PTFE cannot be processed when melted; instead it is processed by paste and ram extrusion or compression molding.

Polyvinyl Chloride (PVC) has an excellent strength-to-weight ratio and superior flame resistance. It is also characterized by its excellent chemical resistance and dielectric properties, good tensile, flexural and mechanical strength, low moisture absorption, exceptional dimensional stability, and good flammability. Rigid PVC is the most common type, possessing a high strength-to-weight ratio, pressure-bearing capability, corrosion and chemical resistance, and low friction loss characteristics. Typical applications include pipes, valves, window frames, and electrical insulation. However, there is concern about their toxicity, both during manufacture and life.

Polyethylene Terephthalate (PET) has excellent wear resistance, a low coefficient of friction, high strength, and resistance to moderately acidic solutions. These properties make it especially suitable for the manufacture of precision mechanical parts which are capable of sustaining high loads and enduring wear conditions. When crystallized, PET has excellent heat resistance but is opaque in this state. When amorphous, PET has excellent clarity, gloss, and toughness.

Table 10.7 compares different resins with their characteristics, including comparative price.

10.6.2 Thermosets

Resins in this family polymerize when catalyzed and/or heated. When set, they will no longer melt if heated; they just degrade. Two-part epoxy polymers used for gluing

TABLE 10.7 Resin Selector Table

Resin Family	Cost	Strength	Stiffness	Toughness	Clarity	UV Resistance	Chemical Resistance†	Important Properties
ABS	xxx	xx	xxx	xxx	no	xx	xx	good appearance, toughness
ASA	xxx	xx	xxx	xxx	no	xxxx	xxx	good appearance, weatherability
Acetal	xxx	xxx	xxx	xxx	no	xx	xxx	stiffness, strength, dimensional stability
Acrylic	xx	xx	xxx	xx	xxxx	xxxx	xx	excellent clarity, weatherability
High Impact PS	x	x	xx	xx	x	x	x	low cost, commonly replaces ABS
Nylon	xxx	xxx	xxx	xxx	no	x	xxx	strength, poor dimensional stability
Nylon 33% Glass	xxx	xxxx	xxxx	xxx	no	x	xxx	high strength, stiffness, temp. resist.
Polycarbonate	xxxx	xxx	xxxx	xxxx	xxxx	xx	xx	clarity, toughness
Polyethylene, LD	x	x	x	xx	xxx	xxx	xxxx	very low cost, chem. resist., flexible
Polyethylene, HD	xx	xx	xx	xxx	xx	xxx	xxx	low cost, medium strength
Polypropylene	xx	xx	xxx	xxx	xx	xx	xxxx	low cost, chem. resist., med. strength
Polystyrene	x	x	xxx	x	xxxx	xx	x	very low cost, clarity, easy to mold
PVC	xxx	xx	xxxx	xxx	xxx	xxxx	xxx	chemical resistance, weatherability

x = low, xx = fair, xxx = good, xxxx = high

†Chemical resistance is used here as a general term. Each plastic has various degrees of resistance to solvents, detergents, petroleum products, alkalines, acids, and other chemicals. Refer to manufacturers' published information for resistance to specific chemicals. For detailed material properties, refer to manufacturers' information or Modern Plastics Encyclopedia. These guidelines are a starting point for designs and should be adjusted depending on the actual materials, processes, and applications.

purposes, handles of pots and pans, rubber used in tires, and silicone polymers are examples of this resin. Most thermoset resins are hard, but some are soft (like natural and synthetic rubbers). Thermosets include epoxy, Phenolic, and polyurethane.

Phenolic resin works well in applications for electrical insulation. Phenolic is a hard, dense material that holds very tight tolerances. It is used in making gears, electrical insulators, terminal blocks, pulleys, washers, spacers, switches, bearings, transformers, machining components, gaskets, industrial laminates, and many other products. Phenolic polymers have excellent dielectric strength; can be machined; are lightweight; resist heat, wear, corrosion, and chemicals; and are mechanically strong and dimensionally stable.

Polyurethanes are among the most versatile materials. Their many uses range from flexible foam in upholstered furniture to rigid foam as insulation in walls and roofs to thermoplastic polyurethane used in medical devices and footwear to coatings, adhesives, sealants, and elastomers used on floors and automotive interiors.

10.6.3 Elastomers

Elastomers are rubbery, soft, and bouncy. Whereas most engineering materials have modulus of elasticity ranging between 1 and 1000 GPa (0.14 to 140×10^6 psi), elastomers' range is 0.001 to 1 GPa (0.14 to 140×10^3 psi). They are capable of returning to the original shape and size with little deformation when a load is removed even if they are stretched many times their original length. Because they are soft, they are also good at clinging to the surface next to them. They are used for making tires, erasers, rubber bands, washers, seals, o-rings, and other similar products. Elastomers include the following:

Acrylic elastomers

Butyl Rubber

Neoprene

Isoprene

Natural rubber

Silicone

Thermoplastic elastomers

Neoprene (polychloroprene) is an extremely versatile synthetic rubber with 70 years of proven performance in a broad industry spectrum. It was originally developed as an oil-resistant substitute for natural rubber. Neoprene is noted for a unique combination of properties, which has led to its use in thousands of applications in diverse environments. Neoprene resists degradation from sun, ozone, and weather; performs well in contact with oils and chemicals; and retains its characteristics over a large range of temperatures.

Silicone oils, polymers, and cured sealants are odorless and colorless, water resistant, chemical resistant, oxidation resistant, stable at high temperatures, and have weak forces of attraction, low surface tension, low freezing points, and do not conduct electricity. These characteristics make them ideal for many applications, from caulking and sealing bathtubs and microwave ovens to breast implants (although they were found to be carcinogenic) and from lubricants and adhesives to gaskets and sealants. Silicone caulking is odorless only when cured. Due to their thermal stability and relatively high melting and boiling points, silicones are often used where organic polymers are not applicable. Silicone is often mistakenly referred to as ''silicon.'' Although silicones contain silicon atoms, they have completely different physical characteristics from elemental silicon.

△ 1	PET	Polyethylene terephthalate
△ 2	HDPE	High-density polyethylene
△ 3	PVC	Polyvinyl chloride
△ 4	LDPE	Low-density polyethylene
△ 5	PP	Polypropylene
△ 6	PS	Polystyrene
△ 7	OTHER	Other plastics

FIGURE 10.2 Common designations for recycling plastics.

10.6.4 Recycling of Plastics

Many plastics are recyclable, and this is an important characteristic to consider. Generally, thermoplastics can be remelted easily and cast or used as raw material for other products. In many cases, it is required that the material used for manufacturing be ''virgin'' or new. Many consumer products used in industry (such as the food industry) are in this category. But many other products can be made from recycled materials, including carpets, cases, and lawn chairs. For example, recycled polyethylene can be used to make many new products, including fiber for polyester carpets and shirts, athletic shoes, luggage, upholstery and sweaters, fiberfill for sleeping bags and winter coats, and automotive parts such as luggage racks, headliners, fuse boxes, bumpers, radiator grilles, and door panels. Figure 10.2 shows the common designations used for recycling plastics, and should be used for promoting recycling and for assisting in their recycling. However, an important issue during the design phase is to consider the recycling of the separate elements, and how they will be recycled. For example, if different parts, whether all plastics of different types or combinations of plastics and other materials such as metals, are assembled together, will it be possible to separate different parts from each other for recycling? This leads to the possibility or the lack thereof of recycling for that part.

10.7 OTHER MATERIALS

Many other materials may be considered for product design and engineering applications, some with specific characteristics that set them apart from others, some novel materials, and some with anisotropic characteristics. These include glass, concrete, stones, composites, biomaterials, and ceramics.

10.7.1 Composites

Composite materials are hybrid and generally made of two or more distinct materials (the reinforcement and the matrix) with different properties that complement each other and remain separate and distinct within the finished product. This means that although mechanically they work together, they are not mixtures. For example, if fiberglass fibers are impregnated with a resin, when loaded, the fibers will carry the tensile loads and thus give the structure its tensile strength, while the resin carries the shear loads and gives the structure its rigidity and shape. Together, they accomplish what neither one of the two can accomplish alone. Another common example of composites is many millennia old and is

FIGURE 10.3 The reinforcement in the composite material may be continuous filaments, whiskers, or particles.

even mentioned in the Bible, when the Israelites were not given straw for their mud bricks. In fact, adding straw to the mud adds structural strength to it in tension and makes it much stronger, although the straw alone would not be a useful structural element.

Many woods and plant stems are composites too. For example, woods are much stronger along the direction of their grain because their cellulose fibers are in that direction. The same is true for stems, where they are strong in one direction and not normal to it. However, the matrix can be removed, separated, or even dissolved by chemicals and the cellulose fibers can be extracted because the two elements remain distinct, although connected. Another common example of composites is concrete. When cement is mixed with water, it hardens exceedingly, but breaks easily. However, when small rocks are added to the mixture, the combined composite is much stronger and far more durable. And yet the rocks can eventually be separated from the cement if concrete is broken. Asphalt is also a composite material of tar and rocks.

In today's world, modern composites are made from strong fibers such as carbon and Kevlar and fiberglass as well as variety of resins. They are exceedingly strong, yet lightweight. In fact, they have become an increasingly important factor in aeronautic and space applications as well as automobile bicycles, and other industries.

Composites may be formed using continuous fibers, short fibers (whiskers), or particles as reinforcement along with the matrix (Figure 10.3). Body panels that are used in automobiles are made with continuous fibers, as are ducts. The mud-brick and straw example mentioned earlier uses short fibers. A specialty concrete made of cement that is embedded with small, tiny spiral inserts is another example. This product has amazing strengths compared to regular cement. Concrete is an example of particles, even if these particles are large.

The filaments used in composites are meant to take the loads directly. As a result, depending on how they are laid out, and based on their angles, the strength of the composite can be significantly different in various directions. For example, if the filaments are laid out in a tri-axial weave, as in Figure 10.4b, the material will carry loads in all three directions, whereas if the filaments are only biaxial (Figure 10.4a), the material will be weak in shear (in between the filaments at 45°). There are many other possibilities for

(a)   (b)

FIGURE 10.4 Two of the many possible lattice configurations for laying filaments in a composite matrix.

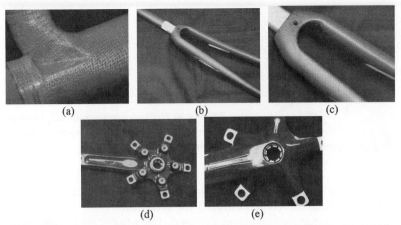

(a)   (b)   (c)

(d)   (e)

FIGURE 10.5   (a) A composite material duct for avionic applications; (b) and (c) a composite material bicycle fork; (d) and (e) a composite material bicycle crank. Notice the inserted metal parts and the direction of the filaments.

laying the reinforcement filament (called a lattice) in the matrix. Design engineers analyze the loads and the response of the composite material to these loads in order to design the composite layout and dimensions.

Figure 10.5 shows a few samples made of composite materials. Figure 10.5a is a composite material duct for avionic applications with Kevlar fibers. Notice the variations in the diameter of the duct and how the fibers are laid out. Figures 10.5b and d are the fork and the crank of a lightweight bicycle. Notice how the carbon fibers are molded into exact shape as well as the direction of the carbon fibers (in c and e). Also notice how metal parts are embedded into the products.

Composite materials demonstrate specific advantages to other materials that are unique to them. In addition to their lower weight and proportionally higher strengths as well as their anisotropic behavior, composites have another unique advantage. To understand this, imagine a cable. If you load a cable, there is a chance that one of the filaments may fail under the load. As soon as this filament fails, it will not carry any load; the load shifts to the remaining filaments that now have to carry somewhat larger values, and thus, are somewhat closer to total failure. However, when a filament in a composite material fails at one point, it is still attached to the resin within the matrix. Since the resin is ductile, by the way of slight expansions, it transfers some of the load to the filament through shear forces. This means that if the filament breaks at one point it will not be completely out of the picture; it can still carry some load through shear forces. Figure 10.6 shows a failed composite material driveshaft. Notice how the shaft was made with both

FIGURE 10.6   This composite material driveshaft failed under load. Notice both circumferential and longitudinal fibers that are supposed to carry torsional and axial loads.

FIGURE 10.7   Crack propagation can be slowed down by the whiskers in a composite material matrix.

circumferential carbon fibers around the perimeter of the shaft to carry the torsional loads as well as longitudinal carbon fibers to carry the axial loads.

The above is true for continuous filaments. But what is the advantage of whiskers and particles? Figure 10.7 shows a matrix with a propagating crack. As the stress at the crack root increases, it will either have to break the whisker, pull it out of the matrix, or separate the matrix from the fiber. All these require additional stress, and consequently, slow the crack propagation. This is in fact equivalent of drilling a small hole at the root of a crack to arrest its growth (a common practice), but happens at a microscopic level throughout the matrix.

Composite materials are often made from certain categories of materials with many varied characteristics. The matrix can be polyester, vinyl ester, epoxy, phenolic, polyimide, polyamide, and others. Continuous fibers are often transformed into a textile material such as a felt, fabric, knit or stitched construction to facilitate the construction process. Reinforcements are often fiberglass, quartz, kevlar, Dyneema, or carbon fiber.

EXAMPLE 10.1

Hydrogen has been identified as a possible source of energy for automobiles. For this, hydrogen needs to be stored at high pressures (5000 to 10 000 psi) inside a capsule. The capsule must be inexpensive, strong, and durable. Since the capsule will be handled and used by a variety of consumers, it must also be tough and resist damage. Composite material capsules are among the contending candidate materials for this purpose since they match most of these requirements.   ∎

10.7.2   Glass

Glass, part of the general family of ceramics, is tough, brittle, and in pure form, very clear. When cooled from melting temperature, glass does not crystallize; it remains amorphous. Adding lead to it will crystallize the compound, thus called crystal. Glass is principally silica, SiO_2, but can be colored by adding different minerals and metals. When, during manufacturing, glass is rapidly air-cooled on the surface (called tempered glass), tensile stresses in the middle create residual compressive stresses on the surface that make the glass about four times as strong. Tempered glass is used in windows, tabletops, and shower doors. However, because there is residual stress on the surface, it will shatter if cut. Glass can be welded by diffusion of melted glass. There is much artistry associated with glass, and its extensive use in household products, industry, and buildings, its low cost and high strength, and its formability make it an extremely versatile product.

Fiberglass is made of very thin fibers of glass, ranging from 0.0001 to 0.01 inches in diameter. It is used for insulation purposes. The modulus of elasticity (E) of glass fibers can be as high as that of music wire and up to six times as high as Nylon-6 (similar to dental floss).

Safety glass, used in automobile windshields is made from two layers of glass firmly held together by a layer of organic material such as polyvinyl butyral. The organic layer keeps the glass parts together if they break, thus preventing the glass from breaking apart.

10.7.3 Ceramics

Up until recently, clay was the primary material for ceramics. However, in recent decades, many new materials have been used to make ceramic products, some of them not found in nature, but with a variety of properties and forms. Ceramics are very tough and wear resistant, but also very brittle. Their atomic bonds are very strong, and as a result, they do not easily deform under load, they do not dissolve in chemicals, and they do not degrade in extreme heat. However, they are very prone to brittle fracture and since they do not yield, they will withstand the load until they suddenly shatter at a stress concentration-causing point, e.g., an internal crack, a void, or other similar imperfections.

Ceramics are used as abrasives, cutting tools, heat shields, electrical insulators, lasing crystals for lasers, artificial bone implant components, and as machine parts where wear resistance against moving parts is desired. They are also used in new applications such as internal combustion engines (cylinder walls, piston tops, turbocharger impeller, and many others) due to their wear resistance, lower weight, and heat resistance. Other uses include bearings, turbine blades, cams, extrusion dies, nozzles, seals, filters, crucibles, substrates, and heat sinks in electrical circuits. Certain ceramics can change their electrical characteristics when exposed to certain chemicals, and therefore, can be used as sensors, electronic switches, memory bits, and other applications.

The process of making advanced ceramics is actually very similar to massproducing regular pottery, where the clay and other particles are mixed with water into a plastic mass. The mass is then shaped by conventional methods such as injection molding into a closed mold, extrusion molding, and slip casting. After it is air dried, it is fired in a kiln at temperatures below its melting point (called sintering). During sintering, the particles are "cooked" together so that most of the voids are removed, forming very strong bonds. As a result, the volume shrinks as well, up to 25 percent during both drying and sintering. For advanced ceramic, the water is replaced with organic polymer binders which are mostly burned out during sintering. Additionally, other additives are added to improve the ceramic's performance against cracking and failure. Advanced ceramics include alumina, silicon carbide, silicon nitride, boron carbide, boron nitride, and tungsten carbide.

As was discussed in Section 10.7.1 regarding composite materials, ceramics can also benefit from embedded fibers as well as intentional microcracks (Figure 10.8). If a crack appears in the ceramic under load, in the absence of mechanisms to arrest it, the crack can easily grow to complete failure. However, if there are fibers embedded in the ceramic or if it is intentionally made with many microcracks, the crack may be arrested by the fibers that take up additional load in tension, or by the microcracks at the root of

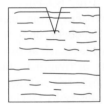 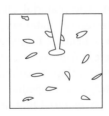

FIGURE 10.8 Fibers embedded in the ceramic as well as intentionally created microcracks in the body can improve the life of ceramics by arresting the crack growth through load-sharing and by distribution of the stresses over a larger area.

the growing crack by distributing the stress over a larger area. The microcracks are purposely created during processing to improve the life of the ceramic.

10.7.4 Biomaterials

Biomaterials are substances that are compatible with human (and animal) bodies. They must not be rejected by the defensive systems of the body as foreign materials; they must withstand the harsh environment of the body; they must be compatible with the requirement of their purpose in their mechanical, electrical, and thermal properties; and they must not create side effects. They should also not be carcinogenic and last a long time. A bone replacement part will have a very different set of required properties than a material used as artificial skin or materials used in an artificial heart.

There are many materials that fit these requirements for different purposes. Titanium has been used for a long time for bone replacements. Many resins are also biocompatible. The trick is to find appropriate properties for the material that will satisfy the requirements.

The Jarvik-7 artificial heart uses a biomer with Dacron mesh, polyurethane, a graphite substrate coated with pyrolitic carbon, titanium, and injection-molded pellethane polyurethane. The materials used for repairing tooth decay include ceramics and cements. A broken bone is secured with titanium and stainless steel plates and screws. However, there are also substrates such as a polyurethane mesh whose surface is flocked with fibrils that can trap blood components which eventually create a smooth collagen-covered surface suitable for use as lining in artificial blood vessels. Another biomaterial mesh, covered with nutritious material was used to grow and multiply skin for transplantation on foot ulcers resulting from diabetes (unfortunately, the company no longer exists). Additionally, new materials are used for direct drug delivery over the patient's skin through diffusion. All these are examples of new biomaterials that have been designed to repair, assist, or replace human organs.

10.8 MATERIAL SELECTION

In order to choose an appropriate candidate material for your application, you will need two sets of data. One is the material requirements such as strength, modulus of elasticity, ductility, cost, weight per unit, and other necessities. The other is material characteristics and behavior data. When these two are matched, an appropriate material can be selected from a list of possible candidates that possess the right characteristic and can satisfy the requirements. The list of requirements is developed by you, the designer, in conjunction with the requirements that come from your problem definition, customers, and design factors (please see Chapter 7). For example, your design may require low density for low weight, but high strength for large load capability. Simultaneously, it may require resistance to chemicals and ductility for machining. You will need to study what needs to be accomplished, what requirements must be met, and what performance and safety needs must be satisfied. The material characteristic, behavior, and data should be collected from manufacturers' catalogs or websites, from handbooks (Mechanical Engineering Handbook, Chemical Engineering Handbook, etc.), from professional societies (ASTM, ANSI, etc.), from your company files (such as test data on previously used materials or newly developed alternatives), or from manufacturers' associations (such as Modern Plastics Encyclopedia). No single resource has enough room for all necessary tables for every material that may be available. Obviously, due to the vast variety of available alternative materials, this is not a simple task and must be done with perseverance and knowledge.

To facilitate this selection process, Ashby et al.[1, 2] have developed an extensive set of charts that relate different pairs of material properties to each other. Using these charts, the designer can compare different families of products as well as individual products within a family relative to each other in order to arrive at an estimate of what may work. You will still need to finalize your selection by comparing data on the particular material you are considering with your desired characteristics, but these charts can facilitate this selection. For example, in the avionics industry, thermal conductivity versus temperature is one criteria for choosing materials for the combustion chambers of jet engines.

Figures 10.9, 10.10, and 10.11 are three samples of material selection charts by Ashby, et al. These charts relate to specific pairs of characteristics such as Young's modulus versus strength, strength versus maximum service temperature, and strength versus density. As you see, families of materials congregate together in common areas, for example, ceramics, metals, foams, polymers, and composites. From these charts, using proper material characteristics, the designer can select families of materials, from which a specific material can later be picked for an application. However, one pair of characteristics is hardly ever enough; you may have to match your selected materials with similar ones for other characteristics, say weight-to-cost ratio or chemical resistance to strength relationship and many more. This means that the selected families of materials must satisfy multiple pairs of variables from multiple charts.

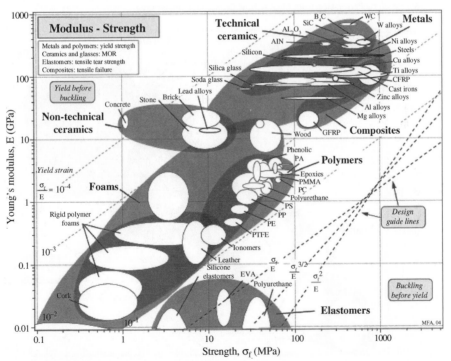

FIGURE 10.9 Pairs of material characteristics are used to categorize families of materials. By comparing the desired values for a design with the material characteristics in these charts, candidate materials can be selected for further analysis and comparison.

(Reprinted by Permission from Mike Ashby, Cambridge University.)

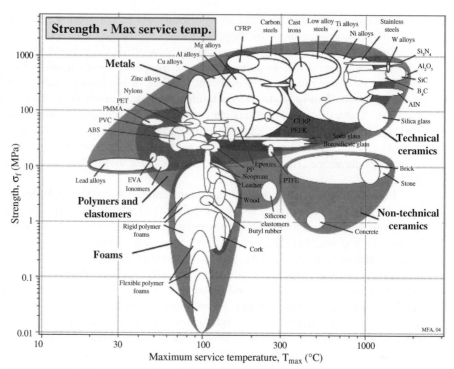

FIGURE 10.10   Material selection chart relating strength to maximum service temperature.
(Reprinted by Permission from Mike Ashby, Cambridge University.)

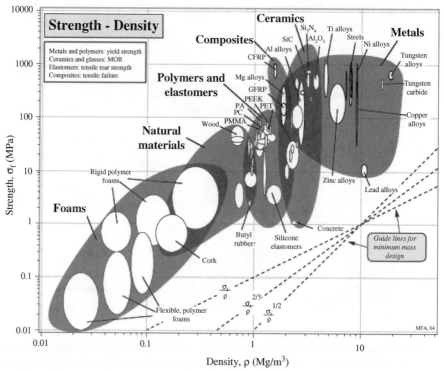

FIGURE 10.11   Material selection chart relating strength to material density.
(Reprinted by Permission from Mike Ashby, Cambridge University.)

Ashby's charts include an extensive list of properties, some mechanical, some aesthetic, some acoustic, including:

Modulus of elasticity vs. density	Expansion vs. modulus of elasticity,
Strength vs. density	Strength vs. expansion
Fracture toughness vs. density	Strength vs. temperature
Modulus of elasticity vs. strength	Modulus of elasticity vs. cost
Fracture toughness vs. modulus	Strength vs. relative cost
Fracture toughness vs. strength	Tactile softness vs. tactile warmth
Loss coefficient vs. modulus of elasticity	Acoustic pitch vs. acoustic brightness
Conductivity vs. diffusivity	Weight vs. abrasion resistance
Expansion vs. conductivity	Stiffness vs. resilience

An extension of the application of these characteristics is to use ratios of pairs against each other. For example, one may compose and use a chart that relates density/yield strength vs. price /density. This relates the weight, price, and strength of a material to each other in the same chart by comparing their ratios. Another example is fracture toughness/density vs. elastic modulus/density. These characteristics are chosen based on the importance they play in the design of the particular product or system. Please see Ashby [1,2] for more charts and how they may be used.

To demonstrate how these charts may be used, let's consider the following example.

EXAMPLE 10.2

Suppose that the design of a machine component with circular cross section in bending requires low weight and high strength. The two equations governing these requirements are

$$m = \pi r^2 l \rho \tag{10.1}$$

where m is the mass, r is the radius, l is the length, and ρ is the density of the material; and

$$\sigma = \frac{Mc}{I} \tag{10.2}$$

where σ is the stress, M is the maximum moment on the part, c is the maximum distance of the outer layer of the part from the neutral axis (usually equals r for circular cross sections), and I is the area moment of inertia of the cross section. Assume that based on our calculations, estimates of these two values are at hand. As you notice, Figure 10.11 relates material strength σ_f with material density ρ. Dividing the stress σ by a safety factor (S.F.) gives us a rough estimate of the required strength for the material. Using Equations (10.1) and (10.2), we can calculate the ratio of these characteristics as

$$\frac{\sigma_f}{\rho} \geq \frac{\pi r^2 l \sigma}{(S.F.)m} \tag{10.3}$$

This equation can be further simplified if we know more about the cross section. However, the notion is that any material we select should have a ratio larger than this value. Figure 10.11 shows "guide lines" for material selection based on various ratios, including ours. Lines parallel to these lines will guide us to materials that have specific ratios. As the line moves upward, the ratio increases. Figure 10.12 shows the same chart with a fictitious guide line drawn parallel to the σ_f/ρ guide line based on our desired values. The width of the guide line is chosen to represent an acceptable range for this ratio. Families of materials that fall within or above this guide line have ratios within our range. The cross section between this guide line and a horizontal guide line of desired strength will allow us to select a particular material. In this case, we may be able to select metals,

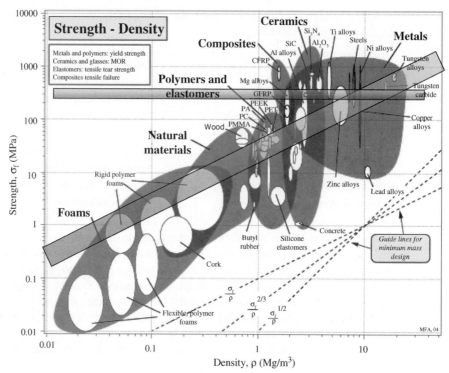

FIGURE 10.12 The application of material selection charts for selecting materials for a specific application. The guide lines are used to relate material requirements to material characteristics.

steels, and zinc alloys. Alternately, a vertical line with a particular value for density may also allow us to pick materials specified by the ratio and density instead. ■

Additional charts may also be consulted for other properties. For example, if the part happens to also be subjected to high temperatures, the thermal properties chart may also be used to further specify acceptable materials. If cost of the material is a concern as well, a chart of strength versus relative cost may be used. Ultimately, the actual material properties must be checked to ensure that the final selection will possess the desired characteristics.

10.9 MATERIAL PROCESSING AND MANUFACTURING

For thousands of years, humans have processed materials to manufacture artifacts and systems, and during all this time they have come up with new, innovative, and more modern methods. Although these methods are influenced by localities, by the local and native materials available to the people, and by people's ingenuity, there are many common methods that have lasted even through today's sophisticated methods. If you visit local manufacturers, third-world countries, and even ''hand-made'' shops of certain communities (who in fact are proud of their heritage of old technologies and traditional methods), you will see the variety of these methods and how ingenious some of them are. However, in our modern world, mass production and the use of modern material processing techniques and automatic machines is extremely prevalent. It is against this backdrop of reality that we talk about material processing. Making a prototype of your design may not necessarily require a machine shop. But the ability to put things together with whatever is available is a skill and an advantage that may set you apart from your competition. In fact, many innovating designers brag about keeping a box of ''stuff'' in their office or room for rapid prototyping. Still, you must also know about major material processing techniques, understand their characteristics and capabilities and limitations, consider their advantages and disadvantages, and incorporate the best practices in your design.

Material processing is not a simple matter though. There are many methods and applications involved that must be integrated into practice. This is why manufacturing engineers are generally part of the design team. Their role is to bring their knowledge of manufacturing techniques into the design process, so that the design will be built with the least amount of effort, at the lowest possible cost, with the best possible quality. To this end, one should not expect that you, as the designer, will know all the details about manufacturing techniques, but it should be expected that you will know enough about these techniques to be a good designer and understand not only the importance of material processing in design, but also its relationship to the design process.

In the following sections we will discuss some very common methods of material processing, but it is expected that you will learn about manufacturing through more extended formal study as well as practice by designing and making things yourself.

10.9.1 Manufacturing in the Global World

There are many statements made by government officials and industry advocates about the global economy and the state it is in, some factual, some speculative. It is very difficult to actually know what is accurate or not, and what is said as fact or prediction, and what is said for one's personal (or company) benefit. What is true is that there are differences between the costs of manufacturing in different countries, and there are differences in the quality, practice, and expectations about these products. It is also true that things change extremely quickly. And it is also true that much of the manufacturing base of the United State has been moved to other countries and even if a company is known as an American company (or any other nationality for that matter), it does not mean that the products the company sells are American (or any other nationality). A Japanese product may come from Indonesia or Singapore, a French product might be Chinese or Korean, and a German product might be Mexican or Brazilian. The communication revolution of the 1990s and early 2000s has changed the way the world is connected, and as such, a product designed in America can be seen simultaneously on a screen in India and be made in China as soon as the data are saved. It is against this background that most design activities are performed in industry.

It is estimated that if the labor cost in the United State is about $30 per hour, it might be around $24 in Western Europe, about $3 in Eastern Europe, $24 in Japan, $2 in China, and $1 in India. However, labor cost is not the only issue; the quality of the product and the productivity of the manufacturing process are also significant. It is true that products made in China and India (and many other countries with lower labor costs) cost less to build. However, it should be noted that the manufacturing processes in many advanced countries are far more productive, and as a result, although products made in these countries involve higher labor costs, the final cost of the product is not proportional to the labor cost alone. A more productive worker produces more and higher quality products in the same given time, thus reducing the overall cost.

The important matter here is that the labor cost in some countries is lower because those workers have a lower standard of living and the cost of living in their native land is lower. If the labor in those countries were to live the same lives as their American and Western European counterparts, they would cost the same. This was true for the Japanese industry too. In the 1950s and 1960s the cost of Japanese products was far lower than the American counterparts. But when the standard of life increased in Japan and their labor cost increased, the Japanese manufacturers moved their manufacturing operations to Singapore, Malaysia, and Korea. So, should we try to retain a lower standard of life in certain countries to keep the labor costs low, or should we expect that as they improve and desire to live a better life their costs will increase the same way? We must not ignore this important issue.

Material processing and manufacturing trends in today's global world can be divided into the following categories:

- Harmonized, sustainable manufacturing
- Precision manufacturing
- Smart manufacturing
- Digital manufacturing
- Micro- and Nano-manufacturing
- Biomanufacturing

Harmonized or sustainable manufacturing relates to finding ways to make manufacturing cleaner, less wasteful, more environmentally friendly, and therefore, more sustainable. Although China boasts about its manufacturing power, the problem of air pollution there is not a joking matter. At present, total GDP of China is about 4 percent of the world, but it uses 31 percent of the world's coal, 7.4 percent of the oil, and 40 percent of the cement. One estimate puts China's sewage contribution into the environment per year at 48 billion tons. If nothing is done to improve the situation, think about the consequences when the population of the world grows at the present rate to 10 billion and beyond.

The trend to improve the situation toward more harmonized, sustainable manufacturing requires that consumers, the governments, the industry, the technology, the resources, and the environment be integrated into one system. The objective should be to

- promote product quality.
- reduce costs by designing more efficiently.
- increase productivity.
- reduce waste, save energy and resources.
- reduce environmental effects.
- recycle and reuse.

When resources are wasted, it has a direct effect on the consumers and the environment. If the technology advances with better productivity, the industry and the consumer and the environment benefit. As was discussed in Chapter 7, these are all parts of the design sphere and should be integrated together as products, systems, and processes are improved and new ones are designed.

Precision manufacturing relates to ever smaller dimensions and accuracies. Nowadays, electron beam lithography and other similar technologies can reach machining accuracies of 10 nm or less. In ultraprecision machining, molecular beam epitaxy (a method of thin film deposition) and similar methods can reach even 1 nm or less accuracy. The ability to machine at these precision levels allows many new applications, processes, and products that are otherwise impossible. There should be an expectation that industry will use these techniques as time passes. Precision casting, precision welding (Friction Stirred), precision forming (Laser Direct Forming), and other precision methodologies should be added to the precision machining.

Smart Manufacturing is intended to be a catch-all phrase for better design, lower cost, less dependence on resources, etc. Smart manufacturing is supposed to foster and support creative design (which has been an emphasis of this book all along), the interactive exploration of design options, worldwide collaboration, intelligent control of manufacturing processes, product quality, and other similar attributes.

Digital Manufacturing relates to the integration of digital systems and control with the manufacturing processes. For example, as will be discussed later, computer aided manufacturing (CAM) is the integration of computer technology with manufacturing. A drawing of a part, made on a computer aided design (CAD) system can be electronically transferred to a numerically controlled (NC or CNC) milling machine for manufacturing. Similarly, the electronic information from a CAD drawing can be used in a rapid-prototyping machine that will make a prototype of the product. We will discuss these later.

Additionally, digital manufacturing also relates to the simulation and modeling of products and systems. This includes modeling the behavior of products under load, in accidents, in high speeds, etc., but also the finite element analysis (FEA) for determining the stresses, strains, deformations, and material behavior under varying temperatures as well as vibration analysis of parts and systems.

Micro- and Nano-manufacturing is for creating or manufacturing structures and systems at the molecular levels. Micro-electromechanical systems (MEMS) and nano-electromechanical systems (NEMS) are expected to allow the creation of sensors, structures, actuators, and even robots at these scales. For example, it is expected that microscale robots would be able to enter the body and repair organs, attack cancer cells, or look for problems.

Bio-Manufacturing relates to regenerative medicine and the availability of methods and techniques that will allow our bodies to replace injured or damaged parts without the need for invasive surgery. For example, imagine that a suitable environment within the body would be created artificially to enable the body to regenerate a joint that is damaged by arthritis. This would eliminate the need for surgery to replace the joint or even medicine to manage it.

10.9.2 Material Processing Categories

Overall, material processing techniques are generally divided into three categories of shaping, joining, and surface treatment, with all their variations in methods and purposes. Shaping includes processes that are used in forming the raw or bulk material into shape, including casting, machining, and material removal. The purpose is to create the part. Joining includes techniques that are used to connect or join components together, including adhesion, welding, brazing, riveting, and fastening. The purpose here is to integrate different parts into more complicated shapes or assemblies. Surface treatment includes many different techniques that are used to change the characteristics of a material (quenching, tempering, and anodizing) and techniques that affect the final appearance of a part (abrasion, polishing, painting, and anodizing). Some of these techniques may be used for multiple purposes while others are hard to categorize. Table 10.8 is a summary of some common techniques for each of the categories.

TABLE 10.8 Common Processes of Major Manufacturing Operations

Shaping	Molding	Injection Molding, Rotational Molding, Blow Molding, Expanded Foam Molding, Compression Molding, Resin Transfer Molding
	Casting	Die Casting, Sand Casting, Investment Casting, Polymer Casting
	Bulk Forming	Shape Rolling, Die Forging, Extrusion
	Sheet Forming	Press Forming, Roll Forming, Spinning, Thermoforming, stamping
	Lay-up Methods	Hand Lay-up, Vacuum Bag, Pressure bag
	Powder Methods	Pressing, Sintering
	Rapid Prototyping	Laser Method, Deposition Method, Polymer Hardening Method
	Machining	Turning, Milling, Routing, Sawing, CNC Devices
Joining	Adhesives	
	Fastening	Sewing, Rivets and Staples, Threaded Fasteners, Snap Fits
	Welding	Hot Gas Welding, Hot Bar Welding, Hot Plate Welding, Ultrasonic Welding, Power Beam Welding, Brazing, TIG Welding, MIG Welding, Soldering, Torch Welding, Spot Welding, Resistance Welding, Friction Welding, Diffusion Welding, Glaze Bonding
Surface Treatment	Printing	Screen Printing, Pad Printing, Cubic Printing, Hot Stamping, In-mold Decoration
	Plating	Vapor Metalizing, Anodizing, Electro Plating,
	Polishing	Chemical Polishing, Electro-Polishing, Mechanical Polishing
	Coating	Painting, Electro Painting, Powder Coating, Enameling
	Etching	
	Texturing	
	Sand Blasting	
	Sanding	
	Grinding	
	Buffing	
	Shot-peening	
	Heat Treatment	

Let's say we consider a family of operations such as shaping. There are different methods of shaping including bulk forming, molding, casting, machining, sheet forming, and more. Each one may be performed in different ways too. For example, casting may be accomplished by investment casting, die casting, sand casting, and others, each with its own characteristics, requirements, costs, and capabilities. Machining may be accomplished by milling, turning, abrasion, and more. Milling machines are operated by an operator, while CNC mills operate automatically under computer control. Joining affords us many choices too: adhesion, welding, fastening, and snapping. Welding includes soldering, brazing, MIG welding, spot welding, hot gas welding, ultrasonic welding, laser welding, diffusion welding, friction welding, TIG welding, and others. Fastening includes using screws and bolts (many varieties), rivets, tongue and grooves, and snaps.

It should also be mentioned here that there are many other industries that use material processing machines and techniques that are not mentioned in the above table and do not even fit these categories. They may use some of the machines mentioned there, but their production and manufacturing is somewhat specialized and as a result, requires specially designed mass production machines. Examples are the paper industry, the food industry, the integrated circuit industry, glass manufacturers, metal manufacturers (such as steel mills, aluminum mills, etc.), and others. Manufacturers of breakfast cereal do not use any of the above-mentioned shaping machines in the production of the cereal, nor do any use joining techniques or the above-mentioned surface treatment. However, they do use specially designed material processing machines that mix, cook, cast, pour, dry, treat, sprinkle, and package the food. A simple product like tomato sauce requires huge machinery that works at amazing speeds, at efficiencies that are unbelievable, but they use food processing machines that are very different from the above too. For example, it is required that all traces of bacteria and spores be removed from the cooked tomato after it is canned (called retort). This needs to be done at 250°C. In order to be able to raise the temperature to this level, it is necessary to place the cans in a pressure chamber. Additionally, if the outside pressure on the cans were not increased, when the sauce inside the can boils, the internal pressure will raise so much that it will damage the can. On the other hand, thousands of cans need to go constantly into and out of the pressurized chamber. So how would it be possible to accomplish this without having to constantly open and close the chamber doors? This is accomplished by passing the cans through a large column of water, behind which the pressure is at the desired level. The cans go through the water column into the chamber, get heated for sterilization, and leave the same way from the other side. This machine is very specialized equipment specifically designed to accomplish this task.

EXAMPLE 10.3

Consider an electric guitar and how it is manufactured. Most guitar bodies are manufactured using computer aided manufacturing (CAM) techniques that use the same processes mentioned above. The body is designed using a computer aided design (CAD) program. The information is sent to a CAM-operated router that cuts the wood and all the necessary grooves and openings for the neck, the electronic components, the bridge, and the pick-ups. Many guitars have veneers for accent or for artistic rendering. The veneers are also cut and later glued to the body. The neck is similarly designed and shaped, mostly with CAM-operated routers and saw blades. Then the body and the neck are sanded repeatedly to completely smooth the surfaces. Next, the surface of the body (as well as the neck, even if not as much) is treated by painting and glossing repeatedly followed by polishing and buffing (on a buffing wheel) to a high gloss. Next, the neck, the bridge, potentiometers, switches, and the pick-ups are joined to the body with screws and the wiring is completed and tested. As you notice, many of the processes mentioned in Table 10.8 are used to manufacture an artistic product like a guitar. ∎

EXAMPLE 10.4

A car body panel is made from already manufactured steel sheets. The sheet is cut and formed by a press at high temperatures, eventually joined to other parts by spot welding, rivets, snaps, or continuous welds, and later, is painted. Some parts may also be treated for specific behavior, for example, to increase their resistance to rusting. In this case, a variety of the above-mentioned processes are used to manufacture the body panel.   ∎

EXAMPLE 10.5

A faucet is composed of many parts made from a variety of different materials. The body is usually cast if made from resins, or cast and subsequently pressed if made of metals. A variety of cutting machines is used to make the seats, cut threads in the body, etc. The body is then sanded, ground, and polished, followed by plating with chromium, nickel, or others. The seat is separately made on a lathe, if brass, or cast, if plastic or ceramic. The handles are similarly made depending on the material. As you notice, in this case too, many of the above-mentioned processes are used to form, join, and surface-treat the product.   ∎

10.9.3   Metallurgical Effects of Forming Choices

One important consideration about different methods of forming, especially for crystalline metals, is the effect on the strength of the product in relation to the way the structure is formed. Figure 10.13 shows the structure of three similar parts made by casting, cutting (machining), and forging. The crystals in the cast part (a) are randomly orientated when it is cast. Part (b) had to be machined from rolled stock. During rolling, the crystals of the material are forced into directional orientation which increases their strength. This grain flow is somewhat similar to the grains of wood that are directional (anisotropic). Thus, when the part is machined (cut) from the stock, it displays better strength, even though the grain flow of the material is disrupted where it is cut. Part (c) is also made from rolled stock. However, the part is forged into shape by heating it (to lower the yield strength) and plastically deforming the material with large forging forces. In this case, the grain flow of the crystals follows the form of the part, and therefore, the part displays the best strength. The same is true for wooden parts. For example, certain parts used in mass-produced furniture are made of plywood or pieces of wood that are formed during manufacture. These parts display superior strength and durability compared to parts that are made from cut-and-glued plywood or wood block. As a result, the choice of the process affects not only the cost of production and the time that it requires, but also the strength and other factors.

In the following sections we will discuss some of these processes and their specific characteristics that are more common and, in general, accomplish a lot of different tasks. For more information about these processes please refer to the list of references provided or to other resources.

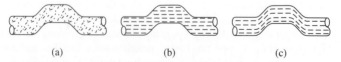

(a)   (b)   (c)

FIGURE 10.13   The effect of forming process on the strength of a part. Casting, machining (cutting) and forging result in different strengths.

10.9.4 Machining and Material Removal

Machining, whether for material removal or not, is one of the major material processing operations. It includes very simple processes such as drilling or sanding to sophisticated and complicated machining such as turning with a lathe or milling, from machining with enormous marble saws, to complex processing with a six-axis forming and material removal machining center. For example, many of the curved surfaces of airplane wings are in fact machined from huge aluminum blocks. These operations require sophisticated multi-axis machines with enormous power and range of movements, as does the making of complicated molds which require multi-axis CNC milling machines.

Material removal may be accomplished in a number of ways. For example, to remove the surface of a material, a sharp tool may be moved over the stationary part back and forth in straight lines while the tool (or the part) moves slightly to one side during return motions such as in a machine called a shaper. Alternately, the part may be moved relative to a tool as in a lathe (Figure 10.14a and b). In this case, the part is held in a chuck and is rotated in front of a tool that is not rotating but moves slowly to advance against the part. Figure 10.14c shows an advanced toolroom lathe.

In a mill the part moves on an X-Y table (if 2-dimensional) or on an X-Y-Z table (if 3-dimensional) while the tool rotates (Figure 10.15a, b, and c). Combining the 3-dimensional movements of the table with the part attached to it and rotating the tool about the roll-pitch-yaw axes while it turns will create a six-axis mill that theoretically can cut any shape. Depending on the tool axis of rotation, there are two varieties of mills: horizontal mills and vertical mills.

Figure 10.16 is an example of a CNC router. A router is similar to a mill, except that the tool is usually smaller and rotates at much higher speeds. It is used for shaping, cutting, and making grooves. CNC routers are also used to cut shapes from sheet material, for example for guitars, in the clothing industry, and in the plastics and wood industries.

Figure 10.17a is an example of a CNC mill interface board for entering programs as well as operation and motion data. CNC mills are powerful and versatile multi-axis machines capable of many sophisticated operations. They are usually equipped with sensitive feedback sensors that allow them to make very accurate motions. Figure 10.17b is the monitor screen of the same mill showing the program as well as data on the axes. Figure 10.17c is a vertical tool turret from which the mill picks up a tool from a collection of different tools as specified by the programmer or the operator. Figure 10.17d is a horizontal

(a)   (b)

(c)

FIGURE 10.14 Material removal with a lathe. The part is held in a chuck and rotated against the tool that moves slowly to advance. (a) Haas TI1, (b) Haas SL10 Applications, (c) A Haas TL-3B toolroom lathe.

(Photos courtesy of Haas Automation, Inc.)

(a)　　　　　　　　(b)

(c)

FIGURE 10.15 (a) A part is milled. (b) A vertical mill with an attachment on the table for holding the part. The vertical mill can be seen above the part. (c) Haas TM-3 toolroom vertical mill with X-Y-Z travels and optional tool changer. The tool changer allows multiple quick tool changes during a task.

(Photo courtesy of Haas Automation, Inc.)

FIGURE 10.16 Haas GR-408 bridge-style router with X-Y-Z travels, fixed bridge, and moving table.

(Photo courtesy of Haas Automation, Inc.)

tool turret which works with a lathe in a similar fashion. As specified by the operator, the turret rotates to line up the appropriate tool against the part.

Figure 10.18 is a three-axis, tabletop, educational CNC mill with capabilities similar to industrial machines. With this simple system, one can create a CAD drawing of the part, automatically create a numerical control (NC) program for machining the part, and then automatically fabricate the part with the mill out of machinable wax, wood, or aluminum.

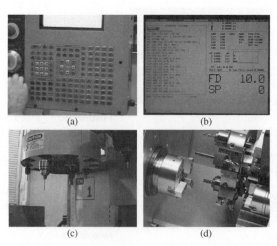

(a)　　　　　　　　(b)

(c)　　　　　　　　(d)

FIGURE 10.17 (a) The interface board of a CNC mill through which an operator enters a program or motion and operation data. (b) The screen of a CNC mill showing the program that is running as well as data about the machine's operations. (c) A tool turret for holding a number of different tools from which the mill picks up tools as specified in the program. (d) A horizontal tool turret for a lathe. The turret rotates to line up the tool specified by the operator with the part.

(Haas TL15 sub main turret. Photo courtesy of Haas Automation, Inc.).

FIGURE 10.18 A table-top educational CNC mill (spectraLight™).

(a)   (b)

FIGURE 10.19 (a) Haas EC-400PP horizontal machining center with six-station pallet pool and a 70-pocket side-mount tool changer. (b) Haas SL-20APL turning center with automatic parts loader and 8-inch hydraulic chucking system for unattended operation.

(Photos courtesy of Haas Automation, Inc.)

FIGURE 10.20 A multi-axis machining center is making a mold.

(Photo courtesy of Haas Automation, Inc.)

Figure 10.19a and b are two examples of a machining center. A machining center is used in conjunction with computer aided manufacturing (CAM) or computer integrated manufacturing (CIM) systems, and may even include robots that load and unload parts or otherwise participate in manufacturing operations. These systems are usually multi-axis, powerful, and accurate machines that can perform many tasks under a central computer control. They usually are equipped with a tool turret that holds a large number of tools that are accessible by the machining center. The appropriate tool is used for each part of the operation.

Figure 10.20 is an example of a tool from a multi-axis machining center making a mold. You may notice how the tool is moved in space in order to follow the 3-dimensional surface. In order to successfully accomplish this kind of a task, the machine should ideally have six axes of motion, three for positioning and three for orientation.

10.9.5 General Machining Operations with a Mill

General machining operations include facing, contouring, drilling, and pocketing, as well as creating a surface of revolution, a ruled surface, and a swept surface. The following is an explanation of these operations.

- **Facing** removes material from across the surface of the piece, usually used for cleaning a rough surface, or to decrease the thickness of the piece. Planers, shapers, routers, grinders, sanders, or milling machine may be used for this purpose.

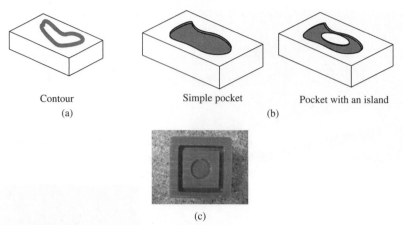

Contour
(a)

Simple pocket
(b)

Pocket with an island

(c)

FIGURE 10.21 A contour and a pocket cut.

- **Contouring** removes material to a specified depth along a specified path line and is used to create a groove, as shown in Figure 10.21a. Contouring is a line cut. It can be accomplished by a router or a mill.
- **Drilling** drills holes in specified positions.
- **Pocketing** removes the material to a specified depth from within a surface area as specified by geometry, as shown in Figure 10.21b. A router or a mill can be used for this purpose too. Figure 10.21c shows a simple part with a square contour and a circular pocket in machinable wax.
- **A Surface of revolution** is created by removing all material within a surface generated by revolving a geometry about the X or Y axes. In Figure 10.22a, when the ½ silhouette of a bottle is revolved about the X-axis, the ½ bottle is cut by the CNC mill. Figure 10.22b is a ½ sphere cut in machinable wax by a CNC mill.
- **Ruled Surfaces** are created by selecting two geometries. The geometries reside on different planes. The material is removed on the surface as a combination of two planes (Figure 10.23).
- **Swept Surfaces** are created by selecting two geometries. One geometry is on the XY plane and defines the swept area. The other geometry is perpendicular to the swept area and defines the material to be removed.

If you are not familiar with machining operations and want to learn, try to locate a machine shop in your school or similar places. Next time you have a project, a need, an idea, or just the desire, try to use the machine shop to make your parts. Many institutions have student-run shops and craft centers. Use these facilities as much as possible. Make your parts out of wood, plastics, metals, foam, or any other material that seems appropriate. Learning by doing is a great asset.

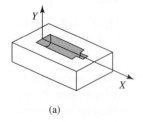

(a)

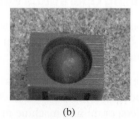

(b)

FIGURE 10.22 Surface of revolution can be cut with a CNC mill.

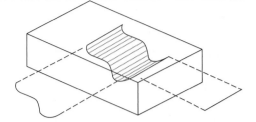

FIGURE 10.23   A ruled surface is created when two geometries in two planes are combined.

10.9.6   Material Joining

Material joining is accomplished with adhesives, screws and bolts, rivets, snaps, or by welding, soldering, brazing, and similar methods. Each method has its own characteristics, associated costs and requirements, as well as mechanical properties. An important characteristic about different methods of joining is whether they are permanent, semipermanent, or may be removed and re-applied easily and repeatedly.

Adhesives and glues create permanent joints. In many cases, the adhesive may be stronger than the two materials it joins together. Adhesives come in many different types, natural and synthetic, water- or solvent-based, soft or hard, etc. They may be used for many different types of applications, including biomaterial applications such as in dentistry and orthopedic replacements. Adhesives may also be applied with computer-controlled glue-laying machines that apply continuous or intermittent beads of glue very accurately. Many polymers act as adhesives.

Welded Joints are also permanent. Welding can be accomplished manually or with automatic, computer-controlled machines. Most automotive welds are accomplished with welding robots, either spot welded or continuous. All different types of welding mentioned in Table 10.8 are common, but require different levels of skill and equipment.

Brazing is with brass or similar softer materials that have a lower melting temperature and flow well when melted. Soldering, a very common techniques used with electronic manufacturing, uses alloys with very low melting temperature. Mass production industries use automatic soldering machines that solder all the connection nodes on a circuit board at once by dipping and then removing the excess solder.

Welding affects the heat-related characteristics of metal parts. If a material is heated to a melting point for welding, but cooled quickly, the affected area may harden. This can have a negative effect on the fatigue life of the part as well as its ductility. You must be careful about the consequences of welding if large forces are involved. Please refer to machine component design references on how to incorporate this into the design of a machine part.

Figure 10.24 shows common weld types and weld joints. Figure 10.25 shows some of the more common welding symbols and what they mean. These are used to communicate the size, type, frequency, and finish of the weld. For more information on other types of welds, please refer to the references provided or see documentations from ANSI and ASTM and other professional organizations.

Riveted joints are considered semipermanent. They can be removed when they are cut (e.g., by drilling) without damaging the connected parts. Rivet joints are assembled by drilling holes, with or without countersink, into the two parts, inserting a rivet into the holes, and hammering the rivet to completely fill the holes and the countersink (if present). Because they are meant to be deformed easily by hammering, rivets are made from very ductile materials. They are very common in assembling the frame and body panels of buses and coaches as well as airplane fuselages. For example, in an airplane it should be expected that the body panel may have to be removed. Riveting allows the removal without damaging the other panels. However, there are hundreds of rivets used

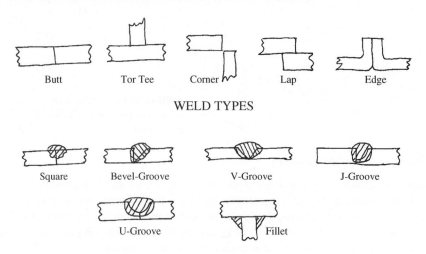

WELD TYPES

WELD JOINTS

FIGURE 10.24 Common types of welds and joints.

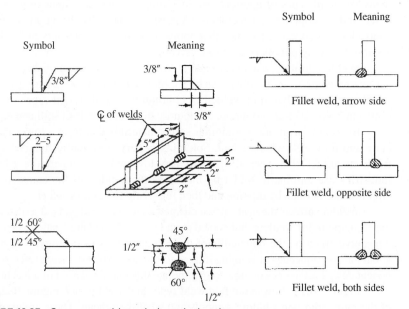

FIGURE 10.25 Common weld symbols and what they mean.

for each panel. All these must be drilled out and removed one at a time. In fact, rivet-removing robots are sometimes used to do this job. In order to accomplish the task, either a vision system is used to look for the rivet heads and direct the robot to it, or the location of the rivet heads from a CAD drawing is fed into the computer database of the robot. Rivets are used in many other applications as well.

Screwed and Bolted joints are considered temporary. This means that they can be removed and re-installed indefinitely (if the parts and the bolt or screw are still intact). However, unlike other methods of joining, bolts can be preloaded. This means that the joint can be preloaded with a compressive force that counteracts a tensile force up to the value of the compressive force before a joint is lost. For example, consider the connection between an engine block and the cylinder head. The large forces of combustion can be completely lost if the joint or connection between the block and the head is lost due to these forces. The preload in the cylinder-head bolts counteracts the separating combustion forces to keep the cylinder head connected to the block with an intact seal. Instead, if a rivet or adhesives were used, this connection could easily be lost. The same is true for a tire rim, where the preload keeps the tire rim attached to the axle even though the bending forces due to the weight of the car tend to separate the two.

Figure 10.26 shows different types of bolts, bolt heads, and drives. There are many other types of bolts with other types of drives, heads, tips, and nuts. For a complete list please refer to manufacturer's catalogs or visit your local hardware store.

Table 10.9 shows the standard Unified National Coarse (UNC) and Unified National Fine (UNF) fastener and thread sizes as well as the required drill sizes for tapping. Although not included in this table, many sizes include Unified National Extra Fine (UNEF) threads as well.

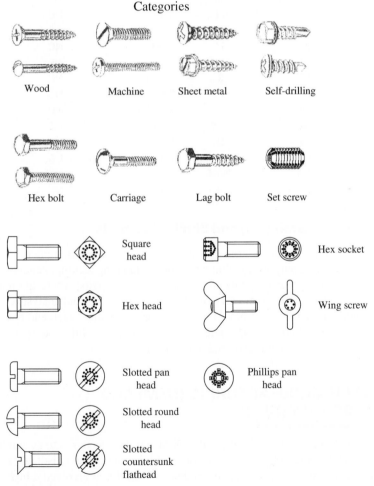

Categories

Wood   Machine   Sheet metal   Self-drilling

Hex bolt   Carriage   Lag bolt   Set screw

Square head   Hex socket

Hex head   Wing screw

Slotted pan head   Phillips pan head

Slotted round head

Slotted countersunk flathead

FIGURE 10.26 Bolt types, heads, and drives.

TABLE 10.9

Thread Size	Pitch Series	Drill Size	Thread Size	Pitch Series	Drill Size
0-80	NF	56	1/2-20	NF	29/64
1-72	NF	53	9/16-12	NC	31/64
2-56	NC	50	9/16-18	NF	33/64
2-64	NF	50	5/8-11	NC	17/32
3-48	NC	47	5/8-18	NF	37/64
3-56	NF	45	3/4-10	NC	21/32
4-40	NC	43	3/4-16	NF	11/16
4-48	NF	42	7/8-9	NC	49/64
5-40	NC	38	7/8-14	NF	13/16
5-44	NF	37	1-8	NC	7/8
6-32	NC	36	1-14	NF	15/16
6-40	NF	33	1 1/8-7	NC	63/64
8-32	NC	29	1 1/8-12	NF	1 3/64
8-36	NF	26	1 1/4-7	NC	1 7/64
10-24	NC	25	1 1/4-12	NF	1 11/64
10-32	NF	21	1 3/8-6	NC	1 7/32
12-24	NC	16	1 3/8 12	NF	1 19/64
12-28	NF	14	1 ½-6	NC	1 11/32
1/4-20	NC	7	1 ½-12	NF	1 27/64
1/4-28	NF	3	1 3/4-5	NC	1 9/16
5/16-18	NC	F	2-4.5	NC	1 25/32
5/16-24	NF	I	2 1/4-4.5	NC	2 1/32
3/8-16	NC	5/16	2 ½-4	NC	2 1/4
3/8-24	NF	Q	2 3/4-4	NC	2 ½
7/16-14	NC	U	3-4	NC	2 3/4
7/16-20	NF	25/64	3 ½-4	NC	3 1/4
1/2-13	NC	27/64	4-4	NC	3 3/4

10.9.7 Pressing, Stamping and Similar Methods

Pressing operations constitute a large portion of many manufacturing plant. Presses are used for many different purposes, including shaping, stamping, cutting, bending, forging, and assembling. In many cases, dies are used for forming parts, where the material is heated to lower its yield point before it is pressed. Careful design and analysis ensure that there is enough material for stretching, that the material remains intact, and that the final outcome is smooth and clean. Applications range from automobile body parts to small inserts stamped into shape with logos and brand names.

10.10 DESIGN FOR MANUFACTURING (DFM) AND DESIGN FOR ASSEMBLY (DFA)

Also known as Design for Manufacturability, DFM signifies the desire to design the product or system while looking forward to the way it will be fabricated. I distinctly remember when as a graduate student I was grading a design project that involved a high-pressure tank that required an attachment at the bottom. One student had designed a thick-plate bottom with a square hole whose depth was only ½ the thickness of the plate. When asked how he would make it, he had no answer. Obviously, casting was not an option, first because high pressure tanks are not made from cast material. Second, the cost of casting such a plate is prohibitive unless it is mass produced. Could a not-through square hole be cut? Perhaps yes, but at what cost and time?

Similarly, design for Assembly (DFA) involves looking into the future to ensure that assembling the parts or components of the system together will be accomplished in the easiest, most economical way. It involves trying to reduce the number of parts as much as possible, and making the assembly of the remaining parts as easy as possible. The fact is that both DFM and DFA go hand in hand; most material processing is followed by assembly. For this reason, many scholars combine the two together into DFMA, design for manufacturability and assembly.

Design for manufacturing and assembly is a methodology used during the design phase that tries to forecast the problems that might arise during material processing and assembly operations, the time that each operation takes, the cost, and whether it can be improved or not. As a result, the designer should expect that by following the DFMA rules and guidelines, manufacturing and assembly of the product will be as easy and cost effective as possible. Considering that most of the future cost of a product (70%) is determined during the design phase, it should be clear why DFMA is so vital.

The following simple guidelines are part of this effort.

- **Try to standardize the parts and the materials used.** If the materials and parts are standard, and better yet, if they are off-the-shelf items, not only will the cost of acquiring the material and the parts be less, it will be easier to work with the material and easier to maintain it in the future (e.g., replacement of broken parts). Standard parts and materials are used by many institutions and industries, and as a result, many suppliers and manufacturers deal with them. Due to more experience with their use and competition for more business, and as a result of consumer feedback, these standard parts are constantly improved. Consequently, they generally have better quality and are less expensive due to competitive pricing by multiple suppliers. Since standard parts can be obtained at any time, there is a better chance of reducing inventories as well. With nonstandard parts, the manufacturer will need to maintain a larger inventory of the same part or material.

   This should be extended to standardizing variations in similar parts too. For example, if your design requires a number of different size and/or type fasteners, see if you can change things to reduce the variations. Reducing two different size screws into one will reduce ordering requirements, inventory, handling, and assembly requirements, but also will eliminate possible mistakes. Figure 10.27 shows how changing the thickness of the base, if possible, can eliminate the need for two different types/sizes of bolts. Can you think of other ways to simplify this?

- **Simplify the design and reduce number of parts.** With a reduced number of parts and a simpler design, both material processing and assembly times will be reduced resulting in lower cost. Using the same part for multiple purposes (e.g., the same size and type of fastener vs. multiple sizes and types) will reduce inventories, minimize mistakes, and simplify assembly.

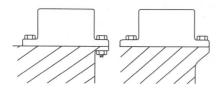

FIGURE 10.27 Simplifying a design to require fewer types of fasteners.

- **Consider methods of fabrication required.** As we have already discussed, alternative methods may sometimes be used to process materials or fabricate a part. You should consider all different possibilities, their associated required time and cost, required machinery, and whether the expertise to work with the process is available to the company or not. If a part is mass produced, you should consider mass production methods. For lower volumes of production, you should consider alternative methods. Additionally, the use of automation techniques, fixtures and jigs, and specialty tools may reduce fabrication and assembly times. DFMA programs point out a number of dos and don'ts about different things such as thicknesses, locations of holes and ribs, etc. that help the designer make decisions.

- **Make the product assembly mistake proof.** For example, imagine that a part must be inserted in one particular way. However, the way it is designed allows the assembler to insert it in other ways as well, even if not desired. You may facilitate this operation by marking the part in one particular way to indicate the correct insertion. However, it will be even better if the part is designed so that it could only be assembled one way—the correct way. In that case, not only have you made it easier for the assembler to quickly do the task, you have also made the design mistake proof. The result will be fewer errors, less correction time, and better quality. Another example is in the handling of parts. Small, thin parts are harder to pick up. Designing the part for easier handling will save time and reduce mistakes. DFMA programs have a list of dos and don'ts that help the designer think about these issues.

- **Consider the penalties for each attribute of every part.** For example, there is a penalty if the part is too small or too big, too light or too heavy, too small or too large, etc. A very small part is hard to handle, but so is a huge part. A part with sharp corner can injure or cut fingers. A very flexible part is difficult to handle. A flat, thin part laying on a flat surface is difficult to grab or lift.

- **Avoid unnecessary requirements.** This includes surface finishes, surface treatments, unnecessary fasteners, and unnecessary tight tolerances (that may not be easily achieved during fabrication).

- **Reduce the number of connections.** This includes reducing the number of parts, but also reducing interconnection between different parts. The common bus of a computer system is an example of this, where through the bus, the data are shared between different systems. Otherwise, think about how many more wires would have to be interconnected between parts. Quick-release connections also facilitate assembly and disassembly.

- **Avoid intermingling of parts.** For example, a part with small holes near another with a small peg can intermingle, resulting in connections or unwanted movements. Springs can intermingle and connect too, especially if their ends are open. Hollow conical objects can stack tightly. (Many cups do the same. Adding small tabs around the walls prevents this.) Figure 10.28 shows a few examples of parts that intermingle.

- **Design for ease of assembly** Reduced number of movements necessary, fewer directions of motions, and fewer parts all result in simpler assembly. For example, consider "vertical stack" (1-axis) assembly where all parts and subassemblies are assembled on top of one another, all from above. This facilitates robotic assembly too, and is used extensively in electronic assembly. Additionally, chamfers, guides, and marks can facilitate positioning, orientating, and insertion operations.

- **Design for better and easier fastening** This can be accomplished by reducing the number of screws, but also by replacing screws with other techniques of fastening,

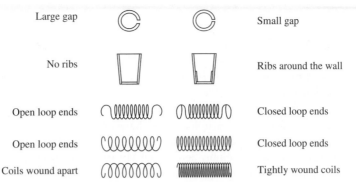

Large gap			Small gap
No ribs			Ribs around the wall
Open loop ends			Closed loop ends
Open loop ends			Closed loop ends
Coils wound apart			Tightly wound coils

FIGURE 10.28 Examples of parts that have the potential to interfere with each other, intermingle, or move each other. The ones in the right column eliminate the problem.

including adhesives, snap fits, and press fits. Minimize the variety of screws as well. Use self-tapping screws and captured washers.

- **Design modular subassemblies.** When a product or system is made from subassemblies, it is easier to assemble, to connect, and to replace. Because each module can also be tested before assembly, the overall quality will be increased as well.

- **Consider automated assembly.** Automated assembly, whether robotic or not, can increase throughput and efficiency and reduce assembly time. You should consider the added cost of automation versus the savings in assembly time and increased quality.

- **Ask relevant questions about the design of every part.** For example, ask whether a component needs to move relative to other parts, be of a material different than others, whether it should or should not be separate from other parts, and whether it is needed in this form or shape. Each question should be answered with a critical review of the part and its characteristics. You should be able to justify your design decisions. If there are alternatives available, why didn't you choose them?

EXAMPLE 10.6

First consider two rotary knobs used on a stereo, one for volume and one for tuning (Figure 10.29). The one that is used for volume must be inserted in a particular orientation to correctly reflect the volume. On the other hand, the knob used for tuning rotates continuously and has no particular orientation. For easier assembly it is desired to design the former in such a way as to simplify correct orientation of the knob for insertion, while the latter should be easily inserted in any direction without a problem. As you see, the time required to assemble two similar parts is influenced by whether or not the part has a particular orientation. Facilitating either requirement may result in different designs for the two similar parts. Now look at the dashboard in your car or your own stereo and determine whether there are any similar requirements for any of the knobs, controls, or dials.

FIGURE 10.29 Two rotary knobs on a stereo, one for volume, one for tuning. Although they look the same, the former has a positional requirement that the latter does not have. This creates different assembly requirements. ∎

FIGURE 10.30 Axes of symmetry influence the ease of assembly based on the requirements of positioning and orientation.

One fundamental way to look at each operation or task during fabrication and assembly is to consider the associated penalties. Each part has a fabrication or processing penalty, handling penalty, and assembly penalty. DFMA programs attach a penalty value to each part based on the information supplied by the designer. As mentioned above, a part that is too heavy is difficult to handle and assemble, but so is a part that is too small. The small part is more difficult to pick up, hold, and assemble, and it may be damaged during handling or assembly. Generally, more complex parts have larger penalties too. More precise alignment and more complicated orientations require more time, and therefore, have a larger penalty as well. As mentioned in Example 10.6, there are times when a particular position or orientation is desired. In that case, the part should be designed to facilitate correct positioning and orientation (dimensional asymmetry). Similarly, there are times when a part may be assembled in any orientation (or position). In this case, geometric symmetries will facilitate assembly. Figure 10.30 shows axes of symmetry for different shapes and how they may be used to facilitate symmetry or lack thereof. The sphere has complete symmetry along all axes, and thus, is easiest to insert if no particular orientation is desired. However, if a particular orientation is desired (e.g., a threaded hole on one side for attachment to another part), it is more difficult to orient. A cylinder has fewer axes of symmetry, followed by a cylinder with asymmetrical ends, and so on. DFMA programs attach a penalty value to each case.

EXAMPLE 10.7

Figure 10.31a shows the original design of a strobe light for an emergency vehicle. Figure 10.31b is the redesigned light. This design improvement resulted in an approximately 90 percent reduction in the number of parts to be assembled, a 70 percent reduction in the number of unique parts to be maintained, and a 75 percent reduction in assembly time. There were additional benefits in reliability, weight and so on.

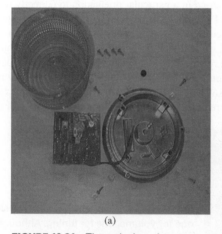

(a)   (b)

FIGURE 10.31 The redesigned emergency strobe light requires 90 percent fewer assembly actions, has 70 percent fewer unique parts, and needs 75 percent less assembly time.
(Courtesy of Prof. Larry Staufer.)

10.11   DESIGN FOR DISASSEMBLY (DFD)

Disassembly is required during a product's life for maintenance and repairs and may be required after a product's useful life for recycling and reuse of parts. Design for Disassembly (DFD) seeks to facilitate disassembly by anticipating it during the design phase of the product. Following problems may be encountered by the disassembler:

- Parts may be difficult to remove because access is obstructed by other parts. In this case the overall product layout may be at fault.
- A part's attachment to neighboring parts can also cause difficulty because some means of attachment (e.g., adhesives) are permanent. A hierarchy of fasteners is sometimes used as a general guideline to remedy this problem; snaps are easiest to remove. Screws are more difficult, and adhesives are the most difficult.
- Detaching or separating a part from its neighboring parts may require too large a force, special tools, or too many interfering motions.
- It may not be clear what needs to be done to disassemble a part, or the sequence of removal may not be clear.

These general guidelines are helpful, but they are not always sufficient. For example, what makes one screw easier to remove than another? Are snaps always easier to remove than screws? And if top-down assembly is not practical for a certain product, how should the parts be laid out? The answers depend on the specific product in question.

EXAMPLE 10.8

Figure 10.32 is the result of applying some of the above-mentioned DFD concepts to a popcorn popper and how it simplified the design as well as disassembly.[19] Specifically, the duct and the motor assembly were simplified to require fewer screws. However, the addition of the tabs in the duct and motor mount facilitate the alignment of the two parts and assembly as well as the disassembly of the parts.

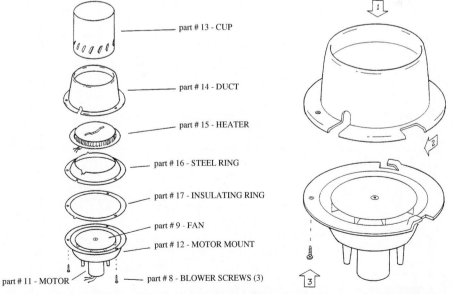

part # 13 - CUP

part # 14 - DUCT

part # 15 - HEATER

part # 16 - STEEL RING

part # 17 - INSULATING RING

part # 9 - FAN

part # 12 - MOTOR MOUNT

part # 11 - MOTOR

part # 8 - BLOWER SCREWS (3)

FIGURE 10.32   Application of DFD methods in simplifying the design as well as disassembly.   ■

Please see Chapter 7 for more information about design for disassembly and its relation to sustainability, pollution, and costs. This material is not repeated here, but it is fundamentally related to this issue. The following procedure is mentioned here for your reference. It describes some fundamental methods and issues to investigate and consider while designing a part for disassembly. For more information, please see Lowe and Niku, 1995.

The following is a summary of a method developed for Design for Disassembly (Lowe, Niku, 1995) This method presents a framework of concepts used to analyze disassembly during the design phase. It describes concepts such as "effort at part," which is the effort experienced by the part; "applied effort," which is the effort expended by the person removing the part; grip/tool advantage techniques, which are used to reduce the applied efforts, disengagement methods, and geometric constraints. It also describes scoring methods for obviousness, path, and force, to evaluate disassembly operations which are encountered during approach, disengagement, and removal of parts. The scores can be multiplied by weighting factors to reflect the relative importance of each category. It also evaluates disengagement methods by considering force, path, integrity of assembly, geometric constraints, tools available, and skill levels. The total effort scores of competing design options weighed against other factors such as increase in manufacturing cost can be compared in a chart. A running subtotal will determine how much disassembly is cost effective for an entire product. The methodology provides for a systematic approach during design phase to consider maintenance requirements for a product and facilitate disassembly of the product at the end of its useful life. Figure 10.33 is a summary of this technique.

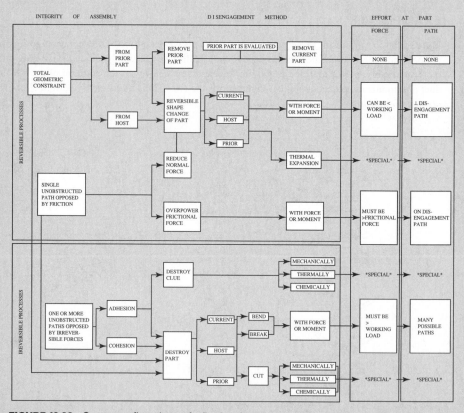

FIGURE 10.33 Summary flowchart of a Design for Disassembly method.

The goal is to reduce the disassembly effort, but there are two kinds of effort to consider: **Effort at part** is the effort which must be experienced by the part for it to be disengaged; **applied effort** is the effort expended by the person removing the part. The two can differ greatly depending on how the part is gripped and what tools are used. For example, a press fitted peg requires a certain force along the removal path to overcome the frictional force which holds it in place. This is the effort at part. But in order to apply the required force, a worker must also apply a squeezing force normal to the removal path to maintain grip. So the applied effort is not the same as the effort at part. Similarly, removing a nut with a long-handled wrench requires less applied effort than removing it with a shorter wrench even though the effort at part is the same in both cases. Analyzing the difference between the two kinds of effort provides another opportunity for design improvement. Because this difference is usually determined by grip and tool use, a "grip / tool advantage" section of the methodology lists techniques which can be used to reduce the applied effort given a predetermined effort at part. The result is that new ideas can be generated using effort at part analysis, improved grip/tool advantage techniques and compared against other designs using applied effort analysis.

Grip/Tool Advantage: An improved "grip" is usually just a reduction in exerted force. For example, it is easier to turn a 5-cm knob than a 1-cm knob because it is easier "to get a grip on it." But it is really an application of mechanical advantage to reduce the force with a longer lever arm. The difference between a grip and a tool is whether the means of mechanical advantage is built into the part or contained in a separate unit. For example, a wing nut and a hex nut both employ leverage to reduce the applied force. But the wing nut has the lever built in, so it is said to be easier to grip than the hex nut. The hex nut's lever is a separate wrench, so it requires a tool.

The first technique is to use mechanical advantage of simple tools (levers). The second technique is to increase frictional coefficient to reduce the normal force required in gripping. As mentioned above, a normal force must sometimes be applied when gripping a part. When the coefficient of static friction is increased, the normal force is reduced so the part is easier to disengage. The third technique is to provide direct application access instead of frictional gripping. The fourth technique is to increase surface area to distribute load. The next three techniques apply to the breaking of parts. Number 5 is to choose easiest failure mode. If a part is subject to occasional large compressive forces in its life, it must be strong enough in compression to withstand those forces. So it will be difficult to make the part fail in compression during disassembly. If the part is subject to only small bending loads, then perhaps bending should be used to break the part during disassembly. The sixth technique listed is to exploit material properties. For example, an extremely brittle part might lend itself to smashing. A highly notch sensitive plastic can be scored with a blade then broken by bending. The use of planned weak points is popular in DFD, so exploiting material properties can be extremely useful. The seventh technique is to consider impact application of force. The part may break more easily when it is impacted than if the force is applied gradually. The eighth technique is to facilitate tool application.

Applied Effort Analysis: Applied effort analysis deals with a part's interface with the disassembler. The goal of applied effort analysis is to estimate the amount of effort required for disassembly while the product is still in design so that the designer can identify disassembly problems early on. But before discussing how applied effort analysis is used, some new concepts which were not required in effort at part analysis must be introduced.

Effort at part is only defined during the **disengagement** of the part; that is while the part is actually being freed from the forces holding it in place. Effort at part is described by only two categories: **force** and **path**. But these two are not sufficient to describe applied effort. Force and path can still be used to describe the physical effort of the worker, but an additional category called **obviousness** is required to describe mental effort. The worker must decide what to do next before moving along any path or exerting any force. So obviousness must be considered in applied effort analysis to reflect the time spent in thought. The location of the disassembly (called **venue**) is also important in applied effort analysis. There are three main venues where the disassembly might take place: factory, scrap yard, and household. Factory workers receive training in the disassembly of the product so that the work is easier for them than for a consumer disassembling the product at home. A consumer may be doing the work for the first time and may not be mechanically inclined at all. Furthermore, the consumer might not own any tools. Even at an all-purpose scrap yard it is likely that the worker would have no experience with the specific product. However, the worker should have some experience with similar products and some degree of mechanical skill. The scrapyard worker is likely to have a full compliment of standard tools available but no access to product-specific tools. However, the scrap yard worker and an owner may disassemble a product only once in a while. As a result, the

time required to disassemble may be less important than in a factory where a worker may have to repeat the operation many times a day and is paid for the work.

A weighted score system similar to a decision matrix is used to assign applied effort scores so that comparisons can be made between disassembly operations. The designer first gives a disassembly operation scores for obviousness, path, and force, making sure that the scores reflect the difficulties encountered throughout all three stages of disassembly. There is subjectivity involved in

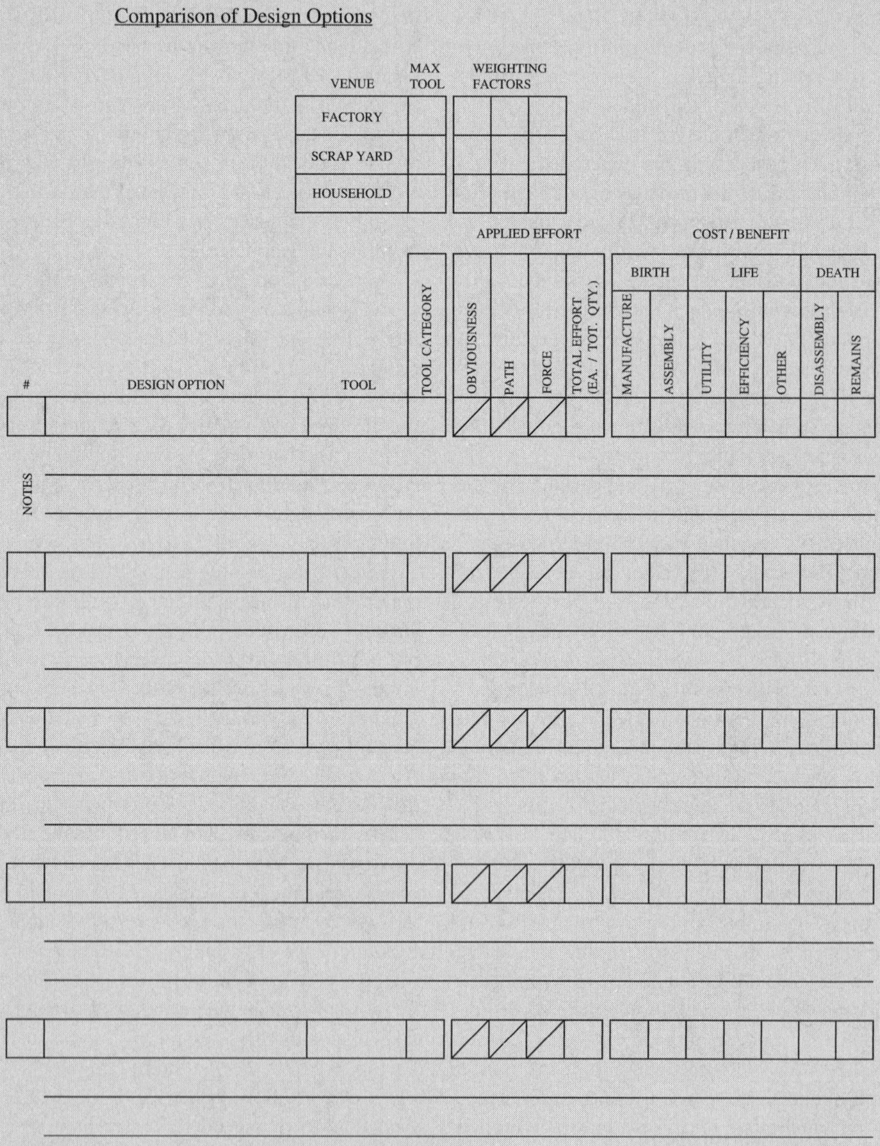

Comparison of Design Options

FIGURE 10.34 Comparison of Design Option form.

examining an operation and assigning a score. However, a guide like the one shown will improve accuracy by ensuring that the most important factors are weighed before scores are given. Next, each of the three scores is multiplied by a weighting factor. Each venue should have a different set of weighting factors. The weighted scores are added to give a total score for applied effort. Higher scores indicate more effort and are to be avoided. The total applied effort score should

reflect the time required for a disassembly operation without necessarily calculating that time absolutely. The score is useful when compared with scores for other disassembly operations.

Figures 10.34 and 10.35 show two suggested chart formats that can be used for this weighted score system to calculate applied effort. The form of the charts can be modified as necessary or used on a spreadsheet. A designer wishing to know whether a particular part is easier to remove when

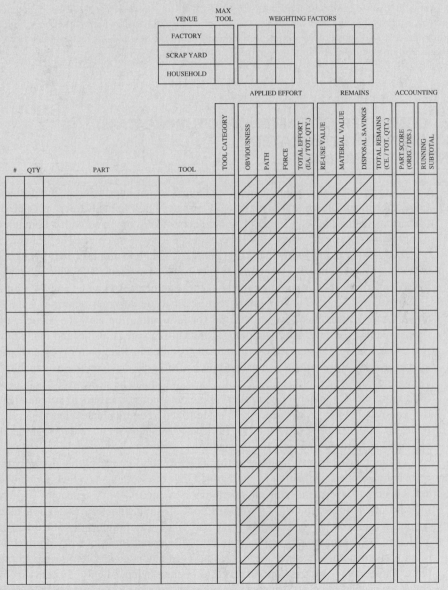

FIGURE 10.35 Complete Design Evaluation form.

it is affixed with a wing nut, breakaway weld, or snaps can use the Comparison of Design Options chart. The options being compared are listed down the left side of the chart and the applied effort is calculated for each as described above. The design with the lowest total effort score is the easiest to disassemble although not necessarily the best design. Parts are listed down the left column in the order of disassembly. For each part, the total effort score is subtracted from the total remains score to give the part score. The part score is then added to a running subtotal to find the extent of disassembly which is most cost effective. For example, the first line of the chart records the cost to dispose of the entire product without any disassembly at all. Then perhaps several parts are removed which cost effort but yield no value. Thus far then, the disassembly would not be economical at all. But the next part removed may have such a high value that the initial disassembly is justified. The running subtotal will rise and fall as more parts are removed. A negative score indicates that disassembly up to that point is not cost effective. The Complete Design Evaluation chart can also be used to spot areas for potential redesign. Any part with a high value that cannot be reached economically should be investigated. There may be a way to move it forward in the disassembly order or redesign a prior part. Any part with a particularly high total effort score should also be considered for redesign.

10.12 COMPUTER AIDED MANUFACTURING (CAM) AND COMPUTER INTEGRATED MANUFACTURING (CIM)

Computer Aided Manufacturing involves the electronic interfacing of the design with a fabrication device in order to automatically execute a task as many times as desired. What this means is that first the design must be electronically prepared with a computer aided design (CAD) system so that it will be understood by a machine. The drawing is then converted into a machine language that can be understood and executed by a controller. This conversion is usually through a numerically controlled (NC) code which is commonly called G&M code (because many of the statements start with G and M commands, as will be seen later). This operation requires that the designer identifies many details about the processes that need to be used for fabricating the part, the number of times (steps) required to do it, advance, plunge, and cutting speeds, paths, and many other factors. A complicated system may require many detailed instruction for correct conversion. However, the code can be downloaded into a CNC-type machine that is capable of executing the code and fabricating the part. The CNC machine needs to be actuated under the control of a computer (or controller) and have feedback sensors to communicate with the controller. A correctly designed system will communicate electronically, even remotely, and therefore, so long as the raw material can be set up in the machine, the part can be manufactured (thus, global engineering).

Computer integrated manufacturing (CIM) is similar, except that not only this process, but also other peripherals are integrated into the system, such that a manufacturing cell, capable of almost independent operation, is created. CIM systems may include loading and unloading systems that work in coordination with the fabrication machines, robots, sensors, conveyor belts, delivery and take-away systems, automatically guided vehicles (AGVs) for the delivery of parts from the storage units or carrying the manufactured parts to storage or other locations, and even automatically retrievable storage systems. These storage systems are designed to operate independently. Systems such as bar codes are used to identify parts that are retrieved from their bins and are delivered to the cells. As long as everything works as planned and everything is in its correct place, these systems function beautifully and without interference. A huge food manufacturing plant in California employs only a few people in its storage units; all bags and boxes of food are stored in its automatic storage and retrieval system before and after processing. The AGVs even deliver the boxes to the trucks at the loading docks.

In addition to material processing and fabrication, CIM ventures into other aspect of manufacturing, including quality, total quality management (TQM), scheduling, simulation of the manufacturing processes, tracking of parts, delivery, storage, and even ordering of components.

Figure 10.36a is a simple 3-axis, tabletop, CNC milling machine. This machine is very inexpensive and simple to use. It can cut machinable wax, wood, and aluminum.

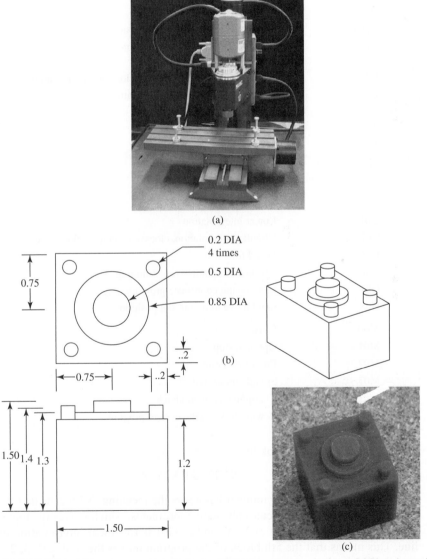

FIGURE 10.36   This part was designed on a CAD system, converted to a G&M code, and made by a CNC mill.

Figure 10.36b is a simple design that was implemented with this mill. The manufactured part is shown in Figure 10.36c. Although this part is extremely simple, it demonstrates the effectiveness of the system. If you have experience in machining, you can imagine how difficult—almost impossible—it is to machine the circular areas by hand. However, a CNC mill can easily and accurately machine this part in a few minutes. But more importantly, it can repeat the same process indefinitely so long as everything is in its place and nothing unexpected happens.

NC Codes   Before a part is machined, every step in the machining process must be defined. The standard language for this purpose is usually referred to as G&M code. The following codes are only a few of the more common characters in the code.

N	Block number
G	Preparatory codes
X,Y,Z	Primary X, Y, Z motion dimension
U,V, W	Incremental X, Y, Z motion dimension for absolute programming
I, J,K	Arc center X, Y, Z axis dimension for circular interpolation
R	Arc radius for circular interpolation
F	Feed rate (in/min)
S	Spindle speed
T	Tool specification
M	Miscellaneous functions
;	Comment line
G00	Rapid traverse
G01	Linear interpolation
G02, 03	Circular interpolation, clockwise, counterclockwise
G04	Dwell time
G90	Absolute coordinate programming
G91	Incremental coordinate programming
G98	rapid move to initial tool position
M00	Pause
M01	Optional stop
M02	End of program
M03, 05	Spindle motor on/off
M10, 11	Clamp/unclamp air chuck
M47	Rewind (restart) the current program

A typical code block may look like

$$N5G90G01X2Y3Z2$$

In this block, every letter or number has a specific meaning. N5 means that it is the 5th block of the code. G90 indicates absolute coordinates, which moves the tool to a point defined by the given values of X, Y, and Z. G01 is for linear interpolation in straight line. This means that the 5th block of the program moves the tool to 2, 3, 2 point in a straight line in absolute coordinates. A G91 (incremental coordinates) instead of G90 would move the tool a distance of 2, 3, 2 from the present location.

EXAMPLE 10.9

Interpret the meaning of the following G&M codes.

N15G1Z-0.2000F9.0; PLUNGE
N16G1Y1.1875
N17X1.1875

Solution Since the coordinate measuring system is not mentioned, it means that it was defined in one of the earlier block. We will assume absolute motion was specified. The first line is block 15 of the code. G1 indicates linear interpolation. There will be a motion along the Z-axis (−0.2) inches below the surface with a plunge speed of 9 inches per minute. Line 2 is block 16, linear interpolation, and moving along positive Y-axis 1.1875 inches. Next is block 17, moving 1.1875 inches in the positive X direction. ∎

10.13 RAPID PROTOTYPING

As the name indicates, rapid prototyping is used to quickly generate models or prototypes of systems and to check their design. Rapid prototyping, for the most part, is not used for production because it is too expensive and time consuming. The principles are not dissimilar to CAM. The design is generated on a CAD system. The model is converted into specific information that can be used by the rapid prototyper for quick manufacturing. Prototypes made with rapid prototyping techniques generally include internal cuts, complicated 3-D shapes, and even parts that move relative to each other. Rapid prototyping machines come in many different forms that work on different principles. However, most of them either slice the part into layers and make each layer, or slice it into paths and follow the paths for manufacturing.

One of the simplest systems that was used for rapid prototyping slices the part into paper-thin layers and cuts the layers on adhesive-back paper. When all the parts are added together, layer by layer, the part is formed. This system is limited in its capabilities, and the strength of the part is limited to that of paper, but it is relatively inexpensive. However, with the advent of newer, more capable machines, this system is not as popular. Figure 10.37 is a sample key that was made with this system.

Another system called Selective Laser Sintering[20] uses a bath of polymer powder that selectively hardens by the application of a laser light. The light is focused in one particular point within the bath, hardening the polymer at that location. By continually moving the focused laser in layer upon layer, and scanning throughout the bath, the part can be shaped.

Another system called Fused Deposition Modeling (FDM) lays a very thin two-part thermoplastic bead (filament) layer upon layer to form the part. The head moves in X, Y, Z directions until the part is finished. The thermoplastic polymer filament hardens as it is

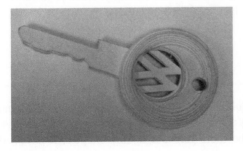

FIGURE 10.37 A rapidly manufactured prototype of a key made with thin slices of adhesive-back paper.

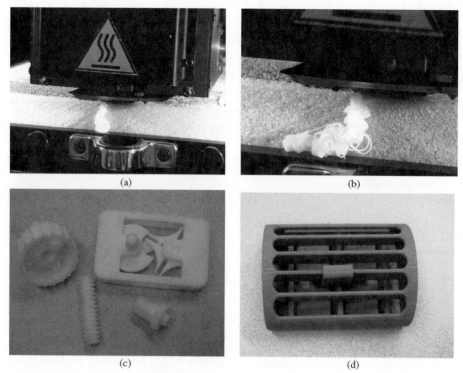

FIGURE 10.38 A Fused-Deposition-Modeling rapid prototyping machine and examples of some parts made with this machine.

cooled immediately after touching the part. Figure 10.38a shows the head of the FDM machine. Figure 10.38b shows the filament collecting in one point as laid by the head. In this system two types of filaments are used, one for the parts that will remain, one for space that is supposed to be removed (these are only used when necessary). The second filament is used only where needed to create temporary support for moving parts, but can be ''washed'' away when placed in an ultrasonic bath. FDM can create models that have moving parts relative to each other all at once, without the need for assembly. Figure 10.38c and d show a few parts made with this system. As you can see, the scotch yoke model has at least three moving parts, as does the planetary gear system. Figure 10.38d shows a small vent with moving vanes and tabs. All parts of each prototype were made together as one piece.

It should be mentioned here that during the design phase, there may be a need for detailed rapid prototyping in order to advance the design concepts. These machines, if available, are excellent aids in this endeavor. However, in many other cases, simple rapid prototyping with paper and cardboard, foam, and pieces of wood may be all you need. We have already discussed this before, but it is worth reminding you that physical realization of ideas should not suffer from the lack of a sophisticated, expensive machine. Get out your box of ''junk'' and model it.

10.14 LEAN PRODUCTION

Lean production is a method of thinking that alters the way production and quality are integrated. It is a fascinating subject that warrants serious attention. Although it is related to manufacturing and appears to be more relevant to this chapter, it is discussed in

Chapter 12, where the subject of quality is first discussed in detail. Please see Chapter 12 for a thorough discussion about lean production.

REFERENCES

1. ASHBY, MIKE F. and KARA JOHNSON, "Materials and Design," Elsevier, Butterworth, Heinemann Publishing, 2002.
2. ASHBY, MIKE F., "Materials Selection in Mechanical Design," Pergamon Press, Oxford, 1992.
3. CREESE, ROBERT C., "Introduction to Manufacturing Processes and Materials," Marcel Dekker Inc., New York, 1999.
4. JAMES P. WOMACK, DANIEL T. JONES and DANIEL ROOS, "The Machine that Changed the World: Based on the Massachusetts Institute of Technology 5-million-dollar 5-year Study on the Future of the Automobile," New York, Rawson Associates, 1990.
5. HSU, TAI-RAN and D. K. SINHA, "Computer Aided Design : An Integrated Approach," West Publishing Co., St. Paul, 1992.
6. REHG, JAMES A., "Computer-Integrated Manufacturing," Prentice-Hall Career & Technology, New Jersey, 1994.
7. BOOTHROYD, GEOFFREY, "Assembly Automation and Product Design," Marcel Dekker, Inc., 1992.
8. BOOTHROYD, GEOFFREY, "Making it Simple: Design for Assembly," *Mechanical Engineering*, February 1988, pp. 28–31.
9. WATERS, FRED, "Fundamentals of Manufacturing for Engineers," University College London Press, 1006.
10. RUBIN, ARTHUR I. and ELDER, J., "Building for People: Behavioral Research Approaches and Directions," U.S. Department of Commerce, Washington, D.C., 1980.

11. JUVINAL, ROBERT C. and K.M. MARSHEK, "Fundamentals of Machine Component Design," 3rd Edition, John Wiley and Sons, New York, 2000.
12. SHIGLEY, JOSEPH E. and L. MITCHELL, "Mechanical Engineering Design," 4th Edition, McGraw Hill, New York, 1983.
13. SHIGLEY, JOSEPH E, C. MISCHKE, and R. BUDYNAS, "Mechanical Engineering Design," 7th Edition, McGraw Hill, New York, 2004.
14. BEAKLEY, GEORGE and E. G. CHILTON, "Design Serving the Needs of Man," MacMillan, New York, 1974.
15. KAHMEYER, LEICHT, "Dismantling Facilitated," translated from Kunststoffe German Plastics, Vol. 12, 1991, pp. 30–32.
16. SIMON, M., "Design For Dismantling," *Professional Engineering*, 11/91, pp. 20–22.
17. FORCUCCI, TOMPKINS, "Automotive interiors: Design For Recyclability," SAE Special Publication n 867, Designing For Recyclability and Reuse of Automotive Plastics, 2/91.
18. STAUFFER, LARRY, Ph.D., P.E., Director, College of Engineering and Engineering Management, Professor of Mechanical Engineering and Engineering Management, University of Idaho, Boise.
19. LOWE, ALLEN and S. B. NIKU, "Methodology for Design for Disassembly," ASME Publication # DE-VOL. 81, March 1995, pp. 47–53.
20. COLLEY, DAVID, "Instant Prototypes," *Mechanical Engineering*, July 1988, pp. 68–70.

HOMEWORK

10.1 Search the Internet and locate manufacturers/distributors for steels and alloys. Find what they offer.

10.2 Search through the Internet or manufacturers' catalogs and find three steel alloys that satisfy the following requirements for strength and hardness:

$$S_y = 45000 \, \text{psi}, \quad \text{Brinell Hardness} \geq 120$$

10.3 Find two metallic alternative materials that have a modulus of elasticity of approximately 30 Mpsi and a tensile strength of between 40,000 and 50,000 psi.

10.4 Which one of the materials mentioned in this chapter has negligible creep throughout its life?

10.5 Use Internet sites such as http://www.matweb.com/search/SearchUNS.asp to learn about metals

and alloys. Investigate what kind of information can be obtained from this and other similar sites.

10.6 Identify at least two species of wood, one hardwood, another softwood, with a minimum modulus of elasticity of 1.5 Mpsi and a minimum tensile strength perpendicular to the grain of 450 psi. Compare their weight and average price. What other characteristics would you consider in order to choose between the two?

10.7 Call your local lumber yard or hardware store and ask about the variety of different plywoods and particle-boards. Then choose two similar-type plywoods with different core materials and compare their price, weight, and strength. How would you choose between these two if you were making a table?

10.8 Design a simple bookshelf for your own use. Compare the price, weight, and strength of a veneered plywood (such as oak), a Melamine white board, and pine boards. What other characteristic would enable you to choose between these three alternatives?

10.9 Search the Internet to locate manufacturers/distributors for polymers and resins. Find what they offer.

10.10 Identify at least two types of thermoplastics that can be used to make a seat for outdoor use. What characteristic will you use to decide?

10.11 Consider a product in your own household that is made mostly or primarily of plastics. Find at least one alternative material that could be used to make it. What criteria did you use to make this decision?

10.12 Write down a list of advantages and disadvantages of the material that is used to make your flooring. How can you improve the disadvantages?

10.13 Find five different types of recyclable plastics and match them to Figure 10.2. Which one of these product types was easier to find? Which one was the hardest to find?

10.14 Select 10 random products made of plastics and look for recycling stamp on them. What percentage of these products do not have a recycling logo stamped on them?

10.15 Investigate your surroundings to see if you can identify any products that use composite materials. If so, try to identify the types of filaments and resins used as well as how they are laid. Can you explain why the fibers are laid as such based on the applied loads?

10.16 If you have access to the charts developed by Ashby in your library, try to use them to select a material based on the provided characteristic ratios.

10.17 Consider two metal machine parts, perhaps from your car or other similar machine elements (e.g., parts of your car's engine, transmission, or suspension). Try to determine what processes were used to manufacture these parts, and whether there are similarities and differences between them. Try to determine why these particular processes were used.

10.18 Consider a simple product like your telephone, toaster, or a printer. List the materials used in its overall construction, how many different major parts are made and assembled, and how they are put together. Would you change any of these methods?

10.19 Pick up a common household product from a thrift store and use it to investigate whether the product was designed for disassembly or not. Try to disassemble it by keeping track of the time required to do so and what steps are required. Next try to devise a plan to improve the design. Can it be made simpler? Can more parts be designed for better disassembly and recycling?

10.20 Interpret the meaning of the following lines of NC code:

N26G2X0.688I0.75J0.75F27
N27X0.8115I0.75J0.75
N32G2X0.562I0.7500J0.7500
N34G0Z0.05; RETRACT

CHAPTER 11

ECONOMICS OF DESIGN

I am not rich enough to buy cheap things.
—Scottish proverb

11.1 INTRODUCTION

In this chapter, we will discuss the relationship between the economics and the design of a product or system. This relationship is a significant factor in the effectiveness and competitiveness or uselessness and failure of the product or system. We will also examine the role a designer plays in this and how the decisions the designer makes can significantly change the final outcome of the product or system.

We will also discuss the basics of engineering economy and time value of money which helps a designer decide between different alternatives based on economic principles.

11.2 VALUE OF A PRODUCT OR SYSTEM

In the market, the worth of a product or system is its value to the customer. In a competitive market, a product that has higher value is the one customers select. What is value? Value is defined as the ratio of what you get in a product or system, including its quality, to what you pay for the product or system, or

$$\text{Value} = \frac{\text{What you get}}{\text{What you pay}} \tag{11.1}$$

Obviously, the more you get for what you pay, the higher the value, thus the higher the competitiveness. Imagine that you are considering two similar products, for example two cameras, two toaster, or two restaurants. You may do some research (or simple comparison) of the quality of the product, the price, the amenities, the looks, the features, or the service each product or system provides to you or costs you. In the end, you will most probably pick the product or system based on its value. The one product or system that has a higher value will eventually be selected. Suppose that between two cars one has a better reputation, better looks, more power, better fuel economy, better service organization, more features, and is more comfortable, but they both cost about the same. The one you will pick will be the one that has all better characteristics at the same cost, because its value will be higher. Similarly, if between two cars, one has fewer features, less power, etc., but it costs much less than another car, so long as it delivers more value to you, you will pick it. There are many times that customers may like to buy an expensive car with more qualities and better characteristics, but they opt for less expensive options because the additional features and luxuries are not worth the cost, thus less value.

The only exception to this rule is what I call "sentimental value." This refers to additional value placed on products or systems based on emotional considerations. For example, if one really likes a particular car, for whatever reason, that car will not be evaluated fairly with the competition, since there is an additional emotional value that changes the perception about that car. In these cases, the customer will pick the product based on the total value, which includes the overwhelming sentimental value (e.g., when customers pay more than the asked price in order to buy a product earlier than others, including electronics toys and hybrid cars). In fact, many manufacturers create sentimental value for their product in order to increase its competitiveness. The "ultimate driving machine" of the 1980s was perhaps less than the lowest-valued automobiles of today, but it was, and still is, referred to as the "ultimate driving machine" then and now.

Therefore, it should be clear that the economically related issues are as important as the rest of the qualities and characteristics of the product. The designer plays a vital, pivoting, and significant role in the final determination of the cost of the product or system, thus affecting its final value. In fact, the designer may also play a significant role in the creation of the sentimental value of a product (as an example, the particular sound of the 45° engines of Harley Davidson has a great sentimental value to motorcycle enthusiasts, all due to its particular design).

In the following sections we will discuss the ways the design engineer plays a role in the final determination of the value of a product or system.

11.3 GLOBAL ECONOMICS MODELS

World economies follow four basic models:

- Traditional
- Controlled (communist)
- Socialist
- Capitalist or free enterprise

Traditional economies are based on production of goods and services by individuals for their own consumption and exchanging their surplus products and services for other things they need, mostly in simple markets. Flea markets and farmers' markets, although enterprising, are variations of this economic system. Traditional economies, although inefficient and small scale, are very common in many parts of the world, especially in smaller towns and villages. However, since this system is not very common in the United State and does not relate to large-scale product design, we will not discuss it in any more detail.

Controlled economies in Communist countries are based on the premise that it is the duty of each individual to do what she or he can for the betterment of the society; and therefore, it is not the individual's welfare that is important, but the whole society's. As a result, the government controls all the affairs of the State, including all economic decisions, all manufacturing and product related decisions, and all financial decisions of the society. Individuals are assigned to the work that is deemed best for their abilities or talents, they are paid what the State decides, the products they buy are controlled by the State through prices and availability of the products, and individuals work and live where the State determines is the best choice for them. Although there are certain positive values in this system in terms of resource management and welfare of the individuals in terms of social and job security, the economic system is too closely tied to the political system of the government, and as a result, it does create many social problems. One basic flaw accorded to this system is that it assumes that people are "good" and they will do what they must for the society. However, human nature is such that unless there is some positive return, individuals will not do their best; if you own your house and you are responsible for what happens to it, you will take care of it much better than if the State gives you a house to live in and you are not responsible for its final value or what happens to it. If you know that your level of job performance will not make any difference in the way you are financially or socially rewarded, you will not do your best. Additionally, in communist systems, the State decides what is to be manufactured, what the prices are, and who can buy the product or not (if not directly, through other controls). Both the lack of competition and the fact that your product will be purchased no matter how competitive it is causes the market to fall behind the competition in other societies and lose its competitiveness. The result is a broken and bankrupt economy. This lack of respect for human nature was the main reason why communism eventually destroyed itself. There are only a few countries in the world today that are still under this economic system, although romantically, many people still believe in it. Furthermore, even in communist countries such as China, the economy is no longer controlled; it is practically capitalism in a controlled society.

Economies in Socialist republics (such as in Europe) are a middle-of-the-road variation between capitalism and communism. In socialist systems, the government is very strong and maintains control over certain parts of the economy such as utilities, transportation, and major national industries. Taxes are generally very high, and individuals cannot get rich as easily. However, the State generally provides more for the social, physical, and financial needs of the individual through more comprehensive health insurance, free education, better unemployment and retirement benefits, and the like. Although there are more regulations, the State does not dictate what products are made and sold, and to whom, and therefore, industries must remain competitive to be able to survive. In that sense, the system is very similar to the free enterprise.

Capitalist or free enterprise economies are based on the freedom of individuals to engage in economic activities as they wish (within the legal limits set by the legislature) in the pursuit of making profit, with which they will support themselves and their families. The premise that work is done to make profit may sound egotistical and selfish. However, if not to make money, why would anyone want to work and produce? And without money, how would one generally provide for his or her family? So long as we do what we do lawfully, and so long as we act justly toward others, the free enterprise accords us the freedom to engage in activities from which we profit in many ways. The control comes from the fact that in free enterprise, one has to compete with others who can do the same as we do. The competitiveness required to stay in the market forces us to be better, more efficient, and produce what customers need and want.

Obviously, all systems have limitations, advantages, and disadvantages. For example, in free enterprise, although we have to be competitive and produce what a customer wants in order to be able to sell our product and make profit, it is also possible to create false need through advertising or other gimmicks, so that we would make and sell products that are in fact not needed. Perceptions can also make a big difference in the way products are sold and presented, sometimes falsely. Therefore, the system is not immune from faults. Similarly, there are many who benefit from the competitive nature of the free enterprise, but many others are left behind, some due to their own fault, many due to lack of opportunities, lower social and educational skills, and disabilities. Whether they should suffer the consequences or not is a big dilemma. On the other hand, although in the free enterprise system it is the individual who is in control, the State does exert a certain amount of control over the economy and the system. For example, through setting the prime rate of interest at which the Central Bank (and other banks) charges other institutions when they borrow money, the state of the economy is controlled. This rate is affected by and affects the rate of inflation and how fast the economy grows. One consequence, for example, is the unemployment rate. In order to control wages the industry would like to see this rate higher. However, this means that somewhere about 6 percent of the workers are always unemployed. Should we care about these individuals?

The free enterprise system is perhaps the most prevalent economic system in the world today. The overriding principles of competitiveness and profit-making create an environment where individuals have to do their best to maximize their benefits. In this system, the designer is a central figure with much influence on the outcomes. Next we will look at the role of the designers in this system, and how they can influence the outcomes.

11.4 COSTS, REVENUES, AND PROFITS

The cost of a product consists of two, essentially separate, parts: direct costs and indirect costs.

Direct costs are the summation of all the costs directly related to the production of the product or system. Therefore, the more products are made, the higher the cost. This, in its simplest form, is a direct linear relationship (although as we will see shortly, the relationship is generally not linear). If no products are made, there is no direct cost. If the production is increased by 100 percent, the direct costs will be increased by 100 percent (Figure 11.1a). So, imagine that you are producing a simple gadget that beeps if your garbage can is too full. Each product requires a sensor, a beeper, a hook, etc. If you increase your production, you will need more parts, more workers, and more machines, proportionally increasing your total cost. However, no production can take place unless there is a support system, thus indirect costs.

Indirect costs refer to the costs incurred as a result of the support system. In order for you to be able to produce even a single product, you have to have a place, an organization, an administration, and a slew of other individuals and requirements that enable you to produce your product. This includes your board of directors, the president, vice presidents, accountants, lawyers, secretaries, janitors, the plant in which you make the product, parking lots, warehouses, and many other elements that cost you, whether you are making a few or many products. In fact, in many cases, even when you are not making any products (as in overproduction, during strikes, long weekends, etc.) you may not produce any products, but you still pay for all of the above (Figure 11.1b). This seemingly fixed cost is independent of your production level. However, in reality, it does vary as production levels change, and we will discuss this in more depth later.

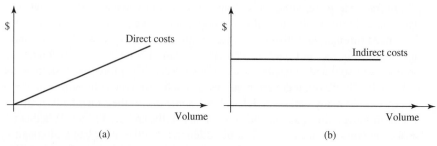

FIGURE 11.1 Direct costs of a product.

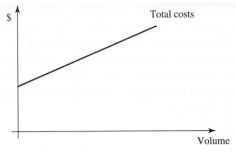

FIGURE 11.2   The total cost of a product.

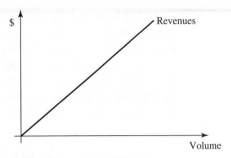

FIGURE 11.3   Revenues as generated by selling your products or systems.

The total cost of a product is the combination of the direct and indirect costs, as shown in Figure 11.2. Obviously, even if there is no production, there are still the indirect costs, but as the production increases, the total cost grows accordingly. This can be represented as

$$C = I + kn \qquad (11.2)$$

where C is the total cost, k is the cost of one additional product (also called marginal cost), and n is the volume of production.

Revenue is proportional to the number of products or systems you sell to your customers. The more you sell, the larger your revenues. In its simplest form, this can be represented as a straight line as shown in Figure 11.3. We will discuss variations to this representation shortly. Revenue can be represented as

$$R = pn \qquad (11.3)$$

where R is the revenue, and p is the price of one unit.

Break-even Point is the number of products or systems that must be sold in order to offset the total cost by the total revenue, thus no loss or profit. To get this value, you must superimpose the cost representation and the revenues representations, as shown in Figure 11.4. The break-even point can be found by setting Equations (11.2). and (11.3) equal to each other, resulting in

$$n_{break-even} = \frac{I}{p - k} \qquad (11.4)$$

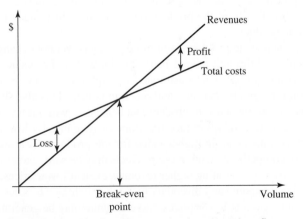

FIGURE 11.4   The break-even point, losses, and profits.

As Figure 11.4 shows, the difference between the costs and revenues is either loss, if the costs are more than revenues, or profit, if costs are less than revenues. At the break-even point, there is no profit or loss. The total revenues are offset by the total cost.

Obviously, different slopes create variability in loss or profit. One hopes that the slope of the revenues is larger than the slope of the costs; otherwise, there would never be a profit. The point is whether these slopes can be changed, and if so, how. We will discuss this in more detail shortly. Profit can be represented by

$$\text{Profit} = (p - k)n - I \qquad (11.5)$$

Although the graph in Figure 11.4 is very simple, it actually teaches us a lot. For example, imagine that we want to increase our profit. One way to do this is to simply increase the slope of the revenue (increase p). This means that we will increase the selling price of our products or systems. Obviously this will generate more income, and there-fore, more profit, so long as we sell the same number of products and systems even at the higher price. However, in real life, as you increase the price, and especially in competitive markets, the product will have less value, and therefore, you may sell fewer items, thus reducing your revenues and profit. In other words, simply increasing your selling price will not necessarily generate more profit, unless at higher prices you still sell about the same number of items. This can only happen if even at higher prices your products still have enough value to be competitive. In other words, you must either have a unique fea-ture, no competitors, or high initial value to be able to compete even at higher prices. This is where the designer's role becomes evident. It is the designer's responsibility to design and create products and systems that are good, have high quality, are needed, and have high value, in order to successfully compete in the market.

Another approach to increasing profit is to lower the slope of total costs (decrease k). This can be done in many ways as well, for example, by excluding certain features from the product or system, lowering the cost of the product by using cheaper raw mate-rial, using more efficient manufacturing techniques, lowering labor costs, and many more. However, once again, the level of profit will only increase if the same number of products is sold. If you find a way to reduce the cost by reducing labor costs or more efficient manufacturing techniques and other similar approaches, without affecting the quality or the value of the product, you will certainly succeed. On the other hand, if you reduce your costs by using inferior raw materials, excluding useful features, or otherwise reducing the value of the product or system, you may no longer be competitive in the market, sell fewer items, and still end up with less profit. Here too, the role of the designer becomes very evident. It is the responsibility of the designer to design the product in such a way to lower the costs, maintain the quality of the product, have useful features and character-istics, and create high value products.

Another approach to increasing profit is to sell more products. Without changing the price, this can be accomplished through having better products at higher value that are competitive in the market and attract more customers, through advertising, or through creating a good reputation for provided services and other amenities. This also shows the role the designer can play in creating a useful product that is needed, with high value.

Yet another approach is to actually reduce the slope of revenues by lowering the selling price of the product, thus creating higher value for the product. As a result, the product may become more competitive, many more products may be sold, and as a result, you may make more profit by yet generating higher revenues even at lower prices. This is the basic logic behind sales, coupons, and discounts for higher volumes of sales. As the seller sells more of a product, even at lower prices, more revenue may be generated with higher profit (Figure 11.5). Many manufacturers and retailers do the same thing. The price

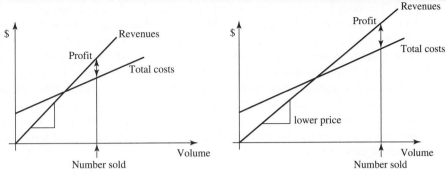

FIGURE 11.5 Effect of increasing volume on profits.

per unit for most products in bigger containers and at larger sizes is less. Most manufacturers and retailers provide discounts for higher volume sales. Many advertise an item by specifying a price for larger counts or bundles, for example, five soda bottles for $5, although they would still sell one bottle for $1.

This brings us to what was mentioned earlier about whether the simple straight-line representation of costs and revenues is accurate or adequate. For instance, as was just discussed, manufacturers may be willing to lower their prices for larger quantities. As a result, the line representing the revenues will have a smaller slope as the volume increases. Similarly, if you are the buyer, as you make more products, your cost of purchasing other materials and components from other manufacturers for each unit will also be reduced. Additionally, as you increase your production volumes, certain costs such as the cost of new machinery are divided among a larger number of products, thus reducing the per-unit cost. As a result, your total cost per unit decreases as your volume increases, reducing the slope for costs and increasing profit margin. On the other hand, as your volume increases, your establishment may reach a saturation point at which you may need to hire more labor, more support staff, may need a larger plant, more machines, etc. This means that as the volume increases, at certain levels, the indirect cost will have a sudden rise in it, reducing your profit. Therefore, increase in profit becomes somewhat stepwise. This is shown in Figure 11.6.

The costs, prices, values, competitiveness, volume, and profits are all interconnected. One cannot arbitrarily change one without affecting the others. The real issue is how to find the right balance between these opposing elements in order to maximize

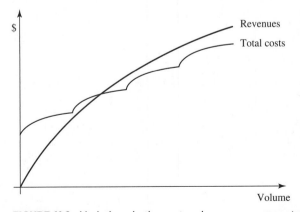

FIGURE 11.6 Variations in the cost and revenue representations as volume increases.

profit. The design engineer plays a vital role in this, having significant effect on the cost, the selling price, the value, the competitiveness, and ultimately, the profit. By creating a well-designed product that is simpler and requires fewer resources, less material, fewer labor, and fewer pieces of equipment, the cost can be reduced. By designing useful features, needed products that actually are beneficial to the user, and by creating good quality products at reasonable and competitive prices, the value is increased. When the product is actually needed, less advertising and fewer incentives are needed to attract customers. And when more products are sold to satisfied customers, more profit is generated over time. In the next section we will investigate in more detail the role the designer can play in each of these issues.

11.5 COST BREAKDOWN OF PRODUCTS AND SYSTEMS

In order to better understand the role a designer plays in the final value of a product or system and to see how the decision the designer makes will affect this final outcome, we will scrutinize the final selling price of a product. For each of these elements, we will look into the role of the designer and his or her influence on the outcome.

The final selling price of a product can be divided into many different categories, and each may include a variety of different elements, all depending on the product. A large system such as a car, produced in this country, may be very different than another product, say a coffee mug, which can easily be made by multiple manufacturers in many countries. The profit margins may also be entirely different. A high-priced, large product such as a car or a major appliance may have much lower profit margin (10–20%) compared to a shirt or a novelty gift item (100–200%). Some of the categories we will discuss may be irrelevant for certain products and very important for others. For example, a simple product such as a cooling vent may require little engineering design, while other products such as a car, an airplane, or a television require significant engineering design. Nevertheless, what is more important is the influence of the designer on each category and on the final outcome rather than the actual percentages of each category and whether one is more or less relevant to a particular product.

The final selling price of a product is the summation of the individual elements shown in Figure 11.7.

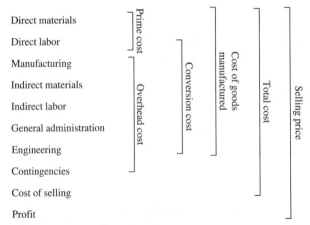

FIGURE 11.7 Elements of the selling price of a product or system.

In most cases, the contribution of each element to the final cost of the product is determined based on experience, on past usage, and by calculation. There are computer programs that assist the designer in cost estimation too. Additionally, most companies have their own cost estimates for each operation and material. However, this chapter is not about cost estimation (which is usually related to manufacturing). In the following sections we will discuss the influence of the design engineer on each element.

11.5.1 Direct Materials

Direct materials are all the raw materials, premanufactured subassemblies, components, fasteners, and other similar materials that are used in each product, and as such, the more products you make, the more of these materials you will need. The direct material category increases proportionally to your production volume. As we discussed earlier, with increased volume, the cost per unit of the direct material may be less as you receive discounts from other manufacturers, but in general, as the volume increases, the cost of the direct materials increases as well.

Direct materials also include packaging. As the volume increases, so does the total number of packages that are needed to ship or to sell the product.

The design engineer has a very direct and significant influence on the total cost of the materials. This can be in the magnitude of the material used, the cost of the material chosen, and the methods of manufacturing that are needed to process the material. With fewer materials, fewer parts, simpler design and manufacturing, and fewer subassemblies required to make a product, the product will be less expensive. This can show up in all stages of design, in many forms. The decisions of the designer will determine how much money is spent on direct materials.

11.5.2 Direct Labor

Direct labor costs include all the costs that are incurred directly for the production or manufacturing of the product or system. As the production volume increases, so does the direct labor needed to accomplish the job. Conceivably, if there is no production, there should be no direct labor cost.

In reality, this is not always true. In some companies, such as some auto makers, the labor is hired on a daily (hourly) basis. This means that based on production levels, the involved labor is called in on an as-needed basis. They use a punch card to be paid based on total number of hours they have worked. In this case, it is relatively easy to calculate the direct cost, since as the need changes, so does the total hours of work by the involved labor. In this method, the labor involved may sometimes not have any work for a stretch of time. As a result, the individuals may be paid at a higher level when they work in order to compensate for the times they may not work. Alternately, when the labor is needed for longer than the regular length of time, they are paid overtime which is higher than normal. On the other hand, many other places hire their labor based on an estimate for constant production, and they are paid a constant (monthly or yearly) salary. In this instance, the labor may do less or more, but gets paid on a constant basis. The concept of direct labor may not accurately apply to this method. However, the concept of direct labor is still valid: The more you produce, the more labor you will need.

The designer has a significant and direct effect on direct labor as well. This effect is through the

- choice of methods of manufacture and how long each operation takes,
- total number of operations that are needed for each product,

- level of difficulty of each operation, and
- the level of expertise of the labor involved and its associated cost.

In an anecdotal example, I remember a story from an old friend whose team had redesigned a washing machine with 3 fewer bolts needed to mount the drum. Although this reduction in number of bolts was simple, it had no negative effect on the quality of the product or its operations. However, they had estimated that this simple change would save money in direct material and in direct labor as it would require a little less time to pick the bolts, insert the bolts, mount the drum, and tighten them. The total saving per machine was in the $1 range. However, since they made about 1 million machines per year, this would translate into a $1,000,000 savings per year, and therefore, that much more profit. Please also refer to Figure 10.31 that shows how the redesigned emergency strobe light required 90 percent fewer assembly actions and needed 75 percent less assembly time, reducing the labor cost (courtesy of L. Staufer).

11.5.3 Manufacturing

The cost of manufacturing is the total cost of machinery and associated peripherals that are used to manufacture products. As the volume of production increases, the cost of manufacturing increases as well. The cost of investment funds used for purchasing and installing machines may either be considered as part of the overhead cost, or more commonly, as part of the manufacturing cost. This is true because the larger the volume of production, the more machinery needed to accomplish the job, therefore a direct cost. On the other hand, if these machines are purchased and installed, but are not used, the company may still have to pay for them, thus an overhead cost.

The designer has a direct and significant effect on this cost as well. If more sophisticated, more expensive, and greater number of machines is needed, the manufacturing cost will increase as well. By designing simpler products that require fewer operations, needing fewer machines and set-ups, and by designing the product with fewer parts, not only the labor costs but also the manufacturing costs will be less.

The example in Figure 10.31, mentioned above, applies here too. With a simple redesign, not only the labor cost was reduced, but so was the manufacturing cost. As another example, imagine that you would require skilled machinists and welders to machine and then weld two pieces of a machine together. Alternately, imagine that these two pieces could be cast with high strength plastics in one piece, automatically, without the need for skilled workers. The manufacturing cost of each unit would be significantly less in both labor and material. However, whether or not the cost of manufacturing is less depends on other factors as well. One factor is the production volume. All castings require a die. A die for plastic casting can in fact be very expensive and may require much lead time for preparation. If the number of products that will be produced is large enough to offset the cost of the die, it may be less expensive to cast the part. If the production volume is low, the cost per unit may be prohibitively expensive (as was mentioned, this is another reason why increased production volume reduces your cost). Another factor is the availability of machines required to do the job. If the company has already invested in casting machines and they are available for manufacturing the new design, the final cost of the part may be reduced. However, if the company needs to invest in new machines, train the workers, and develop expertise in the design of parts with casting, the cost per unit may be very large. Therefore, it is always necessary for the designer to consider the best methods of manufacturing that are available as well as the cost of new methods that are not presently available in the company.

11.5.4 Indirect Materials

Indirect materials are items that are not directly used in the products, and consequently do not significantly change as production levels change. This includes factory supplies, lubricants used with machines, lighting and air conditioning costs, and many other similar items. Some of these items actually vary as production volumes vary, but still are not directly used in the product, and are thus considered indirect material. Depending on the product and the culture of the company, this may be a significant cost or not. However, in general, the designer has little effect on this cost.

11.5.5 Indirect Labor

Indirect labor includes supervisors, foremen, inspectors, factory clerks, janitors, and other personnel that are needed to run and manage the daily operations of the company, but who are not directly involved in the production and manufacturing of the products or systems. As production levels vary, many of these individuals are not affected, unless their jobs reach a saturation level, where additional personnel are needed to continue with the increased levels. Like indirect materials, the cost of indirect labor may or may not be significant. Similarly, the designer's decisions generally do not greatly affect the indirect labor costs.

11.5.6 General Administration

General administration includes most other people that are required to run a business, but who are not directly involved in the production of the products. This includes the management team, including the board of directors, presidents and their staffs, midlevel and lower-level managers, accountants, lawyers, purchasers, clerks, secretaries, janitors, etc. This also includes the cost of buildings, taxes, depreciation, power, and other similar items. Depending on the type of company and accounting practices, what is counted as cost of buildings may vary. For example, some companies consider the cost of renting, leasing, or purchasing the administrative building as administrative overhead cost while counting the plants and factories as part of manufacturing costs, while others consider all building costs as administrative cost. Also, depending on the type of company, this cost can be exorbitantly high or relatively small compared to the size of the company. In the California State University system, not counting the Governor and the Board of Trustees and their staff, who in fact make major decisions about the University, the administration includes the Chancellor, Vice Chancellors, Associate Vice Chancellors, and all their layers of staff, the Campus Presidents, Provosts, Vice Provosts, and their staffs, the Deans and Directors and their staffs, the Department Chairs and department staff, and finally the faculty who teach the students. Similar multiple layers of management exist in industry (as well as the government) all in order that adequate control is placed on the operations of the entity. This may be necessary, but nonetheless, expensive.

In general, the designer has little effect on the level of funds spent on administration, except indirectly. For example, if production of a product requires fewer steps and fewer machines, the total area devoted to the manufacturing of the product will be less, reducing the need for more building, and consequently, reducing the administrative cost of renting or building a plant.

11.5.7 Engineering

Engineering costs include the costs of designing, developing, modeling, prototyping, testing, and setting up the manufacturing processes that are needed to start production of the products and systems. Certain products that are simple may require little engineering design, while others may require significant engineering detail and sophisticated processes. Compare the production of an Allen wrench and a passenger airliner. Obviously, the level of design, engineering, testing, and set-up is extremely different. As a result, the level of engineering costs will also be proportionally different. No doubt engineers can do things to reduce the cost of testing, setting-up, designing, and up-keep of the drawing and details.

11.5.8 Contingencies

Contingencies are just that! They are the means or plans that an establishment has to establish for emergencies in case anything goes wrong. This may be in the form of cash set aside, an insurance policy, or the ability to have credit to borrow money when needed. All of the above cost money, which must be added to the cost of the product.

There are countless things that can go wrong. It may be that the product has a flaw and needs to be recalled, repaired, or exchanged. It may be that due to unforeseen accidents and injuries to the consumers the company is sued and is held liable, with a judgment against the company to pay for both actual and punitive damages. It may be that other disasters, including fire, may strike and damage the plant or otherwise render it useless. Since most of these are unpredictable at best, there is no easy way to know how much to set aside or how much insurance to obtain. Therefore, contingencies can be a very significant consideration for any operation.

The designer has a profound effect on the level of contingencies needed and probability of events that require those contingencies. Excluding natural disasters and general accidents that can happen in any place, most other events that require contingencies, at least as they relate to the products or systems designed by the designer, are affected by the decisions she or he makes.

When the first model of Toyata Camry came out in 1983, the engineers had specified two seemingly identical timing belts from two different manufacturers as acceptable. As a result, either of the two belts might have been installed in any given engine. However, due to differences in manufacturing and quality of the belts, one of the brands would only last about 15,000 to 20,000 miles, while the expected life for timing belts is over 60,000 miles. This caused a high level of dissatisfaction among the owners, and in order to remedy it, the company instituted a recall, where the owners were asked to take the car to the dealers, the mechanic would remove the timing belt cover (a few screws) to look at the brand name of the belt. If it were the acceptable brand, the cover was replaced. If not, the belt would be replaced. To do this, in addition to the costs involved in towing and repairing the cars in which the belts had actually broken, the company had to compensate for the cost of the recall campaign, the cost of sending thousands of new belts to all dealers around the nation, and the cost of inventorying the belts just in case they would be needed. Additionally, the dealers were permitted to charge a minimum ½ hour for checking the brand name of the belt and the actual time needed to replace it. The company had to also deal with the attitude of the customers who were dissatisfied with the quality of the design and the perception they had formed about the company. As you add all these costs for thousands of cars, you will realize how much money was spent by the company to correct a small mistake that the engineers had made. In fact, this is a very minor case, involving no injuries, no lawsuits, and no punitive damage payments. It just required a simple recall and repair.

In contrast, the latch mechanism of the rear door of a chrysler minivan could malfunction if the car were hit in the rear, throwing passengers out of the car, resulting in injuries and death. At one point, the total payments by the company in the form of punitive damages, medical bills, etc. had reached $400,000,000. This is in addition to the cost of recall and repair of the door latches. All this was due to an inappropriate design.

Every year, there are thousands of recalls of products and systems, from cars and major appliances to baby carriages and food items. Most of these can be taken care of during the design. There is really no reason why most of these cannot be provided for as the product is designed or manufactured. They happen because of mistakes, negligence, or lack of consideration for possible scenarios and people's behavior. Please refer to other chapters for more detail about human factors and liability issues.

As was mentioned earlier, to provide for these contingencies, the company needs to set aside money, they need to purchase insurance, or they must have the ability to either borrow or raise cash. Setting aside money in a bank account is costly. It means that the company will have a large sum of money left useless, while they are paying interest for credit. Besides, no one knows how much money may be needed to take care of any particular contingency before it happens. Consequently, there must be additional means of raising funds available to the company anyway. Liability and contingency insurance is also possible, and most companies carry insurance policies for this purpose. However, these policies are very costly and are at times prohibitively expensive. The premiums are based on the total number of products sold, how long they have been on the market, the reputation of the company, whether there have been similar liabilities in the past, and the level of insurance that is desired. In some cases, the liability insurance premiums for certain County governments and municipalities and some branches of medicine (as well as home insurance for areas prone to natural disasters) have become so expensive that these entities and individuals can no longer afford insurance, setting themselves up for disastrous consequences. But in general, it is possible to purchase insurance for unforeseen and unexpected events. This is true for individuals who act as consultant designers or sole proprietors of smaller manufacturing or production facilities. However, in all these cases, the cost of insurance is an additional cost that is added to the total cost of products and systems, raising their final price, lowering their value to the customers. In fact, the reasoning forwarded by attorneys and judges who seek and award punitive damages to victims of accidents is exactly that. They reason that the cost of contingencies required to pay an individual who is injured by a product should be borne by all people who use the product. This is accomplished by passing the cost to the consumers by adding the cost to the final price of the product. The better and safer the design (and thus, the product), the lower these costs. As a result, the design engineer must always consider all unforeseen misuses, abuses, as well as consequences of every design to determine the extent of the dangers involved, the probability of the events happening, and the cost of correcting mistakes and accidents. The earlier in the process a problem is detected and resolved, the cheaper and easier the solution. For example, if either of the two mistakes in the design and selection of the timing belt or the door latches mentioned above were detected early in the design process, it would have been very easy and inexpensive to just change the drawings or specifications and correct the problem before manufacturing processes were set up and before any product was manufactured. If they were detected before any products were shipped, the remedy would have been millions to hundreds of millions of dollars cheaper, with an untarnished reputation. But when the product leaves the company, and especially if injuries or accidents occur, the remedy is much more expensive with lingering effects.

11.5.9 Cost of Selling

Except on rare occasions, it costs money to sell a product. This includes cost of advertising, distribution and shipping, samples and coupons, entertainment, travel expenses for sales personnel, commissions, telephone expenses, postage and handling, freight charges, and cost of attending expositions and shows.

The cost of selling can vary significantly depending on the product. There are products we all buy on a regular basis, some with many competitors, some without, that are never advertised. We buy them because we trust them, we need them, or we do not have a choice. On the other hand, there are other products that have too many competitors, are new, or do not really fulfill any great need. These products have to be advertised regularly to become known, to stay ahead of the competition, or to convince us that we need them. In some cases, when a product is advertised, the competitors' products lose market share just because more people buy the advertised product. As a consequence, the competitors have to start advertising too in order to regain their market share. Advertising is also used to convince individuals that they need a particular product which otherwise could be completely ignored. We, for example, must be told that we are a particular soda's generation and we need to drink one particular soda, although there is absolutely no need to drink soda to begin with. We might enjoy it, but there is no need for it. In these cases, hundreds of million of dollars are spent annually on advertising, a large portion of the cost of the product itself, just to sell it. Now imagine that you have a product that is truly useful and that it fulfils a real need. You may have no need for advertising, or only to the extent that it is needed for introducing your new product into the market. The more competitive your product, the less need for advertising. Therefore, it should be clear that a good, well designed, and competitive product with high value, that is truly needed, could sell itself. The design engineer plays a vital role in the creation of such a product or system.

Certain elements of the cost of selling are similarly affected by how it is designed by the designer. For example, if the product sells well, the company may be able to pay lower commissions to the sales agents and still attract them to the job. There may be no need for free samples, coupons, or sale prices if the product is successful. And there may be less need to go to expositions and shows if the product is well received and highly respected. Other elements of this cost such as handling, postage, and freight charges or distribution costs, although affected by size, weight, delicacy, and other factors, are a necessary part of doing business and should be expected.

11.5.10 Profit

Profit is the ultimate goal of the free enterprise system and the main reason why companies offer their product, systems, and services for sale. The larger the profit, the better it is for the company. However, as it has been discussed so far, profit is greatly affected by other factors and cannot be regarded as independent.

The designer plays a vital role in maximizing the capability of the company to increase its profit. As already discussed, the design engineer affects the material costs, labor costs, and manufacturing cost. The designer also affects the need for contingencies and for advertising costs. All these can directly affect the profit by minimizing the cost. However, the designer can also affect both the total number of products or systems sold, as well as the selling price of the product, thus maximizing profits. Perhaps the best way to judge this is still to look at the value of the product. The more the customer gets for what she or he pays, the better the product.

Some designers feel that they are not good with financial matters, their job is engineering and design, and they do not or should not care about the economics of product design. Many feel that this subject is either beyond them, or they do not have any interest in it. Still, the designer should realize the importance of economics in the life of the product, and regardless of the designer's personal interest, she or he should take interest in economics, make conscious decisions regarding its effects, and keep it in mind during all design phases. It is true that marketing departments usually set the price of a product after surveys are taken and the market is analyzed. It is true that the actual cost of the product depends on how the company can negotiate prices with their own vendors, and it is true that much money is wasted for nonessential needs. But still, it is the designer who enables the company to be competitive, to offer good product to its customers who trust it, develop loyalty to it, and are willing to pay its prices. It is the designer who creates the competitive edge of the company.

11.6 PRODUCT LIFE SPAN

One of the important economic-related decisions the designer makes is the life span of the product. Life span is the planned (or designed) length of the time that the product is expected to last or function properly. For example, many cars are designed for a life of about 10 years or so. This does not mean that cars last only 10 years. In fact, most last much longer. What it means is that after this length of time, the customer is expected to spend much more money on the repair and maintenance of the car, since most parts are designed for this length of time. For example, many parts of a car are designed for fatigue loads. Fatigue analysis requires number of cycles of load to which the car will be subjected. To estimate this number of cycles, the designer needs to estimate the life of a product. Since fatigue analysis is fundamentally based on statistics and the probability of the presence of flaws in the material, the product may fare much better or much worse than the estimated life. However, since the part is designed for the particular life estimated, the manufacturer can expect that many last that long. Many companies warrant their products for specific length of time as well. These are based on statistical data and the probability of certain defects happening within a given time. They usually warrant the product for as long as possible without getting into the time when expected repairs rise in frequency. A few years ago, General Motors increased the time it provided warranty for its Cadillac cars. Many customers perceived the notion that if the company is willing to warrant the car for this long, the car must be of really good quality. Due to this perception, the sales of the car soared enormously.

The life span of a product is related to its initial quality, the planned estimated life, and its function. As the quality increases, the life span increases as well. However, better quality generally means higher production costs (compared to the same design but with lower quality components). Higher quality increases value of the product, but higher costs lower it. So, the designer must find the best compromise between the better quality and higher costs, maximizing the value of the product.

The life span of the product should be estimated based on many factors. For example, toys are generally used for a short time. Children usually outgrow their toys or get bored with new toys and move on to other things. A toy, if designed to last a very long time, will be abandoned nonetheless, even if in good shape. Although it is necessary that the toy be of good quality and be very safe, it does not have to last a very long time. However, a refrigerator needs to work well and last a long time too. No one expects to use a refrigerator for a short time, no one really outgrows it or gets bored with it. Even if one upgrades to a bigger or better refrigerator, someone else may be able to use it. So, the expected life of a refrigerator and a toy should be different. Other factors include obsolescence, need for improvements, economic factors, resources available, and disposal issues. One must consider these factors in choosing the appropriate life span for any product (please also see Chapter 7).

The life span of a product affects, and is affected by, the following factors:

- Obsolescence/ Improvements
- Economic factors
- Disposal issues
- Human and natural resources

Obsolescence/Improvements Imagine buying a computer today. There is a good chance that by next year your computer will be obsolete. Similarly, compare an old gas-guzzling car from the fifties, sixties, or seventies. The car probably had a low efficiency engine, was large and heavy, had no seatbelts, no airbags or antilock brakes, and the seats probably had no head rests (a safety feature to prevent whiplash). It also lacked air-pollution control device as well as many features new cars have (although many of these are still true for modern cars as well). There is no doubt that many would still like to have an old car, precisely for these reasons (that the engines were simpler, that the car was solid, heavy, and safe, that it had no air-pollution control device, etc.). However, this is mostly sentimental and nostalgic. In reality, new cars are modern, safer, have more efficient engines, are less polluting, lighter, and more agile. Owning those old cars today would be difficult because they are obsolete and not as safe, their parts are not easy to find, they require much fuel, and they pollute the air. Most other products are the same as well. Due to improvements made in products, even if superficial, products do advance, rendering the older ones obsolete.

The important issue here is that as products and designs improve, they generally function better and are safer. Then, would it be appropriate to design products to last a very long time? Would we really want to have many old cars, old computers, obsolete power plants, and other outdated products around? Or should we appropriately design the product for a given estimated life so that eventually the product will be replaced by a better and newer version? How long is it expected that the product will be used? Is the product such that it could be given or sold to others, and could it go from generation to generation?

Economic Factors of Life Span Imagine that any product you buy now would last a lifetime. Imagine that your first computer would be so powerful that you would not need another for the rest of your life. And imagine that the first car your parents bought would still be in a great shape and it would be passed down to you, and from you to your children. Sounds great, except that the automobile manufacturer, the computer (and consequently, the software) manufacturer, and all other companies who sell you a product will only do so once (or very infrequently at best). If this were to be true in general, then companies would sell much less, would not have their products replaced, and would have far fewer customers. There are many economic consequences to this. First, many companies would not survive. They would not have enough revenues and would not make enough profit to stay in business. Second, who would buy their products? Everyone would be happy with what they have already bought, or otherwise would pass it on to the next generation. Consequently, there will be far fewer people with jobs, including the designers, workers, and administrators. Consequently, world economies would slow down with lower gross national products and a low level of economic activities. In fact, since most economies are based on consumerism, most would be negatively affected. In other words, it is actually to the benefit of the manufacturers to design and build their product with a planned obsolescence; otherwise they will cause their own doom. Furthermore, companies could not count on repeat customers (that if you are happy with your first purchase, you will buy from them again) because you would not need to buy again. This means that once they sell you the product, they do not have to worry about having you as a repeat customer, thus, not having to worry about quality, customer service, or other related issues. All they have to worry about is to entice you to buy their products once. This will reduce competition, product quality, and importance of company's image. Consequently, the product will actually lose its quality and will not last a lifetime, thus correcting the system. As a result, there needs to be a sensible estimated life for all products to satisfy these competing needs.

Disposal Issues   Another important matter related to the life span is the end-of-the-life issues. Any product whose useful life has ended must be disposed of. The birth of the product is a very important matter, but so is its death. Every product that is sold will eventually die. The more we produce, the more we need to discard. Some products are designed for an extremely short life. A disposable camera is used once (one roll). Then it is discarded (even if recycled and remanufactured). A computer has to be discarded, as are furniture, cars, shoes, etc. What may be done to the product at this time depends on the design and whether the designer considered this point and planned ahead for the disposal issues. Some products can be recycled completely; others cannot. Some can be toxic if disposed of improperly. Some will last hundreds of years before the environmental elements will reduce them to the original compounds. The point is that the more we produce, the bigger the chance of environmental pollution, pollution caused by the production processes we use during manufacturing, and the more waste produced.

Human and Natural Resources   Another issue related to products' life span is the availability of natural and human resources. As production levels increase due to more consumerism, more and more workers may be needed. This means more jobs, higher economic outputs, and more consumerism as more workers consume more products. On the other hand, as more people find jobs, and especially if not enough workers are available for the needed jobs, the workers' demands will increase, affecting salaries, prices, inflation, and economic throughput. As before, these are competing factors and must be considered carefully. Additionally, as more products are manufactured, more natural resources will be needed, requiring more mining and tree-cutting, more pollution as more material is processed, and higher prices due to higher demand. In 2004, China started buying large quantities of aluminum scraps from the United State, creating a large demand for the material, increasing the price of aluminum products many folds. So, as more products are manufactured, natural resources are depleted and prices increase. Therefore, it is important to consider this effect of life span of the products as well.

11.7   ENGINEERING ECONOMY

The discussion above centered on the effects of design decisions on economic issues related to products and systems and the role a designer plays in this issue. The following discussion is about engineering economy, which involves the time value of money, and how different economic situations can be compared against each other. This includes formulating, estimating, and evaluating the economic outcomes of different alternatives. This issue is absolutely important in making decisions about competing options and proposals. In this section we will discuss basic components of engineering economy and their related formulae.

11.7.1 Equivalency

Generally, there may be more than one economic option available to a company (or an individual designer), each with its own characteristics. For example suppose that there are two banks, each willing to finance the resources needed to start production of a new product. Each bank has a different interest rate and payment schedule, one with a fixed rate and fixed payment plan, the other with a variable (but known) rate that changes as more of the principal is paid back. In order to decide which loan is economically better, one would have to compare the total net value of the two options at some point in time, for example the present time, to determine which one is better. As another example, suppose that someone would give you $1,000 today, which if left in a savings account with 5 percent interest, would be worth $1,050 after one year and $1,102.5 after two years. The worth of the $1,000 payment today is equivalent to a payment to you of $1,050 one year into the future, or $1,102.5 in two years. They are equivalent to each other as far as economics is concerned. This is called equivalency.

11.7.2 Time Value of Money

Since money can be invested or left in a savings account that bears interest, it earns more money and grows. Consequently, its value (magnitude, but not necessarily its purchasing power) increases in time. This is called the time value of money. It means that one should expect to receive more money if it is delivered later in time, because one can always invest the money or save it in a bank and thus increase its magnitude. To calculate the time value of money, one would need to consider an interest rate (or equivalent), which is not always easy. Nonetheless, since in most engineering cases it is the relative worth of two competing scenarios that is of concern and not their absolute values, it is not vital to have a particularly accurate interest rate. In most cases, a prevalent interest rate is assumed in order to compare two or more options relative to each other. However, it is crucial to understand the importance of the time value of money and how some money in hand today is better than the same in the future.

11.7.3 Cash Flow, Costs, and Capital Diagrams

In order to simplify the representation of cash flow (revenue, costs, and capital expenditure) at different times, a simple cash flow diagram is used, as shown in Figure 11.8. Each upward arrow represents the receipt of money at a particular time. Each downward arrow is an expenditure. All expenditures at the beginning of the diagram (zero-time) are called sunk costs. Revenues and expenditures during a time period are shown at the end of the time period (usually a year). The diagram is used for visualization and simplifying expression of the situation.

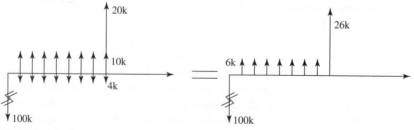

FIGURE 11.8 An example of a cash flow diagram.

EXAMPLE 11.1

A machine costs $100,000 to purchase, generates revenues at a rate of $10,000 per year, and costs $4,000 a year to operate and maintain. At the end of 8 years, its salvage value is $20,000. Figure 11.8 shows its cash flow diagram. ∎

11.7.4 Present and Future Worth

A certain amount of money P at the present time will be worth a different amount F in the future. Assuming that the interest is compounded (the interest remains in the account and accrues more interest), for a total of n intervals at the interest rate of i percent, the future worth will be

$$F = P \times (1 + i)^n \tag{11.6}$$

The factor $(1 + i)^n$ is called compound amount factor and can also be found from tables or dedicated calculators. It is commonly written as $(F/P, i\%, n)$, therefore:

$$F = P \times (F/P, i\%, n) \tag{11.7}$$

Similarly, the present worth of a certain amount of money in the future is:

$$P = F \times (1 + i)^{-n} = F \times \frac{1}{(1 + i)^n} \tag{11.8}$$

The factor $(1 + i)^{-n}$ is called present worth factor and can also be found from tables or dedicated calculators. This factor is also written as $(P/F, i\%, n)$, therefore:

$$P = F \times (P/F, i\%, n) \tag{11.9}$$

In both cases, n represents a time interval or the number of times interest is calculated and added to the principal. For example, if interest is calculated at the end of each year, then n will be the number of years the money is earning interest. However, if interest is calculated each month, since the interest earns even more interest when the intervals are shorter, the effective rate is increased. In that case, the annual interest rate is divided by the number of intervals per year and n is multiplied by the same number of intervals per year.

EXAMPLE 11.2

What is the future worth of an investment of $10,000 in a bank account at the rate of 5 percent, compounded yearly, after (a) 5 years, (b) 10 years.

Solution Using Equation (11.6), we, respectively, get

$$F = P(1 + i)^n = \$10,000(1 + 0.05)^5 = \$12763$$
$$F = P(1 + i)^n = \$10,000(1 + 0.05)^{10} = \$16289$$

This example shows the time value of money. ∎

EXAMPLE 11.3

How much should an individual save in a bank account today at a rate of 10 percent per year in order to have $100,000 in 10 years if (a) the interest is compounded yearly? (b) the interest is compounded monthly?

Solution Using Equation (11.8), with $n = 10$ and $i = 10\%$ for case (a) and $n = 10 \times 12 = 120$ and $i = 10\%/12 = 0.8333\%$ for case (b), we will get

$$P = F(1 + i)^{-n} = \$100,000(1 + 0.1)^{-10} = \$38,554$$

and

$$P = F(1 + i)^{-n} = \$100,000(1 + 0.008333)^{-120} = \$36,942$$

As you notice, when compounded more often, a smaller investment is needed in the same time period to grow into the same value in the future. ∎

EXAMPLE 11.4

A saving of \$50,000 is kept in a bank account at a constant rate of 5 percent for 5 years. Subsequently, an additional \$30,000 is added to the account for an additional 3 years. Calculate the future worth of the investment at the end of 8 years. Interest is compounded yearly.

Solution This can be done by assuming two investments, a \$50,000 amount for 8 years, and a \$30,000 investment for 3 years. We may directly calculate the future worth by

$$F = \$50,000(F/P, 5\%, 8) + \$30,000(F/P, 5\%, 3)$$
$$= \$73873 + \$34729 = \$108,602.$$

Alternately, we may first calculate the present value of both investments, and subsequently, calculate the future worth of the total:

$$P = \$50,000 + \$30,000(P/F, 5\%, 5) = \$73,505$$
$$F = \$73,506(F/P, 5\%, 8) = \$108,602$$

As you see, the results are identical. ∎

EXAMPLE 11.5

Imagine that someone bought a household product for \$1 a long time ago. The same item is now on sale as an antique. Assuming an interest rate of 5 percent, how much money would the original owner have if the money was instead invested in a savings account if the product was purchased (a) 200 years ago, (b) 300 years ago? Neglecting wear and tear of the product, would the product be worth this much as an antique?

Solution The future worth of the saving (today) would be

$$F = \$1(F/P, 5\%, 200) = \$17,293$$
$$F = \$1(F/P, 5\%, 300) = \$2,273,996$$

∎

11.7.5 Recurring Uniform Series Amounts

There are times when uniform amounts A of cash flow recur at the end of equal time periods. For example, there may be uniform income expectations at the end of every month due to interest received for a loan by a bank or due to contracts. Similarly, there may be recurring uniform expenditures such as interest paid for a loan to a bank, maintenance costs, or rent. The present worth value of these recurring amounts may be individually calculated at the present time and summed up, or alternately, a uniform series factor may be used as

$$P = A \times \left(\frac{(1 + i)^n - 1}{i(1 + i)^n} \right) = A \times (P/A, i\%, n) \tag{11.10}$$

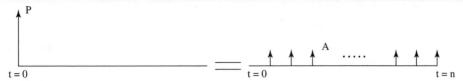

FIGURE 11.9   Cash flow diagram for present worth of a uniform series.

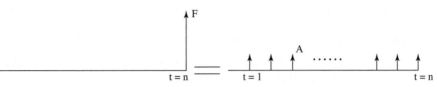

FIGURE 11.10   Cash flow diagram of the future worth of a uniform series.

and conversely,

$$A = P \times \left(\frac{i(1+i)^n}{(1+i)^n - 1} \right) = P \times (A/P,\ i\%,\ n) \tag{11.11}$$

which lets you calculate the necessary recurring amounts for a certain present value. This is called capital recovery. It means that an expenditure at the present time (say, to buy equipment) is equivalent of periodical payments for a certain number of periods. It is very important to notice that there must be a payment at period n as well. In other words, the period n requires a payment as well, which occurs at the end of the period. This is important to remember and to include in your calculations. Figure 11.9 shows the cash flow diagram for the series.

Similarly, the future worth value of these recurring amounts may be individually calculated at a future time and summed up, or alternately, a uniform series factor may be used as

$$F = A \times \left(\frac{(1+i)^n - 1}{i} \right) = A \times (F/A,\ i\%,\ n) \tag{11.12}$$

and conversely,

$$A = F \times \left(\frac{i}{(1+i)^n - 1} \right) = F \times (A/F,\ i\%,\ n) \tag{11.13}$$

that allows you to calculate the necessary uniform amounts over a period of time that have a certain value in the future. Here too, there must be a final payment at the last period. Figure 11.10 shows the cash flow diagram for this series.

EXAMPLE 11.6

The cost of maintaining a certain machine is estimated as $1,000 per year. Calculate the present and future worth of these maintenance costs over 10 years if interest rate is 8 percent.

Solution   The present worth of the expenditures is

$$P = -\$1000 \times [(P/F,\ 8\%,\ 10) + (P/F,\ 8\%,\ 9) + \cdots + (P/F,\ 8\%,\ 1)]$$
$$= -\$1000 \times (P/A,\ 8\%,\ 10) = -\$1000 \times (6.71) = -\$6,710$$

The future worth of the same expenditures at the end of 10 years will be

$$F = -\$1000 \times (F/A,\ 8,\ 10) = -\$1000 \times (14.487) = -\$14,487$$

Please note that if you calculate the future worth of $6,710 at 8 percent for 10 years, you will get the same $14,487. ∎

EXAMPLE 11.7

Calculate the necessary recurring payments to pay back a loan of $100,000 at a rate of 6% in 10 years if (a) payments are made once a year, and (b) if payments are made once a month.

Solution   The uniform payments are

(a) $A = \$100,000(A/P,\ 6\%,\ 10) = \$100,000 \times (0.135868) = \$13,586.80$
  The total payment in 10 years will be $135,868.

(b) $A = \$100,000(A/P,\ 0.5\%,\ 120) = \$100,000 \times (0.0111021) = \$1,110.21$
  The total payment in 10 years will be $133,225.

Notice that the future worth of $100,000 in 10 years at 6 percent, compounded yearly is $179,085. Can you describe the difference between these three values? ∎

EXAMPLE 11.8

Calculate the necessary yearly savings that have to be made for 12 years so that a child will have $50,000 in the bank at college age if the money earns 10 percent.

Solution   The required uniform yearly savings are

$$A = \$50,000 \times (A/F,\ 10\%,\ 12) = \$50,000 \times (0.04676) = \$2338$$
∎

11.7.6   Recurring Nonuniform Series Amounts with a Gradient

This case is similar to 11.7.5, but the payments are not constant and vary with a uniform gradient (constant change) or geometric gradient (constant percentage of change). For example, the maintenance cost for a machine or rent may increase as time goes by. The following equations are used for either case similar to the above:

Uniform Gradient   For uniform gradient series, where the amount varies with a constant gradient, we will assume that the total present worth is composed of two present values, one for the constantly recurring amount, ± the value for the gradient amount, starting at the second time interval (Figure 11.11). If the first cash flow is zero, the constant portion will be zero.

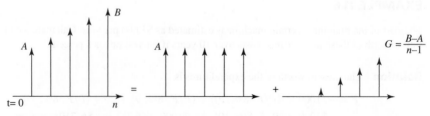

FIGURE 11.11   Uniform-gradient recurring cash flow series.

For interest rate i percent and number of transactions n, the present value of the gradient is

$$P = P_A + P_G \tag{11.14}$$

where

$$P_G = G \times (P/G, i\%, n) = G \times \left(\frac{1}{i} - \frac{n}{(1+i)^n - 1}\right)(P/A, i\%, n)$$

$$= G\left(\frac{1}{i} - \frac{n}{(1+i)^n - 1}\right)\left(\frac{(1+i)^n - 1}{i(1+i)^n}\right) \tag{11.15}$$

and

$$P_A = A \times (P/A, i\%, n) \tag{11.16}$$

Geometric Gradient Series   In this case, the increase or decrease in the periodic values is at a constant percentage g, which is the change in decimal form from one payment to the next. The present value can be calculated from the following two equations:

$$P = A_1\left[\frac{1 - \left(\frac{1+g}{1+i}\right)^n}{i - g}\right] \quad \text{for } g \neq i \quad \text{and} \quad P = A_1\left(\frac{n}{1+i}\right) \quad \text{for } g = i \tag{11.17}$$

The recurring amount at any interval can be calculated from

$$A_n = A_1(1+g)^{n-1} \tag{11.18}$$

EXAMPLE 11.9

The maintenance costs of a machine are expected to be $500 for the first year, but increasing at a rate of $100 per year. Based on an interest rate of 10 percent, calculate the present worth of the maintenance costs for the next 5 years.

Solution   A representation of these costs is shown in Figure 11.12. The maintenance costs increase to $600, $700, etc. each year. The cost for each year can be broken down into a constant $500+ the additional uniform gradient added to it starting at the second year. The total present value of the cost will be

$$P_A = A \times (P/A, i\%, n) = -\$500 \times (P/A, 10\%, 5) = -\$500 \times (3.791) = -\$1,895.5$$

$$P_G = G \times (P/G, i\%, n) = G \times \left(\frac{1}{i} - \frac{n}{(1+i)^n - 1}\right)(P/A, i\%, n) = -\$100 \times (1.8101)(3.791)$$

$$= -\$686.2$$

$$P = P_A + P_G = -\$1,895.5 - \$686.2 = -\$2,581.7$$

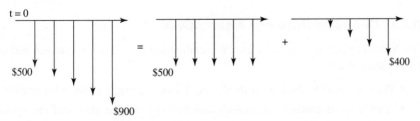

FIGURE 11.12   Representation of uniformly varying periodic amounts.

EXAMPLE 11.10

Repeat Example 11.9, but assume that the cost increases by 20% per year.

Solution In this case, the increase in cost of maintenance is not constant, but increasing at a rate of 20 percent. Therefore, the yearly cost is $600, $720, $864, etc. The total present worth of the cost for 5 years is

$$P = A_1 \left[\frac{1 - \left(\frac{1+g}{1+i}\right)^n}{i-g} \right] = -\$500 \left[\frac{1 - \left(\frac{1+0.2}{1+0.1}\right)^5}{0.1-0.2} \right] = -\$500 \times (5.4505) = -\$2725$$

11.7.7 Shifted Series

There are times when uniform annual revenues or expenditures do not start at the end of period 1 (first year), but at a different time. This is called shifted series. In this case, the same formulae can be used except that the present value is first calculated one period before the first payment, and future worth is calculated at the same period as the last payment. The future or present worth of these values can later be recalculated at other periods for comparison.

EXAMPLE 11.11

A machine costs $100,000 to purchase. It is estimated that there will be negligible maintenance costs for the first three years, and $3,000 per year after that for additional 7 years. Calculate the present worth of the total expenditure if an interest rate of 8 percent is assumed.

Solution Figure 11.13 shows the cash flow diagram of the payments. We will first calculate the present worth of the maintenance costs at the beginning of year 3, then recalculate the present worth of this value at the present, and add it to the purchase cost to calculate the total present worth, as follows:

$$P_3 = -\$3,000(P/A, 8\%, 7) = -\$15,619$$
$$P = -\$100,000 - \$15,619(P/F, 8\%, 3) = -\$112,399$$

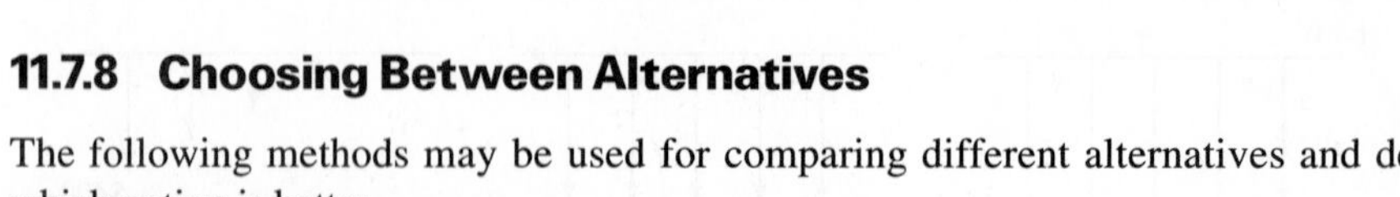

FIGURE 11.13 Cash flow diagram for Example 11.11.

Table 11.1 is a summary of the above-mentioned equations presented for easy reference.

11.7.8 Choosing Between Alternatives

The following methods may be used for comparing different alternatives and deciding which option is better.

- Present worth method converts all cash flows to a single value at the present time.
- Annual worth method converts all cash flows to an equivalent uniform annual series over the plan's life.

TABLE 11.1   Summary of Engineering Economy Conversion Equations

Future worth of a present single payment	$F = P(1 + i)^n$
Present worth of a future single payment	$P = F(1 + i)^{-n}$
Present worth of a uniform series payments	$P = A\left(\dfrac{(1 + i)^n - 1}{i(1 + i)^n}\right)$
Uniform series capital recovery	$A = P\left(\dfrac{i(1 + i)^n}{(1 + i)^n - 1}\right)$
Future worth of a uniform series payments	$F = A\left(\dfrac{(1 + i)^n - 1}{i}\right)$
Uniform series (sinking fund) of a future value	$A = F\left(\dfrac{i}{(1 + i)^n - 1}\right)$
Present value of series with a gradient	$P_G = G\left(\dfrac{1}{i} - \dfrac{n}{(1 + i)^n - 1}\right)\left(\dfrac{(1 + i)^n - 1}{i(1 + i)^n}\right)$
Present value of series with geometric gradient	$\begin{cases} P = A_1\left[\dfrac{1 - \left(\dfrac{1 + g}{1 + i}\right)^n}{i - g}\right] & \text{for } g \neq i \\[4ex] P = A_1\left(\dfrac{n}{1 + i}\right) & \text{for } g = i \end{cases}$

- Future worth method converts all cash flows to a single value at the end of the plan's life.
- Rate of return method determines an interest rate at which the future worth will be zero.
- Savings/investment ratio method determines the ratio of the present worth of the savings to the present worth of the investment.
- Payback period method determines the length of time required to recover the initial investment.
- Capitalized worth method determines a value at the present time that is equivalent to a cash flow pattern that continues indefinitely (using a minimum acceptable rate of return or MARR).

Except for the last two methods, the rest are practically equivalent and will yield the same recommendation when applied to the same options. For example, the present worth method uses the present worth value of each alternative. The one with the largest present worth is the superior choice. Similarly, the option that yields a higher value when future worth is calculated will be superior. These two methods require that all choices be mutually independent (one alternative is not dependent on part of another) and that all alternatives have the same lives. On the other hand, two alternatives that accomplish the same task at different lengths may be compared by the annual worth method. The restriction is that it assumes that alternatives with shorter life can be renewed infinitely up to the duration of the longest alternative. The calculated annual cost is called *Equivalent Uniform Annual Cost* (EUAC).

EXAMPLE 11.12

An investment alternative A costs $100,000 today and pays back $120,000 in 3 years. Alternative B costs $80,000 now and pays back $31,000 each year for 3 years. If interest rate is 5 percent, which investment is better?

Solution   Since these two investments both have the same life (3 years), we will use the present worth method to decide which one is superior.

For A:
$$P(A) = -\$100,000 + \$120,000(P/F, 5\%, 3)$$
$$= -\$100,000 + \$103,660 = \$3660$$

For B:
$$P(B) = -\$80,000 + \$31,000(P/A, 5\%, 3)$$
$$= -\$80,000 + \$84420 = \$4420$$

Clearly, investment B is superior. Similarly, we could have calculated the future worth of these two options, yielding \$4,237 and \$5,118, respectively, indicating that option B is superior.   ■

EXAMPLE 11.13

For manufacturing a product, two alternative machines may be selected. The first option has a life of 10 years, costs \$18,000, and requires \$1,000 per year to run. The second option has a life of 5 years, costs \$8,000, and requires \$1,500 per year to operate. Assume interest rate is 5 percent. Determine which option is superior.

Solution   In this case, since the lives of these two options are different, assuming that we will be able to buy a new replacement for the second option at the end of its life, we may compare them using the EUAC method.

First option: $\text{EUAC}(1) = -\$18,000(A/P, 5\%, 10) - \$1,000 = -\$3,331$

Second option: $\text{EUAC}(2) = -\$8,000(A/P, 5\%, 5) - \$1,500 = -\$3,348$

Based on the above, the first option costs slightly less per year, and therefore, is slightly superior to the first option. Of course, the assumption that we can replace the same machine in 5 years at the same cost is not entirely reasonable. Certain products cost more as time goes by, whereas others cost less. This should also be considered when making a decision.   ■

Another method of comparing different alternatives is Rate of Return, discussed next.

11.7.9   Rate of Return (ROR) and Minimum Acceptable Rate of Return (MARR)

The rate of return is an interest rate that will equate the future (or present) worth to zero. This means that if all cash flows are calculated as the future (or present) worth, the total value of the investment will be zero. For example, assume that an individual invests sum P for a period of time (remember that an initial investment is a negative number) and receives annual payments of A while the investment is kept, plus a sum B at the end of the period (all positive numbers). If you add either the present or the future worth of the initial investment and all payments at the ROR value, the sum will be zero. Normally, the rate of return is the variable that is calculated to make the sum equal to zero and is used to compare different options.

There are two ways to calculate the rate of return, either mathematically by solving the above mentioned equations, or by first assuming two different interest rates, calculating the present or future values, and then interpolating (or extrapolating) between these values repeatedly until the sum equates zero. The mathematical solution is only useful for simple cases; otherwise, the equation can be too complicated. The second method is much more common.

The Minimum Acceptable Rate of Return (MARR) is the minimum rate of return that a firm finds acceptable. All investments that yield a return below this value are deemed not worthy of the risks involved. The MARR may be selected from experience, the minimum discount rate charged by banks, the rate of return of the stock market, or any other rate that the company finds desirable.

To make a decision, the ROR may be compared to the MARR. If the ROR is larger than MARR, the investment is sound. If it is lower, it is questionable or unacceptable. For example, if the prevailing rate of return of the stock market is about 12 percent and this is used as MARR, an investment with a higher ROR will be acceptable.

EXAMPLE 11.14

An initial investment of $100,000 earns $10,000 per year for 4 years and is settled by a payment of $110,000 at the end of the period. Calculate the rate of return.

Solution   Figure 11.14 shows the cash flow diagram of the option. First, we will calculate the total present worth at two rates of 7 percent and 13 percent as follows:

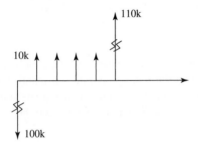

FIGURE 11.14   Cash flow diagram of Example 11.14.

$$P_1 = -\$100,000 + \$10,000(P/A, 7\%, 4) + \$110,000(P/F, 7\%, 5)$$
$$= -\$100,000 + \$33,872 + \$78,428 = \$12,300$$
$$P_2 = -\$100,000 + \$10,000(P/A, 13\%, 4) + \$110,000(P/F, 13\%, 5)$$
$$= -\$100,000 + \$29,745 + \$59,703 = -\$10,552$$

As you notice, one value is positive, one is negative. Interpolating between these two values yields an approximate ROR value of 10.2 percent. We may recalculate the present worth at two new values, close to 10.2 percent to get more accurate result, or simply recalculate at 10 percent to see the result and decide. At 10 percent, the present value will be

$$P = -\$100,000 + \$31,699 + \$68,301 = \$0$$

indicating that the rate of return is 10 percent. We could have also used the future worth with the same result. In this case, there will be an annual cash flow of $10,000 for 5 years, plus a $100,000 return of investment (remember that based on the requirement for future worth of a series, the $110,000 payment in year 5 should be treated as an annual payment of $10,000 plus a sum payment of $100,000 in order to allow us to use the (F/A, i%, n) factor). Therefore,

$$F = -\$100,000(F/P, 10\%, 5) + \$10,000(F/A, 10\%, 5) + \$100,000$$
$$= -\$161,051 + \$61051 + \$100,000 = \$0$$ ∎

11.7.10   Effective Interest Rate

When interest is calculated more often than once a year, and the interest is added to the principal, the effective interest rate is, in fact, more than the nominal value. This is because the accrued interest is added to the principal, and consequently, bears more interest when it is calculated more often at shorter periods. The effective interest rate or Annual Percentage Yield (*APY*) can be calculated by

$$APY = (1 + i)^m - 1 \qquad (11.19)$$

where i is the nominal rate for the period and m is number of periods.

EXAMPLE 11.15

A bank offers a nominal interest rate of 5 percent for Certificates of Deposit, compounded monthly. Calculate the *APY* of the CD. What will the *APY* be if the interest is compounded daily?

Solution   Using Equation (11.19), we get

$$APY = (1+i)^m - 1 = \left(1 + \frac{0.05}{12}\right)^{12} - 1 = 5.116\%$$

and

$$APY = (1+i)^m - 1 = \left(1 + \frac{0.05}{365}\right)^{365} - 1 = 5.127\%$$

11.7.11   Tax Considerations

All entities, whether individuals or companies, may pay variety of taxes, including federal, state, county, and city taxes. Assuming that the federal tax rate is $f\%$ and state (and other municipalities) tax rate is $s\%$, and assuming that the federal tax law treats state taxes as expense (which can be deducted from income), the effective tax rate t will be

$$t = s + f(1 - s) \tag{11.20}$$

Tax is a very important issue, not only in politics, but also in economy. As will be discussed in Chapter 15, one important issue in forming a business is its tax treatment by the government. For example, C-corporations pay taxes on their income before the income is passed on to the stockholders, who in turn, pay taxes on their income, whereas sole proprietorships pay taxes only once.

In engineering economy, one should consider the effect of taxes on the investment as well. What is taxed is generally the profit or income, not revenues or sales. All eligible losses and expenditures are subtracted from income, whereas all revenues and sales are added to it. Therefore, all deductible expenditures are effectively reduced by t percent, and all revenues are reduced by a factor of t percent. Additionally, equipment and investments can be depreciated; and based on a number of different methods and formulae, the cost can be subtracted from income (although depreciation affects income, it has no effect on engineering calculations and deciding about alternatives). Since taxes reduce the final profit, it is important to consider taxes when investment decisions are made.

EXAMPLE 11.16

An entity pays federal taxes as high as 28 percent, with a state tax rate of 7 percent. What is the effective tax rate for the entity in its highest bracket?

Solution   All entities are exempt from paying taxes up to a certain amount of income, and from there, the rate increases as their income increases. Consequently, these values do not represent the total tax rate at which the entity pays taxes, but the effective tax rate at this level. The total tax rate will be lower.

Substituting these values into Equation (11.20) we get

$$t = 7\% + 28\%(1 - 7\%) = 33.04\%$$

EXAMPLE 11.17

The entity of Example 11.16 invests in a machine that costs \$15,000, generates an income of \$5,000 per year, and costs \$700 to operate for 10 years. Neglecting depreciation and salvage value, what is the effective after-tax present worth? Assume interest rate is 8 percent.

Solution Although we can subtract the $700 expenses from the income and calculate the present worth for both, we will do this separately to see their individual effects. Therefore, the total present worth after tax will be

$$
\begin{aligned}
P &= -\$15,000 + \$5,000(P/A, 8\%, 10)(1 - t\%) - \$700(P/A, 8\%, 10)(1 - t\%) \\
&= -\$15,000 + \$33,550(1 - 33.04\%) - \$4,697(1 - 33.04\%) \\
&= -\$15,000 + \$22,465 - \$3,145 = \$4,320
\end{aligned}
$$

As you see, the tax rate affects both the income and the expenses, thus affecting the present worth. Without taxes, the present worth would be $13,853. ∎

REFERENCES

1. OSTWALD, PHILLIP, "Cost Estimating for Engineering and Management," Prentice-Hall, Englewood Cliffs, New Jersey, 1974.
2. DIXON, JOHN and CORRADO POLI, "Engineering Design and Design for Manufacturing," Field Stone Publishers, Conway, Massachusetts, 1995.
3. BOOTHROYD, G., "Assembly Automation and Product Design," Marcel Dekker, New York, 1992.
4. FORCUCCI, TOMPKINS, "Automotive interiors: Design For Recyclability," Designing for Recyclability and Reuse of Automotive Plastics, SAE Special Publication 867, February 1991.
5. KAHMEYER, LEICHT, "Dismantling Facilitated," translated from Kunststoffe German Plastics, Vol. 12, 1991, pp. 30–32.
6. SIMON, M., "Design For Dismantling," *Professional Engineering*, November 1991, pp. 20–22.
7. ADLER, SHWAGER, "Software makes DMFA Child's Play," *Machine Design*, April 1992.
8. http://www.me.utexas.edu/~uer/challenger/chall2.html.
9. WHITE, JOHN A., K. E. CASE, D. B. PRATT, and M. H. AGEE, "Principles of Engineering Economic Analysis,"4th Edition, John Wiley and Sons, New York, 1998.
10. GONEN, TURAN, "Engineering Economy for Engineering Managers, with Computer Applications," John Wiley and Sons, New York, 1990.
11. LINDEBURG, MICHAEL, Editor, "Mechanical Engineering Review Manual," 6th Edition, Professional Publications, San Carlos, California, 1980.

HOMEWORK

11.1 Check the Consumer Product Safety Commission's website and assemble a list of the products for which a recall is issued. Study one of the products in detail to determine the reason the recall was issued.

11.2 Select two similar items that are different in their design. For example, two toasters, two pens, or two bicycles. Compare number of parts, methods of manufacture, and estimated assembly time. How would you improve the inferior design? Can you improve the superior design?

11.3 Select two items of ordinary use, one essential, one nonessential. Find out how these two items are advertised. How many different outlets and methods are used to advertise each product?

11.4 Investigate the methods used by the manufacturer of a new product to introduce it to the market. Do they use the media? News media? Movies? Newspapers?

11.5 If you are involved in a design project in your class, estimate the product's direct and indirect material cost, manufacturing cost, and direct labor cost.

How realistic do you feel your estimate is? Will you have it manufactured in a foreign country?

11.6 What is the future worth of an investment of $50,000 in a bank account in 20 years at the rate of 5 percent, compounded yearly, if the bank charges $25 per year for maintenance?

11.7 Compare the effect of doubling the rate of interest from 5 percent to 10 percent versus doubling the time of an investment from 5 years to 10 years. Which one is more important?

11.8 How much should an individual save in a bank account today at a rate of 5 percent per year in order to have $100,000 in 8 years if (a) the interest is compounded yearly? (b) the interest is compounded monthly?

11.9 A saving of $25,000 is kept in a bank account at a constant rate of 6 percent for 5 years. Subsequently, two additional $10,000 deposits are made to the account in years 5 and 7. Calculate the future worth of the investment at the end of 8 years. Interest is compounded yearly.

11.10 A saving of $10,000 is kept in a bank account at a constant rate of 6 percent for 8 years. Subsequently, $10,000 is withdrawn from the account in year 8. Calculate the present worth and future worth of the investment at the end of 10 years. Interest is compounded yearly.

11.11 The initial cost of a house was $200,000 fifteen years ago. Yearly taxes are about $2,250. Insurance has increased steadily from $200 to the present $900. Ten years ago, the roof was changed at the cost of $10,000. Calculate the actual cost of the house to the owner up to this date if an interest rate of 8 percent is assumed.

11.12 A machine costs $35,000 to purchase. The yearly cost of running the machine increases 10 percent every year, starting from $1,000 in year 1. The salvage value of the machine at the end of 10 years is $5,000. Calculate the present worth of this machine if $i = 5\%$.

11.13 Compare the option of Problem 11.12 with another machine that initially costs $25,000, but has a constant operational cost of $1,500 and lasts 10 years with a salvage value of $1,000. Assuming that this option can be replaced at the same price, which option is better? Assume $i = 5\%$.

11.14 Compare the total payments made to pay back a loan of $50,000 at a rate of 4 percent in 10 years if (a) payments are made once a year, and (b) if payments are made once a month.

11.15 Repeat problem 11.14, but assume that you could save the monthly payments in a savings account for each year before paying back the yearly payments at an interest rate of 10 percent and a tax rate of 30 percent. What option would be better?

11.16 Calculate the necessary yearly savings that have to be made for 12 years so that a child will have $60,000 in the bank at college time assuming the money earns 12 percent if (a) the account is tax free, and (b) a 33 percent federal tax and 7 percent state tax is paid.

11.17 The maintenance costs of a machine are expected to be $1,000 for the first year, but increasing at a rate of $150 per year. Based on an interest rate of 8 percent, calculate the present worth of the maintenance costs for the next 5 years.

11.18 Repeat Problem 11.17, but assume that the costs increase at a 12 percent rate. Also calculate the future worth of the costs in 5 years.

11.19 A machine costs $80,000 to purchase. It is estimated that there will be negligible maintenance costs for the first two years, and $3,000 per year after that for additional 8 years. A major overhaul, costing $10,000 is expected at year 5. Calculate the present worth of the total expenditure if an interest rate of 8 percent is assumed.

11.20 An investment alternative A costs $110,000 today and pays back $140,000 in 4 years. Alternative B costs $90,000 now and pays back $34,000 each year for 3 years. If interest rate is 5 percent, which investment is better?

11.21 For manufacturing a product, two alternative machines are considered. The first option has a life of 8 years, costs $20,000, and requires $1,000 per year to run. The second option has a life of 5 years, costs $10,000, and requires an increasing expenditure of $1,000 + 10\%$ per year to operate. Assume interest rate is 5 percent. Determine which option is superior.

11.22 An initial investment of $10,000 earns $800 per year for 5 years and is settled by a payment of $12,000 at the end of the period. Calculate the approximate rate of return.

11.23 The average price of houses in a city has risen from $100,000 to $500,000 in 20 years. Neglecting taxes, what interest rate would an individual need to earn in order to be able to buy the same house now if she or he had put $100,000 in a savings account 20 years ago?

11.24 For Problem 11.23, calculate the effective rate of return (ROR) if the house was bought 20 years ago at $100,000. Assume a yearly maintenance and tax expenditure of $3,000. If MARR is assumed to be 8 percent, is the investment wise?

11.25 An entity invests in a machine that costs $20,000, generates an income of $4,000 per year, and costs $500 per year to operate for 10 years. Neglecting depreciation and salvage value, what is the effective after-tax present worth? Assume interest rate is 8 percent and the entity pays 33 percent in federal tax and 8 percent in state tax.

CHAPTER *12*

QUALITY IN DESIGN

12.1 INTRODUCTION

The quality of a product is the culmination of many different philosophies, actions, and decisions that are made throughout the design and manufacturing processes. This includes the desire of the institution's top leaders for quality, the buying-in of the individuals within the system, inclusion of practices that increase quality in the design, operations, manufacturing, and shipping of the product or system, and continual evaluation of the effectiveness of the whole operation. Whether we are concerned with a simple or sophisticated product, whether it is technologically advanced or not, and whether it is a manufacturing operation or even an educational institution or the government, the same issues apply. Even in an educational institution there is need for the leaders to set an example for quality and expectations by their directives, decisions, and actions. The faculty must maintain their expectations of quality in their own work and in the work of their students, and the students must be sold the idea that quality matters. Without all three, and without continued evaluations, you cannot expect improved quality. In a manufacturing operation too, without the support and expectation of the top leaders demonstrated by their actions, directives, and expectations, and the drive of the personnel in doing quality work whether in design, in manufacturing, or any other operation, and without the proper inclusion of quality design practices, the product will not be as good as expected. In the mid-1990s a major automobile manufacturer used the motto ''quality is job 1'' to encourage its employees and to convince its customers that quality was important to the company. They used banners on the factory floors and bought advertisements in the media to show their dedication to quality. Of course, mottos do not do it; actions do. Having a few posters and banners around the factory floor will not change the behavior of the workers or the quality and safety of the product. It only demonstrates the supposed buy-in of the management as the first ingredient needed to set the course of the institution, so long as it is sincere. The same must be expected of the lower layers of constituents to try and build in the practices that result in quality products into their thinking, into their systems, and into their operations. Many companies have teams of educators that teach their personnel how to incorporate quality design and manufacturing into their processes, their tasks, their designs, and their decisions.

It has been estimated that production errors and variations are 80–85% related to the processes and methods used, and only 20–15% related to the behavior or expertise of the workers. What this indicates is the importance of incorporating proper methods and processes in the design and manufacturing tasks that yield better results and improve quality.

There is a host of techniques developed over the decades that try to take the guess work out of integrating quality into the product by systematically incorporating scientific and statistical methods into the design and manufacturing processes. These philosophies and techniques allow the designer to design better products with better quality, reduce variations in the product, and make the product more robust and less costly. In this chapter we will look at some of these techniques that enable us to build in quality into the design of the product. Like other subjects, a complete survey and detail treatment of this subject in one chapter will not be possible. You must continue studying this subject in appropriate courses and through dedicated references. However, in this chapter you will learn the basics of this subject, the philosophies behind the concepts, as well as some practical methods that can be used for increased quality, including Taguchi methods, Lean production, Quality Function Deployment, and other similar concepts.

12.2 DEFINITION OF QUALITY

The American Society for Quality (ASQ) defines quality as ''a subjective term for which each person has his or her own definition.'' Quality may be defined by multiple variables that relate to the function, shape, cost, aesthetic, appearance, and life expectancy of a product. But this is not what industry pioneers advocate or consider.

As will be seen later, to most industrial advocates of quality, at least in manufacturing operations, the term relates to reduced variations from the nominal; when parts vary less, the quality increases. This, although an accepted premise, is not necessarily always true. It is not adequate to just have less variation from the nominal to have quality products and systems. It requires that the nominal be of superior quality to begin with before reductions in variations from it increase quality standard.

Let's consider a bookcase you buy and assemble. Imagine that the bookcase is built from inferior particle board that bends under the expected load, with a thin laminate that chips and cracks easily under normal load. If this is the nominal specification for the bookcase, even having extremely small variations will not make the bookcase a quality product. In fact, if variations are large, there is a chance that at least once in a while a better product may be shipped! Otherwise, this means that all manufactured bookcases are close to the low-quality nominal, hardly a quality product. If a television set has a low-quality image, lack of variations from this low quality will not increase the product's quality either. If two mating parts such as a shaft and a bearing are specified with wrong interference measurements, even if the range of variations is extremely small, the shafts and the bearings will not match. However, if the parts are designed as quality parts, with proper dimensions, choice materials, and appropriate characteristics, reducing variations will certainly increase the quality of the products.

So, in this chapter, our discussion about quality revolves mostly around the concept of reductions in variations, mostly as it pertains to variations in manufacturing of products. However, the product must be designed with quality attributes in mind, both in terms of its functional and physical characteristics as well as methods to reduce the variation from the ultimate goal of the product.

It is important to point out here that, in fact, variations from the nominal desired value are not always bad. When the purpose of the task is to maximize the yield, larger variations above the desired value yield better results. For example, in agriculture, if the crop yield is larger than desired (the goal, the nominal value), it will be better (unless we consider the effect of much larger yields that lower the price). If the total sales of a company are larger that the stated goal, it will be better too. Alternately, when the purpose is to reduce an effect, smaller results are better, even if they deviate more from the desired value. For example, in air pollution, customer-wait time, and lost luggage, lower values are better even if farther away from our desired goal. However, in mating parts, we should look for the least amount of variation from one part to another.

12.3 HISTORICAL BACKGROUND

During World War II, W. Edwards Deming who was a statistician and engineer had helped the American industry increase its quality and productivity, even though a great portion of the workforce was poorly trained. After WWII, with the expansion of the markets open to the American goods and the resulting economic boom, American industrialists did not see the need for increased productivity and quality in their merchandise. They were producing products that were popular in the market and were sold easily. They did not need Deming's way of thinking for increased quality.

On the contrary, after WWII, the Japanese industry was in ruins. Their industry was weak and their products unknown in the world. The general perception was that the Japanese manufactured cheap imitations of others' products. In 1950, the Japanese Union of Science and Engineering invited Deming to educate the Japanese industrial leaders on the subject of quality. He told them that with his set of 14-point principles, their industry would turn around in 5 years' time. They did not believe the prediction, but took his methods to heart and surprised him when their industries turned around in 4 years' time instead. Deming became a guru on quality for the Japanese, where winning the Deming Prize is coveted greatly. Japanese products changed significantly for the better over the next few years, to the point where by 1970s, they became the standard against which quality of products from other countries were benchmarked. Products that originally could only compete for their low prices, expecting that they were of low quality, were changed to products that were not necessarily any lower cost than the competition, but at the highest levels of quality. Isn't the same true these days about the Korean and the Chinese products? Their lower cost and quality is benchmarked against higher quality and better products from other places, but also changing fast (the vastly improved Korean automobiles of today are not the same cars of a decade ago). It was not until the mid-1980s that Deming's methods became popular in the United State. The turnaround of the Ford Motor Company in the late 1980s is attributed to Deming and his methods.

Similarly, Joseph Moses Juran, a Romanian immigrant, developed techniques that eventually led to the application of the Six-Sigma principle and lean production. He wrote a quality control handbook in 1951 that became a standard in this field. Like Deming, Juran was invited by the Union of Japanese Scientists and Engineers to travel there in 1954 to introduce the concepts of quality control into the Japanese manufacturing industries.

Interestingly enough, Genichi Taguchi, also a Japanese statistician and engineer, developed more principles and strategies, many of them based on statistical information. However, his methods were first adopted in the United State. Although the Japanese industry used Taguchi's methods as well, he delivered lectures and workshops to countless American institutions. Many concepts such as House of Quality, Total Quality Management, and Robust Design are based on Taguchi's works. His methods also include design of experiments, where the effects of variation on the system are studied, whether caused by changing variables in the process, affected by variations in delivered components, or varied methodologies.

In the following sections, we will discuss these points in order to see how they are interrelated, but also how they can help a design engineer improve the quality of a product or system.

12.4 STATISTICAL QUALITY CONTROL

Although the quality of a product or system is defined in many ways and it can be measured in many ways too, statistical quality control is one of the major ways it is studied and measured. To understand this, let's consider three different products, an automobile, a shaft made on a lathe, and a cup of yogurt. After the product is manufactured, one needs to test it and evaluate whether it is good or not, whether it falls within specification or not, and whether it can be shipped or not. Imagine that every cup of yogurt would be tasted by someone to ensure its quality. Obviously this would mean that the cup, having been tasted, would no longer be fit for shipping. Besides, would it be necessary to taste every cup of yogurt to ensure quality, or would it be adequate to taste one cup out of every, say 100, cups? In that case, the quality control would be based on the faith that if the randomly chosen cup of yogurt is acceptable, that the others in between are also good. Whether this is true or not, we do not know, but we make the assumption. Now imagine that the next cup tested has an unacceptable taste. No matter what the cause, we do not know when this change in taste started and how many are bad. The change may have been gradual, or sudden, but unless we go back and check all of the cups from the last good cup we will not know when the change started.

Now consider the automobile. When the assembly is completed, it is driven for a short distance to ensure that everything is in order, the brakes work, and there are no unexpected noises. This test does not damage the car, and it does not render it useless. Besides, the customer expects that every car is tested for quality and for making sure that it is safe. Would you expect that one car out of every, say 100, cars be tested? This is 100 percent testing. Then what about a shaft, made on a lathe? Like the automobile, it can be tested for correct measurements without damage. But is it necessary to test all shafts as we did with a car? Perhaps not. We assume that random, statistical quality testing and measurement is adequate. This is based on faith that unless something drastic happens (like sudden damage to the cutting tool), system variables (such as gradual tool wear) are slow enough that if we test the product within reasonable intervals, we will be able to catch the variations and remedy them. Still, what if something drastic does happen? Will we find out too late?

The point of this discussion is that in most cases, it is either impossible or too expensive to do 100 percent quality control testing. Our desire should be to test as few items as possible. Against this, we need to ensure that quality is not sacrificed and that we have enough measurements to ensure that we are still within specification. A common way to accomplish this is the idea of using tolerances. We will discuss tolerances next, but we will also see how using tolerances is not the best way to accomplish quality in manufacture. That is where methods forwarded by Deming and Taguchi become prevalent.

12.5 TOLERANCES

When a part is manufactured, no matter what process is used for fabrication or production, there are chances that the part's dimensions may be different than specified. For example, let's say that you need to fabricate a 1-in shaft that is to be assembled into a bearing (a hole of the same size). If we assume that the shaft will be exactly the same size as the hole, the assembler will be able to insert the shaft without difficulty and the shaft will fit perfectly. To begin with, how will we measure that the shaft is in fact 1-in in diameter? To do so, we need to use an instrument, perhaps a caliper. The caliper has an accuracy of perhaps 1/1000th of an inch, assuming that it was manufactured within specification. However, if the shaft is 1.0004 in. will the caliper be able to measure the difference? What this means is that no matter what device we use for measurements, we are limited to the accuracy of the measuring device. The next issue is the variations in the size of the shaft due to manufacturing variations. For example, if the cutting tool used for machining the part wears out, the shaft will become larger as time goes by.

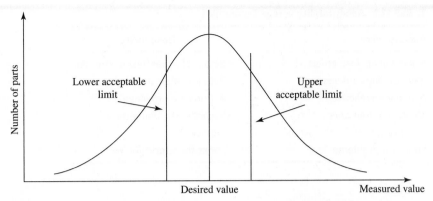

FIGURE 12.1 The bell-shaped distribution of a population of manufactured parts.

As temperatures change, all dimensions change due to thermal expansion as well. This means that the final size of a product can be affected by the temperatures encountered during manufacture, even if these changes are small. Operator errors, differences in materials, and many other factors contribute to this variation in size.

If we are lucky, we will have a shaft that happens to be exactly the same size as the hole. It will fit perfectly without excessive force or friction, and it will not wobble. Now imagine that a slightly smaller shaft is matched with a slightly larger hole. The result will be that the assembly will be loose and wobbly, causing excess noise, wear, and deflections. Conversely, imagine that the shaft is larger and the hole is slightly smaller. In this case, the worker will need to use force to insert the shaft, there will be excessive friction causing heat and wear, and the product will be inferior.

In most cases, if a large number of parts are measured and a histogram is plotted, a natural bell-shaped curve will result (Figure 12.1). Hopefully, most parts are close to the mean value. As the dimension varies more from the middle desirable value, the numbers decrease. Assuming that the bell-shaped curve is a good representation of this population, the mean, standard deviation, and other statistical measurements can be calculated for the population.

To keep these variations under control, tolerances are used to set lower and higher acceptable limits for each product or part such that the largest and smallest counterparts will still function together in an acceptable manner, even if not perfectly desirably. Any part that falls outside of the specified dimensions needs to be reworked or discarded. Obviously, as the tolerance limits are tightened, the parts will be more uniform and more similar to each other (less variation), increasing the quality of the product, but also increasing the cost of production (Figure 12.2). Whether random checking of the parts or 100 percent testing, with tighter tolerances, more parts fall outside of the limits, thus more waste and more time that must be spent manufacturing each part with more care, still increasing the cost. As a result, based on the theory of tolerances, better quality will cost more (as we will see later, this theory is rejected by Deming and Taguchi).

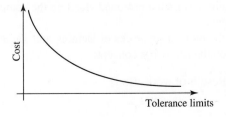

FIGURE 12.2 The relationship between cost and level of tolerances in manufactured parts. As tighter tolerances are specified, the cost will increase.

TABLE 12.1   Acceptability versus Desirability

Acceptability	Desirability
Good, cheap, fast, smart	Better, cheaper, faster, smarter
Relationships are absolute	Relationships are relative
Actions are reactive	Actions are proactive
Meets requirements	Exceeds requirements
Fix or re-do	Improve
Looks for problems	Looks for opportunities

The basic problem with the concept of tolerances is the difference between *desirability* and *acceptability*. To see the difference, imagine a set of wooden stakes. Our desire is to pick stakes that are 10 in. long (desirable), but will accept all that are 10 ± 1 in. and fall between 9 to 11 in. (acceptable). In this case, two stakes that are 9 and 11 in. will both be acceptable, although 2 in. different in size. However, only one of two stakes 8.9 and 9 in. long will be acceptable even though they are only 0.1 in. apart. This dilemma in acceptable and unacceptable differences in specifications and the increased cost for better quality specification lead to lower quality and constant struggle between increased cost and increased quality. A better specification should be to *strive to get closer and closer to the desirable value* and not to have an acceptable range. The difference between acceptability and desirability can be summarized as in Table 12.1.

The desirability mind-set guides the designer to better quality and constant improvement rather than being satisfied with what is.

12.6   WHAT IS QUALITY?

As will be seen later when we discuss Deming's principles, Taguchi Methods, and lean manufacturing, the basic concept of improved quality in relation to cost, and what constitutes quality, is different from tolerances. In other words, in all these methods, the philosophy behind quality is somewhat different than the concept behind tolerances.

As is also discussed in Chapters 7 and 11, the value of a product or system is defined as

$$\text{Value} = \frac{\text{What you get}}{\text{What you pay}} \tag{12.1}$$

This means that, except in sentimental cases, the value of the product is the proportion of what you get from a product compared to what you pay for it. What you get from a product includes many things such as features, looks, comfort, reputation, prestige of the brand, but also quality of the product as a major factor. The more you get for the same price, the better the value. The less you pay for the same quality, features, looks, etc., the higher the value. The value of the product is the ultimate factor of its desirability to the customer. This is why lowering the cost and increasing the quality has such a profound effect on the competitiveness of any product, and thus, any company.

The concept of quality based on the pioneering works of Deming and Taguchi and many others can be summarized based on the following concept:

$$\text{Quality} = \frac{\text{Results of work efforts}}{\text{Total costs}} \tag{12.2}$$

If the focus of the organization is primarily on quality, it tends to increase and as a result, total costs are reduced. When the focus of the organization is on costs, then costs tend to rise and quality falls. This can be seen in many examples. For example, when Ford set overall specifications for the Pinto, the price was set at $2,000, no matter what. The result was that even though the engineers at Ford wanted to add a liner to the gas tank at the cost of a few dollars each to prevent leaks and explosions resulting from rear-end collisions, with the emphasis on costs, it was rejected by the management. On the contrary, as we will see, in lean production, with emphasis on quality, every small defect is cause for stopping production until the problem is solved completely. The emphasis on quality forces everyone to do their utmost to prevent defects at the root, resulting in fewer reworks and rejects, thus reducing costs. When in a university the cost of education is the primary concern, the quality suffers as classes are not offered, unqualified students are admitted, and so on. When the emphasis is on quality, it rises as better qualified students are admitted, more qualified faculty are hired, and classes are taught based on need not enrollment.

12.7 THE CONCEPT OF ZERO DEFECTS

In the early 1960s, the possibility of defective products were expected by the end users. When products were delivered either to a customer, to retailers, or to other organizations, it was expected that some of the products would be defective. At best, the end user would return the defective part for replacement. However, Philip Crosby, who was a quality engineer at that time, coined the phrase ''Zero Defects'' and pushed this new quality standard across the entire industry. In 1979, in his book *Quality is Free*,[9] Crosby summarized his views on quality when he theorized that there are but four ''Absolutes of Quality Management'':

- Quality is defined as conformance to requirements, not as goodness or elegance.
- The system for causing quality is prevention, not appraisal.
- The performance standard must be Zero Defects, not ''that's close enough.''
- The measurement of quality is the Price of Nonconformance, not indices.

The concept of zero defects indicates that all products shipped from a plant must be working properly within the specified tolerances. However, it does not indicate where within the spectrum of acceptable values the product falls. It only indicates that it is defect-free. This model of product management includes the task of dividing a given product into a finite number of parts, with the aim of achieving zero defects at the part level. In achieving the goal of assembling parts with zero defects, it is assumed that if all parts are defect-free, the resulting product will be defect-free as well. The simple working model is that defect-free products result from assembling defect-free parts using a defect-free assembly process. However, there is no real reason why this should be true.

To understand the concept of zero-defect better, let's consider the corresponding definition of quality in a classroom setting. If one were to maintain the same standard of quality as imposed by Crosby, what would be the letter grade requirement for each student if a professor was to strive to achieve zero defects with the performance of his or her class? Would it be all A's as many believe? According to William Bellows, an industry consultant who presented this example, the letter grade requirement for the achievement of zero defects is not A+, or even A; it is D−, which corresponds to a passing grade for everyone. If all students in a class receive a D−, everyone passes the class, thus no defects. Even if all students receive D− and nothing more, there is still zero-defect achievement; no one was rejected. Obviously, this is not what the professor should strive to achieve. In fact, the goal should not even be to achieve a grade of C or even B for everyone. The goal should be a grade of A for everyone. Similarly, this is what a manufacturer should strive to achieve.

12.8 DEMING'S PRINCIPLES

W. Edwards Deming's principles for quality are based on what he called *System of Profound Knowledge*™. His basic premise is that tighter tolerances do not mean higher prices. Rather, as quality increases, the overall cost decreases, and as long as the higher quality is linked to higher prices, the institution will not be able to compete with better quality products at lower cost.[1]

In Prophets in the Dark,[2] a story is told about Frank Pipp, a former executive at Ford, who in the 1960s realized the threat that the quality of cars and trucks made by Toyota imposed on the domestic market in California. He purchased a truck and had his workers dismantle and reassemble it to learn about the vehicle and its quality. They found out that every part was *snap fit*, meaning that they were perfectly matched for assembly, and no external force by a mallet was needed to assemble the truck. This was not true for Ford cars, as based on tolerances, in many instances, parts would not fit and the worker had to use a mallet to assemble the parts. This emphasis on quality had resulted in much smaller variations in parts, providing for complete snap fit. I remember that in 1976, a Datsun B210 owner bragged about his little car having more than 230,000 miles on it, and still running very well.

The following is the Deming's 14-point quality principles that constitute his Systems of Profound Knowledge.

12.8.1 Deming's 14 points

Deming offered the following 14 key principles for management to transform their business effectiveness. It includes:

1. Creating consistency of purpose for the improvement of product and service, with the aim to become competitive, stay in business, and provide jobs.

2. Adopting a new philosophy of cooperation (win–win) in which everybody wins and putting it into practice by teaching it to employees, customers and suppliers.

3. Ceasing dependence on mass inspection to achieve quality. Instead, improve the process and build quality into the product in the first place (this is based on Deming's observation that 80–85% of defects were caused by processes not people).

4. Ending the practice of awarding business on the basis of price tag alone. Instead, minimize total cost in the long run. Move toward a single supplier for any one item, based on a long-term relationship of loyalty and trust. In this case, the supplier will also follow the company's quest for quality, knowing that its survival is not based on competitive prices alone.

5. Improving constantly, and forever, the system of production, service, planning, of any activity. This will improve quality and productivity and thus constantly decrease costs. This means that the goal is not to stay within tolerances, but constantly trying to achieve better quality.

6. Instituting training for skills.

7. Adopting and instituting leadership for the management of people, recognizing their different abilities, capabilities, and aspiration. The aim of leadership should be to help people, machines, and gadgets do a better job.

8. Driving out fear and building trust so that everyone can communicate and work more effectively. The relationship between the management and workers should be friendly not adversary.

9. Breaking down barriers between departments and abolishing competition and building a win–win system of cooperation within the organization. People in research, design, sales, and production must work as a team to foresee problems of production and what might be encountered with the product or service.

10. Eliminating slogans, exhortations, and targets asking for zero defects or new levels of productivity. Such exhortations only create adversarial relationships, as the bulk of the causes of low quality and low productivity belong to the system and thus lie beyond the power of the work force. A banner stating quality is job-1 does not work.

11. Eliminating numerical goals, numerical quotas, and management by objectives. Substituting leadership.

12. Removing barriers that rob people of joy in their work. This will mean abolishing the annual rating or merit system that ranks people and creates competition and conflict. Cooperation rather than competition.

13. Instituting a vigorous program of education and self-improvement.

14. Putting everybody in the company to work to accomplish the transformation. The transformation is everybody's job not a directive from management.

As we will see later, similar principles are embodied in the Taguchi's Methods and lean production and even more recent findings.[3]

12.9 TAGUCHI METHODS

Taguchi Methods[4,5,6] are a set of methods, mostly based on statistics, that are used to specify and design parts with better quality. Some of the methods are philosophical in nature; others are very mathematical. The fundamental concepts of Taguchi methods are based on the following three ideas:

- Quality should be designed into a product or system, not inspected into it. Thus, if a product or system is designed properly, inspections are needed only for verification.

- Quality is best achieved by minimizing variations from the nominal value due to environmental and process factors that are not controllable. Unlike tolerances where all values between a minimum and maximum are acceptable, minimizing variations seeks to achieve the nominal value 100 percent of time.

- The cost of quality is measured by deviations from the nominal value and losses should be measured system-wide. This means that the cost is not just to the manufacturer, but also the customer and the society at large.

Taguchi methods are designed for **off-line quality** and for **on-line quality**. On-line quality methods relate to maintaining and improving quality during manufacture. Off-line quality relates to methods and processes that are used during design of the product or system in order to build quality into the system through design. We will discuss some concepts of on-line quality methods as applied in lean production later. But since this book is about design of products, we will concentrate more on off-line quality.

Off-line quality is applied to both the design of the product and the process of manufacturing it, each one at three levels of system, parameter, or tolerance level. For the design of the product, the three off-line principles are applied as:

- **System design**: Here, the overall system, the product, or subassembly is designed for best quality. This means that the system is designed to minimize variations from the set, nominal goals, although conditions may change. For example, the cooling system of a car is designed to ensure that the temperature remains unaffected by environmental changes. Thus, a thermostat is used to control the flow of coolant between the radiator and the engine block. An external coolant reservoir is also added to ensure that if extra coolant is needed, it will be available, but also that a control system of some sort is present to decide if more fluid should be drawn from the reservoir. A fan that can be turned on and off based on temperature, rather than always being on when driven by a belt, also assists in maintaining constant temperature even at increased environmental variations. The addition of control elements in the system makes it more robust and less affected by environmental effects.

- **Parameter design** relates to the design of the components of the system to ensure that minimization of variations is attainable. Each component should be designed to respond appropriately to the design parameters in order to achieve uniformity of output, even if the input varies. In this case, using the nonlinear characteristics of material and components assists in reducing output variation with respect to input variations. In the cooling system of an automobile, the radiator should be capable of rejecting heat at desired levels, even if the water level varies in it. We will discuss this in more depth later.

- **Tolerance design** is used to create a balance between the cost of components and their variability (quality). There is no question that the cost of components increases as tighter tolerances are specified. However, there is also a cost associated with variations in the components. At this level, tolerances are chosen for the most economical tolerances, a balance between the cost and variability in the product.

  As an example, imagine that the output *(Z)* of a system is related to the outputs of two components (X and Y), such that $Z = kXY^2$ where k is a constant. Imagine that the output variations for both X and Y are 5 percent from nominal. For $X = 1.05 \times$ nominal value and $Y = $ nominal value, the output will be $Z = k(1.05)(1) = 1.05k$ of nominal output value, whereas for $X = $ nominal value and $Y = (1.05) \times$ nominal value, the output will be $Z = k(1)(1.05)^2 = 1.1025k$ of nominal output value. Clearly, it is more important and more cost effective to specify tighter tolerances for Y than for X, assuming that their cost increases similarly as the tolerance is tightened. Otherwise, the increase in component cost must also be considered.

For the manufacture of the product, the three off-line principles are applied as:

- **System design**: At this level, the manufacturing processes that alter a material or subassembly into a more advanced state are chosen or designed. In general, every task may be accomplished with multiple methods. For example, a part may be machined or cast into shape. This step involves selecting the best, most economical methods that produce or advance a subassembly into a higher state with least variations from the nominal goal.

- **Parameter design** involves selecting the operation levels of each selected method such that the variations within the process are minimized. For example, as a part is machined, the tool may wear out. Some tools wear at higher rates than others, thus creating variations in the part. Selecting a method of fabrication that is less prone to tool wear, or alternately, creating a situation where the part is less affected by tool wear will result in less variations in the component.

- **Tolerance design** involves selecting minimum and maximum acceptable levels for the behavior of the process. As discussed earlier, tighter tolerances increase the cost of production, but reduce variations. The engineer should find the most economical value for a balance between cost of production and acceptable variations.

Taguchi Methods revolve around the following points:

1. Total cost to society
2. Continuous quality improvement and cost reduction
3. Reduction in variations, robust design
4. System design, parameter design, tolerance (allowance) design (as discussed above)
5. Design of experiments

In the following sections, we will discuss these concepts.

12.9.1   Total Cost to Society

This is one of the fundamental philosophies of Taguchi Methods. As was mentioned earlier, it is important to not just consider the production cost of better quality products to the company, but the total cost to society, including any losses. For example, when a product is made, it requires resources, energy, time, and effort. These are all costs to society (these could be used for other purposes too. If they were used for this product unwisely, they are lost, costing the society). When the product is discarded, it adds more loss to the society. If it is discarded early in life because the product is no longer needed, has stopped working prematurely, or is of bad quality, the loss to society is even larger. If the product produces pollutants when it is used, it adds more loss to the society. And if the product malfunctions or requires repair, there is even more loss to society, not only due to the expenditure of the owner's money to repair it, but also due to the owner's wasted time away from doing other useful things. Thus, the better the quality of the product, the less loss to society. In other words, according to Taguchi, a better product causes less loss to society because it would require fewer resources, would create less pollution, cost less to make, cost less for the customer to repair and maintain, waste less time, more trust in the company's products, and more profit for the company. Thus, instead of only considering the cost to the company, the designer should consider the total cost to society. In the long run, this will greatly benefit the company too due to customer trust and loyalty. If a customer likes a product and develops loyalty to the brand name, there is a better chance that she or he will be a repeat customer, buying more products from the same manufacturer. So, instead of relying on selling more products with lower life expectancy in order to increase sales, the manufacturer should rely on repeat business due to loyalty to a brand whose products last longer. Every time the manufacturer loses a customer, a loss has occurred that must be added to the total cost. Figure 12.3 shows a typical loss to society curve. As the range of variations in acceptable dimensions is reduced, the cost of production increases while the customer cost decreases. The combination of these two is the total loss to society. This allows the designer to specify the desired values at which the total cost to society is minimum.

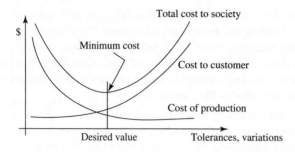

FIGURE 12.3 Total cost to society.

To better understand this, consider the difference between stealing, skipping work, and cheating on taxes, versus murder. In the first cases, one loses and another gains. When a worker skips work, the employer loses (by paying the salary for no return work), while the worker benefits (by getting paid for no work). The net cost is zero. On the other hand, in murder, there is only loss to society. No one gains. Similarly, if a manufacturer's benefit in producing lower-quality products is more than the loss to the customer, there is a net gain in society. If the benefit to the manufacturer is less than the loss to customers, there is a net loss to society. Thus, it is not just the benefit to the manufacturer that should be considered, but also the loss to the customer, and thus the total cost to society, and how to minimize the loss.

12.9.2 Loss Functions

Loss functions represent the loss caused by variations in the product (or part) or variations in the output of a system as the input changes. As was mentioned earlier, for tolerances, an *acceptable* range for a value or dimension is selected. If the part's dimension falls below or above the set limit, the part is rejected; otherwise, it is accepted. However, the *desirable* value for the specified dimension is the value that would best fit the needs. Figure 12.4 shows the lower specification limit (LSL) and the upper specification limit (USL). Any part that falls outside of these limits is discarded, thus a loss. Otherwise, if the part is within the tolerance limit, no matter how close to or far from the desired value, it is acceptable and creates no loss. However, according to the concept of the total loss to society, the goal is to get as close to the desirable value as possible, where the loss to society is the minimum. At any other point, the loss is more, even if within the tolerance limits. As you notice, with this approach, even if a part is within tolerance (acceptable), it may still create a loss if it is not at the desirable nominal value. Equations 12.3 and 12.4 represent the loss for each case. The goal should be to constantly strive to find ways to improve the quality of the product by reducing the loss through *reducing variations in the product*. This premise leads to the application of other Taguchi Methods and philosophies.

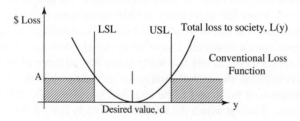

FIGURE 12.4 Loss functions for tolerance-specified design and total-loss-to-society based design.

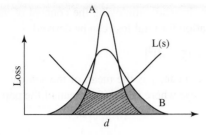

FIGURE 12.5 Two distributions with different variances have different losses.

For tolerances:

$$\text{Loss} = \begin{cases} A & \text{if} \quad y < \text{LSL} \quad \text{or} \quad y > \text{USL} \\ 0 & \text{if} \quad \text{LSL} \leq y \leq \text{USL} \end{cases} \qquad (12.3)$$

For total loss to society:

$$\text{Loss} = \begin{cases} k(y-d)^2 & \text{if } (y-d) > 0 \\ 0 & \text{if } (y-d) = 0 \end{cases} \qquad (12.4)$$

where k is a constant that must be found for each case, y is the dimension of any given part, and d is the desired dimension.

Equation (12.4) represents a second-order polynomial of the difference between the actual value y and the desired value d (square of the error), based on the first term in the Taylor series. This establishes a measure of the dissatisfaction of the customer with the variations (lack of quality) of the product.

Now consider two distributions A and B of similar products as shown in Figure 12.5. Each distribution demonstrates the variations in the product's variable from the desired nominal value for N samples. Distribution A is narrower with a larger mean (nominal) value, while distribution B is wider, with a smaller mean. Obviously, distribution A is more desirable because the variations from the nominal value are less. The area under the Loss function is the related loss for each distribution. Therefore, the losses related to this distribution are also smaller than the losses related to distribution B. This is why narrower distributions will increase quality and decrease total loss.

EXAMPLE 12.1

Assume that the desired dimension for a part is $d = 1$ in. The selected tolerance for this dimension is ± 0.001 in. It has been determined that the cost (loss) for a part outside of this limit is $1. Determine the loss function for this part. Also determine the associated cost to society for a part that is at 0.9995 in.

Solution   Substituting these values into Equation (12.4) we get:

$$L = k(y-d)^2$$
$$1 = k(0.999 - 1)^2 \quad \rightarrow \quad k = 10^6$$
$$L = 10^6(y-1)^2$$

The loss associated with a part at $d = 0.9995$ is: $L = 10^6(0.9995 - 1)^2 = \$0.25$. ∎

Using Equation (12.4), the loss associated with each part based on its particular dimension can be calculated. Of course, for a limited number of parts, this is manageable. But for large product populations such as in mass production, it is impossible to calculate the loss associated with each particular part. Instead, employing the concept of variance and mean of a population the following equation for total loss can be derived:

$$L = k[S^2 + (\bar{y} - d)^2] \tag{12.5}$$

where S is the standard deviation around the mean, $\bar{y}$ is the mean of the sample population, and d is the desired dimension. For the case where $\bar{y} = d$ (the mean of the population is the same as the desired value), $L = kS^2$.

EXAMPLE 12.2

For the example 12.1 above, assume that the mean value of a 1000-part sample is 1.00 in. with a standard deviation of 0.00025. Calculate the average loss per part and total loss for the whole population.

Solution   Using Equation (12.5) we get

$$L = 10^6[0.00025^2 + (1 - 1)^2] = \$0.0625 \text{ per part.}$$

The total loss for 1000 parts will be

$$L = 0.0625 \times 1000 = \$62.5$$

As you notice, in a population of 1,000 parts with different values, the average loss per part is different from a specific part with a known dimension such as in Example 12.1. ∎

EXAMPLE 12.3

Repeat Example 12.2, but assume that the mean value of the population is 1.0002 instead.

Solution   Substituting the corresponding values into Equation (12.5) we get:

$$L = 10^6[0.00025^2 + (1.0002 - 1)^2] = \$0.1025 \text{ per part.}$$

The total loss will be

$$L = 0.1025 \times 1000 = \$102.5$$

As you notice, since the mean value of the distribution is no longer the same as the desired value, both the loss per part and the total loss have increased. Therefore, it is desirable to not just minimize the variations (standard deviation), but also keep the mean value as close as possible to the desired value. ∎

Now imagine that the tool which is used to manufacture the part wears out. As a result, the mean value also changes over time. Allowing the tool to wear out for a longer range will increase the change in the mean value. However, correcting the tool to reduce variations requires stopping to manufacture, which results in a loss as well. Assuming that W is the width of wear allowed, we can substitute a corrected value for standard deviation as a function of W:

$$S^2_{\text{corrected}} = S^2_{\text{nominal}} + \frac{W^2}{12} \tag{12.6}$$

Using this equation, one can calculate the allowable width of tool wear for a desired loss, or calculate the loss associated with a range of tool wear.

EXAMPLE 12.4

For Example 12.2, calculate the corrected standard deviation and associated loss if a tool wear width of 0.0001 or 0.0003 is allowed.

Solution   Substituting these values into Equations (12.5) and (12.6) for each case we get

$$\text{For } W = 0.0001 \qquad S_{\text{corrected}} = \left(0.00025^2 + \frac{0.0001^2}{12}\right)^{1/2} = 0.0002517$$

and

$$L = 10^6[0.0002517^2 + (1-1)^2] = \$0.0634 \text{ per part.}$$

$$\text{For } W = 0.0003 \text{ we get} \quad S_{\text{corrected}} = \left(0.00025^2 + \frac{0.0003^2}{12}\right)^{1/2} = 0.0002646$$

and

$$L = 10^6[0.0002646^2 + (1-1)^2] = \$0.07 \text{ per part.}$$

As you notice, the loss increases as the tool-wear width is allowed to grow, even if the mean value is kept at the nominal dimension. This increase in loss can be compared to the cost of correcting the tool more often in order to minimize the total loss. ∎

Now consider the effect of changes in the mean and standard deviation in the population. As you may expect, changes in the mean and standard deviation of the population will result in different values of total loss. The influence of each variation can be calculated using the above mentioned equations. For example, consider the three pairs of distribution in Figure 12.6.

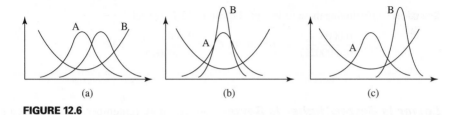

(a)   (b)   (c)

FIGURE 12.6

In part (a), two distributions A and B are similar but the average value of A is closer to the nominal value. In this case, A is clearly better. In case (b), the two distributions have similar average value, but their variances are different. In this case, B has smaller variations and is clearly better. In case (c), the distribution of B is narrower but off to one side while A is more distributed but closer to the nominal value. In this case, it will be necessary to compare the two distributions and calculate the total loss related to each case and decide which one is better. This shows that reducing the variations is not automatically better if the average is not close to the nominal value.

EXAMPLE 12.5

Referring to Figure 12.6, the standard deviation and mean values for three pairs of populations are given below. For each pair, calculate the loss per part and compare.

(a) $S_A = 0.00025$, $\bar{y}_A = 1.00$ and $S_B = 0.00025$, $\bar{y}_B = 1.0002$

(b) $S_A = 0.00025$, $\bar{y}_A = 1.00$ and $S_B = 0.00015$, $\bar{y}_B = 1.00$

(c) $S_A = 0.00025$, $\bar{y}_A = 1.00$ and $S_B = 0.00015$, $\bar{y}_B = 1.0002$

Solution   Case A for all three pairs is the same as in Example 12.2 and yields a value of $L = \$0.0625$ per part.

For (a), distribution B, the loss is: $L = 10^6[0.00025^2 + (1.0002 - 1)^2] = \$0.1025$.

For (b), distribution B, the loss is: $L = 10^6[0.00015^2 + (1 - 1)^2] = \$0.0225$.

For (c), distribution B, the loss is: $L = 10^6[0.00015^2 + (1.0002 - 1)^2] = \$0.0625$

As you see, for case (a) the loss per part is $0.0625 versus $0.1025 as the same distribution's mean value increases away from the nominal value. In case (b) the loss is $0.0625 compared with $0.0225. In this case, the narrower distribution with smaller variations at the same mean value causes much less loss to society. However, in case (c) even though the distribution B is far less varied, since its mean value is off the nominal value, it causes the same loss as A. ■

Process Capability Index One way to measure quality using the distribution of the population is called *process capability index* C_p which can be calculated as

$$C_p = \frac{\text{Tolerance}}{\pm 3S} \tag{12.7}$$

In comparison with the same tolerance specified for a part, as the standard deviation S decreases, C_p increases.

EXAMPLE 12.6

The tolerance for a part was specified as ± 0.0005 inch. As a result of improvements in manufacturing, the distribution of the part was improved such that its standard deviation changed from 0.00025 to 0.00015. Calculate the improvement in its process capability index for ± 3 sigma.

Solution Substituting these values into Equation (12.7), we get

$$C_p = \frac{0.0005 \times 2}{6(0.00025)} = 0.67 \quad \text{and} \quad C_p = \frac{0.0005 \times 2}{6(0.00015)} = 1.11$$

■

Lower Is Better/Higher Is Better As you may remember, we discussed situations where lower is better as well as when higher is better. In these situations, achieving mean values that are lower or higher is better than striving to achieve the nominal value. For example, reducing pollution or wait-time is better than remaining on a nominal goal. Similarly, increasing efficiency or life expectancy above the nominal goal is advantageous. The loss to society for these situations can be expressed as follows (Figure 12.7).

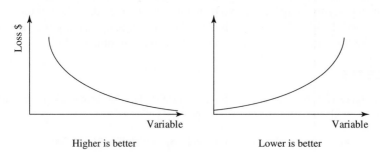

Higher is better   Lower is better

FIGURE 12.7 Loss functions for lower-is-better and higher-is-better cases.

	Loss for an Individual Part	Average Loss per Part in a Population
For lower is better	$L = ky^2$	$L = k[S^2 + \bar{y}^2]$
For higher is better	$L = k(1/y^2)$	$L = k(1/\bar{y}^2)[1 + (3S^2/\bar{y}^2)]$

12.9.3 Continuous Quality Improvement

Figure 12.5 demonstrates the effect of improvements on total loss. As the variance decreases due to improvements in manufacturing as well as processing, the total loss decreases as well. The manufacturer should strive to continually decrease the variance and get closer and closer to the nominal value. This devotion to continuous improvement should continue indefinitely.

Continuous improvement is the hallmark of many corporations as they strive for improved quality. As their quality improves, the end-users' expectations increase as well, demanding better quality products. Other companies that do not follow suit will fall behind. In fact, the same buzz-word of *continuous quality improvement* is applied to the curriculum at the university level too, even if objective measurements are not used for its application.

12.9.4 Reduction of Variations and Robust Design

As was seen above, the goal of the designer should be to reduce variations in parts in order to increase quality and minimize the total cost to society. Robust design refers to methods of design that make the product less vulnerable to disturbances and variations in the input. In other words, the product functions properly under conditions that vary and when disturbances are present. For example, consider a battery. If the battery supplies constant voltage regardless of outside temperatures, age of the battery, the variations in the materials, etc., then its design is more robust. A television set should have a good picture whether in a metal-frame building or not, whether in a stormy weather or not, and whether new or a few years old. A robust system is expected to be insensitive to these external perturbations and continue to function properly and satisfactorily even as external variables fluctuate. External variables include ambient factors, environmental factors, variations in end users, age-related changes, and many more.

One way to achieve this is to take advantage of the relationship between different variables and their level of influence on the behavior of the system. For example, if a variable has a more pronounced effect on the output of the system, it is more important to try to control that variable than another variable that has a less severe effect on the output. For example, consider the volume of a container. Imagine that this volume determines the accuracy of another function that is very important to a task. Thus, we should want to make sure that the volume is accurate. The volume of a cylindrical container is $Vol = \pi r^2 h$ where r is the radius and h is the height of the container. If the radius of one container is $1.001r$, the output of the system, namely volume, will be $vol = \pi(1.001r)^2 h = 1.002(\pi r^2 h)$. If the height is $1.001h$, the resulting output will be $vol = \pi r^2 (1.001h) = 1.001(\pi r^2 h)$, which is less than the first case. Hence, in this case, the container's volume will be less varied if the radius of the container is controlled with better care.

Another way to make systems more robust is to take advantage of nonlinearities in material properties. As an example, consider the nonlinear response of a part such as a spring to variations in input, for example, the displacement (stretching) of the spring under a tensile load (Figure 12.8). As shown, the variations in output of the spring (displacement) to the same variations in the input (tensile load) at different levels will be significantly different. If the designer chooses to use the spring within the range where the output is more saturated (flat), the variations in the output will be less, resulting in a more consistent output. If this spring, under this level of loading, were used in a device,

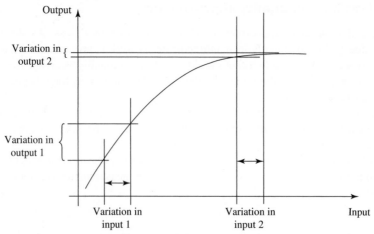

FIGURE 12.8 Application of the use of nonlinear response of materials to reducing variations in parts and improving the quality of the product.

say a check valve, even if the input to the spring varied much, the variations in the output of the spring would be little, thus reducing variations in the product. The nonlinear output response of a transistor can similarly be used to reduce variations in output. Other specific characteristics of materials and parts may be used similarly to decrease variations in products, and thus, improve the quality of the product.

The following is an actual example of how these concepts were applied to the design and manufacture of a real system and how it was improved.

12.9.5   Case Study: Pratt & Whitney Rocketdyne's Rocket Engine Hardware

In the mid-1990s, engineers at Pratt & Whitney Rocketdyne in Los Angeles, California, were challenged to deliver rocket engine hardware of exceptional performance, quality, and cost to extremely demanding customers. The primary customer was the U.S. Air Force, which was in need of a new launch vehicle for use as a satellite delivery system. Secondary customers included a growing community of providers of commercial satellite systems. As one of the premier rocket engine companies in the world, Pratt & Whitney Rocketdyne is well known for its design and development of liquid rocket engines. Engines and applications include the F-1 engine for NASA's Saturn launch vehicle, the Space Shuttle Main Engine, the RS-27 engine for Boeing's Delta II rocket, and the MA-5A engine cluster for Lockheed-Martin's Atlas II rocket.

Faced with both military and commercial demands for high performance and high-quality rocket engines delivered at low cost, the engineers at Pratt & Whitney Rocketdyne used their growing understanding of Taguchi's quality loss function to design and deliver the needed hardware systems for the RS-68 engine (Figure 12.9, Reference 10). Among the many resulting applications, a simple example is included here to demonstrate the dramatic paradigm shift by the engineers in moving from part quality to system quality. Specifically, this application involved the brazing of a component of the combustion chamber, which is a round, flat plate, approximately 30 inches in diameter and 1 inch in thickness. The design of this plate (Figure 12.10) includes 628 holes, roughly ½ inch in diameter, in which a hollow post is secured in

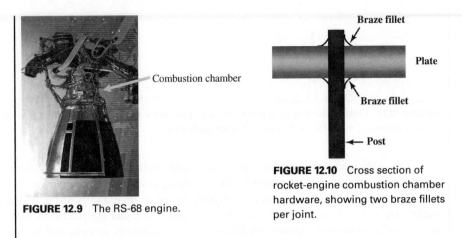

FIGURE 12.9   The RS-68 engine.

FIGURE 12.10   Cross section of rocket-engine combustion chamber hardware, showing two braze fillets per joint.

place in each and every hole with a thin foil ring of braze alloy. Ideally, the braze foil flows uniformly in the gap between each hole and post, and forms a resulting 360° fillet at the top and bottom interfaces of the plate with the post, a total of 1256 brazes. The presence of the braze fillet assures the secure bond of each tube within each hole, leading to the high structural integrity of this hardware.

As discussed above, striving to achieve 100 percent good parts (zero-defect mentality) is counter to Taguchi's approach to quality loss function mentality. This situation defines the starting point when engineers on the RS-68 Integrated Product and Process Team (IPT) reviewed the quality of the brazing process for similar combustion chamber hardware designs—hundreds of holes in a plate with a post brazed in each hole.

A standard way to document the quality of the brazing process for hardware such as this design is to record the total number of successful 360° braze fillets after each brazing cycle. Ideally, all fillets across the plate, both top and bottom, would be achieved after the first braze cycle. If so, there would be no need for subsequent second and third brazing cycles, each of which adds significant costs and schedule delays for hardware deliveries and requires additional re-work space and equipment for second or third brazing cycles.

The IPT's review of Pratt & Whitney Rocketdyne's historical success in brazing similarly designed plate-and-tube hardware revealed a first-pass brazing success rate of between 50 and 85 percent, well below the desired goal of 100 percent. As a result, the team was faced with a significant challenge to design a product and its manufacturing process in a way that all 1,256 fillets in this assembly would be realized in one brazing cycle.

Next, the IPT began to collect data to document the machining process for producing holes in the 30-inch plate and the grinding process for producing the outer diameter of each post. In doing so, they knew that manufacturing engineers for both the plate and posts were employing statistical methods to manage the variation in each process by documenting the mean and standard deviation of the hole diameters and the post-outer diameters for a variety of manufacturing processes for each one. Furthermore, they were aware of the potential advantage of adjusting the average value and standard deviation for each. Doing so would require process changes, possibly even significant process improvement, depending on the severity of the demanded process changes.

FIGURE 12.11 Distributions of hole diameters and post outer diameters in a 1-inch plate from four hypothetical manufacturing processes for each part.

Given the awareness of the role of statistical methods to manage variation in the noted process features, the next step for the IPT designers was to define the specific mean value for the manufacturing process for both the plate hole and post-outer diameter; namely the nominal target values for each. The question under consideration by the IPT was "Given fabrication options for both the hole-machining process and the grinding process for the outer diameter of the post, which process for each do we use and, where do we locate the mean nominal value for each as well?"

In the spirit of thinking together, the IPT designers on the combustion devices team collected process data for the hole diameters and post outer diameters from their manufacturing counterparts. Figure 12.11 provides representative examples of the type of data they collected, each showing the process data for several different processes. This data, along with proprietary experimental data on the optimum gap size for the given design (plate, posts, and braze alloy), were used to select a set of processes and target nominal values for hole machining and post grinding. This would enable them to manufacture the posts and the holes with the least amount of variation from the target nominal value. Through a concerted effort to manage the variation in the outer diameter of the posts and the diameter of the holes, IPT members predicted that all 1,256 fillets could be achieved in one braze cycle; and, they were right. The never-before-heard-of result of 100 percent first pass braze quality was achieved with the first hardware set. For a projected production volume of one hardware set per week, the savings in cycle time was well worth the extra effort required to manage the process variation with respect to the target values for both the diameter of the 628 holes and the outer diameter of the 628 posts. Although more time and effort were required to manage the variation of these dimensions in this manner, the added attention to detail eliminated the need for far more expensive second- and third-braze cycles. From a systems thinking perspective, the significant savings realized by Pratt & Whitney Rocketdyne for this remarkable level of braze quality was well worth the investment of time to study and document the hole machining and post grinding processes. To date, 100 percent first pass braze quality results have been repeated on a routine basis, results which have made it unnecessary to obtain a second braze furnace. According to one of the manufacturing engineers involved in this effort, "It looks like we can repeat this process over and over until the cows come home."

12.10 DESIGN OF EXPERIMENTS

When the design of a system involves many different variables, it is essential that these variables are chosen properly. Otherwise, the system may not be optimal or even useful. For example, consider the compressor unit in a refrigerator. It includes a motor, a compressor, an interface between the two, a condenser, an evaporator, and others. Each one of these parts must be selected appropriately in order for the system to function properly.

At a more detailed level, assume that the connection between the electric motor and the compressor is a v-belt and pulley system. The design engineer has a large number of different choices available for the simple belt-pulley system. As the size of the belt, environmental conditions, torque, size of pulleys, center-to-center distance, and other variables change, the performance of the system will change too. Thus, the engineer needs to pick a set of values for all these variables that collectively work together satisfactorily.

One way to do this is to perform experiments with a variety of different values for each variable in order to measure the best performance from the set. Doing so may require a tremendous number of experiments as many different values for every variable are tested for best combination. The total number may be as high as the full-factorial number. This means that all possible combinations between all variables and their values are tested. As an example, assuming that there are 7 variables, each at two levels (values), there will be a total of 2^7 or 128 experiments needed to test all combinations. Although this has the potential to reveal the optimal values for all variables if enough tests are performed, and although this has been common in agriculture, it is obviously too expensive and time consuming. An alternative to this is to reduce the sheer number of experiments needed through a one-factor-at-a-time approach,[11] in which only one factor is changed and the consequences, and their severity, are measured to determine the effects on the result. Later, another factor is changed and its effects are measured. The process continues until all variables are determined. Although this reduces the total number of experiments needed, the potential exists that combinations of effects and interactions between different variables are missed or not measured unless they are included as well.

Another approach is to use the Taguchi Methods. In this method, statistically oriented tables called *Orthogonal Arrays* are used to determine the effects of variable, their importance, as well as optimal levels, all with relatively few experiments. Furthermore, orthogonal arrays also reveal interactions between different variables or lack thereof. As an example, imagine that you want to develop a new recipe for cookies. The list of ingredients may include 5 different items (flour, eggs, oil, sugar, and nuts). Obviously, variations in each ingredient will affect the outcome. For example, if the amount of flour is changed (varied), the quality of the cookie will also vary (e.g., the thickness and diameter). The same is true for other ingredients. In order to develop the best cookie, you may have to bake a very large number of experimental batches to figure out the best combinations of ingredients, their magnitudes, and their conditions (a heavenly experiment for cookie-lovers). Alternately, with orthogonal arrays, a minimum number of experiments will help you determine the best combinations of elements and their effects on the outcome. Similarly, the effects of variables such as temperature, time in oven, and quality of raw materials on the cookie can also be determined with similar series of experiments (even more cookies to eat). Considering the same example as above for a system with 7 variables, each at 2 levels, the application of orthogonal arrays will reduce the total number of needed experiments to only 8. What is nice is that regardless of the nature of the problem, the same approach may be applied to other design problems, whether they be engineering problems or not. So, for example, the same may be done to determine the effects of the similar variables (temperature, time in oven, and quality of raw materials) on the strength of aluminum during heat treatment. This means that whether we speak about cookies or heat treating aluminum, the process of measuring quality due to variables using Taguchi methods is the same. It is the process that is important, not what is measured. Another example might be the measurement of the effects of variables such as pressure and texture on the ability of a clutch to transfer torque. A few tests of these variables will allow the designer to determine both the effects of each variable as well as interactions between them.

$L_4 (2^3)$

No.	1	2	3
1	1	1	1
2	1	2	2
3	2	1	2
4	2	2	1
	a	b	a
			b
Group	1	2	

(1) 1 —— 3 —— 2

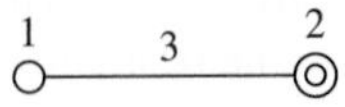

FIGURE 12.12 An $L_4(2^3)$ orthogonal array (with permission from the American Supplier Institute, Dearborn, Michigan).

Orthogonal Arrays Taguchi's orthogonal arrays assist the designer in implementing experiments and analyzing the data for optimum results. There are many different arrays available from different sources such as the American Supplier Institute. What follows is a sample of some of the more common arrays that will be used in the subsequent examples and how to use them. Please see reference 12 for other arrays.

Figure 12.12 is a sample of an $L_4(2^3)$ orthogonal array. In this notation, L_4 means there will be 4 experiments involved in the process with 3 variables, each at 2 levels (values). The three variables may be variable A, variable B, and the interaction between them AB. Variables A and B will have two values each. Thus, the four experiments will be A1-B1, A1-B2, A2-B1, and A2-B2, or simply 1-1, 1-2, 2-1, 2-2.

As you notice on the left side, each row relates to one of the experiments, a total of four. The three columns on the right relate to the relationship between the variables, listed on the top row as 1, 2, and 3. If we assign variable A to column 1 and variable B to column 2, column 3 will represent the interaction between the two variables AB. The numbers in the rows on the right side represent the order of experiments. For example, row 1 relates to values A1 and B1 or 1-1. Row 2 represents A1 and B2 or 1-2. The bottom row also shows the variables and the interactions.

You also notice a linear graph on the right side. The graph shows the variables and the possible interactions. In the case of $L_4(2^3)$, a simple case with only two variables, the graph shows only one possible interaction, namely AB (1-2). As we will see later, other arrays with more variables provide for more interactions (as you notice, these arrays have two sets of numbers 1, 2, etc. which can be confusing. Later, we will use letters and numbers for this purpose to clarify notations).

Figure 12.13 is an $L_8(2^7)$ array, with 8 experiment runs, 7 variables, each at 2 levels. The variables may be A, B, C, AB, AC, BC, and "others," including ABC. The eight experiments will be 1-1-1, 1-1-2, 1-2-1, 1-2-2, 2-1-1, 2-1-2, 2-2-1, and 2-2-2, as shown in the table. However, please note that although column 1 is assigned to A (or 1), and column 2 is assigned to B (or 2), it is column 4 that is assigned to C (3) and not column 3. To see the reason, please look at the linear graph in Figure 12.13. You will notice that nodes 1 and 2 are connected by line 3, which is an interaction between 1 and 2 (A and B). Node 4 is another variable, with lines 5 and 6 representing interactions between 1 and 4 (or A and C) and between 2 and 4 (B and C). Alternately, the other linear graph on the right may be used for assigning variables to columns.

Referring to columns 1, 2, and 4, you will notice the relationship between variable values for each experiment. For example, row-1 has 1-1-1; row-5 has 2-1-1; etc.

L₈ (2⁷)

No.	1	2	3	4	5	6	7
1	1	1	1	1	1	1	1
2	1	1	1	2	2	2	2
3	1	2	2	1	1	2	2
4	1	2	2	2	2	1	1
5	2	1	2	1	2	1	2
6	2	1	2	2	1	2	1
7	2	2	1	1	2	2	1
8	2	2	1	2	1	1	2
	a	b	a	c	a	b	a
			b		c	c	b
							c
Group	1	2		3			

(1)
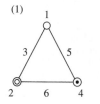

(2)

FIGURE 12.13 An L₈(2⁷) orthogonal array (with permission from the American Supplier Institute, Dearborn, Michigan).

For each case, a performance criterion must be selected for measurement. This criterion is the value that is measured as the result of each experiment, for example the strength of a metal under consideration, or the size of a cookie, or the life of a v-belt. These measurements will be used subsequently to determine the effect of each variable and its severity. For each case, the average of the value measured for a variable at two levels is calculated and is later compared to the alternative. For example, to calculate the effect of variable A (in column 1) in $L_8(2^7)$, the results of experiments 5, 6, 7, and 8 (related to the 2-values) is averaged and compared to the average of experiments 1, 2, 3, and 4 (the 1-values). Similarly, for variable B in column 2, the average of results of experiments 3, 4, 7, and 8 (2-values) is compared to the average of experiments 1, 2, 5, and 6 (1-values).

These results are then compared (subtracted) in order to determine which variables have a larger effect on the outcome of the experiments. The larger the number, the bigger the effect. Additionally, these results may be graphed in order to determine the relationship between different factors, and whether their relationship is positive (interaction) or negative (reaction).

Figure 12.14 is an $L_{12}(2^{11})$ orthogonal array. This array is a very common, but also a special array. One important point about it is that all interactions are eliminated from the columns. Thus, it allows for 11 variables to be tested at once, each at two levels. With the full factorial number of experiments, the total number of experiments would be 2^{11} or 2,048, whereas this array requires only 12 experiments, a tremendous saving of resources and time.

$$L_{12}\,(2^{11})$$

No.	1	2	3	4	5	6	7	8	9	10	11
1	1	1	1	1	1	1	1	1	1	1	1
2	1	1	1	1	1	2	2	2	2	2	2
3	1	1	2	2	2	1	1	1	2	2	2
4	1	2	1	2	2	1	2	2	1	1	2
5	1	2	2	1	2	2	1	2	1	2	1
6	1	2	2	2	1	2	2	1	2	1	1
7	2	1	2	2	1	1	2	2	1	2	1
8	2	1	2	1	2	2	2	1	1	1	2
9	2	1	1	2	2	2	1	2	2	1	1
10	2	2	2	1	1	1	1	2	2	1	2
11	2	2	1	2	1	2	1	1	1	2	2
12	2	2	1	1	2	1	2	1	2	2	1
Group	1					2					

The $L_{12}\,(2^{11})$ is a specially designed array, in that interactions are distributed more or less uniformly to all colums. Note that there is no linear graph for this array. It should not be used to analyze interactions. The advantage of this design is its capability to investigate 11 main effects, making it highly recommended array.

FIGURE 12.14 An $L_{12}(2^{11})$ orthogonal array (with permission from the American Supplier Institute, Dearborn, Michigan).

Another useful and common array that we will use later is $L_9(3^4)$, with 9 experiments, 4 variables, each at three levels (Figure 12.15). The benefit of this array is that it allows for three levels for each variable thus enabling the designer to investigate for optimal values. For example, if the effects of a variable have already been tested between a lower and higher range, this array will allow the designer to test a value

$$L_9\,(3^4)$$

No.	1	2	3	4
1	1	1	1	1
2	1	2	2	2
3	1	3	3	3
4	2	1	2	3
5	2	2	3	1
6	2	3	1	2
7	3	1	3	2
8	3	2	1	3
9	3	3	2	1
	a	b	a	a
			b	b²
Group	1		2	

(1)

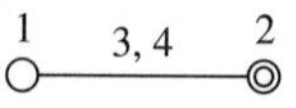

FIGURE 12.15 An $L_9(3^4)$ orthogonal array (with permission from the American Supplier Institute, Dearborn, Michigan).

in-between and determine what value in the range may yet render the optimal value. This test can be repeated for even more accurate values until the designer is satisfied with the final results.

Procedure for the Design of Experiments   In order to design experiments the designer may do the following:

1. Choose a performance criterion that will be measured as a measure of quality. For example, the spread (thickness or diameter) or shelf-life of a cookie, the strength of heat-treated aluminum, the transferred torque by a clutch, and applied torque on a bolt before it strips can all be measures of performance. This measure is used for comparing the effects of variables as they vary.

2. Select the variables, their values, and their level. This will determine what variables will be used in the experiments, at what levels, and with how many variations. For example, if the oven temperature is used as one variable, how many different values will be tested, and at what level (say 300 and 350 °F).

3. Based on the number of variables and their levels, select an appropriate orthogonal array for interactions and number of experiments.

4. Perform experiments and measure the performance criteria selected in 1 above.

5. Calculate the effect and importance of each variable as well as interactions between the variables on the performance.

6. Based on the results from the above, refine experiments by repeating as needed in order to get more accurate results.

We will use the following examples to demonstrate this approach.

EXAMPLE 12.7

The department of Food Science and Nutrition at Cal Poly has a pilot plant laboratory in which students learn the process of making food as it is done in industry. Among other products, students also make salsa. To prevent the salsa from spoiling and to meet government regulations, it is necessary to ensure that the pH of the mixture remains below 4.5, preferably about 3.4. In order to achieve this, the pH of the mixture of all ingredients but vinegar is measured. Next, vinegar is added until the desired pH is achieved. However, if the original pH of the mixture varies, a varying amount of vinegar must be added to achieve the desired pH, thus changing the final taste of the salsa. In order to reduce variations in taste, it is desired to maintain the uniformity of the pH level of the mixture before vinegar is added.

The pH level of the salsa mixture is affected by the characteristics of the ingredients, but also by their magnitude. When the ingredients are added according to the recipe, the accuracy of the measured amounts will affect the pH of the mixture. In order to reduce variations, we would like to determine which ingredient has the most effect, and if so, how severely.

The ingredients of the salsa mixture include tomatoes, tomato juice, onions, fresh peppers, herbs, salt, lemon juice, spices, and vinegar. In order to test the effect of different ingredients, samples of salsa were made in the Food Analysis Laboratory and the pH of each mixture was measured. For each sample, a fraction of the recommended weights of each ingredient were used. The same amount of lemon juice was added to all samples.

As mentioned above, we will select a performance criterion, variables, levels, and orthogonal arrays and will finally calculate the effects using the obtained results from lab tests.

Part I   In this part, we will consider the effects of only two ingredients (variables), each at two levels. This will require a total of 2^2 or four tests. For this part, the magnitude of tomatoes and peppers were selected as variables. As discussed earlier, we can use an $L_4(2^3)$ array.

Performance criterion: pH of the mixture

Variables: A: Tomatoes at two levels, (1) 40 gr, (2) 65 gr

    B: Peppers at two levels,   (1) 6 gr, (2) 9 gr

Interactions: AB between A and B

Number of experiments: 4 (2^2): Use orthogonal array L4 (2^3)

    [3 variables at 2 levels] Results are shown in Table 12.2.

TABLE 12.2   Results of the pH Values for Example 12.7, Part I

	A	B	AB	Result, pH
Experiment 1	1	1	1	4.25
Experiment 2	1	2	2	4.29
Experiment 3	2	1	2	4.20
Experiment 4	2	2	1	4.21

To calculate the effects, we will sum the values related to 1s and 2s in the table and subtract, as follows:

$$\text{Effect of factor A:} \quad (4.21 + 4.20)/2 - (4.29 + 4.25)/2 = -0.065$$

$$\text{Effect of factor B:} \quad (4.21 + 4.29)/2 - (4.20 + 4.25)/2 = 0.025$$

$$\text{Effect of factors A\&B:} \quad (4.29 + 4.20)/2 - (4.25 + 4.21)/2 = 0.015$$

Results: The magnitude of the tomatoes (A) has the bigger effect on the pH. Controlling it will also have the most effect on performance. The negative sign indicates that increasing the value decreases the result. The magnitude of the peppers has less effect. There is some interaction between the two. Although this result is useful, it is not adequate. We may need to investigate in more detail.

Part II   As you noticed, we only considered two ingredients in Part I. In this part we will add a third variable to see whether the other major ingredient is also important, and if so, how much. However, with three variables, more experiments will be needed. We will follow the same process as before to select performance criteria, variables, levels, and their effects.

Performance criterion: pH of the mixture

Variables: A: Tomatoes at two levels, (1) 40 gr, (2) 65 gr

    B: Peppers at two levels,   (1) 6 gr, (2) 9 gr

    C: Onion at two levels,   (1) 7 gr, (2) 10 gr

Interactions: AB, AC, BC, and ABC or other (please see Figure 12.13 graph 1).

Number of experiments: 8 (2^3): Use orthogonal array L8 (2^7) [7 variables at 2 levels] Results are shown in Table 12.3.

Please note that the major variables A, B, and C are listed in columns 1, 2, and 4 (not 3). Thus, experiment 3, for example, relates to levels 1,2,1 for tomatoes, peppers, and onions respectively, at 40, 9, and 7 gr each, whereas experiment 7 relates to levels 2,2,1 with 65, 9, and 7 gr each. Thus, you should be able to tell what values were used for each variable in each experiment. Next, we will calculate the average of the pH values obtained for 2-values and 1-values and subtract. Hence, for each column, the corresponding values will be:

TABLE 12.3   Results of the pH Values for Example 12.7, Part II

	1 A	2 B	3 AB	4 C	5 AC	6 BC	7 Other	Result, pH
Experiment 1	1	1	1	1	1	1	1	4.25
Experiment 2	1	1	1	2	2	2	2	4.28
Experiment 3	1	2	2	1	1	2	2	4.29
Experiment 4	1	2	2	2	2	1	1	4.35
Experiment 5	2	1	2	1	2	1	2	4.20
Experiment 6	2	1	2	2	1	2	1	4.24
Experiment 7	2	2	1	1	2	2	1	4.21
Experiment 8	2	2	1	2	1	1	2	4.26

$$\text{Effect of A} = (4.26 + 4.21 + 4.24 + 4.20)/4 - (4.35 + 4.29 + 4.28 + 4.25)/4 = -0.065$$
$$\text{Effect of B} = (4.29 + 4.35 + 4.21 + 4.26)/4 - (4.25 + 4.28 + 4.20 + 4.24)/4 = 0.035$$
$$\text{Effect of AB} = 0.020$$
$$\text{Effect of C} = 0.045$$
$$\text{Effect of AC} = 0$$
$$\text{Effect of BC} = -0.010$$
$$\text{Effect of column 7} = -0.005$$

Results: The magnitude of tomatoes has the most effect on performance, followed by onions and peppers (this is interesting because originally it was thought that onions may be more important than tomatoes. This shows how these experiments can clarify false assumptions and misconceptions). The interaction between A and B is also somewhat important. However, other interactions are small and can be ignored for now. The ABC interaction or ''others'' is also small, indicating that other ingredients can be ignored for now.

Part III   In this part, we will consider three levels for each variable. This will help us obtain a more accurate, more optimal value for each variable. To do so, we will select the same three variables, each at three levels, requiring 9 experiments.

Performance criterion: pH of the mixture

Variables: A: Tomatoes at three levels, (1) 40 gr, (2) 53 gr, (3) 65 gr

B: Peppers at three levels, (1) 6 gr, (2) 7.5 gr, (3) 9 gr

C: Onion at three levels, (1) 7 gr, (2) 8.5 gr, (3) 10 gr

Interactions: AB only. Others are small or insignificant.

Number of experiments: We will use orthogonal array L9 (3^4) [4 variables at 3 levels] Results are shown in Table 12.4.

Ordinarily, as shown in Figure 12.15, columns 3 and 4 would both be effects of 1 and 2. This means that if we assign A to column 1 and B to column 2, columns 3 and 4 will be effects of AB. However, if we assign A to column 1 and C to column 2, columns 3 and 4 will be effects of AC, which is zero. Thus, we can assign B to column 3 and leave column 4 for ''others.'' The results of our tests are summarized in Table 12.4.

The averages for the 1-values, 2-values, and 3-values for each variable are:

Tomatoes: 4.33, 4.32, 4.30 (average of 1-values, 2-values, and 3-values)

Peppers: 4.31, 4.31, 4.34

Onions: 4.32, 4.30, 4.33

TABLE 12.4   Results of the pH Values for Example 12.7, Part III

	1 A	2 C	3 B	4 Others	Result, pH
Experiment 1	1	1	1	1	4.32
Experiment 2	1	2	2	2	4.29
Experiment 3	1	3	3	3	4.39
Experiment 4	2	1	2	3	4.35
Experiment 5	2	2	3	1	4.32
Experiment 6	2	3	1	2	4.31
Experiment 7	3	1	3	2	4.30
Experiment 8	3	2	1	3	4.30
Experiment 9	3	3	2	1	4.30

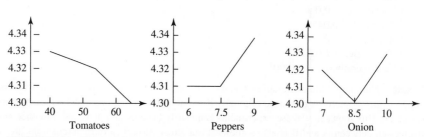

FIGURE 12.16   Plots of the results of experiments of Example 12.7, Part III.

Figure 12.16 is a graph of these results. When multiple variables at multiple levels are investigated, these graphs assist in visualizing the best results. Since for the salsa we strive for lower pH, we should look for levels that result in lower pH values without negatively affecting the taste by significantly altering the recipe. Based on these results, level 3 for tomatoes (65 gr), lower levels of peppers (6 gr), and level 2 for onions (8.5 gr) should give us the lowest pH without sacrificing the taste (we are still within the original range of the recipe). These small values can now be translated to the original bulk rates for the entire production.

One final comment about these experiments. As you probably noticed, part I is a subset of part II. One could either start with part I or part II. Part II requires more tests, but yields more results too. One can start with part I to get a direction, or simply start with part II. However, part III might not have been possible without part II because without prior knowledge about the interactions between A and C or A and B, we would not have been able to assign the variables as shown. We could only do this because interactions between A and C were found to be zero. In fact, originally, the tests for part III were planned and performed at the same time as parts I and II to save time. In that, A, B, and C had been assigned to columns 1, 2, and 3, respectively. Later, after the interactions between AB and AC were discovered, it became evident that these assignments would not work; C had to be assigned to column 2 and B to column 3. Hence, the tests for part III were repeated. ■

This is not a complete treatment of the power or capabilities of the orthogonal arrays, but an introduction. Please refer to the listed references for more information.

Procedure for the Application of Orthogonal Arrays to Product Design   Orthogonal arrays can also be used for the determination of a product's design parameters. For example, imagine that a product has three major

components whose values must be determined. Each component may be exposed to external sources of environmental and functional effects, called noise. In this case, we will need to determine the optimal value of each variable when exposed to each noise. Assuming that there will be need for N experiments to determine the optimal value for variables at each noise value, and there will be need for M experiments to determine the effects of each noise factor, the total number of experiments needed will be N × M experiments. This can become a huge undertaking. The process to do this is as follows:

- Define the design variables and their ranges
- Choose an appropriate array for design variables with N experiments
- Define the noise factors and their ranges
- Select an array for noise variables with M experiments
- Specify the performance criterion or result
- Arrange the two arrays around an M columns by N rows matrix
- Perform N × M experiments to determine the results of variables
- Compute and record "signal-to-noise" ratio S/N for each row based on the following:

  - When nominal value is best:

$$S/N = 10\log(\mu^2/s^2) \tag{12.8}$$

  where μ is the mean value and s is standard deviation of results for each row.
  - When smaller is better (e.g., cost):

$$S/N = -10\log\left[\left(\frac{1}{n}\right)\sum_1^n r_i^2\right] \tag{12.9}$$

  where r_i is each result and n is the number of experiments in each row.
  - When bigger is better (e.g., efficiency):

$$S/N = -10\log\left[\left(\frac{1}{n}\right)\sum_1^n (1/r_i)^2\right] \tag{12.10}$$

EXAMPLE 12.8

Imagine that two major parts of a jet engine are assembled together through a series of bolts. The designer is faced with a choice of number of bolts, size of bolts, and their grade (ultimate strength of the bolt material) as well as external environmental effects. The ultimate goal is to select these variables to maximize the life of the product. To aid the designer, the following is suggested:

Design variables:

        (A) bolt size A_1, A_2

        (B) number of bolts B_1, B_2

        (C) grade of bolts C_1, C_2

        Interactions as specified by the orthogonal array

        Array chosen: L8 (2^7)

Performance criterion: product life

Noise variables:

(E) Temperature $\qquad$ T_1, T_2

(F) Presence of chemicals $\quad$ low, high

Interactions as specified by the array

Array chosen: $L4(2^3)$

Results: A total number of $8 \times 4 = 32$ experiments will have to be performed. The experiments will be based on a table made up of the two selected arrays above, as shown in Table 12.5. The table will be filled with the results of each experiment, r_i.

TABLE 12.5   Combining the Two Arrays for the Design of an Experiment for Example 12.8

							EF	1	2	2	1	
			Noise factors		$\rightarrow$		F	1	2	1	2	
					$\rightarrow$		E	1	1	2	2	
												S/N Ratio
A	**B**	**AB**	**C**	**AC**	**BC**	**ABC**						
1	1	1	1	1	1	1						
1	1	1	2	2	2	2						
1	2	2	1	1	2	2						
1	2	2	2	2	1	1						
2	1	2	1	2	1	2						
2	1	2	2	1	2	1						
2	2	1	1	2	2	1						
2	2	1	2	1	1	2						
$\uparrow$	$\uparrow$		$\uparrow$									

Design variables

The results r_i from this table will be used in Equation (12.10) to calculate the signal to noise ratio S/N for each row for "bigger is better." The S/N yielding the largest number will be the best choice.

Similar to Example 12.7, additional tests may be devised for optimal values. ∎

12.11   SIX-SIGMA

Six-Sigma relates to a methodology for the inclusion of quality in the design and manufacture of products. In general, sigma refers to the standard deviation of a population. In statistics, standard deviation is defined as

$$\sigma = S = \sqrt{\frac{\sum x^2}{N}} \tag{12.11}$$

where sigma (σ or S) is the standard deviation, x is the individual deviation of one data point from the mean of the population, and N is the number of data points in the

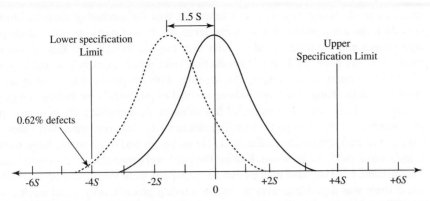

FIGURE 12.17   The Six-Sigma production goals are to achieve a process capability index of larger than 2 which results in fewer than 3.4 defects per million parts.

population. Standard deviation is a measure of distribution of data points. 99.73 percent of all data points are within $\pm 3S$.

As discussed earlier, Equation (12.7), repeated here, shows process capability index C_p. This is a measure of the capability of the process to produce parts that meet design specifications as set by the designer:

$$C_p = \frac{\text{Tolerance}}{\pm 3S} = \frac{\text{Tolerance}}{n} \tag{12.7}$$

where n is a range other than $\pm 3S$. The larger the process capability index C_p, the higher the possibility that the part will meet the design requirements. Many companies set C_p equal to 1.33, which translates to $n = \pm 4S$ (or more correctly, to smaller S for the same tolerances). Now assume that the production standard is set at $\pm 3S$. As Figure 12.17 shows, this appears to be satisfactory, as practically most of the products fall within specifications. However, in general, the assumption is that manufacturing processes may shift as much as $\pm 1.5S$. The discrepancy between the shifted population and the desired tolerances can be as much as 0.62 percent, or 6,200 parts per million, which is unsatisfactory. Setting C_p equal to 1.66 which translates to $n = \pm 5S$ will reduce defects to about 200 parts per million. Top quality companies set their process capability index C_p equal to 2 or more, which translates to $n = \pm 6S$. This yields product defects below 3.4 parts per million, a lofty goal which sets these products apart from their competition. This methodology is referred to as Six-Sigma and requires extremely low variations in products, the same concept as advocated by Deming and Taguchi.

12.12   LEAN PRODUCTION

One of the major attributes of the Japanese methods of production is quality of the products. This level of quality has been achieved by the Japanese industry as a result of their application of the quality control methods advocated by Deming as well as Taguchi, but also the philosophies of lean production advocated by large companies such as Toyota and others. The methods of lean production have now been implemented in the United State as well as parts of Europe, improving the quality of their products.

Lean production was first started in Toyota and is also referred to as the Toyota Production System. It came about when in 1950 Eiji Toyoda, a young engineer and a member of the family that owns Toyota visited Ford's Rouge plant to learn about

American productivity. One of his observations was the method of manufacturing body parts. In large mass production systems, large batches of parts are cut and stamped by large presses and are inventoried for later use. Toyoda thought that this system was not practical for his plant, as Toyota's production level was extremely low and he could not afford to buy many large presses. He, together with his production counterpart Taiichi Ohno, set out to change the large volume method of production by finding and perfecting a method to change dies in a matter of few minutes, thus enabling the system to produce large numbers of different parts in small quantities with only one press. What they noticed was that this had two positive effects: (1) He no longer had to inventory large numbers of manufactured parts for later use. Rather, the parts were manufactured in small quantities when needed. (2) They also noticed that because parts were manufactured in small quantities, if there was a problem with the manufacturing process, only small number of parts would be defective, allowing easy correction. This saved the company much loss. These observations became a basis for developing a method of production that is dubbed *Lean production.*

Lean production,[7] as opposed to common mass production, is based on the philosophy that quality must be integrated into the process of production. To understand the difference between the two, let's first review the concept of mass production as proposed by Henry Ford, and as perfected by the thousands of companies that have used it since then. In mass production, the idea is that the job of production is divided into small portions, where each person performs a part of it, and as the product goes through an assembly line, individual parts are assembled together to create the whole. Each person works on the part for which she or he is responsible and should try to do it as directed.

However, this individual has no role on the rest of the process and neither receives, nor sends, any feedback to the rest of the assembly line. This method was first used by Henry Ford to produce thousands of inexpensive cars, replacing job-shop, hand-made cars that were expensive and difficult to make. In order to be able to do this, unlike the individually produced cars, in which parts were manufactured and fitted individually, it was necessary to mass produce the parts. Tolerances were used to specify the minimum and maximum acceptable values for each dimension so that they could be assembled together. However, as we have already seen, mating parts that happen to be at the extreme values of their range of tolerances still pose a problem. This is what was discovered by Frank Pipp's team at Ford[2] when the Toyota truck was found to be all snap-fit, whereas Ford's trucks and cars were not; Ford's workers needed to use mallets on a regular basis for assembly. Variations in tolerances caused fitting problems, even if within their acceptable range.

Additionally, depending on the quality of manufacturing processes and the level of quality control in the factory, it is possible that some parts may fall outside the acceptable range of tolerances, but not detected. These parts will eventually find their way to the assembly line. This happens when statistical methods are used for quality control instead of 100 percent inspection, where, only a sample of produced parts is measured for accuracy. In this case, if, say a lathe, produces faulty shafts that are not detected, these shafts will eventually get to the assembly line. When parts come to the workers at their station on the assembly line, the worker will try to assemble the faulty parts into the product without success. However, the basic principle of mass production is speed. Thus, when a part does not fit, the particular task cannot be finished. In this case, the product is removed from the assembly line, moved into a ''re-work'' area, and is tagged for hand fitting or completion later, a very costly endeavor. Alternately, the product may continue on the production line, assuming that at the end of the line, the mistake will be caught and corrected. The worker will not stop assembly to find why the parts do not fit, thus providing no feedback. The work goes on as scheduled, even if many incomplete products are partially assembled and sent to the re-work area. Stopping the assembly production line is the responsibility of the senior management only. Each time the assembly line is stopped, there will be need for overtime work to compensate the loss in production, thus adding to the loss.

Most production assembly lines that do not use lean production principles have large re-work areas dedicated to finishing the partially assembled products with skilled workers who adjust, re-work, and hand-fit parts. This is in fact a very expensive and time-consuming task. However, this is a basic premise of mass production with assembly lines; the job cannot stop for every small mismatch or for every part that does not fit or is of inferior quality. Speed is the key factor in lowering the cost, and thus, the assembly must go on as fast as possible.

In lean production, the idea is that everyone in the assembly line has total control on the assembly. In other words, if due to any reason, a part at any station in the assembly line does not fit, the worker stops the whole production until the source of problem is found and corrected. This provides instant feedback to the earlier stations that are producing faulty parts. They stop production and immediately look for the cause of problems. The assembly stops until the source is found and it is fixed, changed, or corrected. The result is that there are practically no incomplete products, and almost no re-work is required.

Initially, this method of production seems to be very drastic with potential to cause chaos in the assembly line. Imagine that for every part that does not fit or is not acceptable, the whole production process would simply stop. Further, imagine that any worker on the assembly line could instantaneously stop a factory's operation because a bad part is found (in fact, a cord is provided above each station, where the worker can pull the cord to stop production). How could the assembly line ever work with this chaos? However, when everyone knows that they are totally responsible for the whole production to go smoothly, everyone feels that they must produce the best parts. One part that one person produces in one station will be judged immediately by all the succeeding stations. No one would want to be the source of this stoppage. Thus everyone will do their best possible work to ensure that no part, not even a single part, will leave their station or their unit in unacceptable condition. Consequently, the quality of the whole product increases drastically. On the surface, this seems to be an extremely competitive and stressful environment in which to work, contrary to one of the Deming's principles. Deming stresses the importance of a friendly environment, abolishment of merit salaries, and cooperation rather than competition. However, the idea here is not to judge individuals, but the processes. Hence, if a part is unacceptable, it is not the person who is at fault; it is the process that causes the problem, and thus, the person in charge will try to immediately fix the problem. In this environment, workers cooperate with each other to improve the quality so as not to have to stop assembly, not because their salary is in jeopardy. Thus, they have the power to make suggestions to the engineers and managers for improvements, they take part in decision making, and they have pride in their accomplishments. The result is improved quality and reduced cost.

The idea of lean production has roots in Deming's way of thinking, where slogans like quality is job 1 is not needed because everyone is automatically responsible for quality anyway. If you go back and look at the Deming's 14 principles as well as Taguchi's Methods, you will find many of their philosophies and ideas in line with lean production and its consequences (that will be discussed next).

Lean production applies to product development as well. Its basic premise is simplicity and effectiveness. During the process of evaluation, every activity must be scrutinized to determine whether it adds value to the overall development process or not. If not, it is modified or eliminated. For example, suppose that the traditional method of product development in a company is to have the design be reviewed by different departments.

This will add to the cost and time-to-market since one department will not be able to review the design until another is finished. Communication between departments will eliminate this hurdle, shortening the time-to-market and reducing the cost. Concurrent engineering is one concept that helps this too.

Lean production has a snow-balling effect on all other aspects of production as well. This way of thinking affects final cost, suppliers, warehousing, workers, customers, labor relations and unions, and many others. In the following sections we will discuss some of these consequences.

12.12.1 Effects on Quality and Cost

As was discussed above, lean production requires that every part and subassembly be in such a desired state that, during succeeding assembly and production operations, workers and assemblers will encounter no problem in putting their parts together and advancing the state of the product. This requirement forces every part to be at its peak level of quality. When every part and subassembly has a high level of quality, the quality of the product will also improve. Since the quality of the product improves, there will be less re-work and corrections needed, reducing production costs. Since customers will be happy with a higher quality product, they will have a higher level of loyalty to the company and will return for more business. The consequence is a lower total cost to society, benefiting both the manufacturer and the society.

In common production systems the assumption is that problems are random events. Thus, if a part does not fit or an assembly is defective, the next one should be fine and free of defects. Thus, there is no need to stop production on the account of one bad or unacceptable part. Lean production assumes that if a part is bad, the next ones are probably the same, thus production should stop until the problem is fixed, and it must be done now not later. Thus, when a problem is found and production is halted, the whole production team comes around to help fix the problem.

Womack et al.[7] talk about the U.S. and European automobile manufacturers and lean production. As of the writing of their book, most European (and American) manufacturers had not adopted this method, thus they had extremely large re-work areas, as much as 20 percent of the production area. In some cases, some European manufacturers bragged about the fact that their "craftsmen" were hand-fitting their cars. In fact, this means that they had problems with manufacturing quality, and consequently, their craftsmen had to hand-fit the parts that could not be assembled during regular assembly operations. In some plants, as much as 25 percent of production time is spent on re-work. Womack et al. also mention the disastrous situations that Volkswagen and Renault encountered when they used common methods of production in their assembly plants in the United State and Mexico. For example, Renault bought American Motors in 1979 with the hopes of establishing a low-cost presence in the United State. However, using the same methods of production, they could not improve the quality of their products, and consequently, suffered loses. In 1987 they sold their operations to Chrysler at a fraction of their cost. However, Honda and Toyota have successfully opened production plants in the United State with the same quality automobiles. This shows that it is not the workers but the process that affects the quality.

12.12.2 Effects on Suppliers

In a common production system, parts are either manufactured in-house or they are subcontracted to other companies who manufacture parts or complete subassemblies.

These parts and subassemblies are then assembled together in the final production section into the final product. In many industries, including the automobile industry, this is a very common practice. In these cases, the drawings and specifications are drawn up by the engineers and designers at the company, and subsequently, are sent to bidders for pricing. The pricing is usually contingent on a set maximum number of defective parts, usually per 1,000. The company with the lowest bid wins a contract. This process is usually handled by the purchasing department of the company through their buyers.

This process has at least three problems associated with it (we will discuss other problems next). One is that since the lowest bid is granted for production, the company who is making the part or subassembly tries to present the lowest bid to win, and thus, may sacrifice quality in favor of lower costs. As long as the maximum number of defective parts delivered is lower than the number in the contract, the quality no longer matters. The supplier tries to make more money by delivering the lowest, and thus apparently cheapest, quality. There is no incentive for improving the quality of the part or subassembly. In fact, if the price of the part increases due to improved quality, the suppliers may lose their contract with the company.

Second, since lower cost is the driving factor, the buyers at the company constantly look for lower bids. In reality, in most cases, the buyer's year-end bonus is related to how much money they save the company by finding lower bids from alternate suppliers. This seemingly positive cause for competition is actually a detriment. Companies do not try to improve the quality of their products to win a contract; they try to lower the costs. There is no incentive for better quality. The consequence is that the supplier, who has won this round of bids, may lose the bid the next time around, or in the absence of a long-term contract, lose the contract any day. In that case, there is still no incentive for the supplier to improve the quality of their part or subassembly by investing in better production systems, machines, or skilled workers. They may lose their contract any day, so why should they spend money investing in quality improvements? Their interest is in lowering costs not improving quality.

Third, there is generally no long term relationship between suppliers and the final assembling company. As soon as a less expensive supplier is found, the relationship is severed. Therefore, the culture of the company is not transferred to the suppliers, their engineers and management do not know each other, and their relationship is only based on contracts.

In contrast, the same basic philosophy of lean production can be extended to the subcontractors and other suppliers. Imagine that the company would instead look for suppliers that follow the same rules of lean production. Instead of looking for the lowest bidder, the company will look for suppliers that deliver the best-quality parts and subassemblies for their price. Thus, it is the quality standard that results in contracts not prices. The buyers of the company will not simply switch to new cheaper suppliers to save a little money. They look for best product quality. However, since they do not blindly follow price, they end up in long-term relationships with their suppliers. The suppliers know that their contracts are safe, so long as their product is improving. Since they are not worried about losing their contract, they dare invest in new technology, better equipment, and improved techniques in order to improve their products. Their investments will be paid off by having long-term contracts, by continuing to supply parts, and by reducing their employee turnover and training costs.

In most cases, there is a continuous and friendly relationship between the company and their suppliers. It is in the best interest of the company to improve the quality of their suppliers, and the suppliers do strive to improve the quality in order to keep the company happy. In contrast with the first case where the specifications and drawings are sent to the suppliers, in lean production the suppliers are generally a part of the design team as well.

Considering the fact that suppliers are much more experienced in their own business, their expertise is a major addition to the team.

The story of Ford and Firestone tires, mentioned earlier, is a good example. Ford and Firestone had a long-term relationship for close to a century. When during the Ford Explorer fiasco both companies blamed each other for the problems, and neither one accepted the fault, this relationship was severed. Ford had reduced the tire pressure specification of the Explorer without Firestone's consent in order to lower the center of gravity of the car and to reduce the danger of rollover. The tires were manufactured with low quality with an increased chance of exploding because of increased heat production due to low tire pressure in hot whether.

12.12.3 Effects on Warehousing and Supplies

One of the consequences of lean production and lean supplier chains is that parts and components may not, and cannot, be ordered in large quantities and be warehoused. Large orders and warehousing are contrary to the basic philosophy of lean production that if there is a problem with a part, production must stop until the problem is fixed. If a large quantity of parts or subassemblies is ordered and warehoused for eventual use, and if it is discovered that a part or subassembly is defective in one respect or another, what will be done with all the inventories and warehoused parts? In addition, the assembler can either stop production completely and wait for the supplier to find the problem, devise a fix, and reproduce the parts and deliver for another inventory or produce partially completed products that are kept in re-work area for eventual completion, or continue making defective products. None of the above is satisfactory, and none matches the basic premise of lean production. Instead, the idea is that parts and subassemblies are produced and used in succession, such that if there is a problem, it can immediately be solved and remedied. Consequently, it is necessary that only enough parts be delivered to the assembly line to satisfy production for a very short time (in many Toyota plants this is about a 1-hour supply). Another batch must be delivered as soon as the present batch is used. At Toyota, this concept of *just-in-time* and *day-to-day* delivery is called *kanban*. In practice, a container system is used, in which parts are delivered to the assembly line. The container is returned to the previous step of production from which it came, signaling the need to produce another batch while the next container is used. Thus, just-in-time production is implemented by need.

On the contrary, in common mass production systems, the company may purchase large quantities of manufactured parts and keep them in inventory and use the parts as time goes by. Seemingly, this enables them to buy the parts in larger quantities and keep their purchase price down. However, as mentioned above, in addition to the cost of inventory which is high, if the purchased parts are defective, large numbers of products will remain incomplete requiring large re-work areas, the whole production may come to a halt, or defective parts will be produced. In lean production, subcontracted parts from suppliers are purchased continually, sometimes delivered many times a day. In this case, if one batch is defective, it can be quickly reported back to the supplier, who in turn, will correct the production and will supply desirable parts to the factory. Thus, not only the cost of inventory is eliminated, the assembly is far less sensitive to defective supplies too. On the other hand, in this case, the factory is much more sensitive to the lack of responsibility by the suppliers. In other words, since there are no extra parts available in the system, if the supplier does not deliver his parts on time, the factory will stop production. This includes the possibility of strikes by the suppliers' workers. As a result of this dependence on the suppliers' quality of work and the treatment of their workers, the factory can expect that the suppliers will also follow the same rigorous lean production methods, increasing their own quality. Otherwise, they will lose their contract with the factory. This results in a domino effect, strengthening relationships between suppliers, assemblers, and factory workers.

Another problem with large orders in common mass production systems is lack of response due to fluctuations in delivery requirements. Suppose that an order is received for a large batch of parts and is subsequently delivered by the subcontracting company. What should the company do next? Should it wait until a new order is received before continuing to produce more? Should it assume that another order will be received in due time, and thus, continue production and warehouse the parts for future delivery? Should it stop production until an order is received, at which time the production starts in a hurry? What should it do with the workers during these times? Should it lay off the workers, or should it continue to pay them until the next order is received? Can new skilled workers be hired quickly and trained for production? As the need for production varies in time, adjustments must be made which can affect both profits and quality. However, in the lean production supply chain, production is steady and continuous and small variations can be dealt with easily. Thus, the supplier can rely on a uniform delivery schedule and plan ahead for best quality and maximum profit. There is no inventory to keep and pay for, there is no need to wait to receive orders, and there is no need for constantly laying off or hiring workers. There is no question that the just-in-time delivery puts a lot of stress on the supplier since the production of the next assembler is dependent on its delivery. But on the other hand, a uniform and continuous delivery schedule allows the supplier to plan ahead and ensure reliable delivery of quality parts and subassemblies with a predictable schedule.

12.12.4   Effects on Workers

The domino effect of consequences stemming from lean production continues downstream from suppliers to inventories and supplies and to workers. Lean production affects the philosophy behind the relationship between the assembler and supplier and the supplies. Similarly, it has a fundamental effect on the treatment of workers and the relationship between the management and the workforce. As was mentioned earlier, the friendly versus adversarial relationship between the managers and workers is one of the fundamental points in Deming's 14-point philosophy as well.

As mentioned, one of the consequences of lean production is that parts and subassemblies are manufactured and delivered just-in-time. This means that there is a very small inventory of parts available for just-in-time assembly. There needs to be a fresh batch delivered just-in-time for the next interval, sometimes within the hour; otherwise, production will stop. The result is that there is very little, almost zero, tolerance for non-delivery of the next batch or container. The supplier must continually deliver parts and subassemblies to the assembler. Then what will happen if workers go on strike due to unsatisfactory conditions in the workplace? What will happen to the stream of parts that must be manufactured and delivered if workers stop working? The result will be disastrous. A few years ago UAW members went on strike in part-manufacturing plants in the United State. A few days later Chrysler assembly plants in Canada ran out of parts and had to stop work until the strike ended.

The response on the part of employers is then to try and keep the workers happy and satisfied. They must create a family oriented environment of mutual respect and shared values. The whole enterprise must be treated as one big family working in unison. Workers are part of the system, not just in theory, but in practice too. They take part in the day-to-day decisions about the production and supplies, they are part of production teams, their views are respected, and they are treated as part of a family. The result of this is that the adversarial relationship that exists between management and workers in mass production systems is replaced with one that is based on respect and shared values, leaving no need for unionization. Lean production companies do not have unions, even in the United State. There is no need for them, and workers do not ask for one.

Additionally, workers are normally employed for life. Not only is the employer expected not to lay off workers, the workers are expected to remain with the company for life too. This is contrary to the normal relationship in mass production systems, especially in the United State, where workers at all levels leave one company for another for more pay or better benefits, and companies encourage this by hiring from their competitors. Hiring employees for life results in more care about the future of the company and its well-being, long-term relationships among workers and management, and a sense of security. Additionally, a system of seniority-dependent salary structure discourages transfers from one institution to another; as a worker leaves one employer for another, his or her salary will be lowered due to lower seniority in the new firm.

The final result of this improved relationship is improvements in quality of the products the company produces.

12.12.5 Economic Effects

The obvious benefit of lean production systems is improved quality of products produced by the enterprise. In a long-term point of view, this revitalizes the economy, increases market share, increases the export capability of the company, develops a satisfied and loyal customer base, and establishes the company as a leader in its particular industry. No one expected that Toyota would achieve its lean production success overnight. In fact, at the beginning of this endeavor, production was halted constantly. But as quality improved, the total number of work-stoppage occurrences decreased significantly. Today, there is hardly any need for halting production because problems are resolved immediately when they occur and before they become significant. Even though workers still have their cord to pull and stop production, they hardly even need it. In 1955, Toyota had an almost zero market share in the U.S. auto market. Today it is a different story. The economic effect of lean manufacturing on the Japanese industry has been enormous and cannot be disputed. Others are catching up.

It should be mentioned here that there is no intention to praise or criticize any car manufacturer. All major manufacturers have their own strengths and weaknesses and design and manufacturing successes and failures. Just recently, the case involving the gelling problem of Toyota and Lexus engine oil has surfaced with potentially devastating results.

There is much more to lean production. For more information about its history, the beginnings, and the state of global auto manufacturing, please refer to References 7 and 8.

REFERENCES

1. TRIBUS, MYRON, "Deming's Way,"*Mechanical Engineering*, January 1988, pp. 26–30.
2. KEARNS, DAVID, D. NADLER, "Prophet in the Dark," Harper Business, New York, 1992.
3. COLLINS, JIM, "Good to Great," Harper-Collins, New York, 2001.
4. RANJIT, ROY, "A Primer on the Taguchi Method,"Van Nostrand Reinhold, New York, 1990.
5. TAGUCHI, GENICHI and E. ELSAYED, T. HSIANG, "Quality Engineering in Production Systems,"McGraw-Hill, New York, 1989.
6. ROSS, PHILLIP J., "Taguchi Techniques for Quality Engineering,"McGraw-Hill, New York, 1996.
7. WOMACK, JAMES P., D. T. JONES, and D. ROOS, "The Machine that Changed the World," Rawson Associates, New York, 1990.

8. MORGAN, JAMES M. and J. K. LIKER, "The Toyota Product Development System," Productivity Press, New York, 2006.

9. CROSBY, PHILIP B., "Quality is Free," McGraw-Hill, New York, 1979.

10. http://www.boeing.com/defense-space/space/propul/RS68.html.

11. FREY, Daniel D. and E. M. GREITZER, "One Step at a Time," *Mechanical Engineering*, July 2004, pp. 36–38.

12. "Orthogonal Arrays and Linear Graphs: Tools for Quality Engineering," American Supplier Institute, Inc., Center for Taguchi Methods, ASI press, Dearborn, Michigan, 1987.

13. DIXON, JOHN R., and C. POLI, "Engineering Design and Design for Manufacturing: A Structured Approach," Field Stone Publishers, Massachusetts, 1995.

14. REHG, JAMES A., "Computer Integrated manufacturing," Prentice-Hall, New Jersey, 1994.

HOMEWORK

12.1 Assume that the desired dimension for a part is $d = 5$ in. The selected tolerance for this dimension is ± 0.001 in. It has been determined that the cost (loss) for a part outside of this limit is $6.00. Determine the loss function for this part. Also determine the associated cost to society for a part that is at 4.9993 in.

12.2 For Problem 1 above, the mean value of a batch of 500 samples is 5.0005 in. with a standard deviation of 0.0002. Calculate the average loss per part and total loss for the whole population.

12.3 Repeat Problem 2, but assume that the mean value is 1) 5.0003 and 2) 5.000. Compare the results.

12.4 Repeat Problem 2, but assume that the standard deviation is 0.0001 in. Compare the result with Problem 2.

12.5 For Example 12.2, calculate the corrected standard deviation and associated loss if a tool wear width of 0.0002 is allowed.

12.6 Assuming that the value of constant $k = 1.05 \times 10^6$, for a tool wear width of 0.00015 in., find the required nominal standard deviation that should be expected if the loss per part is not to exceed $0.05. The average value of the parts is the same as the nominal value.

12.7 The elastic constant for a spring used in a product is selected to be $2 \pm 0.01 \, \text{N/mm}$. The cost of adjusting the device for parts outside of this range is $2 per machine. Determine the loss function for this part. Also calculate the cost to society for a spring that is measured to be at 2.005 N/mm.

12.8 Assuming that a batch of 1,000 springs of Problem 7 have a mean value of 2.003 N/mm and a standard deviation of 0.002, calculate the average loss per part and total loss for the whole population.

12.9 Repeat Problem 8, but assume that the mean value is 2.00 N/mm. Compare the results.

12.10 Consider the springs of Problem 7. As a result of improvements in manufacturing, the distribution of the spring constant was improved such that its standard deviation changed from 0.003 to 0.002.

Calculate the improvement in its process capability index. Can the company achieve ± 6-sigma quality? If not, determine the standard deviation that must be achieved to ensure ± 6-sigma quality.

12.11 In order to investigate the effects of major ingredients in a cookie recipe on its shelf-life the following experiments were conducted:

Assumptions: No preservatives. Shelf life affected only by fresh ingredients used.

Performance criterion: Rancidity (days)

Variables: A: Oil used, two types, (1) shortening,
(2) vegetable oil

B: Eggs, two levels, (1) 1 egg,
(2) 2 eggs

Interactions: AB between A and B

Number of experiments: 4 (2^2): Use orthogonal array L4 (2^3) [3 variables at 2 levels]

TABLE PROBLEM 12.11   Rancidity Tests

	A	B	AB	Rancidity, days
Experiment 1	1	1	1	14
Experiment 2	1	2	2	13
Experiment 3	2	1	2	11
Experiment 4	2	2	1	10

Determine the effects of each variable, what is more important, and whether the results make sense.

12.12 In order to investigate the effects of major ingredients in a cookie recipe on its shelf-life, the following experiments were conducted:

Assumptions: No preservatives. Shelf-life affected only by fresh ingredients used.

Performance criterion: Rancidity (days)

Variables: A: Oil used,   (1) shortening,
two types,   (2) vegetable oil

B: Eggs, two levels, (1) 1 egg, (2) 2 eggs

C: Nuts used   (1) no, (2) yes

Number of experiments: 8

Rancidity test results in days:

1,1,1	14	1,1,2	12
1,2,1	13	1,2,2	10
2,1,1	11	2,1,2	8
2,2,1	10	2,2,2	6

Use an appropriate array and calculate the effects of each variable as well as possible interactions between them.

12.13 The strength of an experimental cable in torsion is related to the number of strands used and the helix angle of the strands. To determine the effects of each variable, a series of test were performed with the following results. Choose an appropriate array and calculate the effects of each variable and possible interactions between them.

Number of strands:	(1) 15	(2) 21
Helix angle:	(1) 15°	(2) 30°
Strength in lb-ft:	1,1	100	1,2	80
	2,1	110	2,2	95

12.14 The strength of an experimental cable in torsion is related to the number of strands used and the helix angle of the strands and possibly to the two suppliers who supply the material used to manufacture the cable. In order to determine the effects of each variable and the role the raw material plays in the final quality of the cable, a series of tests were performed on the cable, as shown below. Determine the effects of each variable and possible interactions. Also determine what might be done if an optimal value for each variable is sought.

Number of strands:	(1) 15	(2) 21
Helix angle:	(1) 15°	(2) 30°
Suppliers:	(1) JD	(2) ST
Strength in lb-ft:	1,1,1	100	1,1,2	95
	1,2,1	75	1,2,2	78
	2,1,1	120	2,1,2	116
	2,2,1	100	2,2,2	103

12.15 Consider a product with which you are familiar. Select an appropriate number of variable, levels, and a performance criterion for the product. Pick an array that would appropriate and determine the tests that will be needed in order to study the variables and their effects. Write down the characteristic values of each variable for each experiment.

CHAPTER 13

DESIGN AND PRODUCT LIABILITY

Murphy's Law: If anything can go wrong, it will.

Murphy's Law of Jelly Bread: If a jelly-bread falls down on an expensive carpet, it will invariably fall with its jelly face on the carpet.

—*Unknown source*

In the case above, the fault is with the person who put the jelly on the wrong side of the bread.

—*Danny Kaye Show*

13.1 INTRODUCTION

Products and designs, like all other aspects of personal and professional life, carry a certain liability if they fail to perform as indicated or expected. The liability may be as simple as providing an opportunity for the user to return a product or ask for the repetition of a service, or may be as expensive as a judgment in favor of the plaintiff with exorbitant punitive damages. In all cases, the designer should consider product and design liability as a design issue that must be addressed during the design phase.

Product liability has changed significantly during the century. According to the *Mechanical Engineering magazine*,[1] there used to be a time when there would be an ax hung on the wall near rotating machinery in manufacturing plants, just in case someone's hair, clothing, or body parts would be caught in the machines and needed to be chopped off; the managers did not want to shut down the whole plant (which in those days were run by overhung shafts, pulleys, and belts, driven by a central motor) in favor of saving a worker. During the latter part of the nineteenth century, some 400 workers in a hosiery factory were locked into the factory every day until the end of the day, when the manager would open the door to let them out.[2] Imagine what would happen if the factory caught fire. Today it appears that almost any accident is someone else's fault, and they are liable for it. We have gone from one end of the spectrum to the opposite end, from having the individual responsible for everything that happens to him or her to holding a manufacturer liable for all damages and injuries. What we hear in the news are insane judgments and awards, illogical cases and arguments, and emotional charges against others. We have gone from severing someone's arm that is caught in rotating machinery to holding a fast-food restaurant liable for the spilled hot coffee on a woman's legs when she holds the hot coffee between her legs while driving, blaming the restaurant for providing coffee that was too hot. However, we must realize that unlike sensational news, real liability cases have forced manufacturers and designers to be more careful, to design and manufacture safer and better products, and to respect the customers' and the users' abilities and limitations. We have already seen the effects of human factors in the safety of products. We have also seen how good products function better, are more efficient, and are safer. Therefore, it is unfair to only see the negative aspects of the system in the United States without considering the positive effects it has had on product safety. Still, there is no question that at times, and in certain cases, things have gone awry, and that as a result of unreasonable product liability suits, many manufacturing plants and companies have been shut down and individuals ruined. However, as design engineers, we must understand the issue, incorporate the necessary elements into our designs, and produce better and safer products.

In this chapter, we will study some of the basic issues, definitions, and requirements that relate to product and design liability. It is impossible to predict how a product will be judged in a court, and whether a jury will find the behavior of the designer acceptable or not. Nonetheless, the designer's behavior and the decisions that she or he makes may be judged in a court. Therefore, the designer must have a basic idea about what is expected of him or her.

13.2 PRODUCT LIABILITY: HISTORY AND BACKGROUND

Historically, the marketplace was governed by "Let the buyer beware" (Caveat Emptor), or "What you see is what you get." In early markets, the designer and manufacturer of most household and agricultural products was also the seller of the product. The buyer was responsible for understanding what the product was, and would assume the responsibility of whatever happened to the product or from it, unless there was an actual contract between the seller and the buyer (called privity). If anything happened resulting in damage or injury, the only remedy was based on the contract between the seller and the buyer. The privity provided for claims against the seller, based on the contract and what he could afford. This, in fact, is true even today in many societies; when you buy a product, you assume the responsibility for the product. There is no return policy if you do not need the product or if you are not satisfied with it.

As industry advanced, larger companies with mass production capabilities were formed. Since they could not directly deal with the consumers, distribution systems consisting of distributors and retailers were formed. As a result, privity (direct contractual relationship) no longer existed between the buyer and the manufacturer or the distributor, but with retailers who generally had nothing to do with the performance or safety of the product. As a result, with the absence of privity with the manufacturer, the original concept for product liability could not work; the buyer could only bring a complaint against the retailer, not the manufacturer.

In 1916, an individual named MacPherson brought a suit against the Buick Motor Company because the wooden spokes of his car failed and injured him. The court determined that the manufacturer, although not in direct privity with the consumer, was at fault, because they had manufactured a product that was unsafe. With this ruling, the shield of privity was removed when a product was unsafe, enabling the users to sue the manufacturers instead of retailers.

In 1961, a Ms. Henningsen bought a new car from Bloomfield Motors, Inc. and drove it for a few days before the steering mechanism suddenly failed and injured her. A suit was brought against the manufacturer. The suit was based on the premise that she was not knowledgeable enough about a complicated product such as a car to examine it in detail to determine whether it was safe or not before she purchased the product (let the buyer beware or caveat emptor). As a result she had to rely on merchantability of the product, implied by it being offered for sale. The car had a defect and should not have been sold. The court ruled in favor of her, and thus, the shield of privity was removed from all warranties, and the concept of merchantability was enforced. After this case, it is implied, even if not expressed, that a product offered for sale is safe and that it works (implied warranty).

This concept was later interpreted as a case of strict liability in tort, meaning that if a manufacturer placed a defective product into the market which could cause damage or injury, the manufacturer was strictly liable for the consequences of the defect, regardless of privity, warranties, disclaimers, etc. Causing wrongful harm to another person or his or her property is a civil wrongdoing and is called tort (as opposed to contract violation), and the legal theory is called strict liability in tort.

In 1963, there was a definite shift from contract law to tort law as a result of a case (Greenman vs. Yuba Power Products, Inc.) in which a homeowner used an attachment in connection with his powered wood-working tool enabling him to use it as a lathe. During normal operations, the wooden piece flew out of the attachment and severely injured him in the head. Although the case was brought against the manufacturer based on negligence and breach of warranty, the court decided that the correct theory was strict liability. The decision was based on the belief that the purpose of product liability is to ensure that the costs of injuries caused by products are borne not by the injured, but by the manufacturer (called *Caveat venditor*, or let the seller beware). The effect is that this cost to the manufacturer is eventually borne by all buyers through higher product costs either as insurance premiums or direct costs figured into the price of a product.

In the case of May vs. Columbian Rope Co, a new rope severed from a scaffolding, plunging May into the ground. The defendant could not point to a defect in the rope. However, the court ruled that regardless of what caused the rope to sever, the manufacturer of the rope was at fault, as the case spoke for itself (*res ipsa loquitor*). This became the basic theory for what is called absolute liability, in which, regardless of fault or cause, the manufacturer of a product that causes injury or damage is held liable for the product.

Later cases have reduced the role of contributory negligence on the part of the user from liability cases. As an example, the courts have ruled in favor of a person who was injured when he used a lawn mower as a hedger by turning it sideways and using it to hedge bushes. There is no doubt there will be further changes in the liability laws as time goes by.

13.3 THE LEGAL SYSTEM OF THE UNITED STATES

The United States government consists of three branches, each with a different responsibility:

- Legislative: *This branch consists of the Congress, representing the people of the United States, and has the responsibility to create and enact the substantive laws that govern the nation, including taxes. The substantive laws relate to the statutes or changes in statutes that create, define, and regulate the States' and the citizens' responsibilities and rights.*

- Judicial: *This branch consists of the United States courts and includes the Supreme Court, the 11 circuits of the U.S. Courts of Appeal, and the 94 U.S. District Courts. The judicial branch has the responsibility to interpret the laws and the statutes.*

- Executive: *The executive branch issues orders and regulations at the State and Federal levels. The administrative branch has broad regulatory powers over industries, including utilities, communication systems, transportation systems, and others. It provides oversight through agencies such as the Food and Drug*

Administration (FDA), Occupational Safety and Health Administration (OSHA), and Environmental Protection Agency (EPA) and their regulatory laws.

The law system of the United States and most other English speaking countries is *common law system* while most other countries have a *civil law system.*

A civil law system is based on constitutions and statutes. The law is interpreted by scholars and not in courts. As a result, it does not change as the law is applied to cases in courts, but only if the constitution or the statutes are changed.

The common law system is based on the customs of people. Originally, what was the custom of the people became the law. Although the constitution and the statutes are bases of the law, they are interpreted in courts as they are applied to cases. Through a mechanism called *Restatements of the Law*, court decisions become a part of the law, setting precedents for other courts to follow. Hence, the constitution, the statutes, and the interpretations in court cases constitute a hierarchy of sources of law. Court cases mentioned in Section 13.2 are examples of this. When in 1916 the court ruled against the application of privity in the McPherson vs. Buick Motor Company product liability case, the ruling became a basis for all similar cases, changing the interpretation and the application of the common law. Scholars claim that the eventual explosion of the product liability cases is the direct result of the common law system and its ability to change as time goes on. For example, between 1974 and 1985, a mere decade, the number of product liability suits filed in federal courts increased from 1579 to 13,554, a 758 percent increase.[7]

Civil and Criminal Laws   The law can also be divided into criminal and noncriminal (civil) laws. In criminal law, the government brings a suit against an individual as treason, felony, or misdemeanor. These are acts against society and are punishable by fines or imprisonment. This includes murder, robbery, and other criminal activities. In civil law, a civil action is brought by one entity (a person, a corporation, or government) against another entity to enforce his private right such as breach of contract or duty, and is punishable by monetary damages or a court order. Product liability suits are civil actions and not criminal. Therefore, when an entity is found guilty in a product liability suit, only monetary awards are granted and corrective actions are ordered, but no one will be imprisoned unless a criminal activity can be proven.

13.4   MAJOR ADMINISTRATIVE LAWS RELATED TO PRODUCT LIABILITY

There are many laws and regulations brought about by government agencies such as the EPA, FDA, ICC (Interstate Commerce Commission), IRS, OSHA, and CPSC (Consumer Product Safety Commission), as well as nongovernment agencies that create codes and standards, such as ASME, SAE, etc. These agencies create laws, regulate, investigate, make judgment, and enforce their laws. Product liability is also subject to these rules and regulations.

There are three major regulatory agency/administrative laws that are directly related to product safety and liability. Two of these are related to the work environment (employer/employee relationship) while the third is related to other environments such as home and schools (consumers).

- Workmen's Compensation: Provides compensation for a workman who is injured in the workplace, including medical costs, lost wages, and future lost wages, but not compensation for pain and suffering or punitive damages. It also provides for the cost of retraining the injured if the individual cannot continue with the same job due to the injury. It also protects the employer from liability. This means that the injured employee may not sue the employer for other damages. However, it does not protect third parties from liability, including the manufacturer, seller, and designer of the product that caused injury. This in fact is a form of no-fault insurance, paid for by the employer, to ensure a certain level of security for both the employer and the employee. However, in recent years, the premiums required for this service have skyrocketed and have become a significant campaign issue in states such as California.

- OSHA (Occupational Safety & Health Act/Administration, 1970) has the aim of ensuring ''so far as possible, every working man and woman in the nation, safe and healthful working conditions and to preserve our human resources.'' Under the law, employers must provide their employees with a safe working environment and protect them from their own carelessness. OSHA has created a significant body of regulatory laws that relate to the work environment, handling of hazardous material, number of hours of work and conditions permitted, noise levels, and many more.

- Consumer Product Safety Act & Commission (CPSA/CPSC) was created in 1972 to ''protect the public against unreasonable risks of injury associated with consumer products'' in a nonworking environment, such as home, school, recreational facilities and others. CPSC has also created a significant body of regulatory laws that relate to consumer products, including toys, lawn mowers, ladders, and power equipment. All these laws and regulations are to be considered while a product is designed. Please refer to the documents provided by the regulatory agencies related to these regulations for details.

13.5   SOME BASIC DEFINITIONS OF TERMS

The following definitions of terms related to the product liability are provided as a basis for our further discussions. These definitions are used to develop theories about liability as they relate to safety, injuries, and responsibilities.

Tort: A wrongful act or failure to exercise necessary care.

Liability: An obligation to compensate for or correct a wrongful act or its consequences such as damage or injury by an entity that is found to be responsible.

Duty of Care: The expectation that everyone would exercise care to prevent injury or damage to property of others whenever possible.

Reasonable Care: The degree of care exercised by a prudent person.

Great Care: The degree of care exercised by a very prudent and cautious person.

Negligence: Failure to exercise reasonable care necessary to prevent injury or property damage to others.

Contributory Negligence: Negligence of an individual that contributes to his own injury or damage or to his own property regardless of the negligence of a third party.

Privity: A direct contractual relationship between two parties. If *A* and *B* have a contract and *B* and *C* have a contract, there is no privity between *A* and *C*.

Warranty: A statement by a manufacturer or seller or the expectation by the public about the fitness of a product and its safety.

Expressed Warranty: Written or orally expressed statement of warranty by a manufacturer or seller about the fitness of a product and its functions and its safety.

Implied Warranty: The expectation, implied by law, that a product offered for sale is safe and that it will perform its function.

Strict Liability in Tort: Strict liability in tort is a legal theory that the manufacturer of a defective product is responsible for the injuries or damages that the product causes, regardless of negligence.

Absolute Liability: A theory that assumes that the manufacturer of a product is liable for damages and injuries caused by the product regardless of defect or negligence.

Restatement of Laws: Restatement of Laws is the collection of the comprehensive understanding of the rulings and interpretations of the courts when a sufficient body of cases has been decided upon to significantly add to the total understanding of the law. Based on the common law system, restatements become an integral part of U.S. law.

One important point to notice is the way duty of care at different levels is defined. The distinction between the different levels of duty of care, reasonable care, and great care is the individual's prudence. One is an ordinary person, one is a prudent person, and the last is a very prudent and cautious person. However, there are no examples of who this person might be. Is the judge presiding over a case in court a prudent person (or the standard to go by)? Are the members of the jury standards of prudence? Are the attorneys? It is unclear who the role model for the design engineer should be. As a result, it is not clear whether the standard of care exercised by an engineer is adequate or not until it is judged in court. For example, pilots and bus drivers are used as examples of individuals who are required to exercise great care. In a span of about three years, at least two pedestrians were struck at intersections and killed by bus drivers in the city of San Luis Obispo, with one additional injury case and 10 non-injury accidents (in 2002, there were 7 fatal and 542 injury accidents caused by school buses, 36 deaths and 1572 injuries by other buses, and 15 deaths and 1498 injury collisions by emergency vehicles in California alone, as reported by CalTrans). We have also heard numerous stories about drunken pilots and captains and sleepy drivers. Obviously, these are not very good role models for the design engineer.

13.6   BASIC THEORIES OF PRODUCT LIABILITY LAW

The theories of product liability law are based on the following four concepts:

- Warranties
- Negligence
- Strict liability
- Absolute Liability

Warranties are based on contract law, while negligence, strict liability, and absolute liability are based on tort law.

13.6.1   Contracts

A contract is a legal and binding agreement between two parties or entities for whose breach the law provides a remedy. The mere *sale* of the product constitutes a contract for this purpose and is defined as "the total legal obligation that results from the parties' agreement." An expressed warranty and implied warranty are parts of a contract. When a warranty, either implied or expressed, is not honored, the contract is breached. For example, an implied warranty states that a product, by merely being offered for sale, is reasonably safe for use. If it damages someone's property, a contract is broken which must be remedied.

13.6.2 Tort Law

Tort is a wrongful act or a failure to exercise due care resulting in injury or damage to property. Tort law establishes standards of human behavior and duty, for whose breach the law provides a remedy in the form of compensation to the individual who suffers losses because of dangerous or unreasonable conduct of others. For example, negligence occurs when an individual acts with less care than a reasonable person with no intent to cause harm (otherwise, it will be a criminal act).

13.6.3 Expressed Warranties

Expressed warranties are the written and oral statements that manufacturers and sellers make regarding the safety, fitness, and features of a product that influence the sale. According to the Uniform Commercial Code "any affirmation of fact or promise made by the seller to the buyer that relates to the goods and influences the sale, including catalogs, pictures, advertisements, brochures, etc., any description of the goods that influences the sale, and any sample or model that influences the sale" constitutes expressed warranty. Therefore, all information in catalogs, technical data sheets, samples, advertisements, commercials, and descriptions about the product is included in the warranty. If a car is shown to be at the top of a mountain, and if the manufacturer claims (even indirectly) that the car can climb the mountain, the car better be able to do it, or that can be a breach of the contract. If a consumer asks for a product to clean a floor, and the sales clerk in the store makes a recommendation, the product should be able to do the job without damaging the surface, or it can be construed as breach of contract. In either case, the manufacturer or the store are responsible for the breach and must provide a remedy (for example, by providing a refund for the returned product or paying for the damages). False claims and false advertising have been a problem for a long time and government agencies have created many rules to govern them. However, exorbitant claims about products and services are still prevalent in commercials.

13.6.4 Implied Warranties

Implied warranties are related to merchantability and fitness. Courts hold that the mere act of offering a product for sale implies that the product is safe and has at least minimum qualities expected from such a product. This is called *merchantability*. Fitness as an implied warranty results when a specific recommendation is made by a representative of the manufacturer or seller in response to a situation described by the purchaser.

The fact that a product is offered for sale indicates that it is safe and fit for its stated purpose and that it actually works and performs its functions. There is no need for the manufacturer to specifically express this, and the consumer does not need to look for indications that this is true. For example, an electrical device is assumed to be safe and not electrocute the user. There is no need for the manufacturer to have a sign on the product indicating that the product will not electrocute the user. However, if by any chance the product does electrocute the user, the manufacturer *cannot* claim that since this was not specifically expressed, there is no liability on the part of the manufacturer. The safety of a product is implied, and it is expected that it be safe; so it is for fitness. It is implied that when a product is merely offered for sale, it works and it performs its functions satisfactorily. The law provides automatic implied warranty for safety and fitness of the product.

There is some relief from possible breach of warranty by using disclaimers. However, courts tend to subject disclaimers to very strict interpretations in case of economic loss, and hold them of no value if the product causes personal injury.

13.6.5 Negligence

Negligence is failure to exercise due care. Negligence may happen in many different stages of the product's life (and we will discuss this later). However, in terms of product liability, negligence refers to the lack of care in the design, manufacture, shipping, or installation of a product, causing the product to have an inherent flaw or defect. To prove negligence, the plaintiff will need to prove that the product has an inherent defect and that the defect was the cause of harm, injury, or damage. However, if a product has an inherent defect that is due to its design or method of manufacture or installation, all products manufactured by the same technique and based on the same design will have the same inherent defect. In other words, if another similar product is purchased, it will have the same flaw or defect with the same potential for injury or damage. As an example, imagine that a stool is designed and manufactured such that the seat can separate from the frame if a heavier person sits on one side of it. Imagine that this is due to the improper way the seat is attached to the frame such that under heavy load on one side of the seat, the screws holding the seat can come off. In that case, all stools manufactured by the same technique will have the same defect, and there is the potential danger that any one of them may fail under similar loading conditions. In this case, the design is flawed, the product is defective, and there is grounds for negligence.

Obviously, there is no intent on the part of the manufacturer or the designer to design and build a defective product; otherwise, it could be considered a criminal act to offer for sale a product that has the potential to deliberately injure someone, unless the product is intended for this purpose, e.g., a gun. The designer and the manufacturer believe that the product is safe, and they have done whatever they could to insure that the product is safe. Still, if the product they offer for sale is defective, they may be liable for negligence. Please notice that in this case, it is the designer or the manufacturer that are negligent and at fault, not the product (since the product is functioning as designed or as manufactured).

13.6.6 Strict Liability in Tort

Strict liability in tort relates to the responsibility of the manufacturer of a product whose product has caused injury or damage to property of others, and that he must provide a remedy for the damage or the loss. In strict liability, the conduct of the manufacturer or his care is irrelevant (negligence is not important and not an issue). Rather, if the product is defective and unreasonably dangerous, the manufacturer is strictly liable. Here, the product is at fault, not the manufacturer, but the manufacturer is liable for remedy.

In strict liability, the assumption is that regardless of whether or not the designer and the manufacturer are at fault, the product they manufactured has a defect. This means that they have performed their duty to care to their best ability, and that the line of product they have is not necessarily faulty. Nevertheless, a product they manufactured and offered for sale was defective, and it caused injury or damage to property. This means that there is no need for the plaintiff to prove that there was negligence on the part of the designer or the manufacturer, or that all the products of the same nature and design are defective. All that is necessary is to prove that the product in question had a defect, even a random defect, not related to the design or methods of manufacture. And yet, the product caused injury or damage. In our preceding stool example, imagine that the design of the product is perfectly fine, and that similar stools have performed satisfactorily in all situations and under all different loading conditions. However, due to some other reason, one stool had a defect. This can be grounds for a product liability suit.

The difference between the negligence and strict liability is that in negligence, the conduct of the designer and the manufacturer is the issue, so they are at fault. The resulting product, though flawed, is not the issue, since all the products of the same nature are similarly flawed. In strict liability the product that has caused the injury or the damage is at fault, not the conduct of the designer or the manufacturer, and therefore, there is no need to prove that all the similar products have the same flaw. What is needed is to prove that the product in question (the one used by the consumer) was the cause of injury or damage, regardless of the conduct of the designer and the manufacturer or what other similar products may do. Obviously, this is a much easier burden to prove and is much preferred by trial lawyers. Many legal scholars contend that the product liability crisis is caused by this theory. It is estimated that in 1965, there were about 100,000 product liability suits in the United States, including the state and local municipalities.[4] It is also estimated that in 1985, there were in excess of 2.5 million cases.

These theories are not mutually exclusive, but each represents a different case. For example, under the theory of negligence, the plaintiff's negligence will provide a strong defense for the manufacturer (they are both negligent). Under the strict liability, contributory negligence of the plaintiff is generally not valid in most jurisdictions (it is the product's fault). However, due to ''comparative fault,'' proportional to the contributory negligence of the user, his recovery from the manufacturer may be reduced.

13.6.7 Absolute Liability

In absolute liability, the manufacturer is liable for any damage caused by his products as long as the product is the proximate cause of injury. In this case, neither the conduct of the designer and the manufacturer (negligence) nor the defect of the product (strict liability) is the issue. The point is that the product was the proximate cause of the injury or damage (even if the user contributed to the cause by being negligent in the use of the product) and the manufacturer is absolutely liable for his product. Since this theory blames the manufacturer for even the contributory negligence of the user, it is still not favored by many courts but has the potential to become popular in product liability litigation.

13.7   WHO MAY BE HELD LIABLE

Based on the *Restatement of Torts, Second Edition*:[3]

- *One who sells a product in a defective condition unreasonably dangerous to the user or consumer or his property is subject to liability if*
  - *— the seller is engaged in the business of selling such a product and*
  - *— it reaches the user or consumer without substantial change in the condition.*
- *These rules apply although*
  - *— the seller has exercised all possible care in the preparation and sale of the product and*
  - *— the user has not bought the product from or entered into any contractual relation with the seller.*

As was mentioned earlier, when a sufficient body of cases has been decided upon in the courts to significantly add to the total understanding of the law, scholars collect the comprehensive understanding of these rulings into a volume called *Restatement of Laws*, which based on the common law system, becomes an integral part of the U.S. law. The above requirements for holding an entity liable in a product liability suit are from the *Restatement of Torts, Second Edition*. One important indication of liability is that the seller must be engaged in the business of selling such a product. This means that if you are in possession of a product that later you sell to someone else, if anything happens to the new user, you will not be responsible for it because you are not in the business of selling that product (unless you deliberately removed safety devices or made changes to render the product unsafe). However, this also means that the manufacturer, the distributor, and the retailer may all be held liable for a defective product since they are all in the business of selling it.

The second part is very important. Imagine that you are the manufacturer of refrigerators and you buy your compressor from another manufacturer and use it to make the refrigerator. Now assume that you install the compressor into your refrigerator without changing its design or specifications, but the product fails later and it is determined that the compressor is defective. In this case, since you have not changed the design or specifications of the compressor and that you are using it according to the specified conditions, the original manufacturer of the compressor will be held liable. This is because, as stated, the compressor reached the user or consumer without substantial change in its condition. Alternately, assume that in order to increase the perceived efficiency of the compressor, you make changes in its condition, either by addition or subtraction or changes, and the installed compressor is later found to be defective (causing injury or property damage). In that case, the original manufacturer of the compressor will not be held liable because you changed the product he sold to you substantially enough to change its condition. You will be liable. The same will be true if a product is changed substantially enough by a retailer so as to change its condition. Then the manufacturer may not be held liable.

You notice that the second part of the statement about who may be held liable refers to the two famous cases that were discussed earlier. One case is related to the fact that privity or direct contractual relationship between a manufacturer and a consumer is not a necessary condition for liability. The second case is that even if the manufacturer has exercised all due care in the design and fabrication of the product, he may still be held liable if the product injures someone or damages property, thus the theories of negligence and strict liability.

13.8 ORIGINS OF PRODUCT DEFECTS AND LIABILITY

Product defects and liability may stem from the design, manufacturing and materials, packaging and installation, warnings and labels used with the product, or the disposal of the product.

13.8.1 Design

The design of a product is perhaps one of the most vulnerable yet important aspects from which product liability may result. This is also why it is so important for the designer to be familiar with this issue and to deal with it during the design phase, not when the product has been released.

As an example, imagine that a product includes a load-carrying member with relatively large loads on it. Also imagine that the part is designed with a groove in it so as to cause a large stress concentration factor at the groove. If, due to negligence, the stress concentration factor is not appropriately accounted for, the part may fail under load, causing injury or damage. In this case, the design is flawed, the product is defective, and there is grounds for negligence. This simple example shows how the design of a product may affect the safety of the product. However, there are hundreds of actual examples of this simple case around. For instance, the gearshift lever of the 1970s Ford Capri had a groove in it which would fail due to fatigue loading, separating from the gearbox amid driving. The trunk-door latch of the Chrysler minivans of the late 1990s would fail if hit in the rear, providing an opportunity for the passengers to be thrown out (this cost the company hundreds of millions of dollars to correct the design flaw and pay for punitive damages). The mistaken design of the engine pylons of a Lockheed 1011 jetliner broke due to fatigue, crashing the airliner with its hundreds of passengers. The incorrectly specified timing belt of the 1983 Toyota Camry would only last about 16,000 miles, far below the (implied) expected 60,000 miles, requiring that the company change it at their own expense. The Ford Explorer of the late 1990s had an inferior Firestone tire that would burst when heated, but the problem was severely exacerbated by the decision Ford Company made recommending a lower tire pressure, thus further heating the tires. However, the reason for the lower recommended tire pressure was that the center of gravity of the car was designed too high and made the car unstable. Ford tried to lower the C.G. by lowering the tire pressure. But the problem was in the design of the car and should have been solved by correcting the design and not by lowering the tire pressure. In 2004, the U.S. government asked Chrysler to recall 600,000 Dodge Durango SUVs because a ball joint could separate and the tires could fall off. Also in 2004, Chrysler recalled nearly 900,000 minivans for electrical problems that could prevent airbags from deployment (which had caused injuries too). Examples abound in industry, sometimes with catastrophic results. All these examples indicate the importance of engineering design decisions that sometimes are simply overlooked, mistaken, or ignored. When these mishaps occur, other engineers may be called upon to testify for or against the product as an expert witness, judging the actions of the designer. Documenting the design steps, calculations, tests, and verifications is as important as the design itself when the product is judged in court for its safety.

Liabilities may result from the following aspects of the design of a product:

1. **A concealed danger is created by the design.** This means that either the design itself has flaws and inadequacies that create an inherent danger, or the elements of the product are inadequate and dangerous. The above-mentioned examples about the engine pylons, the door latch, and the gearshift lever are good examples of this. The fact that a part was designed without adequate strength is a concealed danger.

   In July 1999, a jury initially awarded \$4.9B to a family whose GM Chevy Malibu's gas tank had exploded. Apparently, GM knew about the danger but did not do anything to correct it. The award was later reduced to \$1B, with GM appeal pending.

2. **Needed safety devices are not included.** When there is knowledge that safety devices are needed to make a product safe, but the design of the product lacks the necessary device, the product will be considered dangerous. Consider the microwave oven. Originally, microwave ovens had no safety door latches. Unfortunately, ignorant consumers abused the oven in many different forms, causing grave damage to themselves. In some cases, some consumers used the microwave oven to warm

their cold hands by placing their hand in the oven and turning it on. Later, manufacturers added a switch to the door such that the oven would stop if the door was opened. Unfortunately, some consumers would press the button by hand while the door was open in order to turn on the oven. When this was discovered, the oven was equipped by an internal switch that engages when the door is closed and a latch presses the switch inside the body of the oven. Obviously, this can still be overridden by enterprising consumers, but it is clearly a malicious act. What is important is that when the manufacturers recognized a danger and the need for a safety device, they needed to act. The failure to act in response to a known danger would be negligence. A similar case can be made about washing machines. In the past, washing machines had no safety switch on the top door. Unfortunately, small children were severely injured when, mesmerized by the speed of the spin cycle, they would get their hands into the spinning drum. Recognizing the danger, a safety switch was added to the door to stop the spinning drum when the door is opened. However, in the absence of a braking mechanism, by the time a drum stops, it may still injure the person. Therefore it is necessary to have a brake mechanism on the drum to stop it immediately if the door is opened, or there must be a simple centrifugal latch that prevents the door from opening when the drum is rotating.

3. **Design involved materials of inadequate strength that failed.** This is evident when a member fails under load.

4. **The designer failed to consider unsafe conditions due to abuse or misuse that are reasonably foreseeable.** This is a very important point about product liability. Notice that it is expected that the designers consider unsafe conditions caused by both the misuse and abuse of the product. To understand why this is so, can you honestly say that you have never used a knife as screwdriver, or you have never stood on a chair or desk to reach up? Both are examples of the misuse or abuse of common products that are reasonably foreseeable. Still, many student desks on the campus of Cal Poly are designed such that if one would stand on the table, it would tip over, throwing the person onto the ground. The abuse of the microwave oven and the washing machine by ignorant or underage consumers are also foreseeable misuses, as is the use of the tip of chain saws, which are not to be used for cutting. For these, it is expected that the designer provides adequate safety devices. The fact that a lawn mower can create a hazardous situation if the engine is left running unattended is also foreseeable, but it is not an abuse or misuse. It is just a danger that is created when the user may let go of the lawn mower due to variety of reasons, including due to being injured by flying objects from the lawn mower. For this reason, it is required by law that lawn mowers be equipped with a braking mechanism that will immediately shut down the engine if the handle is released. However, it is hard to believe that someone would turn a lawn mower to its side and use it as a hedger, as someone did, injuring himself. This may not be reasonably foreseeable (although the plaintiff won the case). Consequently, it is important that the designer considers all possible abuses and misuses of the product and provides for them. We will discuss the economic ramifications of this issue shortly.

Please note that it is unreasonable for a designer to assume that warnings and labels are solutions to design defects. Where unreasonable danger of a product can be cut by redesign, the courts may decide that the manufacturer does not have immunity from liability just because warnings were provided. Imagine that a label would be affixed on a washing machine warning little children about the danger.

13.8.2 Manufacturing and Materials

Another source of product liability is in the selected materials and methods of manufacturing that can create hazards or inadequacies. For example, once again consider a product made of mild steel with a loaded, grooved member and stress concentration factor. Mild steel can relatively easily yield locally at the high stress areas, redistributing the stresses. As a result, it is safe to assume a lower stress concentration factor. However, imagine that during the manufacturing, the member becomes very hot, and in order to speed the process of manufacturing, it is quenched in cold water, thus hardening the material. The simple hardening of the material may prevent the yielding of the material at the groove, causing it to break under load. In this case, although the material originally chosen was appropriate for the stresses present, the manufacturing technique used to make it has changed the material property and has created a hazard. Another example of the hazards created during manufacturing is the Firestone tires mounted on Ford Explorers of the late 1990s. A major problem with the tires, which caused them to explode when hot, was the bubbles that were created in the tire because the material used was inappropriate and old. In fact, there are reports that the workers at the Decatur plant were given an awl to burst the bubbles during manufacturing. The problem cost both companies hundreds of millions of dollars and their reputation.

13.8.3 Packaging, Transportation, Installation, and Application

Imagine buying a glass table and upon opening the container discovering that the glass top is broken. Would you not hold the manufacturer liable for the damage? Obviously, the original design and manufacture of the product was perfectly fine and had no role in this damage. However, the packaging of the product was flawed, causing damage. In fact, this can be even worse if the broken glass injures someone. Clearly, in this case, the source of liability is the packaging of the product that causes damage, or possibly, injuries. Now imagine driving in a highway behind a truck hauling automobiles to dealers, when an improperly restrained car falls off of the truck and hits your car (as happened to a relative of mine in New York). In this case, the automobile itself has no flaws, the manufacturing is sound, and the product is safe. However, due to the way it is transported, the manufacturer (or the transportation company) is liable for the damage or injury to others. The Exxon Valdez oil tanker that hit a reef off the Coast of Alaska in Prince William Sound was also perfectly fine. It was due to the Captain's improper behavior that the accident occurred. Still, Exxon was held liable.

Installation of a product is another source of liability. Imagine that a wall mounted air-conditioning unit is improperly mounted on a wall and falls and damages property or injures someone. Once again, the product was designed properly, its manufacture was flawless, and it was packaged and transported properly, but its installation was improper and flawed. The party (whether the manufacturing company or another contractor) is responsible for the damage and injuries.

The same can be extended to the application of a product. If the product is recommended for an application in a wrong manner, or if the product is not proper for the recommended application, the manufacturer or the party recommending it will be liable for its applications.

13.8.4 Warnings and Labels

According to one story, some time ago, perhaps in the 1980s, a teenager tried to get high by inhaling the propellant of a spray can. He put the spray can into his mouth and inhaled the gas, but died as a result of the toxins. His mother sued the manufacturer of the can, accusing them of negligence, because there were no labels on the can warning that inhaling the gas can be fatal (now a requirement). On the opposite side of the spectrum, during the early 1990s, the manufacturer of a baby walker sent warnings to all registered customers asking them to remove all labels, including warning labels, from the walker. They had found out that the labels were not affixed to the walker securely enough, such that the toddlers could peel them off, put the labels in their mouth, and choke. These examples show that it is necessary that there are enough warning labels on products about potential hazards, especially if the hazard is not very clear. The warning must be clear and easy to understand, intelligible, adequate, and placed appropriately, but it should also be secure and last the life of the product.

If a product has an obvious danger that is common knowledge, warnings are not necessary (such as a knife or a gun). When dangers are inherent, concealed, or known by manufacturers but cannot be remedied without substantially changing the product's cost or utility, warnings are necessary. However, although warnings help both manufacturers and users against risks, they are not substitutes for obvious dangers which are known by the manufacturer and should not be used as a simple way of dealing with dangers when redesign of the product can eliminate the danger without substantial cost increase or decrease in utility of the product.

In July 1990, the U.S. Supreme Court ruled that labels on packs of cigarettes will not protect companies from lawsuits.

13.8.5 Disposal and the Environment

Another aspect of a product life that can create hazards and liability is the issue of the disposal of the product after its useful life. If a product can create hazardous materials when discarded, the manufacturer may be liable. This has become an important issue with the popularity of computers and monitors in the last decades, and the desire to update this equipment on a regular basis and discarding the old equipment. Nowadays, many dumpsites and thrift stores do not even accept computers and monitors, and there has been much discussion about how to recycle computer equipment and whether the manufacturers are responsible for their discards or not. The same can be true with other products, including aluminum cans and glass bottles, cars, mattresses, and many others.

Additionally, if a manufacturer uses fabrication techniques or materials that pollute the air, water, or the environment in general, it may be liable (although the products they manufacture may be sound and without flaws). There are many environmental laws created by OSHA, EPA, and other governmental agencies that govern the production, handling, and disposal of hazardous materials. To this we must add recycling and discarding issues as well.

13.9 THE UNREASONABLY DANGEROUS PRODUCT

The standard for determining negligence is what a *reasonable* person may or may not do under the circumstances. A reasonable person whose behavior is the standard of expectation is an imaginary person. Consequently, it is difficult to define or guess what this person may or may not do under certain circumstances. The reasonable person, however, is not an ordinary person, a juror, a judge, the attorney(s), your professors, or clergy, but is a prudent and careful person. However, this has to be determined by a court of law. In other words, until someone's behavior is contested in the court, it will be impossible to assess whether the individual has been reasonable or not.

To determine if a product is unreasonably dangerous, consider the following aspects. The combination of the severity and importance of these issues can help you determine the possible acceptable danger level:

1. **How useful the product is:** If a product is useful, customers may tolerate a higher level of hazard associated with it. For example, cars pollute the environment and kill tens of thousands of people every year. However, unlike other products, the drivers are generally blamed and not the product. This is because we all need cars so badly that we cannot stop their production.

2. **Whether safer replacement products are available:** If another product similar to the one in question can do the same function but is safer, it becomes the expected standard for the safety of this product as well. For example, when a manufacturer places a guard at the tip of a chain saw, rendering it safer, the state of the art changes. Other manufacturers who do not add similar safety devices may become liable. This is because it has been shown that the product can be safer and there is really no reason why other products similar to it should not follow suit. This is what has occurred with the automobile industry as well. When certain cars started installing air bags (first for the driver, then on the passenger side, then on the side panels) and antilock brakes, others followed suit, making these safety devices practically standard equipment in all cars. The same is true with gunlocks.

3. **How likely the injury is:** If there is a chance of injury but it is very low, the necessity of safety devices and its importance may be different than when the chance of injury is high. The higher the likelihood, the more important it is to provide for safety of the product.

4. **How serious the injury may be:** Imagine that a product may have the potential to cut the user's finger during operation. This is very different than the potential to electrocute the user, crash and kill hundreds of passengers, or the danger that the user may fall and break his neck. Every product can inherently have some danger in it. But both the likelihood as well as the severity of the injury determine the importance of the danger.

5. **How obvious the danger involved is:** Before consumers were educated about the dangers of microwave ovens, many had no idea that the microwave ovens could be very dangerous. They would attempt to remove the food while the machine was on Some consumers even tried to warm their hands by placing it in the oven and turning it on. In fact, new issues regarding handling of food in the microwave ovens have recently surfaced of which most consumers are not aware. On the other hand, it is obvious that a power tool such as a chain saw is a dangerous product. Still, it may not be obvious to many that the tip area of the saw can get caught in a nail or other hard material in the wood, causing it to recoil and severely injure the user. When the danger is not obvious, there is more urgency to add safety devices to the product and place warning labels on it in order to inform the consumers of the potential dangers.

6. **Can the risk be eliminated without impairing the utility of the product?:** A knife is meant to be sharp and its function is to cut. A gun is meant to be a weapon, and its function is to injure or kill. An iron is meant to be hot, and its function is to transfer the heat to the fabric. All these products are inherently dangerous. Completely removing the danger will render them useless with no utility. However, a light bulb's function is to produce light. The fact that an incandescent bulb gets hot because of the way it is designed is an unnecessary by-product. A drill motor's utility is to provide a powerful torque that can be used for drilling holes and turning screws or other similar functions. But designing the trigger switch to run most of the length of

the handle, preventing the user from letting go unless the other hand holds the motor, is a flaw. For these examples, we already know that there are alternatives that are better and less dangerous. The design engineer must assess whether a product can be made safer without severely reducing its utility and usefulness, and if so, what the best compromise may be. In general, it is expected that necessary safety devices be added to a product unless they render the product non-competitive in the market.

7. **The cost of making the product safer:** Long ago, it was determined that adding a third brake-light to the back of cars at a higher level than normal would help prevent accidents. This was true when the driver in a first car in the row would brake, but the driver in the second car would not pay attention. A third driver behind would not know that the first driver was stopping, and together with the second car, would run into the first car. The higher brake light would warn the drivers behind the second car about the actions of the first car, decreasing the likelihood that chain accidents would occur. The cost of adding the third light was estimated to be about $10 per car during manufacturing. However, while adding many unnecessary gadgets to the car automobile manufacturers resisted this safety feature for years so as to not make their cars more expensive. In general, it is expected that safety devices should be added to a product unless the added cost will make the product non-competitive in the market.

8. **Are consumers willing to pay for a higher-priced but safer product:** When consumers realize that added safety features do help, they generally are willing to pay for them. The perceived added safety of Sport Utility Vehicles, due to their bigger size, is one of the reasons why these cars have become so popular, even if they are more expensive, require much more gas, and are inherently less stable and roll easily (according to the National Highway Traffic Safety Administration, published in 2004, the chance of rollover is between 8 percent and 16 percent for most cars and minivans and between 13 percent and 28 percent for most SUVs, with 34.8 percent for Ford Explorer). Certain brands of cars that boast safety are also more expensive. Nonetheless, the design engineer should determine how willing the public is in paying for added safety features or reduced utility of a product. An example of this is a device called Stop-Saw, which by measuring the current going through the saw blade, can immediately stop a table-saw blade if it comes into contact with a body part. However, it is estimated that the cost of this safety device could be as high as the cost of the saw itself, doubling its total price. This significant added cost must be weighed against the benefits it provides, as well as the willingness of the consumers to pay for additional safety devices. It cannot be expected that the manufacturer adds safety devices to the product at so high a cost as to make the product non-competitive.

13.10 REDUCING PRODUCT LIABILITY RISK

Genuine attempts by the design engineer to address the safety issues to manage hazards is among the main countermeasures that can be taken against future product liability suits. This means that, as was already discussed in Chapter 7, product liability must be one of the design elements that are considered during the design phase of the product, and to ensure that attention is paid to the standards and codes, hazard analysis, warranties, human factors, and other such factors.

The following activities may help in this task:

1. **Include safety as a primary specification throughout the life of the product.** This can be accomplished by considering all the different phases of the life of the

product, from design to fabrication, to storage, to shipping, to installation, to application, to consumption and utility, and to discarding or recycling.

2. **Design to nationally recognized standards.** These include the required governmental codes as well as engineering codes that may or may not be required, but are certainly useful and expected. A partial list includes EPA, CPSC, OSHA, ASME, SAE, Military, and ANSI standards (the ASME pressure vessel code is so complicated and involved that there are books written about how to use the code).

3. **Integrate design considerations into the product and communicate with others completely and accurately.** Design considerations were discussed in Chapter 7. Make sure that those and other relevant design considerations are integrated into the design of the product, including safety consideration, human factors, fabrication techniques, and required codes. However, also ensure that others who will deal with the product after it is designed are part of the design team. This will increase the chance of a safer and better product.

4. **Select materials and components of acceptable qualities and standard deviations.** This should be considered when the product is designed and components are selected and specified. You also need to ensure that the products that are delivered to you by your suppliers are tested for compliance with your specifications. Quality assurance for the supplied materials can be as important as selecting the proper material.

5. **Test the product under realistic conditions that represent true utility and aging.** Products behave differently under different conditions, and depending on their exposure to the natural elements, they age differently. If a plastic material can degrade under the sun or high humidity, losing its strength, it can create a hazard if this is not taken into consideration. Similarly, if a product can behave differently under different temperatures or lighting or other environmental conditions, both its utility and its safety may be jeopardized.

6. **Document the history of the product development and all activities that were undertaken.** As in patenting, if you are called to testify about a product, or if the product is investigated by outside entities, these records will determine whether the company was negligent or not. Keeping accurate and complete records will allow you to defend your actions and decisions, will show why certain actions were or were not taken, and will determine what the results of your actions were.

7. **Murphy's Law applies. Consider worst-case scenarios.** According to Murphy's Law, if anything can go wrong, it will. This means that you must consider all possible misuses and abuses, as well as natural mishaps, that may occur and create safety problems or hazards. You should not assume that people will use your product only the way you have planned. If they can use it in other ways, they will. If anything can go wrong, it will. Do consider the worst-case scenario, the worst that can happen to, or as a result of, your product. Make sure that there are remedies, solutions, or safety devices to counter the worst case(s).

8. **Work with the advertising department to guard against overstatement of product performance.** This is important because whatever is claimed through commercials and advertisements is considered an expressed warranty, and the institution will be liable if anything happens as a result of exaggerated claims.

9. **Use warning labels on the product when appropriate.** However, beware that labels do not guard against hazards or against liability suits. Labels must be adequate, clear, safe, and understandable to individuals who do not read or speak the language, and they must last the life of the product.

10. **Determine all service or maintenance necessary to keep the product safe and advise the user.** However, allow for the possibility that the consumer may not necessarily perform the specified maintenance routines as scheduled. This should not create a blatant hazard.

11. **Ask for reports and feedback from sales and service personnel and dealers about complaints or problems with the product.** The quicker you get this information and correct the problem, the less the liability. However, do not use this approach as a substitute for quality design and production. It is many times more expensive to change a design or correct a mistake after the product is fabricated and leaves the plant than it is to correct it during the design phase, especially if the defect results in an actual injury or damage to property.

13.11 FAILURE MODE AND EFFECT ANALYSIS (FMEA)

Failure Mode and Effect Analysis (FMEA) is used to determine in what ways a system or its elements may fail and what will happen to the system and its environment if it does fail in each of the modes. Therefore, FMEA involves both how a system may fail and what happens when it fails. The information thus developed is used to devise techniques or solutions to either prevent the failure or to include remedies to circumvent the effects. This makes the product safer and more reliable, and by reducing subsequent problems, it reduces product development time. Documentations developed during FMEA may also be very useful in legal proceedings if injuries occur and legal proceedings result. FMEA analysis may also be used in developing maintenance procedures and intervals.

The springs used in overhead garage doors as counterbalance have the potential to fail due to fatigue. If this happens, because they are in tension and can uncontrollably fly around when they snap, the springs may hit a person or a car that is close and severely injure the person or damage property. Proper design of the spring may either postpone or completely eliminate the chance of fatigue failure. Nevertheless, a telescoping bar (usually two interlocking u-shaped bars) or a cable is inserted inside the spring to prevent the failed spring from flying uncontrollably and hitting a person or a car, thus minimizing its effect.

Internal combustion engines use a timing belt or chain to transfer the crankshaft motion to the camshaft in a coordinated fashion. If the engine is designed with a timing chain, since a timing chain generally has a much better life than a belt, the chance of failure is significantly reduced. However, if due to economic considerations, a timing belt is used, it is inevitable that the belt will eventually fail. Therefore, the owner should be advised to have the belt replaced at recommended intervals. Additionally, the damage to the engine can be minimized if the engine is designed to be a noninterference type (the piston and the valves will not run into each other). This means that although the engine stops and the timing belt must be replaced, no other part of the engine will be damaged. Otherwise, the damage to the valves and pistons can be substantial.

Although engineers have always been wary of failures and their consequences, and therefore, have always looked for ways to minimize danger, a formal methodology for analyzing failure modes was not devised until the 1960s during the space program. At this time, several professional societies began to publish methodologies for failure mode and effects analysis.[8] One of the earliest ones was Society of Automotive Engineers' Aerospace Recommended Practice ARP926, published in 1967 entitled ''Fault/Failure Analysis Procedure.''[9] In 1974 another document called ''Procedures for Performing a Failure Mode, Effects and Criticality Analysis,'' MIL-STD-1629 was published by the military. Subsequent publications by Ford, the military, SAE and others have made FMEA a standard procedure for analyzing failure modes and their effects.

Examples of failure modes abound. For instance, a switch may chatter, remain open, partially open, remain closed, or partially closed. A cable may be pinched, may break, fray, or stretch. A valve may remain open or closed, may leak, or wobble. A spring may stretch, break, yield, or fatigue. An operator may get tired, fatigued, may apply a wrong procedure to a wrong item, a wrong procedure to the right item, a right procedure to a wrong item, perform too early, too late, or not perform and not notice. A tank may explode, leak, rust, collapse, roll, or fall. A toaster may catch fire, may burn the item, may become too hot, and may smoke. In addition, each component of it may also fail in different modes.

FMEA is used at two different levels:

- *Functional level:* At this level, the function of the product or system is analyzed. For example, the air speed of an airplane is measured from the ram effect of air into a small, hollow tube called a Pitot tube that is located outside of the airplane on the wing or fuselage. As the speed changes, the pressure in the Pitot tube changes as well. This pressure difference is used to measure the air speed. However, due to extreme cold temperatures during flight, ice may form on the tube and prevent it from working properly. The result is that the gauge will show no airspeed. The remedy is then to add a heater to the Pitot tube to melt the ice.

- *Hardware level:* At this level, individual components are analyzed and the effects of each failure mode are considered and remedies are included. In the above example, the failure of each component of the Pitot tube system would be analyzed. For example, what would happen if the tube is accidentally bent or plugged? Although the consequence may be similar to freezing, the solution required may be very different.

13.11.1 Terminology

First, let's define the following terminology which is used in FMEA:

- **Fault:** This is the inability to function properly.
- **Fault Indication:** The method by which the fault is detected, e.g., visual, audible, odorous, console display, and printout.
- **Fault Tree:** A pictorial representation of the combinations of faults and failures that can result in a system failure. If probabilities are added to the fault tree, it is possible to estimate and evaluate the probability of ultimate system failure.
- **Failure:** The inability of a system or component to perform its intended function.
- **Failure Mechanism:** The process involved in the cause of failure. It indicates the process of failure, e.g., fatigue failure.
- **Failure Mode:** The way in which a system or component fails. It determines how the part fails, e.g., as a result of fatigue failure, a part breaks into two pieces, whereas the result of an explosion is the shattering of a container into many pieces. The consequences of these two modes are very different.

- **Bottom-up Analysis:** This means that the analysis of failure modes and their effects starts at the lowest level of components, eventually to subsystems, and then to the whole system.

- **Compensating Provisions:** Design provisions or actions by operators that mitigate the effects of a failure. These provisions mitigate the effects, but may not prevent the failure (remember the noninterference engine design in case a timing belt fails).

- **Criticality:** A relative measure of the impact of a failure mode on the mission objective. The failure of a component such as the heat-resistant skin of the space shuttle will destroy the vehicle, and therefore, the mission.

- **Criticality Analysis:** A procedure for ranking the combined influence of the severity of a failure and the probability of it happening.

- **Single Point Failure:** Failure of an item that is not compensated by redundancy or by other means which results in an adverse system failure, e.g., failure of the engine in a single-engine airplane.

- **Fail-safe:** A concept that a device must be designed to still be safe even if it fails. In this case, it may stop operating, but it is still safe.

- **Interface:** The means through which different parts of a system are interconnected.

- **Interface FMEA:** An analysis of how failure modes and effects of different hardware and software subsystems affect each other through signals, cables, wires, hydraulic lines, etc.

- **Verification:** The process of proving that a system complies with the formally established requirements.

- **Validation:** The process of confirming to accepted engineering principles.

13.11.2   FMEA and the Design Process

FMEA is applied throughout the design process, and therefore, it goes hand-in-hand with the design process. FMEA's planning is accomplished during conceptual design, when the system is designed as a whole. During the initial detail design of the components, functional analysis of FMEA is accomplished. Interface analysis of the effects of failures between different subsystems, as well as detailed analysis and updating of functional analysis of the system, is accomplished during the detailed planning and design of the system. When the design is evaluated and tested and verified, FMEA analysis is also verified. FMEA continues as field analysis when the product is shipped and used. Figure 13.1 shows the schematic of this relationship.

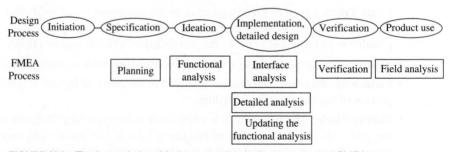

FIGURE 13.1   The interrelationship between the design process and FMEA.

UNION CARBIDE'S BHOPAL, INDIA DISASTER

Union Carbide established a plant in Bhopal in 1969 to manufacture fertilizers. One of the intermediate products used in the manufacture of carbaryl is Methyl Isocyanate (MIC) gas. When mixed with water, MIC produces much heat with increased pressure. On December 3, 1984, a holding tank with 43 tons of stored MIC overheated and released the toxic gas into the surrounding area when water entered the tank. Although originally the plant was outside of the city of Bhopal, a pseudocity of shacks with a large population had subsequently been created by the workers around the plant. The heavier-than-air toxic gases killed and injured thousands of individuals. Some are still suffering from cancer, breathing problems, blindness, and more.

There have been numerous investigations by Union Carbide, by the government of India, and by others. Although still unclear how, water had entered the holding tank and had mixed with MIC, which created the hazardous situation. The plant had been designed with multiple layers of safety devices, but most of these systems were either inadequate or were not operating. The cost-cutting measures had been introduced in 1982 when the demand for carbaryl fell and the company lost money. For example, it is speculated that alarms were not working, there was only one manual back-up working, the flare tower and the gas-scrubber had been out of service, and the refrigeration system, designed to inhibit the volatilization of MIC, was left idle. Many other safety problems are also cited by different investigators. The problems were not just economic; there were language, cultural, and work ethic differences as well. A sister plant in the United States never had a serious problem.

As you notice, multiple layers of safety devices were included in the system. As far as FMEA is concerned, one could see that failure modes were considered and remedies were included. However, the safety devices were not operating at the time of the accident, and consequently caused one of the worst industrial accidents in history.

13.11.3 Application of FMEA

The following steps may be taken in order to accomplish FMEA:

- **Gather information about the product or process under investigation:** You will need to obtain necessary information, drawings, and specifications that define the product. You will also need to establish what it is that you are trying to analyze; whether the system, all of it, a part of it, or components and subassemblies.

- **Determine functions involved:** Investigate all the functions and components that need to be assessed and analyzed. Establish the relationship between different components and their integrated functions (interfaces). In this step you should also define the scope of your analysis. You need to define the purpose as well as the level of detail desired. FMEA may be performed, for example, to determine mission success or the cost of repair. It may be performed for the entire system, a subsystem, an assembly, or particular components.

- **Identify potential failure modes:** For each component or function, determine all the possible failure modes.

- **Determine causes of failure:** As was mentioned earlier, the cause of a failure is the mechanism that creates the conditions under which the failure occurs. For example, fatigue loading may be the cause of a failure. The result is that a part will break in two. However, the cause of a failure as a result of stretching (yielding) is overload. In the Union Carbide's Bhopal plant, the mixing of the water and MIC was the cause of excess heat, pressure, and leaks.

- **Determine effects of failures:** For each failure mode, determine the effect and the consequences. In most cases, the consequences have a domino effect as one failure affects another component. For example, if the key holding a flywheel on a shaft fails, the flywheel may become loose and be dangerous and out of control, hitting other components. Additionally, the motor driving the flywheel may accelerate uncontrollably because there is no longer any load on it.

- **Determine methods of detection used:** You need to determine how the failure may be detected. In some cases, the failure results in nonoperation, explosion, or other visible and clear consequences. In other cases, it may not be clear that a component or subassembly has failed. For example, if the contacts in a switch fail, there may be no indication alarming an operator that the component has failed, and even if the system's operation is affected negatively, it may not be clear to the operator what the cause is. A leak in a valve may not be detected until too late, unless the failure is reported and indicated to an operator, who will take action to remedy the failure.

- **Determine the severity of the results and assign risk factors:** FMEA involves both the determination of the failure modes and their effects as well as the probability and severity of the effect. Please also see Section 13.9. The following criteria have been used for the military, the aerospace, and automobile industries:
  - **US MIL-STD-1629 criteria for 4-level severity ranking:**[10]
    - ○ **I: Catastrophic:** A failure that can cause death or system loss.
    - ○ **II: Critical:** A failure that can cause severe injuries or major property damage or mission loss.
    - ○ **III: Marginal:** A failure that can cause minor injuries, minor property damage, or delay in a mission or its degradation.
    - ○ **IV: Minor:** A failure that is not serious enough to cause injuries or property damage, but will require unscheduled maintenance or repair.
  - **SAE ARP1834 criteria for 3-level aerospace industry severity ranking:**[11]
    - ○ **Critical:** A failure of a function of a system or any design error that would prevent the continued safe flight and landing of an aircraft.
    - ○ **Essential:** A failure of a function or a design error that would reduce the capability of the aircraft or the ability of the crew to cope with the adverse operating conditions.
    - ○ **Nonessential:** A failure or a design error that will not significantly degrade the aircraft's capability or crew ability.
  - **SAE J1739 criteria for 10-level automobile industry severity ranking criteria:**[12]
    - ○ **10: Hazardous, without warning:** The failure affects safe vehicle operation and/or involves noncompliance with government regulations without warning (such as loss of braking fluid).
    - ○ **9: Hazardous, with warning:** Same as above, but with warning.
    - ○ **8: Very high:** A failure that leaves the vehicle inoperable with loss of primary function (such as engine failure due to a broken timing belt).
    - ○ **7: High:** The vehicle remains operable, but comfort or convenience items may be inoperable, causing discomfort (such as loss of air conditioning).
    - ○ **6: Moderate:** Vehicle remains operable, but comfort or convenience items operate at reduced levels, causing customer dissatisfaction.
    - ○ **5: Low:** Same as above, but at reduced level, causing some dissatisfaction.
    - ○ **4: Very low:** Cosmetic defect in finish and fit, squeaks, and rattle. Not conforming to specifications. Customers notice the defect.
    - ○ **3: Minor:** Same as above, defect noticed by average customer.

○ **2: Very minor:** Same as above, but minor enough that only discriminating customer would notice.

○ **1: None:** No effect.

Barring any other numbers to use, you may assign the following numbers to indicate the severity of a failure mode:

$$9 = \text{Severe}$$
$$3 = \text{Moderate}$$
$$1 = \text{Low}$$
$$0 = \text{Unlikely}$$

- **Determine the probability of failure:** In addition to the severity of failure, it is also important to determine the probability of failure (please also see Section 13.9). Table 13.1 shows a qualitative probability interval set[12] that is used in the automobile industry to assign ranking to failures based on the probability of their occurrence.

    Criticality Index is often defined as the product of the severity and probability indices. The higher the criticality index, the higher the priority for change.

- **Implement solutions:** Depending on the results of your analysis, if the consequences are unacceptable, if the severity of the failure is high, or if its probability is elevated, you will need to take proper action to remedy the problem. You may need to add safety devices, change the design, recommend operator-initiated actions, or a slew of other solutions.

- **Assess the results:** After solutions are implemented you will need to assess the results. You will need to verify that the action has resolved the problem or improved the situation and that it complies with formally established requirements. You will need to validate your results to accepted engineering principles.

13.11.4 Fail-Safe Design

The concept of fail-safe design is an important concept in managing failures. This concept requires that a system be designed in such a way that even if a critical component of the

TABLE 13.1 Qualitative Probability Intervals Used in the Automobile Industry[12]

Probability of Failure	Probability	Ranking
Very High	0.5 (greater than 1 in 2)	10
Failure is almost inevitable	0.33 (1 in 3)	9
High	0.125 (1 in 8)	8
Repeated failure	0.05 (1 in 20)	7
Moderate	0.0125 (1 in 80)	6
Occasional failure	0.0025 (1 in 400)	5
	0.0005 (1 in 2000)	4
Low	0.0000667 (1 in 15,000)	3
Relatively few failures	0.00000667 (1 in 150,000)	2
Remote: failure unlikely	0.000000667 (1 in 1,500,000)	1

system fails, it still remains safe. I already mentioned that a restraining bar or cable is inserted in a garage-door spring in order to ensure that even if the spring fails, the system remains safe. In this case, although the garage door will no longer operate properly, it will remain safe. A single engine aircraft is not fail-safe; if the engine fails, all power is lost. Alternatively, if there are multiple engines, even if one fails, the other engine(s) will allow the pilot to safely land the aircraft, even though performance is compromised and the flight is uncomfortable. Due to this important consideration, the majority of airplanes, even military aircraft, have multiple engines. In fact, many systems include redundant assemblies and components in order to avert consequences of failure and make the system fail-safe. Another example is the fail-safe braking system, alluded to in Chapter 4, designed by George Westinghouse. He realized that when air brakes of locomotives failed, the train would be completely uncontrollable in downhill motions. Instead of a braking system that required air pressure to be applied, he designed a brake system that would be normally on, and instead, required air pressure for disengagement. Hence, if air pressure were lost, at least the brakes would automatically stop the train, therefore fail-safe. The same concept is still used in trains, trucks, buses, and other similar systems. In fact, to prevent electric robot manipulators from falling limp if the power is lost, they have similar normally-on brakes on each joint to make them fail-safe. Instead, electric power is applied to disengage the brakes and allow the arm to move. All lawn mowers have a normally-on brake as well in order to make them fail-safe in case the operator releases the handle of the lawn mower in an emergency.

13.11.5 Single-Point Failure

Single-point failure occurs when the failure of just one assembly, subcomponent, or component causes complete failure of the system. As mentioned earlier, the failure of the engine of a single-engine aircraft will cause the failure of the whole system. Similarly, the failure of just one heat shield tile in the space shuttle caused a catastrophic failure of the whole system. To ameliorate this problem, one or a combination of the following may be used:

- Use redundancy. Redundant assemblies or components may completely prevent the failure of the system until the failed assembly or component is replaced or fixed.
- Consider a complete design change. When the design of the system is changed, it may resolve the single-point failure problem.
- Use equipment that is extremely reliable. The increased reliability of each component or system increases the reliability of the entire system. However, more reliable components and systems can be very costly.
- Perform frequent maintenance operations and replace parts and assemblies on a regular basis.
- Reduce or eliminate external stresses on the system for which it is not designed.

Each remedy requires additional cost and/or time. Therefore, it is important that the designer performs a cost/benefit ratio to determine which one of these remedies is more beneficial. However, remember that sometimes cost alone is *not* the factor. When the lives of users or others are in danger, the cost is a secondary issue. In the 1970s, Ford had calculated that their cost of settling death and injury cases as a result of Pinto fuel tank explosions would be less than the cost of resolving the design problem (such as adding a rubber insert in the tank to prevent spillage of fuel when it was punctured). However, the courts added significantly to the liability of Ford for punitive damages, forcing it to stop production of Pinto and paying huge sums to the individuals who were injured or to the families of those who died as a result.

13.11.6 Application of FMEA

A table similar to Table 13.2 may be used to apply FMEA to a product or system. You may use commercial programs, download forms from a variety of sources, or make your own table using any commercial program like Excel. Table 13.3 is a blank form that you may use for your applications.

The example included is for reference only. It is not a complete analysis of a system, but to demonstrate the process. The product is divided into subassemblies, and then to components. For this example, only some failure modes of each assembly or component and their effects are considered, and remedies are suggested. The severity and probability of each failure are also estimated, and a criticality index is calculated. The larger this number, the more important it is to resolve the problem.

13.12   CODE OF ETHICS FOR ENGINEERS

As was discussed in Chapter 7, the National Society of Professional Engineers has developed a Code of Ethics for Engineers that governs the ethical behavior of engineers. Similar codes of ethics exist for other professions, including but not limited to, physicians. Designers, whether engineers, industrial designers, architects, or any other profession, should read, understand, and regularly consult the Code whenever necessary in order to ensure that their decisions are ethical.

The codes, in addition to being a primary source for ethical decisions, can serve as a guide to the designer in ensuring that safety considerations are addressed during the design phase of the product. As designers and engineers, we are required by the Code to "hold paramount the safety, health, and welfare of the public." Adhering to this requirement will also aid the designer in defending his decisions if his actions are questioned in the courts.

Please refer to Chapter 7 or the NSPE website for the complete text of the Code of Ethics for Engineers.

13.13   STANDARDS AND CODES

The following is a partial listing of organizations that can be considered primary sources of standards. Depending on what you design and the requirements of your products, there may be other codes and standards that you must consider:

American National Standards Institute (ANSI)

American Society of Mechanical Engineers (ASME)

Society of Automotive Engineers (SAE)

Institute of Electrical and Electronics Engineers (IEEE)

Underwriters Laboratories (UL)

American Society for Testing and Materials (ASTM)

National Fire Protection Association (NFPA)

National Safety Council (NSC)

TABLE 13.2 Example of the Application of FMEA to a Product

System	Subsystem	Assembly	Component	Failure Mode	Failure Cause	Failure Effects	Severity	Probability	Criticality	Remedy, Possible Solution
Floor lamp	Light fixture	Halogen Lamp	Lamp	Burns out	Normal use	Non-operation	1	3	3	Must be replaced by the user
				Burns out	Oil from touching	Non-operation	1	5	5	Warn user to not touch
				Too hot	Normal operation	Can hurt the user	5	5	25	Cover with tempered glass
		Lamp holder	Brackets	Rust	Moisture	Lost contact	1	2	2	Use plated contacts
				Becomes loose	Repeated application	Lost contact	1	2	2	Use spring washers
			wires							
			screws							
		Protective Glass	Glass							
			Brackets							
		Switch	Knob							
			Potentiometer							
	Dimmer switch	Dimmer	Resistor	Gets too hot	Dimmer effect	Dimmer burn out	3	10	30	Add heat sink to cool it
			Capacitor							
	Base	Base	Base	Rolls over	Too light	Fire hazard	9	7	63	Heavy, wide dimensions for stability
			Attachment	Loosens	Lamp may fall	Fire hazard	9	3	27	Secure connection with pin
	Cable									

TABLE 13.3 FMEA Blank Form

Project #: _____
Project name: _____
Analysis By: _____

Sheet #: _____ of _____
Date: _____
Prepared by: _____
Approved by: _____

System	Subsystem	Assembly	Component	Failure Mode	Failure Cause	Failure Effects	Severity	Probability	Criticality	Remedy, Possible Solution

Failure Mode and Effects Analysis

Food and Drug Administration (FDA)

Federal Trade Commission (FTC)

Occupational Safety and Health Administration (OSHA)

National Electric Code (NEC)

Consumer Product Safety Act (CPSA)

Environmental Protection Agency (EPA)

13.14 WHAT PROTECTION DOES THE DESIGNER HAVE?

In general, if you work for a company, the company will have the primary responsibility for product liability. If the product that the company sells causes damage or injury, and if it is sued for negligence, for breach of warranty, or for strict liability, it will be the company's responsibility to defend itself. If it loses, the company will also have to pay the awarded amount and the attorney fees. You, as the engineer participating in the creation of the product, may have to cooperate with the company's defense, and obviously, it is also possible that you may be called as a witness, in effect helping the plaintiff. However, based on the "big pocket" theory, the entity that has the most resources will end up paying for the damages.

On the other hand, if you are a consulting engineer or the owner of your own business, then you will be responsible for the liabilities that stem from your products and services. In that case, it is absolutely necessary to have liability insurance. For this, you must consult a professional insurance agent in order to determine what is best for you. The premium you pay does depend on the number of products sold, how long the product has been on the market, what your yearly gross income is, as well as a liability multiplier which is related to the danger level of your product or services (mostly based on statistics). However, you must be aware that the cost of liability insurance for all sectors of the economy has skyrocketed in the past couple of decades. In fact, in certain professions, including certain branches of medicine, as well for some governmental agencies and city and county governments, the premiums have gone up to the point that these entities can no longer afford liability insurance. In some cases, they have stopped practicing their profession, or they do not carry any insurance, leaving themselves open to much danger (case in point, in 2004, the cost of liability insurance for a party with 250 guests in a community center, serving wine, for 6 hours, was quoted as $770. The cost of renting the community center itself, including all chairs, tables, and services, was $650).

REFERENCES

1. HUNTER, THOMAS, "Design Errors and Their Consequences," *Mechanical Engineering*, June 1990, pp. 54–58.
2. HADDOCK, DORIS, with DENNIS BURKE, "Granny D: A Memoir," Villard Books, New York, 2001, p. 188.
3. THORPE, JAMES F., and W. MIDDENDORF, "What Every Engineer Should Know about Product Liability," Marcel Dekker, Inc., New York, 1970.
4. MURR, LAWRENCE, "What Every Engineer Should Know about Material and Component Failure, Failure Analysis, and Litigation," Marcel Dekker, Inc., New York, 1987.
5. WANG, JOHN X., and M. L. ROUSH, "What Every Engineer Should Know about Risk Engineering and Management," Marcel Dekker, Inc., New York, 2000.
6. HUNTER, THOMAS A., "Design Errors and Their Consequences," *Mechanical Engineering*, June 1990, pp. 54–58.
7. CORTES-COMERER, NHORA, "Defensive Designing: On Guard against the Bizarre," *Mechanical Engineering*, August 1988, pp. 40–42.

8. "Recommended Failure Modes and Effects Analysis (FMEA) Practices for Non-Automobile Applications," *Society of Automotive Engineers*, SAE ARP-5580, July 2001.

9. "Fault/Failure Analysis Procedure," *Society of Automotive Engineers*, Aerospace Recommended Practice, ARP926, September 1967, ARP926A, November 1979.

10. "Procedures for Performing a Failure Mode Effects and Criticality Analysis," US MIL-STD-1629 (ships), November 1974; US MIL-STD-1629A, November 1980; US MIL-STD-1629/Notice 2, November 1984.

11. "Fault/Failure Analysis for Digital Systems and Equipment," *Society of Automotive Engineers*, Aerospace Recommended Practice, ARP1834, August 1986.

12. "Potential Failure Mode and Effects Analysis in Design (Design FMEA) and Potential Failure Mode and Effects Analysis in Manufacturing and Assembly Processes (Process FMEA) Reference Manual," *Society of Automotive Engineers*, Surface Vehicle Recommended Practice, J1739, July 1994.

HOMEWORK

13.1 Pick a consumer product or a power tool. Study the product in detail. Write a report about its shortcomings, problems, and safety issues. Suggest solutions for the problems. Can safety devices solve the problems?

13.2 Consider a chainsaw. Suggest a way to make the chainsaw safer by providing a system to prevent the user from using the tip of the chain saw for cutting. The design should make it impossible for the consumer to easily override or remove the safety device without seriously altering the saw or rendering it useless or inoperative.

13.3 Assume that due to an accident, a product has injured a user. Make up a case study. Then choose three teams and a judge. One team should act as the plaintiff and the prosecuting attorneys. The other team should act as the defendant and the defending team. The third team should act as the jury. The judge will ensure that the proceedings will go according to a set of rules or common sense. In this mock trial, you should consider all facts and assumptions that could be made by both teams and argue for them. This will help you understand how different people with various points of view will see the same facts and arguments differently.

13.4 Arrange for a mock trial of an industrial or consumer tragedy such as the Ford-Firestone tire episode. Do the trial as in Problem 3 above. Study the case as available on the Internet and extract as much information as you can find. As in real life, each team may find different information, making the case even more educational and interesting.

13.5 Compare defects or failures you may have experienced with your own car to the recommended list provided by the SAE J1739 Criteria for 10-level automobile industry severity ranking criteria.

13.6 Make an FMEA table and apply the analysis to a product of your choice. Select a familiar product in order to be able to recognize the components, the failure modes, the effects, and the severity and probability of each effect. Include suggestions about how to deal with each failure mode and its effects.

13.7 Apply the FMEA to your class project. Consider all failure modes and their effects, their severity, and their probability. Include solutions to remedy each failure mode or to make your design fail-safe.

CHAPTER 14

INTELLECTUAL PROPERTY PROTECTION
PATENTS, TRADEMARKS, COPYRIGHTS

14.1 INTRODUCTION

By now you probably realize that you are a much better and more creative designer, and as such, you can better solve problems and perhaps, even come up with new ideas for products and systems. This chapter explains how to protect your intellectual properties, from the beginning of the project until it is registered in the form of a patent, a copyright, or a trademark. We will also discuss nondisclosure statements and how to document your design activities.

14.2 PATENTS

Patents are used for the protection of new products, machines, processes, new compositions of matter, and new uses for or improvements of the above. There are three types of patents: utility patents, design patents, and plant patents. Plant patents are issued for new plants that are created asexually (not from a seed). Since this is unrelated to our discussion in this book, we will not consider plant patents at all. Design and utility patents are discussed below.

14.3 WHAT IS A PATENT?

A (US) patent is a grant by the (US) government of property right to the inventor, issued by the U.S. Patent and Trademark Office (USPTO). The patent gives the patent-holder the *right to exclude others from using, making, offering for sale, selling, or importing, the patented item in the United States.* Unlike popular belief, it does not give the inventor the right to use, make, offer for sale, sell, or import the invention. It only excludes others from doing so.

  The above has further implications too. Imagine that someone has just invented a gadget that makes your CD player play and digitally record two programs simultaneously. Also imagine that the player still has a valid patent. The new patent issued to the inventor of the gadget will exclude the player patent-holder from using the new gadget, but it does not give the new inventor the right to use, sell, make, or import a player. He can only make and sell the gadget for which he holds a patent. Therefore, without either permission or an assignment from the CD player patent-holder, he will not be able to implement the new invention on a player. Similarly, imagine that someone has just been issued a patent for a new drug. The patent will exclude others from using, making, selling, or importing the drug in the United States. However, it does not give the inventor the right to sell, make, or use the drug until such time as all other regulations and requirements such as clinical testing, FDA approval, etc. have been met.

Another reason for this distinction is that, unlike popular belief, although the patent is issued in the name of an inventor, the inventor may have no right to use the patent if the right is sold or assigned to a third party. In that case, the inventor is himself excluded from using, making, selling, or importing the invention.

The primary reason behind patents is to encourage economic activities by providing a sense of protection to the inventor, and to allow the individuals and institutions to feel confident enough to invest in the new intellectual property without fear of being copied and forced out of the market by competitors. This additional economic activity will increase the GNP of the nation, and will generate additional taxes (which is the ultimate goal of the government in issuing the patent). Understanding this principle will help you better understand the reasoning behind certain regulations and laws related to patents that will be discussed later.

14.4 UTILITY PATENTS

Utility patents are what most people refer to as patents. They are the most powerful, most useful, and most expensive type of patents. Utility patents remain in effect for 20 years from the date of application, so long as the maintenance fees are paid on time. Certain other factors such as a provisional patent application can have a minor effect on the total effective length of a patent (adding up to one year), but in general, the start date is when the application is first submitted to the United States Patent and Trademark Office. Since most applications take less than 3 years before a patent is issued, the new process is better than the old 17 years from the date of issue. As a result of Uruguay Round Agreements Act, the new law went into effect on June 8, 1995, satisfying international agreements and to end a practice called "submarine patents," in which delaying tactics were used to postpone the date of issue of a patent until the industry related to it matured. At that point, when the patent was issued, the patent holder would take legal action against anyone who had used the "not yet patented" idea.[4] A utility patent may be extended by the Commissioner for up to 5 years beyond the normal 20 years for pharmaceutical inventions only in order to compensate for marketing delays due to regulatory agencies. Other patents can only be extended for certain circumstances (such as delays) as provided by law. Otherwise, patents may not be extended beyond the normal 20 years. Later, the American Inventor's Protection Act of 1999 decreed that all patent applications must be published within 18 months of filing. Therefore, all applications from that date are published by the USPTO even if not granted yet.

The usefulness of a patent is related to how powerful the patent is, but also to how it is written. A badly written patent with limiting definitions and descriptions in its claims or with claims that are incomplete and too narrow can render a patent useless. This is why it is important that you consult professionals for writing and executing your patent application.

The United States is a first-to-invent country (versus first-to-file). This will be discussed in more detail later and in relation to international patents.

14.5 REQUIREMENTS FOR UTILITY PATENTS

Requirements for utility patents include statutory classes, usefulness, novelty, nonobviousness, and ownership. We will discuss each of these requirements in the following sections.

14.5.1 Statutory Classes

A utility patent must be one of the following five statutory classes: Manufacture; Machine; Process; Composite of matter; New use for or improvements of the above. If the subject does not fall within one of these five classes, a patent will not be issued. This is one way to understand the difference between a copyright and a patent, for a written novel does not fall within any of these classes. Neither does a method of doing business, which is not patentable. An abstract idea in itself is not patentable either, unless it is reduced to a machine, a process, a manufacture, or a composite of matter.

A manufacture, a machine, a process, or a new composite of matter may individually and independently or collectively be patentable. Each refers to a different part of the process of new creations (inventions).

A *manufacture* is a product or device that has not been in existence before. As an example, consider a new bolt with a new feature. Imagine that this new bolt is designed such that it could accept a nut, be used like a bolt, and be loaded axially and still remain laterally flexible. Whether a new process, a new machine, or a new material has been used in this product or not, the manufacture itself can be patented.

A *machine* refers to a device that performs a certain task and which can be used to make other machines or products, whether new or not. Assume that the new bolt mentioned above requires a new machine to produce a new type of thread for the bolt. In that case, not only the product (manufacture), but also the machine that produces it may be patented. Still, it is also possible that the new machine may be used to produce known, existing products that cannot be patented. Nevertheless, the machine can still be patented.

A *process* refers to a new method of creating a machine or a manufacture. For example, a new method of manufacturing the bolt by depositing layers of polymers over each other and hardening the material by heat or ultraviolet may have been used. This process may be used to create existing products as well. In either case, the new process can be patented. The chemical process of making a new drug falls under this category.

A *composite of matter* refers to new materials that have not existed previously. Once again, imagine that the polymer that is used in the above-mentioned process is new, such that it will harden with exposure to ultraviolet light. Whether it is used for a new product such as the new bolt or not, and whether it is used in a new machine as described above or not, and whether the process is new or not, the new composite of matter can be patented. New drugs fall under this category as well and are patentable.

It is very important to realize that a composite of matter requires a chemical reaction. Hence, in most cases, simple mixtures do not qualify for patents. This can be better understood by realizing that in most cases, mixtures retain the characteristics of their individual ingredients. For example, soda is composed of water, sugar, colorings, flavorings, and carbon dioxide. The mixture is fluid, it is sweet, it will have the color of the added food coloring, it will taste the same as the added flavors, and it will fizz with the carbon dioxide, all similar characteristics to the original ingredients. Therefore, sodas cannot be patented. This is precisely why the exact mixture of brand-name sodas are still kept as trade secrets. In reality, with available processes today, it is very easy to determine the exact mixture of the ingredients in a soda. Anyone can also use, make, sell, or import the exact mixture as well, since it is not patented. But there is a chance that the individual may break trade-secret laws by doing so. However, the brand name of the product is registered as a trademark, and may not be used by others. Consequently, although the soda can be copied exactly, it may not be sold under the same brand name. Nevertheless, when you mix two chemicals that react in such a way that the result is something new that does not have the same characteristics of the original ingredients, the new material can be patented. This is true whether the product is the result of simple mixing of other materials that have reacted chemically, or due to an elaborate process with multiple steps. An exception to the above is when the mixture does not chemically react, but still has characteristics different from the ingredients. As an example, concrete is a mixture of sand, cement, and water. However, it is not fluid, it is not a powder, and it is not grainy. New alloys fall under this category as well and can be patented.

New uses for or improvements of the above indicates that if you make improvements on an existing patent, the improvements can be patented, as can any new uses you may find for an already existing patent. You will not have the right to use the product, but you can patent the new use for it.

14.5.2 Utility

An invention must be useful; otherwise, there is no need to exclude others from using it if the invention is not useful. This goes back to the point made previously about the reason behind patents. As was mentioned earlier, patents are issued to encourage the inventor in disclosing it and using it, so as to increase GNP, economic activities, and tax collection. If an invention is not useful, it will not create new economic activities, and thus, there is no need for it.

Based on this requirement, the following are not patentable:

- Unsafe new drugs because they cannot be safely used.
- Whimsical inventions.
- Inventions that can only be used for illegal purposes, because no one is allowed to use them.
- Immoral inventions (although not enforced).
- Nonoperable inventions because they do not function. Perpetual machines fall under this category and are not patentable.
- Theoretical phenomena/ideas. You must solve a problem, not just find it. Hence, the idea must be reduced to practice.

In fact, there was a time when the patent office required that the invention be submitted for examination. However, as the patent office ran out of room within a very short time, the requirement was lifted. It is now assumed that the invention is in fact made and it works, although it is not required that the invention be submitted.

Based on the Atomic Energy Act of 1954, inventions that are solely used for the utilization of nuclear weapons are not patentable either. This is primarily for national security reasons.

14.5.3 Nonobvious

An invention must be nonobvious. This means that simple variations made in the invention, e.g., dimensions of a patented item or materials used, are not sufficient for an improvement in a patent to warrant a new patent. The subject matter for which a patent is sought must be sufficiently different from the *prior art* to make it nonobvious to a person having ordinary skills in the subject area related to the invention. If a person with this level of skill could think of this variation, it is obvious. This is one of the reasons why patent drawings do not have dimensions. It is assumed that changing dimensions is not a sufficient reason for noninfringement.

14.5.4 Ownership

The patent is always issued to the inventor, even if the inventor has assigned all the rights to another entity. Therefore, a corporation who employs the inventor cannot apply for, nor is it issued, a patent. The employee must apply for the patent even if the company owns the right and is paying for all the costs associated with the patent.

Many corporations and employers require that before employees start working for them they sign a contract assigning all future rights to any patents the employees receive to the company, and to cooperate with the employer in applying and securing patents for their inventions. In these cases, the patent will still be in the name of the inventors, but all the rights to the patent are assigned to the company. You must always be prudent and carefully read the contract before you sign it, because the contract may severely limit what you may do in the future, even after you leave the company.

If more than one person has contributed intellectually to the invention, the patent must be issued in their collective names. It is illegal for an individual from a group to apply for a patent without the others. However, each individual must have contributed intellectually to the invention. Therefore, if someone makes a model based on your drawings, the contribution is not intellectual. Similarly, someone who only provides financial support is not a joint inventor and cannot be joined in the application as an inventor. But if someone contributes an idea that becomes a part of the invention, that person has contributed intellectually, and thus, is entitled to be part of the patent application.

If the right to the patent is not already assigned, then all the patent holders have an equal and similar right to the patent. Every one of the patent holders has the right to use, make, sell, or import the invention. The right to the patent may also be sold by any one of the patent holders, but not exclusively. In other words, any one of the patent holders may sell the right to use, make, sell, or import the invention to any other party or institution, but other patent holders can do the same simultaneously. In most cases, companies will not enter into an agreement with only one individual patent holder in order to prevent legal challenges, especially if they are interested in exclusive right to the patent.

In order to prevent future problems when more than one person is involved in a patent, it is advisable to make an agreement between the involved individuals at the beginning. The agreement should identify who is to be included in the patent application, who will pay for the costs, how the individuals will cooperate if a patent is granted, how they will enter into agreements with other parties, and how the royalties will be divided among them. Without such agreements, there can be tremendous conflicts in the future if and when a patent is granted.

Obviously, this requirement also means that a person who has not invented the subject matter may not apply for a patent, even if there are no patents. This includes inventions that may or may not be patented in foreign countries. As a result, even if there is an invention in another country that is not patented in the United States, another individual in the United States may use, make, sell, or import it, but may not try to patent it under his own name.

14.5.5 Novelty

The last requirement for a patent is that the subject must be new (novel). This means that the invention must be sufficiently different from other inventions or other items in common use to be considered a new item. A patent application requires that the applicant search all the issued patents since the beginning (prior art) to ensure that the invention is not similar to patented or commonly used items. The USPTO will do the same. The applicant will need to describe and reference the prior art in the application and explain how the new invention improves upon the old and why it is useful.

Additionally, the law specifies that the invention cannot have been in use, described in a publication, be public knowledge, or offered for sale more than one year before a patent application is submitted. As a result, if the inventor describes the invention to others, makes a sample and asks others to test it, or offers any samples for sale, she or he will have only one year to apply for a patent. Obviously, no one else may apply for the patent either before or after the one-year limitation, but even the inventor may not apply for a patent after one year. This is especially important for students in colleges and universities. Many students invent things in their design classes or as part of their theses and projects and must publish their design in final reports that become public. Note that immediate family members and colleagues who work on the same project are not considered public and the one-year rule does not apply to them. International patents have other requirements regarding this issue. We will discuss this later.

14.6 TO PATENT OR NOT TO PATENT: THAT IS THE QUESTION

There are a number of important issues that must be considered in order to determine whether the inventor should apply for a patent or not. These issues can be complicated and difficult to analyze, may require resources, time, and energy, and may involve legal, financial, technical, and marketing issues that cannot be answered by individuals who are not trained or experienced. In the following sections some of these issues are discussed. Nevertheless, you should consult with professionals before deciding.

14.6.1 Patent Search

The first step in determining whether to apply for a patent or not is to make sure that the above five requirements are met. Otherwise, there is no need to continue with spending much time and resources to apply for a patent that will not be granted. This requires that you make sure that the invention, although novel and nonobvious in your mind, is actually not already patented. This is primarily accomplished by performing a patent search. Patent search involves checking through all the issued patents and to make sure that the same invention has not been patented. The search may be done by individuals, by professionals, or both.

The USPTO examiner will also do his/her own patent search before a patent is issued anyway. So why should you do it as well? The reason is that by the time you apply for a patent, you may have spent many thousands of dollars, which will be completely wasted if you do not get a patent. So before you do so, it is strongly recommended that you do a search and ensure your invention is patentable.

Full text copies of all patents since 1976, full image of the first page of all patents since 1790, as well as full text copies of all patent applications since March 2001 are on the web site of the USPTO, and may be accessed by any individual. To do this, go to http://www.uspto.gov and search for patents. The search may be performed on titles, abstracts, inventor names, patent numbers, issue dates, claims, assignee names, application dates, and many other fields. Copies of any patents may also be ordered from the USPTO for a small fee. Since this search is so readily available to everyone, it is highly recommended that you do your own preliminary search to see if any patents similar to your invention can be found. Obviously you cannot assume that the content of any given patent may be understood well enough to determine whether your invention interferes with or infringes upon any other patent, especially if the invention is highly technical. In that case you will need the services of a professional to interpret the claims in a patent against your invention. However, this preliminary search is so easy and inexpensive that you should definitely consider doing it.

Other venues for a personal patent search include searching through the documents and publications provided by the USPTO, including the published patents in the Patent and Trademark Depository Libraries (PTDL) in major public libraries and the Official Gazette of the Patent Office. You should contact the patent office or your local library for the nearest depository of patents and Gazettes.

The next step in a patent search is to request a professional agent to do a similar search. In most cases, your attorney contacts professional agents who specialize in patent searches. However, a fee is required for this service. Nevertheless, in most cases, this fee is small compared to the cost of preparing and submitting a patent application. Therefore, it is still highly advisable to perform a thorough and professional patent search before you embark on applying for a patent.

Obviously there are millions of patents, and no one would want to read every patent to find a possible interference. Fortunately, all patents are classified by subject matter, permitting easy reference to patents in a particular technology or area. U.S. Patent Classification System contains over 400 classes and over 136,000 subclasses. Additionally, the USPTO publishes the Index to the Patent Classification System, a publication that contains an alphabetical list of approximately 65,000 common informal headings or terms that refer to specific classes and subclasses used to categorize patents. This index can be used to search for particular technologies, classes, and subclasses related to your invention.

14.6.2   Patenting versus Trade Secret

Another important consideration in deciding whether your invention should be patented or not relates to its value and how to protect it. A patent gives you the right to exclude others from using your invention for 20 years from the date of application in return for disclosing all the details about it. During this time period, others will have the full knowledge of your invention, but cannot make, use, sell, or import it. However, after the 20 years are over, anyone can use your invention without having to pay royalties or infringe on the patent. In other words, after the 20-year period, the patent will be worthless. Additionally, your competitors may even try to circumvent your patent by creating new improvements on your invention and patenting those improvements, or by creating their own inventions. In both cases, your patent may be less valuable as time goes by. On the other hand, imagine that you have an invention that is not easy for others to copy if they do not know the details of how you do it, the processes that are used, or the technological advancements you have utilized in manufacturing it. All they have available is the product itself. In that case, you may decide to not patent the invention, and thus forgo disclosing its details to others, and instead, treat it as a trade secret. As long as a trade secret is maintained successfully, your invention will not lose its value, and you may continue to dominate the market. Many products enjoy this status, and this is why industrial espionage is such a prevalent crime in industry. Many common drinking sodas are still trade secrets, although they could not be patented anyway. Even if they were patented, they would have been worthless a long time ago. However, as a trade secret, they still enjoy their value. In this case, a few select individuals in a central plant prepare the common extract and then ship it to the bottling companies that add water and carbon dioxide to it and fill the bottles (called bottling plants).

Trade secrets are appropriate for special cases when the technology of the product, the process, or the manufacture is such that others will not be able to copy it without detailed information. This may not be possible for many common, but still valuable, inventions that can be easily copied if a patent does not exist. Another option for these inventions is to take the risk of manufacturing a large quantity of your invention, then flood the market with it. By the time your competitors can make their own design and offer it for sale, you will be the only choice and will dominate the market, will establish your market share, and may enjoy good income. When your competitors enter the market, you will have to compete with them in price, quality, and name recognition. In fact, in some cases, a product may not even be patentable, but this strategy may still work. An example of this is the dancing flowers that appeared in the market a few years ago and were sold for over $30 initially, but in less than a year their price was reduced to about $2.

14.6.3   Cost of Applying for a Patent

Another important consideration in whether to apply for a patent or not is the cost involved in the preparation and applying for a patent. By the time you pay for a patent search, fees for the preparation of the patent application by an attorney, submission fees, as well as maintenance fees, a patent may easily cost $15,000 to $20,000. If you think your invention has a good chance of returning your investment, the patent will be worth the risk. Otherwise, you should consider the benefits of the patent versus the cost of it. There are thousands of patents each year that do not earn any royalty or benefit. You should also consider a phenomenon called ''not invented here (NIH).'' According to this theory, many companies have little interest in buying the rights to patents that are not obtained directly by their employees. However, this is not necessarily true for every company, and there are countless examples where a patent obtained by individuals has even been embraced by multiple large companies (e.g., antilock brake system and airbags).

14.6.4   Direct Licensing

Although risky, it is also possible to try to license your intellectual property to an interested enterprise without a patent. You must certainly try to properly protect your intellectual property before disclosing it to anyone by obtaining a solid non-disclosure statement (which will be discussed later). There may be a chance that other institutions may license your intellectual property and try to get a patent on their own, with your cooperation of course. Clearly, your cost and risk are lower, but you will most probably receive lower royalties as well.

Most experts do not recommend that inventors use the services of establishments that advertise obtaining patents for individuals at low prices or for no fees.

14.7   PROCESS OF PATENTING

Assuming that you have already decided that you want to patent your invention and you have done a thorough patent search to ensure that, at least to the best of your knowledge, your invention is patentable, the next step is to prepare a patent application. Patent applications must follow a standard format. This information is available from the USPTO website and from off-line resources.[1,2,3] The application may be prepared either by individuals, including you, or by professionals, although in general, USPTO requires that the application be filed by a registered patent agent. In most cases, it is highly recommended that you seek the assistance of a professional patent attorney for the preparation of the patent application. Although you may (and should) do as much as you can yourself, and although the patent attorney will require your participation in preparing the application, it is important to file the application properly. First, if you do the preliminary work of writing about your invention, thinking about its applications, why it is different from other inventions, and what improvements your invention has over other products, you will save much money in attorney fees. Shorter time spent on your application results in smaller fees. Second, your preliminary work will enhance your application and will make it stronger since your responses to the attorney will be more accurate and well thought-out. However, a patent is only as strong as its contents and what is claimed. Therefore, it is very important to write the patent application and the claims in such a form that will give it credibility to stand examination and trial by the USPTO, the market, and the public, and to prevent others from easily bypassing your claims and rendering it useless. This will be further discussed later.

Patent applications usually include (a) a written document called specifications, which includes the description of the invention and the claims, as well as an oath or disclaimer; (b) a drawing, if necessary; and (c) the filing fee.

Most patent applications require drawings. Patent drawings are neither engineering drawings with dimensions nor perspective drawings. They have their own standard and format. The drawings are supposed to inform the reader about the invention, but they are not supposed to restrict the invention by dimensions, by specific material, or by specific ways of assembly, unless these are an integral part of the claims. It is also strongly recommended that you seek the assistance of a professional to prepare your drawings (if you have a patent attorney, she or he will arrange for the preparation of the drawings).

Each application received by the patent office will be examined for adhering to the rules and for completeness. The accepted application is then assigned to an examiner in the particular area of technology that relates to the application and is examined in the order it is received. The examiner will perform a thorough patent search through the U.S. patents, foreign patents, and available literature, to determine whether there is any interference between the claims of the application and any other existing patents. The decisions made by the examiner are transmitted to the applicant or his submitting attorney of record through a notice called "action." If there is any interference, the adverse action or the objection to the application will state the reasons why a patent is not issued, and information about references is given that may help the individual decide how to proceed with the modification of the application, or withdrawal. If claims are found to be obvious or lack novelty, they can be rejected too. It is very common that in most applications, some or many claims may be rejected. Lack of rejection of claims usually means that either the intellectual property is unusually strong or the claims are too narrow.

If the decision is made to continue, the application may be amended or modified, and is subsequently resubmitted with a written request for reconsideration. Each request must clearly and distinctly indicate where the examiner has erred in his judgment, or must clearly indicate why the modified claims are fit to be patented in reference to the prior art and the examiner's objections. A mere statement that the examiner has erred is not sufficient. The application will be reconsidered based on the provided information, and claims are allowed, objected to, or denied. The second office action is usually the final decision. In this case, the applicant's response is usually limited to an appeal to the Commissioner. Further amendments are not allowed.

If an adverse final decision has been made, or if the applicant does not agree with the examiner who insists on rejecting the application, an appeal may be filed with the Board of Patent Appeals and Interferences in the USPTO. This Board consists of the

Commissioner, Deputy Commissioner, the Assistant Commissioners, and administrative patent judges. Each appeal is usually heard by only three Board members. If the decision made by the Board is still adverse and the applicant is unsatisfied, an appeal may still be taken to the Court of Appeals for the Federal Circuit, or a civil action suit may be brought against the Commissioner in the U.S. District Court in the District of Columbia. The court will either affirm or reverse the decision made by the Board. Each appeal requires additional fees.

If the patent application is found allowable during examination or as a result of appeal actions, a notice of allowance is issued to the applicant or the attorney of record. Usually within three months, an issue fee is due. If the fee is not paid in time, the application is considered to be abandoned. When the fee is paid, the patent is issued as soon as possible depending on the volume of printing at hand. On the day the patent is issued, it also becomes public.

The patent office receives close to 5 million pieces of mail each year, with over 300,000 patent applications. At this writing, USPTO has close to 3,000 patent examiners. This means that on average, each examiner may be responsible for 60–70 patents per year. As a result, it is possible that a patent may stay at the USPTO for a long time (perhaps over a year) before any action is taken.

14.8 CLAIMS

Claims are the essence of any patent application and are effectively what a patent is about. In fact, what the patent protects is only what is stated in the claims section. The description portion of the application does not provide any protection, and does not exclude others from using, making, selling, or importing the invention. Therefore, if the invention or any parts of it are not claimed in the application, there will be no protection.

The way claims are written is extremely important. If the claims are written with narrow definitions, the patent may be useless because others can easily extend the idea into their own methods of accomplishing the same thing and not interfere with the claims. On the other hand, if the claims are too broad, they may interfere with other patents that are similar, and thus, the patent may become unacceptable during examination. As a result, it is strongly recommended that the inventor seek the help of professionals to write the claims as broadly as possible in order to encompass as much as possible, and still be specific enough to be patentable.

Claims are also used for the analysis of infringement issues. In order to prevent others from misinterpreting or misrepresenting your claims in a court, it is important to word the claims with sufficient reference. The following claims from a patent illustrate this point:

What is claimed is:

1. A fastener, comprising: a shank covered with a spiral set of teeth; and a flexible core material running internally to said shank and along the length of said shank; and coupling means for connecting said flexible core material to said shank at both distal ends of said shank so that said fastener is laterally flexible along its length.

2. The fastener of claim 1, wherein said coupling means is welding.

3. The fastener of claim 1, wherein said coupling means is pinching.

4. The fastener of claim 1, wherein said flexible core material comprises innermost wires grouped and bound with wires spirally wound in opposing directions.

As is shown, each part of the sentence in the first claim is referred back to other parts by the word ''said'' in order to ensure that no one will claim that it refers to other things. The next three claims are referred back to the first one. The remaining claims will continue to refer back to other ones for the same reason. Since this is not the way normal text is written, it is difficult for nonprofessionals to ensure that the claims will be written adequately. The consequence may be that a prudent attorney may be able to later challenge your patent and render it useless by misinterpretation or misrepresentation. Kirk Teska[5] describes how the interpretation of the words in a patent regarding razors eventually ended up in the Court of Appeals between the rivals Gillette Co. and Energizer Holdings Inc. In this case, the patent owned by Gillette claimed a razor with a group of first, second, and third blades, each progressively positioned further outward from the blade below it. The Quattro razor by Schick has four blades, or depending on how you interpret it, two ''second'' blades. The point here is not to investigate who is right, as even the courts have not settled the issue yet. Instead, the point is to realize the power of words and the interpretation of such in a patent.

Claims are also used to restrict the utility and usefulness of a competitor's patent. Many large companies file for a large number of claims (or individual patent applications) that relate to a legitimate patent owned by their competitor. The result of this action is that the competitor will not be able to improve the original patent, or at times, even deviate from the exact claims of the patent. This will stifle the usefulness of the original patent and will effectively choke it and may cause it to quickly become obsolete. Therefore, it is important that the claims be as broad as possible and encompass as many variations as possible.

14.9 APPLICATION AND MAINTENANCE FEES

Fees are required for many different aspects of the application process and the maintenance of a patent. The fees related to the submission of a utility application include, but are not limited to, the basic filing fee ($150 for small entities and $300 for large entities as of 2006), a fee for independent claims in excess of three, a fee for claims in excess of 20, a fee for multiple dependent claims, and a fee for non-English specifications. Additional fees are required for the issuance and for the publication of a patent. Other miscellaneous fees also apply for other purposes such as continued examination, for expedited examination of a design application, submission of informal disclosure statement and many others.

Fees for large entities are generally twice as much as the fees for small entities. Small entities include individual inventors, nonprofit organizations, universities, and small businesses.

Maintenance fees must also be paid at 3.5, 7.5, and 11.5 years for a patent to remain in effect. If fees are not paid, the patent is considered abandoned and the subject becomes public. The maintenance fees in 2006 were respectively $900, $2,300, and $3,800 (a total of $7,000) for large entities and half as much for small entities.

As you notice, maintenance fees become higher as time goes by. As was mentioned earlier, the main reason for the issuance of patents is to encourage economic activities that generate income, and thus, a larger tax base. A patent that is not commercialized either by the inventor or through licensing is useless. The maintenance fees are supposed to encourage the inventor to commercialize the patent and generate income. If there is no income from a patent, and if the patentee is not willing to pay the fees, then the patent becomes null and everyone has the right to exploit it. If the patentee has exploited the patent and has earned income, she or he should be happy to pay the fees to maintain the patent rights. This is also why the fees increase as time goes by. Even if by 3.5 years a patent is not yet commercialized, there is still hope that it may be. By 11.5 years, there is probably little chance that a patent will be exploited for the first time. Therefore, the increased maintenance fee will discourage a patentee from keeping the patent in effect for 20 years without exploiting it.

No maintenance fees are required for design patents (can you guess why?).

14.10   INTERFERENCE AND DILIGENCE

The United States is a "first to conceive" country. This means that if two individuals or groups of individuals independently file for substantially the same patents, the individual or group who was first to conceive the idea will be entitled to receive the patent. This is called interference, and in these cases, an interference proceeding is initiated by the USPTO to determine who gets the patent. The same may happen between a patent application and the claims in an issued patent if the patent was not issued more than one year prior to the date of application. However, USPTO will not involve itself with interference between two issued patents, for which a court must decide. About 1 percent of applications become involved in interference proceedings.

Each party will have to submit documents proving the dates of "the conception of the invention" and "reduction to practice." Date of invention is when the methods or ideas about the invention were first completed. Actual reduction to practice relates to the date when the invention was actually made into physical form. Constructive reduction to practice relates to the date when the patent application is submitted to the USPTO. The applicants must show proof of the dates for the above. Otherwise, the date when the patent application is submitted will be considered as the date of invention. The inventor who proves to be the first to conceive the invention and the first to reduce it to practice will be considered to be the first inventor and entitled to the patent. In general, matters are much more complicated and need to be carefully analyzed by the USPTO. One complicating matter is diligence.

In many cases, the diligence of the inventors becomes a serious issue as well. If an inventor conceives an idea and reduces it to practice before another inventor, his diligence will not play a role. However, imagine a case when an inventor, without a good reason, delays the work of reducing the idea to practice or applying for a patent. If another inventor has conceived a similar idea later than the first inventor, but he reduces the idea to practice and applies for a patent, the second inventor will be entitled to the patent. This is true because he has shown more diligence. As was mentioned earlier, the main purpose for issuing a patent is to increase economic activities and generate more income. Someone who is more diligent in reducing an idea to practice and to patenting it may also be more diligent in utilizing the patent and exploiting the invention and generating more income, and consequently, taxes. This is why it is so important to keep acceptable documentation about your invention. We will discuss this in detail later.

14.11   PROVISIONAL PATENT APPLICATION

A provisional patent application is a simple but effective way to protect an invention for up to one year while an actual patent application is prepared. It allows the applicant to file the utility patent application up to one year later while enjoying the following benefits:

- The provisional patent application establishes a filing date for the later full patent application. In case there will be an interference proceeding between this application and another application, the provisional application will have established an earlier date. The 20-year life of the patent starts from the date the actual application is filed.

- In many countries, it is required by law that an invention not become public in any form before an application is filed. The United States law allows for up to one year of public disclosure of an invention before a patent application is required. The provisional patent application provides for both. It establishes proof that the patent was first filed before becoming public, but also allows the applicant to disclose the patent to the public up to one year before the actual application is filed.

- It is required that an applicant submit a patent application before the invention is designated by "*patent pending*" marking. The provisional application also provides for this designation.

- Since a provisional patent application provides a basic protection mechanism for inventions up to one year, and since the 20-year life of a patent starts from the date a nonprovisional patent application is filed, it is practically possible to extend the life of an invention by one year. However, you must make sure that an attorney will help you with the requirements.

A provisional patent application consists of a description of the invention with enough detail to ensure that the invention is fully described. However, no oaths or declarations are necessary. There are also no claims in a provisional patent application. There is no need to cite references either; thus, no patent search is required. The provisional patent application is *not* examined by the Patent Office. USPTO will only examine the provisional application for required content and filing fees, will send a notice to the applicant or his designee, and will file the application for one year. If no other action is taken by the applicant to submit a nonprovisional application before the one-year timeframe is over, the provisional patent application will be abandoned (it will not be made public).

Neither provisional nor nonprovisional patent applications provide for any right to the applicant to exclude others from using, making, selling, or importing an invention until the actual patent is granted. In that sense, a provisional application is limited in its protection as well. However, if and when a patent is issued, the right to exclude others will become effective.

Provisional patent applications are very inexpensive and individuals can submit their own applications easily. However, you must make certain that all information about your patent is included in the application, as no new information may be alluded to in the nonprovisional application that refers to the provisional application. In other words, if you would like to connect the nonprovisional application to the provisional application and enjoy its benefits of earlier filing date, foreign patent filing requirements, claims, etc., the basic material in the provisional should be essentially the same as in the nonprovisional. This does not mean that they must contain the same language and content, but all essential information about the invention must already be in the provisional application.

The filing fee for a provisional patent application as of 2003 is $80 for small entities and $160 for large entities. Provisional patent applications may not be filed for design patents.

14.12 INFRINGEMENT OF PATENTS

Infringement of a patent involves making, using, importing, selling or offering for sale any patented invention that is not yours in the United Sates and its possessions and territories. This includes manufacturing a patented invention in the Unites States for exclusive export or for sale to other countries where the invention is not patented. It also includes importing the invention into the United States from other countries where the invention is not patented, even if it is manufactured there. However, if an invention is not patented in other countries, you cannot exclude others from making, using, selling, or importing it into and between those countries. The United States government may use any patented invention without permission of the patent holder, but the patentee is entitled to fair compensation.

The Patent Office will not enforce the patents. This means that if another entity makes, uses, sells, or imports your invention without your permission, the patent office will not take legal action against the entity. You must take action by first informing the entity that they are in violation of your rights. If they do not stop with their infringement, you will need to take legal action against them in a Federal court in order to enforce your patent rights. The final decision will be made in the court as to whether or not the other entity is infringing upon your patent. However, this can only be done if you have a patent. An application for a patent, and consequently, patent pending status, is not enough to enforce a patent. Others may continue to use, make, sell, offer for sale, or import, any invention that is not yet formally patented. However, they will have to stop as soon as the patent is issued. They will have to stop all actions no matter how much is spent in investments, how many products are already made, or how good their product is. Still, you may not claim any damages for their use of your patent before the patent was issued. Obviously, the other party may also try to prove your patent invalid or prove that they are not infringing your patent (based on interpretations of the claims). Either party may appeal the final decision of the court in higher courts.

14.13   PATENT MARKING AND PATENT PENDING

When a product is manufactured in association with a patent or an assignment by the patentee, the product must be marked with the word ''patent number'' and the numbers of the patents that are involved. Otherwise, the patentee may not be entitled to any recovery of damages from an infringer unless the infringer has been duly notified and he has ignored the notice.

If a patent application is submitted, the applicant may mark the products that involve the invention with ''patent pending'' or ''patent applied for.'' These two, or similar markings, have no legal value; there is no protection unless a patent is granted. However, these markings can forewarn others that a patent application has been submitted and that as soon as a patent is granted they will have to stop using, making, selling, or importing the article, even if they have invested and manufactured the article.

It is unlawful to either mark an article falsely with a patent designation and number, or marking with ''patent pending'' and ''patent applied for'' or similar designations if no patent application has been submitted. As was mentioned earlier, filing a provisional patent application allows the use of the above-mentioned markings.

14.14   INTERNATIONAL PATENTS

Patents are granted by individual governments for their own territories. This means that if you receive a patent in the United States, it is only valid in the United States and its territories and possessions. This also means that the exclusion of others from making, using, selling, or importing your invention is only valid in the United States and its territories and possessions. Hence, others can legally manufacture, sell, import, or use your invention in all other countries and between all other countries. This also means that even if someone makes your invention in other countries, no one but you (or your assignee) may import the invention into the United States.

If you would like to have your invention protected in any other country, you must apply for a patent in that country as well. And if you would like this right in multiple countries, you must apply for patents in all those countries. Since each country has its own patent laws and associated fees, you will have to apply for a patent in other countries based on their own requirements, regulations, and fees. In short, there are no international patents yet.

Many countries require that a patent application be submitted before the invention is made public in any form. This is different than in the United States, where you may make your invention public for up to one year before you must apply for a patent or relinquish your right to apply. Therefore, it is very important to remember that if you plan to apply for a patent in other countries, you should be careful about making your invention public by presenting it to the public, by offering it for sale, or by publishing an article about it. (In the case of educational institutions, this is true even for a report written for a class if the report is made public such that others find access to the information about the invention, for example, by filing it in the library.) One very effective way to circumvent this problem is to submit a provisional patent application before making your invention public. This way, not only will you establish priority of date in the United States, you will also be protected if you later decide to apply in other countries. The provisional patent application will establish that you have applied for a patent before making your invention public, and this is enough for those countries.

Other differences between different countries include the requirement in some countries that the patented invention be manufactured in that country within a certain period of time, usually 3 years. Otherwise, the patent will be void, or the patent holder must assign the right to the patent to anyone who applies for a license.

There are two treaties that have had some positive effects on this process and have created some uniformity and commonality between the laws of different countries. One is called the *Paris Convention for the Protection of Industrial Property*, which is adhered to by over 140 countries, including the United States. The Paris Convention provides that each signatory country will guarantee that the citizens of all other countries who apply for a patent or trademark in these countries will have the same rights and protections as their own citizens. It also provides for the *right of priority,* which means that the applicant has the right to apply in all other countries within a certain period of time after the first application in any country. The date of the first application in any member country will be used as the date of application for all other subsequent applications. This means that the applicant will have priority over other applicants who may file in other countries after he has applied in the first country but before he has a chance to apply in other countries. This is also why a provisional patent application will protect the applicant from making an invention public before a nonprovisional patent application is submitted.

Another treaty, known as the *Patent Cooperation Treaty,* is adhered to by over 90 countries including the United States. This treaty provides for a centralized filing procedure and a standard application format, making it much easier to apply for multiple foreign patents.

It is important to know that under United States law, it is required that you obtain a license from the Commissioner of Patents and Trademarks before you apply for a foreign patent if the invention was made in the United States. The filing of an application in the United States automatically constitutes a request for the license. The approval or denial of a license is made when a filing receipt is issued to the applicant. If you want to first apply in another country, or if you apply for a foreign patent within 6 months from the date of your U.S. patent application, you must request for this license.

The United States does not discriminate against foreign applicants. The same requirements are applied to those applicants as domestic applicants, regardless of the laws of those countries. However, no patents will be issued to foreign patent holders if their application to the USPTO is made later than 12 months from the date when the foreign patent was issued.

14.15 DOCUMENTATION AND RECORD KEEPING

It is essential that an inventor keeps complete and accurate record of all activities that relate to the process of conception and reduction to practice of the invention. This is especially important if at any time it becomes necessary to provide evidence of ownership, time of conception, or diligence, in matters relating to interference or contesting against claims in a court.

As we have discussed previously, in case two inventors apply for the same patent independently of each other, the patent office will declare interference. The patent will be issued to the person who first conceived the idea, unless lack of diligence can be proven. In that case, one who may have conceived the idea later than the first one, but who diligently reduced the idea to practice and who has diligently applied for a patent may be entitled to the patent. Therefore, it is essential that the inventor be able to prove, beyond any reasonable doubt, that he was the person who first conceived the idea, and to show that he diligently worked to reduce the idea to practice. If one person takes longer than another to reduce an idea to practice due to reasonable delays such as illness, other work that takes precedence, and responsibilities beyond his or her control, she or he will still be considered diligent. However, this must be documented to be acceptable.

The most common and acceptable way to keep record is to keep a notebook, in which, regular entries of all activities are kept, witnessed, and signed off. To do this, you must have a notebook or a diary, preferably one with bound paper and all pages numbered, so that no new matter may be added or removed later. All activities relating to the invention, starting with the conception of the idea to the reduction to practice must be entered into the notebook on a regular basis, even daily. The patent office and patent attorneys recommend that all entries be read by at least one, but preferably two, witnesses, who will sign and date the entry with ''I have read and understood the above.'' The witnesses may neither be your family members, nor your colleagues who are part of the same project or invention. Obviously, this means that you must have witnesses who have enough expertise to read your entries, sometimes in small parts, which may be very technical, and understand them. Finding such witnesses can be very difficult. It can also jeopardize your invention, since many will read your documents and will understand them. Still, it is important to keep the records acceptable to the courts.

You must not leave any blank areas in your notebook without crossing them out. This will prevent suspicion that material may have been added to the notebook later and will ensure that the notebook will be a trustworthy document in the court. If you add entries to the notebook later than the date the activity occurred, you must indicate so in the notebook and have it witnessed with the current date. These minimal efforts by you will indicate to the court that you are honest, and so must be your notebook.

14.16 LICENSES AND ASSIGNMENTS

A patent is personal property that may be sold, assigned, transferred, mortgaged, be given to heirs in a will, or licensed. In fact, in the U.S. tax return system, patent royalties are claimed on the same form as rental property. Therefore, a patent holder may either use the right of exclusion afforded by the patent himself, or may assign it to others exclusively or nonexclusively. As part of the original application, the applicant must declare whether she or he already has an assignment agreement with another entity or not. Consequently, if you work for someone or a company, or if you have already sold your interest in a patent to another entity, the assignment will be part of the application. However, note that even if you have sold or assigned all of your interest to another entity, the patent application must be in *your* name and the patent will be issued in *your* name.

Many companies require that a prospective employee sign a document assigning to the company all the rights to all patents that result from their work in the company. This is understandable, since the company hires engineers and designers to produce worthy inventions for the business to survive. However, you must be careful about these documents, since at times, the company will claim rights to patents that you may obtain outside of your employment, from your previous work, or even future work. If necessary, *you* must exclude specific personal projects and patents or patent applications from this document and make sure that the company agrees to the exclusion.

The assignment of the rights to a patent may be exclusive, nonexclusive, partial, mortgaged, or for limited territories. In exclusive assignments, the assignee has the exclusive right to exclude all others from using, manufacturing, selling, or importing the invention, as the original patent holder did. In a nonexclusive assignment, the assignee gains the right to use, make, sell, or import the invention, but understands that the same right may be assigned to others by the original patent holder. Therefore, he will have to compete with others doing the same business. In partial assignments, the assignee gains a partial right to the patent such as 25 percent, 50 percent, etc. This means that others who may gain the right as well are limited to the remainder of the right only. This way, the assignee will have a better control on the level of competition. The patent right may also be mortgaged, which means that it is assigned for a limited length of time, after which the assignment is revoked. In partial territory assignments, the right to a patent is given only for certain territories, whether exclusive or nonexclusive. This is especially useful when the assignee may not be able to cover large territories, the distribution system may be inadequate, or the product is large and shipping is a problem. In each case, the royalties gained by the assignment may vary greatly. The more rights to a patent are assigned to an entity, the larger the income. Obviously, the total income is directly related to the value of the patented item, and its potential for income generation. However, a more exclusive or complete assignment will warrant larger income. It is also possible to structure the agreement in many different ways. For example, it is possible to assign the right to a patent for a lump sum at the beginning, without any future royalties. In this manner, the patent holder receives an income, regardless of whether or not the invention will be successful. The assignee is taking a risk, assuming that he will be able to generate income from the patent in the future, and is willing to pay for it up front. If the patent is successful, he will enjoy larger income if he is not paying royalties in the future. However, if it is not successful, he has lost his initial investment. This is also a good way of receiving income from a patent if there is a danger that the assignee may decide to not use the patent (e.g., if the patent is bought to "kill the competition" by not using it). On the other hand, the agreement can be for the payment of royalties as the patent is used. In this case, it should be expected that the initial payment to the patent holder might be very small, with the expectation that there will more income in the future. Obviously, if the patent is valuable and successful, both the inventor and the assignee will enjoy larger incomes from it, and vice versa. Always be careful about the agreement you make with others and the way a contract is written. You should consult a professional about contracts to ensure that your income from a patent assignment is protected.

A concept, sometimes referred to as "sunflower," suggests that instead of trying to sell a patent to a big corporation, one may try to assign the rights to a number of small entities who effectively choke the bigger company's ability to succeed. As a result, the large company will eventually become interested in licensing the technology. Many automobile parts were introduced into the mass market the same way. Small manufacturers would introduce the innovation into their product as a competitive edge against the large manufacturers, who eventually were forced to adopt the same systems.

As was discussed earlier, when more than one person is engaged in an invention, they all will be named in the patent, and they all have the same rights, no matter how small their contribution. It is also possible that the right to the patent may be shared between the inventor and another assigned entity. In both cases, each individual patent holder has the right to use, make, sell, or import the invention, but also to assign it to others nonexclusively. Hence it is absolutely important to have an agreement in place among the group alluding to the future rights and responsibilities of each individual before the process of invention begins.

14.17 SAMPLE PATENT

Pages 500 and 501show parts of a sample patent. Only the first and the last pages are shown. The name of the inventor is deliberately removed. It shows the format of a patent, the assignment, the title of the patent, the dates, references, the abstract, the description, the drawings, and the claims. Notice how the different portions of the patent are numbered for easy reference. Obviously the length of a patent is related to how complicated it is and how many claims are included. There are thousands of useless, and even funny, patents around. You may check the USPTO website for other examples.

14.18 CONFIDENTIAL DISCLOSURE STATEMENT

There are many instances when you may have to disclose to others proprietary information, including information about your invention, without making it public. Examples include disclosing your invention to an interested party for partnership or to attract sponsors and investors, disclosure for the purpose of raising money to continue with your invention before a patent application is filed or a patent is issued, or if you desire to treat your ideas as trade secrets. To ensure that your disclosure will not be treated as having made the invention public (which starts your 1-year limit for applying for a patent and endangers your foreign patent potential), and to reduce the risk of others stealing your idea, it may be necessary to make your disclosure confidential. By signing a confidential disclosure statement before you disclose your information, the person or entity to whom you divulge the information will assert that they will treat the information with confidentiality and will not disclose it to anyone else who does not need to know about the information. Obviously, it must be made clear that if the entity has already known about this information, or if they receive it independently of you from other sources, that they will not be bound by the disclosure statement. All documents given to the entity must be clearly marked as confidential as well. Most entities will insist that there be a time limit to the confidentiality of the information unless it is a trade secret.

The following is an example of a confidential disclosure statement. You must adapt the form to your needs before using it. The advice of a professional is recommended if your disclosure is a serious matter.

US006955513B2

(12) **United States Patent**

(10) **Patent No.:** **US 6,955,513 B2**
(45) **Date of Patent:** **Oct. 18, 2005**

(54) **FLEXIBLE FASTENER**

(75) Inventor:

(73) Assignee: **California Polytechnic State University Foundation**, San Luis Obispo, CA (US)

(*) Notice: Subject to any disclaimer, the term of this patent is extended or adjusted under 35 U.S.C. 154(b) by 0 days.

(21) Appl. No.: **10/065,564**

(22) Filed: **Oct. 30, 2002**

(65) **Prior Publication Data**

US 2003/0180117 A1 Sep. 25, 2003

Related U.S. Application Data

(60) Provisional application No. 60/366,165, filed on Mar. 21, 2002.

(51) Int. Cl.^{7} **F16B 35/02**

(52) **U.S. Cl.** **411/382**; 411/392; 411/424; 411/438

(58) **Field of Search** 411/378, 392, 411/424, 438, 411, 901, 902, 383, 384

(56) **References Cited**

U.S. PATENT DOCUMENTS

36,014 A		7/1862	Meissner
240,780 A	*	4/1881	Smith 74/458
370,136 A	*	9/1887	Goddu 411/411

(Continued)

FOREIGN PATENT DOCUMENTS

FR	958192	3/1950
GB	572218	9/1945

OTHER PUBLICATIONS

PCT/US03/08140. International Search Report. May 5, 2003.

"Professor, university pursue provisional patent for flexible bolt", by Colin Hester, Mustang Daily–a publication of CalPoly San Luis Obispo. vol. LXVI, No. 123, (May 3, 2002).

Brochure, "Producing Superior Results From Concept to Completion", The Deshler Group Inc., Deshler, Ohio; Amanda Bent Bolt Co.

Catalog; Simpson Strong–Tie Co. Inc., Copyright 1998, pp. 14–17.

Catalog. "Stow Flexible Shafts annd Flexible Couplings", 9th edition, The Stow Manufacturing Co, Binghamton, New York.

Website; Amanda Bent Bolt; www.amandabentbolt.com, (Jan. 9, 2001).

Website; AristoTechnics, Inc.; www.artsotechnics.com, (Jan. 11, 2001).

Primary Examiner—Flemming Saether
(74) *Attorney, Agent, or Firm*—Thomas F. Lebens; Sinsheimer, Schiebelhut & Baggett

(57) **ABSTRACT**

This invention relates generally to a flexible fastener for coupling members. In a preferred embodiment, the flexible fastener may be used to couple members that are non-parallel, non-aligned, or in specific instances when selective compliance in a member is desired. The present invention is directed generally to a fastener that is laterally flexible along its length, comprising a shank covered with a spiral set of teeth and a flexible core material running internally to the shank and along the length of the shank, wherein the shank and the flexible core material are coupled at both distal ends of the fastener. In a preferred embodiment, a means for imparting rotational movement to the fastener is attached at a distal end. The preferred means are a bolt head and a screw head.

4 Claims, 7 Drawing Sheets

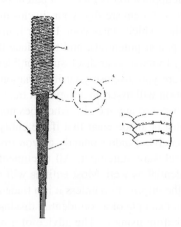

US 6,955,513 B2

| 7 | 8 |

geously cut from a square threaded bolt as shown in FIG. **14**. Threads (spiral set of teeth **3**) can also be produced by rolling, extrusion, casting, and all manners of computer aided manufacturing such as fused deposition, stereo lithography, and 3-dimensional printing. Besides the spiral set of teeth **3** (i.e. thread) conformations shown in FIG. **12**, FIG. **13**, FIG. **14**, FIG. **15**, FIG. **16**, FIG. **17**, and FIG. **18**, in another embodiment, the threads may also be comprised of a compression spring with proper pitch and wire diameter to match the flexible core material's **4** diameter. In any case, the pitch and diameter of a spiral set of teeth **3**, compression spring, or threads should be commensurate with the pitch and diameter of the flexible core material's **4** pitch and diameter as well as the pitch and diameter of a given nut to be used in conjunction with the flexible fastener **1**, so that the nut may be advanced on it. The threads may be fashioned in numerous versions with interchangeable features such as placement and orientation of protrusions, shape and curvature of threads, inclusion or exclusion of interlocking features, and pitch and diameter of threads.

As is previously described, the flexible fastener is useful in numerous situations. The flexible fastener is especially advantageous to easily and inexpensively couple non-parallel and non-aligned members. Additionally, the flexible fastener is also useful in various situations where selective compliance is desired. Among the many situations where the flexible fastener may be advantageously used are: construction, seismic applications, robotics, machine applications, and flexible power transportation. All references cited herein are incorporated by reference.

What is claimed is:

1. A fastener, comprising:
a spiral set of teeth, wherein said spiral set of teeth have a convex portion and a concave portion and wherein the convex portion smaller than said concave portion; and
a selectively compliant core material running internally to said spiral set of teeth; and
coupling means for connecting said selectively compliant core material to said spiral set of teeth at both distal ends of said spiral net of teeth no that said fastener is laterally flexible along its length.

2. The spiral set of teeth of claim **1**, wherein said convex portion of a lower layer of spiral teeth fits into said concave portion of an upper layer of spiral teeth.

3. A fastener, comprising:
a spiral set of teeth, wherein said spiral set of teeth have a convex portion and a concave portion and wherein the convex portion is smaller than said concave portion; and
a selectively compliant core material running internally to said spiral set of teeth and along the length of said spiral set of teeth; and
coupling means for connecting said selectively compliant core material to said spiral set of teeth at both distal ends of said spiral set of teeth so that said fastener is laterally flexible along its length; and
means at a distal end of said spiral set of teeth for imparting rotational movement thereto.

4. The fastener of claim **3**, wherein said convex portion of a lower layer of spiral teeth fits into said concave portion of an upper layer of spiral teeth.

* * * * *

MUTUAL CONFIDENTIALITY AND NON-DISCLOSURE AGREEMENT

This Agreement is made and entered into this ___ day of _____, 20___, by and between _____, having its principal place of business at _____ (hereinafter "_____") and _____, having its principal place of business at _____ (hereinafter "_____").

RECITALS

WHEREAS, _____ and _____ intend to enter into discussions and/or to perform services and functions for the purposes of doing business or possibly doing business with each other relating to **[INSERT SPECIFICS OF TRANSACTION HERE]** (hereinafter the "Transaction"); and

WHEREAS, each party may disclose to the other certain business and technical information (whether oral, in writing, in machine readable or other tangible form) concerning their respective interests and activities which the disclosing party deems proprietary, including but not limited to, financial information, trade secrets, know-how, product formulae, processing procedures and equipment, standards and specifications, product samples, product development plans, proposed products and services, business plans, customer lists, prices, market and sales information and plans, **[INSERT ADDITIONAL SPE-CIFIC CONFIDENTIAL INFORMATION HERE]** and any non-public information which concerns the business and operations of a party to this Agreement (hereinafter "Confidential Information");

NOW, THEREFORE, in order to induce _____ and _____ to disclose such Confidential Information to each other, and for other consideration, the receipt and sufficiency of which is hereby acknowledged, the parties agree as follows:

1. Confidentiality of Information

Each party agrees to receive the Confidential Information in absolute confidence. Each party agrees that it will not distribute, disclose or disseminate any Confidential Information of the other party in any way to anyone, except only to its employees and contractors who need to know the Confidential Information and to its financial, legal or other advisors who are advising such party regarding the Transaction, or as required by law. Each party agrees that its disclosure of Confidential Information to its employees, contractors and/or advisors who have such a need to know shall be limited to only so much of such Confidential Information as is necessary for an employee, contractor and/or advisor to perform his/her function. In consenting to the disclosure of Confidential Information to any third party, the disclosing party may require such third party to sign a confidentiality agreement similar to this Agreement. Each party also agrees that, until the parties agree to publicize the Transaction, it will keep confidential the Transaction and the discussions of the parties relating thereto.

2. Permitted Uses

The receiving party agrees that it will only use the disclosing party's Confidential Information for the purpose of analyzing, negotiating, and/or providing services or functions concerning the Transaction.

3. Standards of Care

Each party agrees that it will treat the Confidential Information of the other party in the same manner it treats its own Confidential Information that it does not wish to disclose to the public, but in all events each party agrees to use at least a reasonable degree of care to protect the Confidential Information of the other party.

4. Inapplicability of Restrictions

There shall be no restrictions under this Agreement with respect to any portion of the Confidential Information which:

    (a) is known to the receiving party or any affiliated company of the receiving party at the time of its disclosure to the receiving party;

    (b) is or becomes publicly known through no wrongful act of the receiving party or of any affiliated company of the receiving party;

    (c) is received from a third party without breach of the restrictions contained in this Agreement;

    (d) is independently developed by the receiving party or any affiliated company of the receiving party;

    (e) is furnished to any third party by the disclosing party without a similar restriction on the receiving party's rights;

    (f) is approved for release by the disclosing party; or

    (g) is required by the Public Records Act or otherwise by law, court order or a governmental agency to be disclosed.

5. Ownership

All Confidential Information delivered by either party to the other pursuant to this Agreement shall be and remain the property of the disclosing party. All such Confidential Information, and any copies thereof, shall be promptly returned to the disclosing party upon written request, or, at disclosing party's option and instruction, destroyed. Additionally, upon written request by the disclosing party, the receiving party shall certify in writing to the disclosing party that, to the best of receiving party's knowledge, all originals and copies of any Confidential Information that were used or possessed by receiving party have been returned to a designated officer of disclosing party.

6. Term

The restrictions and obligations contained herein shall continue for a period of ___ years from the date of this Agreement.

7. Enforcement

The parties agree that money damages would not be a sufficient remedy for any breach of this Agreement and that the parties shall be entitled to seek injunctive relief or remedy to prevent any breach or threatened breach of this Agreement. Such remedy shall not be the exclusive remedy for any breach of this Agreement, but shall be in addition to all other rights and remedies available at law or in equity.

8. Miscellaneous

(a) This Agreement shall be binding upon the parties, their successors, and assigns. Neither party shall assign this Agreement or any Confidential Information received from the other party pursuant to this Agreement without the other party's prior written consent.

(b) Nothing contained in this Agreement shall be construed as granting or conferring any rights by license or otherwise in any Confidential Information disclosed to the receiving party.

(c) This Agreement shall be governed by, and interpreted in accordance with, the laws of the state of California.

(d) The furnishing of Confidential Information hereunder shall not obligate either party to enter into any further agreement or negotiation with the other or to refrain from entering into an agreement or negotiation with any other party.

(e) This Agreement constitutes the entire agreement and understanding of the parties with respect to the subject matter of this Agreement. Any amendment or modification of this Agreement shall be in writing and executed by duly authorized representatives of the parties.

BY: _____   BY: _____

_____   _____
                (Print Name)   (Print Name)

Title: _____   Title: _____

Date: _____   Date: _____

14.19   MOST NUMEROUS RECENT PATENT HOLDERS IN THE UNITED STATES

Each year the USPTO announces the top ten private sector patent recipients. Listed in Table 14.1 are the ten corporations who in recent years received the largest number of patents. It is interesting to notice not just the large numbers involved, but that many companies continue to lead in total number of patents issued to them each year. It is also interesting to notice how many of the companies are foreign. The U.S. government is also among the entities that receive a large number of patents each year. Think about the cost of preparing, applying,

TABLE 14.1   Top 10 Private Sector Patent Recipients in Recent Years

	1995		1996		1997		1998
IBM	1383	IBM	1867	IBM	1724	IBM	2657
Canon	1087	Canon	1541	Canon	1381	Canon	1928
U.S. Gov.	1026	Motorola	1064	NEC	1095	NEC	1627
Motorola	1012	NEC	1043	Motorola	1058	Motorola	140
NEC	1005	Hitachi	963	U.S. Gov.	935	Sony	1316
Mitsubishi	973	Mitsubishi	934	Fujitsu	903	Samsung	1304
Toshiba	969	U.S. Gov.	923	Hitachi	903	Fujitsu	1189
Hitachi	910	Toshiba	914	Mitsubishi	892	Toshiba	1170
Matsushita	854	Fujitsu	869	Toshiba	862		
Sony	754	Sony	855	Sony	859		

	1999		2000		2001		2002
IBM	2756	IBM	2886	IBM	3411	IBM	3288
NEC	1842	NEC	2021	NEC	1953	Canon	1893
Canon	1795	Canon	1890	Canon	1877	Micron	1833
Samsung	1545	Samsung	1441	Micron	1643	NEC	1821
Sony	1410	Sony	1385	Samsung	1450	Hitachi	1602
Toshiba	1200	Micron	1304	Matsushita	1440	Matsushita	1544
Fujitsu	1192	Fujitsu	1147	Sony Hitachi	1363	Sony	1434
Motorola	1192	Matsushita	1137	Mitsubishi	1271	GE	1416
Lucent	1152	Hitachi	1036	Fujitsu	1184	HP	1385
U.S. Gov.	983	Mitsubishi	1010			Mitsubishi	1373

maintaining, and enforcing such a large number of patents each year in time, effort, and fees. It is also interesting to think about why some companies seek so many patents. In many cases, the patents issued are not real inventions that will be directly used for business, but are for encircling and choking the competitors' patents. No statement of guilt is made against any of the companies mentioned here though. Nevertheless, experienced attorneys from large law firms have repeatedly alluded to this practice in national conferences about patents.

According to the Patent Office, in 2001 the USPTO received more than 325,000 patent applications and granted over 170,000 utility patents.

14.20   BAYH-DOLE ACT

In 1980 the Bayh-Dole Patent and Trademark Act Amendment of 1980 created a uniform patent policy among the many federal agencies that fund research. In short, before this, if the federal government funded university or other national laboratory projects, the outcome was public. However, the Bayh-Dole act allowed the funded agencies to retain the right to the project, gain patents, and license it to others for commercialization. As a result, there has been a tremendous growth in technology transfer between universities and the private sector, resulting in countless new products, systems, and processes. By the year 2000, there were more than 330 U.S. and Canadian universities engaged in technology transfer, with gross licensing income of over $1.26B. Nevertheless, the federal government retains the right to license the invention to a third party, without the consent of the patent holder or the original licensee, if it determines that the patent is not made available to the public on a reasonable basis.

14.21 DESIGN PATENTS

A design patent is issued for aesthetics and the shape of a product and not its function. It only protects the design of the invention and not its function. An example of a design patent is the shape of an automobile, or any part of it. For example, if the shape of an automobile's fender is patented, no one else may make or sell it other than the patent holder. This is used to prevent other manufacturers from copying a design and from making the same part for their own use or to sell to others. However, as soon as the dimensions or the shape is changed, the patent will cease to protect the invention. Design patents are granted for 14 years from the date of issue (not the date of application), and no maintenance fees are required. The specification part of the application is usually short, and only one claim covering the design of the invention is permitted. Otherwise, the process and requirements for design patents are very similar to utility patents. The cost of preparing design patent applications and the fees are much lower than utility patents as well, and the applicant only has to show that the design is original and different from other designs. Consequently, it is usually much easier to get a design patent.

14.22 COPYRIGHTS

Copyrights are used for the protection of other intellectual property such as writings, paintings, drawings, poems, musical and theatrical expressions, movies, and software. Copyrights, unlike patents, are not used to exclude others; instead, copyrights prevent others from copying the intellectual property of another person. Copyrights only protect the particular expression of the work—not its content. As an example, all the material in this chapter is from the U.S. patent law and codes and is not originally developed by the author. However, the book is copyrighted, which means that the way the information in this chapter is expressed is copyrighted—not the content of it.

Copyright protection is automatic. No one is supposed to copy another person's work without permission. Anyone may simply state in his or her work that the work is copyrighted. However, it is necessary that each and every copy of the work be marked as copyrighted with the symbol © or the word copyrighted, with the date of first publication, and the name of the copyright owner.

You may formalize the copyright protection by sending a copy of your intellectual property to the copyright office, with the associated fee, and request copyright. There will be no attempt by the copyright office to ensure that the work is in fact original. The fact that you apply for copyright indicates that the work belongs to you. You will be responsible if this is not true.

Like patents, copyrights are property, which can be assigned or sold. Since a patron may commission a work by someone, the right to the copyright will belong to the patron and not the originator of the work. As in patents, there should be a valid contract with necessary royalty speculations when copyrights are assigned.

For all work created by independent authors after 1978, the term of a copyright is the life of the originator plus 50 years. Prior to 1978, the term was 28 years, plus possibility for another 28 years. Corporate works are protected for 75 years from the date of publication or 100 years from creation, whichever is shorter.

The USPTO only deals with patents and trademarks. Copyrights are registered by the Copyright Office of the Library of Congress. Please contact the Copyright Office or visit their website at http://www.copyright.gov for more information. Additionally, please see Copyright Basics (Circular 1) and Copyright Registration for Computer Programs (Circular 61) of the Copyright Office, Library of Congress, for detail information about filing requirements, location of the symbols, and addresses.

As was mentioned earlier, software is copyrighted. However, if particular software is written for a specific machine or process, such that the working of the machine is related to the software, and the software's work is directly related to the machine, it is possible to obtain a patent for that software. This is a complicated matter, and if it relates to your work, you should consult a professional patent attorney to see if it should be copyrighted or patented.

14.23 TRADEMARKS

Trademarks are used as means for identifying an entity such as a business, and to uniquely identify goods and merchandise or services provided by an entity through a name, a symbol, a logo, a device, a mark, or a design. Many companies use both a name and a symbol to identify their products and their business. A trademark is generally associated with a product, whereas a servicemark is used for services offered by a company and not a product. A company engaged in both may use the same name as a trademark and as a servicemark. Depending on the strength of the company, the trademark may be an invaluable asset for the company. To be able to claim a trademark, it is necessary that the symbol be affixed to the product or the container of the product, and for the product to be sold (intrastate, interstate, or international). Similarly, a servicemark must be used as a sign of a service business. Otherwise, as long as the symbol or the name is not a registered trademark, any symbol or name can be claimed by a business as trademark or servicemark. In interstate and international cases, a business can request registration from the USPTO. After 5 continuous years of use of the trademark, it becomes incontestable. There is no need to federally register a trademark if your business is intrastate, or limited to a region. However, another person who may innocently use the same name or symbol for the same business or goods in another area cannot be forced to stop the business at a later date if you later decide to federally (through USPTO) register your trademark. In fact, the other person may be able to exclude you from doing business in his territory. Therefore, it is recommended that you register your trademark or servicemark through USPTO.

The owner of the business who has a trademark or servicemark can stop any other business that uses the same registered symbols in the business, including importation of goods from other countries that bear the same name or symbol. A few years ago, the Sony Corporation forced a Los Angeles restaurateur named Sony to change the name of the restaurant even though this business was completely different, and even though the restaurant owner had used the name for decades. As the name or the business of the company grows, the value of the trademark will also increase, especially when customers readily recognize the name or the symbols used. This is why companies defend their trademarks to the best of their ability and worry about the perception their customers have about the name of the business.

A trademark may be a single word (Google, Dole), letters or numbers (E-bay, IBM, 7-11), a slogan (Just Do It), a design (the symbol used by AT&T or Apple computers), a made-up name (TravAlong, The Haul of Fame), a color (UPS's brown), product shape (the specific shapes used in some computer games), product's container shape (liquor or perfume bottles, Top Ramen soup package), building shapes (McDonald's building shape), sounds (the music used in the Tonight show, news programs, Harley Davidson engine sound), or even fragrances (Dial soap).

It is possible that there may be overlap in patent, trademark, or copyright protection. For example, if a soda bottle's shape is patented as a design patent, the protection may last up to 14 years. However, the shape may also be registered as a trademark, with protection that may continue as long as the bottle is still used by the company. The picture of Tigger in the "Winnie the Pooh" shows is protected by a copyright, but also as a trademark, it has additional, and in fact longer lasting, protection. However, there are also fundamental differences between trademarks, copyrights, and patents. As we have seen, a patent protects the function and a method of doing something. Copyright protects original works of expression. Trademark protects symbols and other specific trade-related items that are used for the recognition of a business or product.

For more information about trademarks and servicemarks, please refer to (2) or to http://www.uspto.gov/web/offices/tac/tmlaw2.pdf. Additional information may be obtained from Basic Facts About Trademarks from the USPTO.

REFERENCES

1. "General Information Concerning patents," U.S. Patent and Trademark office, updated regularly.
2. KONOLD, WILLIAM, BRUCE TITTEL, DONALD FREI, and DAVID STALLARD, "What Every Engineer Should Know About Patents," 2nd Edition, Marcel Dekker, Inc., 1988, New York.
3. PRESSMAN, DAVID, S. ELIAS, "Patent It Yourself," 10th Edition, Nolo Press, Berkeley, California, 2004.
4. SIEGEL, R.P., "Down but Not Out," *Mechanical Engineering*, October 2004, pp. 44–47.
5. TESKA, KIRK, "Does 4 Include 3? ," *Mechanical Engineering*, January 2006, pp. 32–33.

HOMEWORK

14.1 Perform a patent search for a favorite product that you use every day. Are the patents utility type patents or design patents? Describe how the product has evolved with new patents.

14.2 Perform a patent search for one of your own ideas. Determine whether your idea is novel and useful enough to be patentable. Are the patents related to your idea design patents or utility patents?

14.3 Develop and write a confidential disclosure statement for your idea.

14.4 Write a simple patent application for your idea.

14.5 Write a few distinct claims for your idea.

14.6 Find five patents owned by a large corporation and study them lightly. Can you determine whether the patents are for composition of matter, manufacture, process, machine, or a new use of the above? Can you determine how useful any of the patents may be?

CHAPTER 15

ENTREPRENEURSHIP AND INNOVATION

"It must be remembered that there is nothing more difficult to plan, more uncertain of success, nor more dangerous to manage than the creation of a new order of things. For the initiator has the enmity of all who would profit by the preservation of the old institutions, and merely lukewarm defenders in those who would gain by the new ones.

—Machiavelli, 1469–1527

15.1 INTRODUCTION

As mentioned in Chapter 1, in a report delivered to President Bush in December 2004, the National Innovation Initiative's 21st Century Working Group[13] defined innovation as "the intersection of invention and insight, leading to the creation of social and economic value." Entrepreneurs take an innovative idea and turn it into a successful business that provides economic opportunities for them, their investors, their employees, and the nation.[2,3,4,5]

Some innovations are gradual and incremental and occur over time. Others are discontinuous and create disruptive technologies. For example, until Xerox copy machines came into the market, most copying and duplication was done through chemical processes (e.g., using alcohol for transferring ink onto the paper) or by photographic processes (e.g., use of chemicals in a tray to change the nature of substances smeared over the paper). The discontinuous innovation (a real invention) of Xerography changed the previous methods completely and created a disruptive technology that drove the old systems out of the market. However, since then, thousands of minor innovations, both by Xerox Corporation and other competitors, have changed the original idea to the products we have today. Each innovation adds to the value of the product and creates competitiveness. However, they are continuous and gradual. Another example is the digital data storage devices. Tape storage units gave way to floppy disks, to hard disks, to Zip disks, and to Flash memory. Each one was a discontinuous innovation that made the previous technology obsolete. However, improvement in each device was continuous and gradual. Televisions, telephones, printers, and many other products have a similar history (have you seen a dot-matrix printer lately?). How many discontinuous innovations can you name in the automobile industry since the early 1900s? Was the Internet not a disruptive innovation too?

Harold Evans[1] draws the following lessons on innovation:

1. Inventors invent it; innovators improve it. Carlson invented Xerography. Countless other innovators have made it into the machine it is now.

2. Doubt others' statements of fact. Nicolai Tesla was given a long lecture by his professor on why Tesla's AC-motor idea would not work. He went home and made the first AC motor.

3. If things do not work the first time, there is always a next time. If it works the first time, it is probably too simple. Learn from your mistakes, then they are not mistakes; they are lessons.

4. New things are not always accepted until later. The first Ford Taurus looked strange to a lot of people. Nowadays, most cars have the same streamlined features.

5. Success is intoxicating. After WWII, the Big-3 US car manufacturers thought they had the best cars. They thought no one else could catch up to them. They did.

6. Partnerships help. As we have seen in the previous discussions, teams and groups can do more than individuals, especially if they complement each other's characteristics. A business oriented person and a technical person together can complement and help each other in more ways than not.

7. Do it because there is some good in it for others. If the net result does not add value to someone's life, it is not worth doing.

15.2   INNOVATION IN A GLOBAL BUSINESS WORLD

In this era of the 21st century, most everything is affected globally, whether it is in politics, the economy, businesses, markets, or innovations. An idea may be seeded in one country, be designed in another, manufactured in yet another, and be sold somewhere else. A product may be designed for local, regional, national, or global markets. A car designed in Japan must be adapted to markets in the United States or Europe or the Americas. Innovations are no exception. Innovations are similarly affected by global issues, global economies, and global markets. Language barriers, cultural barriers, economic capabilities, and expectations must be integrated with innovations, whether discontinuous or gradual, in order to affect markets. A small innovative juice shop in San Luis Obispo, created as a result of a senior project in the College of Business at Cal Poly, eventually became the international chain known as Jamba Juice. It is a great example of entrepreneurship at a local level that became a large and successful business, even though the product was a simple, but innovative, concept. Many others with expertise in corporations and global business were also involved in making this small entrepreneurial endeavor into what it is today. We will discuss the role of these contributors to the success of a business later.

15.3   ENTREPRENEURSHIP AND ENTREPRENEUR

Entrepreneurship is the act of taking an innovative idea and making it into a successful business. In reality, even if your business is not successful you may be an entrepreneur, since an entrepreneur is a person who is willing to take a risk by taking an idea, an innovation, or invention, and trying to make a successful business out of it. The motivation may be different for different entrepreneurs, whether it is financial success, recognition, challenge, or the desire to do something good to help others. But being an entrepreneur requires the ability to take risks, work hard, be resourceful, and be innovative. Not everyone is naturally entrepreneurial. But everyone can learn to be one.

The following are five key concepts about entrepreneurship:

1. Great opportunities are rare. When you find one, pursue it. Many opportunities are time sensitive as well; if you wait, they may become obsolete.

2. Networks are more than their individual components, as groups are more than the sum of their individuals. Build networks.

3. No matter what else you do, be a salesman too, including selling yourself and your relationships with your network. Confidence in what you do is an essential component. Others must see your dedication, energy, and knowledge of the endeavor.

4. Live proactively. Reactions come too late. By the time U.S. auto manufacturers reacted to the competition, they permanently lost market share.

5. Respect the abilities of others and celebrate them. Recruit good help at all levels to help you be successful. Retain talent, hold on to them, but set them free to do what they do best, and reward performance.

Most technology driven entrepreneurships start from academic research, government labs, corporate R&D, or ''dining room tables'' (individual entrepreneurs). Academic research is mostly licensed to corporations or to individuals through technology transfer offices of the university. Government labs also license their intellectual property to private institutions or individuals for commercialization. Corporate R&D is a huge source of innovations that support corporations in maintaining their competitiveness and development of new products, and are funded and used by corporations. Individuals are the most prolific source of innovative enterprises. Individuals usually start small, and although the majority of start-ups fail financially, many become successful corporations. However, even though these businesses fail financially, they are the backbone of innovative ideas and free enterprise economy, and an invaluable experience and educational lesson for those who dare attempt it.

The path between an innovative idea and ultimate success is not a straight line, but like a minefield. You have to negotiate your path within a maze of dangerous, potentially fatal moves, any one of which may destroy your institution. But when you are through, you will be rewarded handsomely. The maze includes personal, financial, emotional, legal, and health risks. During the journey, countless assumptions are made about the market, the value of the product or service, customers, competition, economy, prices, and many other factors. Because assumptions are just that and not facts, and because entrepreneurs usually have limited experience (and cannot afford professional advice and management at the beginning), and since there are few resources available to the entrepreneur, entrepreneurship is a risky venture. The chances of success are generally less than 3 percent. Therefore, in entrepreneurship, risk reduction is often a more effective approach than planning for success.

An entrepreneur is a passionate innovator, a visionary, a manager, a risk taker, a ''bull dog,'' and a leader. She or he needs to have an innovative idea that makes the future business competitive. It may be a discontinuous innovation (an invention), or a gradual innovation. But there must be some advantage upon which the business is built. The entrepreneur is a visionary who looks at the future as she or he wants to build it, not as an imitator. She or he has to also be a manager who manages the available resources, but is resourceful enough to create needed resources. The entrepreneur must be a risk-taker, as entrepreneurship is risky. No one really knows whether an innovative idea will take hold, or whether a product or business will be successful. Much risk is involved in entrepreneurship, but the rewards are proportionally satisfying as well. The individual must be a ''bull dog'' too, a go-getter, someone who won't let go, a persistent hopeful who will not give up. But she or he must also be a leader who can lead his or her "people" into this adventure, the people who work for him or her, the helpers, the bankers, the creditors, and the customers. Above all, the entrepreneur must be passionate about the idea, and must be able to infect others with his or her passion.

15.4 THE ENTREPRENEURIAL BRIDGE

Entrepreneurship involves taking an innovative "technology" into market. Here the term technology does not mean a technical product, but an innovative idea, whether technical concept or product, marketing technique, a new business idea, or any other technique of accomplishing an idea. Innovation involves the development of the technology. Entrepreneurship is what you do with the technology. Joe Boeddeker[10] likens entrepreneurship to building a bridge. The bridge connects two essential entities of technology and market. You must have something of value to offer your customers, and there must be customers that need your technology. Without either the technology or the market, there is nothing to connect to. The bridge provides a connection between the technology push and the market pull (technology needs a market, the market needs the technology). However, a bridge stands on pillars of needed resources that include financial resources, human resources, managerial resources, physical resources, technological advantages, and many more, as depicted in Figure 15.1. The entrepreneur is the visionary person that realizes this connection, gathers the resources, and builds the bridge.

Managing entrepreneurial resources involves:

- Organizing and planning (goals, commitments, schedules); identifying, soliciting, engaging, and motivating people who can help.

- Financial matters (budgets, loans, raising funds, financial and tax records).

- Marketing and selling ideas and products (goals and projections, presenting ideas and plans, direct selling to customers, creating distribution systems and sales force).

- Management (overseeing employees and the activities of the business).

FIGURE 15.1   The Entrepreneurial Bridge. [Courtesy of J. Boeddker[10]]

15.5 RISKS AND REWARDS

Entrepreneurship involves many risks and many rewards.[7,8,9] So does going to college, getting married, and having children. And yet, many people take the risks for the anticipated benefits. Like other major life decisions, you must decide what is important to you, what your level of comfort is in risk-taking, what your priorities are, and whether entrepreneurship is for you or not. The risks include:

- **Emotional risks:** A new adventure requires many decisions and involves successes and failures. It becomes like a child, an important part of one's life. Emotions run high, letting go is difficult, and at times, one sacrifices other essentials for the success of this one. You must be emotionally strong, and must learn how to deal with these issues when they arise.

- **Stress risks and hard work:** Starting a new business is tremendously involved, requires hard work, and puts the person under a lot of stress to accomplish, to be successful, and to not lose everything. I have known many individuals who quit from high-paced businesses and moved into smaller communities just to reduce the stress in their lives.

- **Financial risks:** Since the majority of new businesses fail, an entrepreneur must be prepared to take a financial loss. In most cases, a personal financial involvement by the entrepreneur is inevitable, even if most of the finances are loans and investments by others. Therefore, there is always a great risk of financial loss with any entrepreneurial endeavor.

- **Health risks:** The hard work and constant stress associated with starting a business and dealing with the financial aspects of an enterprise puts a person in health risk as well. Stories abound in the business world about people who have ulcers, get nervous breakdowns, and are always tired. Of course this does not mean that everyone will suffer from health risks, but there is no doubt that many do.

- **Time risks:** Entrepreneurship requires time and effort, sometimes to a point that negatively affects the people involved. No time is left to spend with your family or children, to participate in daily affairs with your family and friends, or take a holiday. Additionally, if you fail, you will have lost years of your life.

But the rewards are great too:

- **Financial rewards:** When an enterprise is successful, it undoubtedly brings financial rewards with it. This is one of the most prevalent motivations that drive individuals into entrepreneurial activities, a chance to be rewarded handsomely, sometimes beyond imagination.

- **Recognition:** A successful entrepreneur is generally recognized as a person with great abilities, strong personality, and expertise in many fields, including business. Others ask the entrepreneur for advice, the individual is recognized as a leader in his or her field, and at times, the entrepreneur becomes famous and a celebrity (with increased chance of expectations for philanthropy).

- **Fulfillment:** That the work is accomplished, that a service is provided, that jobs are created, that wealth is increased, that others are helped. For some, this feeling of satisfaction may be even stronger than the desire to be recognized or become rich. Others even reject the latter, and are satisfied only by the former.

- **Experience gained is invaluable:** Many entrepreneurs continue to start new businesses and be involved in other ventures. The experience gained may be a tool that is useful for life.

You should prioritize your desires and your tolerances and decide whether entrepreneurship is for you or not. In Chapter 4 we discussed the application of the "worst-case scenario" method to overcome fear. Similarly, you may consider the worst-case scenario approach to determine whether it is acceptable to you or not. Is the worst thing that can happen, an eventual failure of the business, acceptable to you, despite the cost (your money, time, energy, and stress)? Will the possible rewards, whether financial or otherwise, be valuable enough to you to make the risks acceptable? These answers will help you decide whether to go forward or not, and if so, in what form. We will discuss possible different approaches later.

15.6   MOTIVATIONS AND OBJECTIVES

One of the early decisions you need to make is about your motivations and objectives. These will help you decide why you may want to start your own entrepreneurial work, and what you expect from it. You make these decisions based on a number of factors, including the risks and rewards, but also the climate, the market, your capabilities, and your willingness and level of tolerance to get involved in new endeavors.

The birth of a new product or service may be divided into three phases of invention/innovation, product prototyping, and production. This not only applies to a physical product, but also to services and other business models. For example, Amazon.com was a new business model based on an innovative method of selling books on the Internet. There was no new physical product, but a new method of selling products. The innovative phase was to come up with the method of doing business. For a new engineering product, the initial design phase of developing a new product or improving it with innovations is similarly the first phase. The second phase relates to prototyping, testing, evaluating, and presenting the prototype. To do this, you may need additional resources, both technically and financially. You must somehow secure these resources in order to successfully perform prototyping and testing. The third phase, also called scaling or maturation, is the actual production of the product or presentation of the service to the customers. This may require many more resources, many workers and employees, production facilities, and a distribution and sales force. You must ensure that there are enough resources to spend until the product generates revenues and becomes profitable.

However, it is possible for the entrepreneur to only be involved in parts of this process, at different levels, with varied roles. For example, the entrepreneur may license the innovation to someone else and just receive royalties or compensation (phase one only). In that case, the role of the entrepreneur in the further development of the product or service is minimal, albeit with less reward. The entrepreneur may continue involvement in a consulting capacity after licensing. In this case, she or he will continue more intimate involvement with the innovation and its destiny. One may continue to be involved at a technical level only and remain with the development of the product without dealing with the management issues by leaving those to experts. And finally, one may remain intimately involved in all aspects of the endeavor including fundraising, management, hiring, day-to-day running of the business, and so on. Many technical entrepreneurs have more interest in the technical issues related to their invention or innovation and leave the business related issues to experts. By determining your motivations and objectives, you can decide about the level and capacity of your involvement in the business.

15.7 SHOULD YOU START A BUSINESS?

This is not an easy question. There are countless encouraging stories about successful entrepreneurs as well as countless horror stories about failures. Business authorities spell out many different reasons for this, from the strength of the innovation and intellectual property to the business climate to marketing to management styles and to the personality of the entrepreneur. This chapter cannot present a simple recipe for this decision either. You must consider many varied factors before you decide whether you should start your own entrepreneurial business, to license the innovation to other entities for development and profit making, or something in between. A simple toy, made up of a laser-light source and a number of mirrors and prisms, devised by a group of students as a class project, made a profitable and successful, albeit small, business, still operating from a "garage" after 3 years. Another innovative idea, developed as a class project, involving minute, metallic, spirals that were added to cement to strengthen it became a successful business, receiving an award from the National Collegiate Innovators and Inventors Alliance (NCIIA) in 2006. Will yours do the same?

Experts suggest that you must answer "yes" to the following six questions if you are planning to start your own business:

1. Will this business satisfy most of my wants and desires?
2. Do I have, or can I develop (or hire), the skills and capabilities required for this business to be successful?
3. Is the current economic condition and outlook favorable for this type of business?
4. Is there a viable market for this product or service?
5. Do I have or can I obtain the financial resources necessary to fund this business until it can pay for itself?
6. Am I, and is my family, willing to accept the emotional and financial risks and personal commitment associated with starting this business?

When a new, innovative product or service is developed, there are a number of different options available to the designer/innovator:

- Contact the technology transfer department of your institution (if it exists). As discussed in Chapter 14, there may be license and right-to-use issues involved with projects that are performed in universities and companies. In such cases, the rights to the idea (patent) may belong to the institution, and as such, they will try to license the product or system (patent) to others, or they may help you create a business. Even in the absence of rights and license issues, technology transfer departments usually help the designer in many forms.
- Contact a patent lawyer. A patent lawyer can help you decide whether you should patent the idea or not, and how much it costs. They can also help you with licensing issues and writing contracts. In most cases, initial consultation with a patent attorney is free.
- Make sure it is protected and documented before you disclose any information to the general public, for example by showing the product to others that are not involved in the design and implementation of the idea.
- Incorporate. Incorporation may have a significant effect on your income, liability, and tax liability. Consult a business attorney.

- Decide what role you want to play in the business. This includes:
  - **Licensee:** You only receive royalties, but are not directly involved in the business.
  - **Consultant:** You may be a consultant to the new business. In this capacity, you will have an influence on the future of the product, and may even be compensated in addition to your royalties.
  - **Owner:** Start your own business. You will have full responsibility for the well-being of the business, but will also have full benefit if successful. Remember that being an innovator and being a good CEO are not the same thing.
  - **Give it up to others:** Let others use it for free.

If you decide to start your own entrepreneurial business based on your innovative idea about a product or service, you will need to develop a business plan and then work on implementing it. The following sections present guidelines and information about issues related to this.

15.8 MARKET RESEARCH

In order to eventually create a business plan and convince others to invest in your business, you will need to first research the market in which you intend to enter. In general you should have professionals do your market research. This is different from marketing your product for sales and the creation of market share. Market research involves determining what may be the market for your product or system and what potential exists for your success. Many consulting institutions exist for this purpose. However, you can accomplish much on your own by doing preliminary market research to determine where you stand, what the potential is, and whether you should spend more time, energy, and money on additional market research. Many governmental and business institutions gather and publish information about business climate, customer demography, economic data, and statistical information that can be invaluable in this market research. For effective market research you must

- **Understand your customer/consumer.** Consumers should have a need that will be satisfied by your innovative product or service. Who are the consumers? What need is being addressed by your product or service? How many potential consumers are there? What segment of the society constitutes your consumers? Are your consumers individual buyers? Are they other original equipment manufacturers who will use your product in their systems? Is your potential consumer base international? Are they family members, professionals, physicians, technicians, or do it yourselfers? Are your consumers also your customers, or will your consumers buy your product from your customers who buy from you and sell to them? In many cases, the individual or institution who is your customer and buys your product will actually sell it to their own customers, who are the real consumers of your product. Do you directly sell to them, or do you have to satisfy and attract the attention of a corporate buyer who will decide how to sell your product to the real consumer?

- **Determine your competition.** Are there other products in the market that will compete with yours? What is their market share? How is that particular business doing? Unless your product is a totally new concept, if there are no competitors, it might indicate a lack of need, and presumably, lack of market potential.

- **Benchmark against your competitors.** What is your competitive advantage over other products or systems already available to the customer/consumer? Why should anyone want to seek your product or use your service? How will your product compete with the existing product? Why do you believe you will be able to capture any significant market share? Do you have a determinable economic advantage? What is your value?

- **Identify trends.** Our society is greatly influenced by trends and what is fashionable at any given time. However, remember that trends can change and shift swiftly, a popular product or brand may go out of fashion or favor quickly, and others will replace it immediately. This can work against you if you plan your business and the associated products according to this trend. It can also help you if your product changes the trend by successfully challenging the existing products. Check the industry, product sales reports, demographics, and cultural and societal trends. Also talk to industry experts, potential customers, distributors, suppliers and retailers, online data sources, trade associations, trade show attendees, and government regulators. Also check census data. The census data is rich with information about customers, their location, their distribution, and their economic vitality in each region. The data can also help you establish future trends.

- **Forecast the market.** You also have to identify whether a market, even if presently large, is diminishing, steady, or growing. If the market is growing, competition and capturing some market share may be easier. If the market is diminishing, it may be more difficult to compete, unless you have a very innovative and disruptive innovation.

- **Determine channels of distribution.** You should determine how you will get the product or service to the customers. Will it be sold through sales agents? Direct mail? Advertising? Dealerships? Internet mail order? The available channels of distribution or lack thereof can have a significant effect on your ability to distribute your product and service to your customers.

- **Think through pricing.** The final price will affect the competitiveness of your product or service as will its innovative features and competitiveness. To the cost of manufacturing you must add profit, distribution cost, advertising costs, insurance, dealer/retailer profits, as well as contingencies. Will the product still be competitive and display enough value to the consumer to be desirable?

- Use the following resources for your market research. There are many more; some are free of charge, some are not free.

○ USPTO.gov	○ LexisNexis.com
○ SEC.gov	○ Google search on the subject
○ Corptech.com	○ Trade associations
○ Catalogs.google.com	○ Product catalogs
○ Thomasregister.com	○ Dunn and Bradstreet
○ Hoovers.com	○ ProQuest
○ Switchboard.com	○ Census data

15.9   BUSINESS PLAN

A business model defines your business goals.[11,12] In fact, starting a business and doing entrepreneurial work is the *problem* for which you are trying to find a solution, a plan, or a strategy, and as was discussed in Chapter 5, the business plan is the actual definition of what is to be accomplished and how you will accomplish it. The business plan states what needs to be done, who will do it, what the purpose is, the timeline, the resources needed, the strategy for securing the resources that are not available, and the final result sought. The fact that you need to write a business plan before you can talk to an investor or a bank or a creditor attests to the necessity of it, but also how useful it can be. Without a business plan, no one knows what the final goal is, and how you intend to accomplish it. Additionally, having to produce a business plan forces you, the entrepreneur, to think through all the steps, the end game, the process, the people and resources involved, and the methods you will take. In order to be able to produce a business plan you will *have to* do the necessary preliminary analysis, just as in technical problem solving. You will have to identify all the details of the business by thinking, researching, and defining all necessary components needed to accomplish it. The business plan *is* your problem definition. A business plan should answer the following questions:

- Who are the customers and/or consumers?
- What is the competitive edge of your innovative product or service?
- Why may a customer/consumer choose your product and service over the competition?
- How does the customer/consumer make decisions about your product?
- How will the product or service be priced?
- How will the product be distributed, and how will it reach the customers/consumers?
- How will you reach the customers/consumers, and how will you attract their attention?
- What is your economic advantage?
- How will you produce your product or service, and what is the associated cost?
- How will you keep your customers and consumers?
- What are the risks and rewards associated with your enterprise?
- What resources are needed to accomplish the task?

Guy Kawasaki, author of The Art of the Start,[6] states the following general points about starting a new business as an entrepreneur:

- *Make meaning, not money. Your main objective should not be to get rich, although that may be a consequence. The main objective should be to make a difference in someone's life, by providing a solution to a need. That will make meaning for someone else who will be willing to pay for it.*

- *Make a "slogan" for your purpose. You should be able to describe your business and your intention in only a few words that clearly state the purpose.*

- *Define a business model, specifically describing your business and your intentions. Keep it simple too.*

- *Niche yourself. Find your innovative, competitive edge that differentiates you from the rest.*

- *Follow the 10/20/30 rule of presentations. 10 slides, 20 minutes, 30 point font. This will limit the amount of information you are able to communicate or present, thus, you will have to cut out the fat and present the bare essentials that matter. Brevity is important.*

- *Hire people who believe in your product or service, who will be your ambassadors, who will infect others with your product too. How will any customer feel when your employees use the competitors' products? Hire people who are better than you. Do not be afraid of them, as they will help you move up with them.*

- *Make it easy for the customer. There really should be no need for an instruction manual (unless it is rocket science!). When you buy a product, do you like reading a thick manual even before you test it?*

- *Propagate the good. Ask people what they like about your product and do more of that. Do not ask what is wrong with it, but what is right with it. Allow people to test your product and run with it to get hooked.*

- *Be a mensch. Help people, pay back the society, be part of your community. It still matters.*

15.10   DEVELOPMENT OF A BUSINESS PLAN

A typical business plan should contain the following sections. Obviously, you should adapt this list to your requirements as needed:

- **Title Page**
- **Table of Contents**
- **Value Proposition:** This section describes the value of your proposition and what the proposition offers others such as your investors.
- **Executive Summary:** This is a short (1–3 pages) summary of the rest of your plan, usually used by managers, executives, banks, venture capitalists, and others who may not have the time to read through a plan. It should present the basic premise and the value of your proposal.
- **Business Plan Body:**
  - **Business Opportunity:** What your business proposes to accomplish.
  - **Products:** What is your product or service, its variations, characteristics, and value?
  - **Markets:** Describe the markets you believe are available to you for your product, the demographics, the size, the competition, the professions, and other detail.
  - **Operations:** Describe the operations that will be needed to establish the business and develop your product or service, your marketing operations, day-to-day operations, budget control, accounting, projected income statement, balance sheet, cash flow, and other details within the business.
  - **Management:** List the management team that will work with your enterprise, their qualifications, experience, and roles.
  - **Financials:** Describe financial resources available as well as what will be needed. Also describe how the flow of money will be controlled, where the money will come from, and how it will be spent.
- **Appendix and Exhibits:**
  - **Technology status:** Describe the status of the technology without your product and how it might be affected by your innovation.
  - **Market Analysis:** Include information you have gathered about the market.
  - **Venture development timeline (project plan)**: Use common methods of showing your proposed timeline to accomplish each task.
  - **Management credentials:** Include the credentials of your management.
  - **Financials:** Include details of the available finances and your expectations.
  - **Equity distribution:** Include details of equities already distributed. This is important information to anyone else who may want to invest in your enterprise.

15.11 PRESENTING YOUR BUSINESS PLAN

Most entrepreneurs will have to present their business plan to others who will invest in their proposition, who might be interested in joining them, or who will fund its operations (such as banks). A professional presentation, especially in the business world, is crucial. In addition to attire, and content and organization of your presentation, the way you speak in public is also extremely important. While you make your presentation, remember that truthfulness and honesty are also extremely important. At the same time, remember that you must make a positive impression on the audience, and you need to convince them of your worth. Therefore, you need to listen, respond, and convey your enthusiasm and confidence in your abilities to others. The following are some basic guidelines to follow:

- **Your objectives:** What do you want to accomplish through this presentation? What is the goal you want to achieve?

- **The audience:** Consider the audience, their expectations, who they are, and what their background is. Your presentation to a banker might be different than to a venture capitalist or to a potential customer. What they are interested in learning is also affected by who they are. You should adapt your presentation accordingly.

- **Importance of brevity:** As Guy Kawasaki mentions, remember the 10/20/30 rule (and remember this is a rule, not a law): 10 slides, 20 minutes, 30-point fonts. This means you must be brief, concise, and to the point. Always practice to ensure your timing is correct.

- **Public speaking:** This is a very important part of any presentation; to deliver your point in a way that makes a positive impression. Remember the following:
  - **Show your passion:** If you are not interested, why should others be?
  - **Eye contact:** Make eye contact with the audience, do not stare at one person, and do not stare at your slides or the ceiling. This is a sign of shyness and lack of competence.
  - **Tone of voice, loudness:** Speak loudly and clearly. Your tone of voice should be dynamic and pleasant like music, not monotone and boring like noise.
  - **Crutch words:** Make certain that you do not have crutch words like ''ah, ehm, you know,'' etc. Clicking your pen, stomping your foot, and banging on the lectern are physical crutches. Practice them out.
  - **Accents:** Many people have accents, whether regional or international. You cannot change your accent. But if you speak completely and clearly, accents are generally ignored by the audience.
  - **Length, Q and A's:** Keep track of time and allow for questions and answers.
  - **Videotaping practice session:** Practice many times, and time yourself. Videotape yourself and watch for crutch words, hand movements, and other behavioral patterns.

- **Main points of the plan:**
  - **Summary:** Present an outline and a summary of the plan first in order to engage the audience and to let them know to what they should look forward.
  - **Market:** Describe the market size, condition, and state of the market.
  - **Your solution:** Describe your solution and how you intend to implement it.
  - **Your team:** Mention who is on your team and their qualifications.
  - **Funds and their use:** Describe the financial situation, including what is already available and what is needed, and how it will be spent.
  - **Mention expected and potential rewards:** Others who will get involved make their decision based on the reward they will expect. Describe what they may realistically expect.

- ○ **Summarize:** Summarize (not repeat) the plan.
- ○ **References:** Cite references and the sources of your data and information to add credibility to your presentation.
- ○ **Tell them what you want them to do:** Do not assume that they should figure this out on their own. You should tell the audience what you expect.

15.12 LEGAL STRUCTURE OF A BUSINESS

In the United States, businesses can choose from a variety of legal structures that best suit them. The type of business structure affects taxes as well as liability of the owner(s) and shareholders as well as "pass-through" of the income/profits. One of the basic and fundamental questions about taxes is whether the establishment is taxed by the government for its profits first, before any income is passed through to the shareholders, or the establishment is exempt from taxes, but the recipients pay taxes on their income from the establishment. Consequently, the business is either taxed twice or once. However, the establishment and the owners have different levels of liability against the establishment and their creditors. It should be mentioned here that in general, the legal structure of a business may later be changed when it no longer suits the needs of the people involved.

The laws of the United States are among the most favorable for entrepreneurs. They allow for a variety of systems under which a business can operate, and they provide room for growth and for profit making. In fact, unless the business engages in tax fraud or illegal practices, there is little liability risk for the involved individuals. Although most start-ups fail, the society admires people who take a risk and start new businesses, and in most cases, the economy is driven by small entrepreneurs who start these businesses and grow them to maturation.

The U.S. law allows for the following types of ownership and business structures, each with a different ownership, liability, management, tax treatment, and source of capital. The following is a very brief summary of these structures. You must consult professionals before you choose a business type and start your business:

- Sole proprietorship
- Partnership
  - ○ Limited partnership (LP)
- Corporation
  - ○ S-corporation
  - ○ C-corporation
- Limited Liability Company (LLC)

Sole Proprietorship   A sole proprietorship is an unincorporated business that is owned by one individual. It is the simplest form of business organization to start and maintain. The business has no existence apart from the owner. Its liabilities are the proprietor's personal liabilities. She or he undertakes the risks of the business for all assets owned, whether used in the business or personally owned. The income and expenses of the business are included on his or her own tax return. All investment comes from the owner, but all income (and liability) goes to the owner as well. The company is not taxed since all income is passed through to the owner. In turn, she or he will treat the income and expenses as personal income and expenses.

Partnership (General)   A partnership is the relationship existing between two or more persons who join to carry on a trade or business. Each person contributes money, property, labor, or skill, and expects to share in the profits and losses of the business.

A partnership must file an annual information return to report the income, deductions, gains, losses etc. from its operations, but it does not pay income tax. Instead, it passes through any profits or losses to its partners. Each partner includes his or her share of the partnership's income or loss on his/her tax return.

An unincorporated organization with two or more members is generally classified as a partnership for federal tax purposes if its members carry on a trade, business, financial operation, or venture, and divide its profits. However, a joint undertaking merely to share expenses is not a partnership.

Limited Partnership   A limited partnership (LP) consists of two or more persons, with at least one general partner and one limited partner. The general partner has unlimited personal liability, while the limited partner is liable only to the extent of his/her investment. This is usually used to raise investment funds. For example, a general partner may collect investment money as "shares" to buy a commercial building. The general partner is responsible for running the partnership, while the liability of each limited partner is limited to their investment. In return, the general partner has the right to manage or even sell the building and divide the proceeds, but limited partners cannot sell the building. Venture capital funds usually use this model as well. In most cases, this model does not apply to an entrepreneurial start-up company. Limited partnerships usually pass their income through to the partners; therefore, the partnership is not taxed. The partners are individually taxed based on their income from the partnership.

Corporations   A corporation is a legal entity which exists completely separately from the individuals involved. This separation gives the corporation unique powers which other legal entities lack. The extent and scope of its status and capacity is determined by the law of the State in which it is incorporated.

Investors and entrepreneurs often form joint stock companies and incorporate them to facilitate a business. Since this form of business is now extremely common, the term corporation usually refers to these entities.

Limited Liability Company (LLC)   A limited liability company (LLC) is an entity formed under State law by filing articles of organization as an LLC. Unlike a partnership, none of the members of an LLC are personally liable for its debts. An LLC may be classified for Federal income tax purposes as if it were a sole proprietorship, a partnership, or a corporation. If the LLC has only one owner, it will automatically be treated as if it were a sole proprietorship, unless an election is made to be treated as a corporation. If the LLC has two or more owners, it will automatically be considered to be a partnership unless an election is made to be treated as a corporation. Many "angel" investors prefer this model, although it can be more costly to set up.

S-Corporation   An eligible domestic corporation can avoid double taxation (once to the corporation and again to the shareholders) by electing to be treated as an S-corporation. Generally, an S-corporation is exempt from federal income tax other than tax on certain capital gains and passive income. On their tax returns, the S-corporation's shareholders include their share of the corporation's separately stated items of income, deduction, loss, and credit, and their share of non-separately stated income or loss.

C-corporation   A C-corporation is a separate taxpaying entity for the purposes of federal income tax law. To the extent a C-corporation has taxable income, it must pay the corresponding tax. If a C-corporation has losses, the losses must offset the corporation's

TABLE 15.1 Advantages of Different Business Structures

Advantages of Pass Through	Advantages of Separate Taxable Entity
Owner's personal income is sheltered from losses if the owner materially participates in the business (not passive income)	Lower tax bracket on retained earnings
No limits on the deductibility of the owner's income	Tax-free employee benefits for owner employees
Double taxation is eliminated	
No double taxation when the business is sold	

TABLE 15.2 Liability and Taxation for Different Business Structures

	Liability of Owner	Taxation
Sole Proprietor	Unlimited	Pass through
Partnership	Unlimited	Pass through
C-Corporation	Limited	Separate taxable entity
S-Corporation	Limited	Pass through
Limited partnership	General partners: Unlimited Limited partners: Limited	Pass through
Limited Liability Company	Limited	Pass through

income. None of the profits, losses, or other tax items of a C-corporation are passed through to its stockholders or used on their personal income tax returns.

A C-corporation has the potential to cause double taxation on income. For example, assume a C-corporation owned by 10 stockholders of equal shares has income of $1,000 and it pays a federal income tax of $300 on that income. The C-corporation then pays each stockholder a dividend of $70, the amount of its after-tax funds divided by the shares. The stockholders must now report the $70 of income and pay a second or double tax on the $1,000 earned by the corporation. If each stockholder's marginal tax rate were 28 percent, each stockholder would pay $19.60 in federal income taxes. Together, the corporation and its stockholders pay $496, which is almost 50 percent of the $1,000 earned by the corporation.

Table 15.1 summarizes the advantages of pass through versus separate tax entity. Table 15.2 summarizes the liability of the owner versus taxation for different types of business structures.

15.13 FINANCING YOUR ENTREPRENEURSHIP

Every business requires funds to start, operate, grow, pay salaries, and function, before it makes enough revenues and eventually profits to remain sustainable. Securing adequate initial funding to start a business is a crucial step in this process. On the one hand, securing funds, especially from others who are not familiar with you, your abilities, and your work ethics, and who may be completely unfamiliar with the value of your innovation, idea, or intellectual property, is very difficult. Not everyone is willing to invest in unknown and unproven businesses. Therefore, expecting to be able to secure funding quickly and easily is naïve. However, it should not be a sign that your entrepreneurial idea is not worth the effort. On the other hand, the ability to secure funding suggests that your idea is worthwhile. The fact that someone else is willing to invest in you and your idea is an excellent sign that you are on the right path.

The funding level needed to start a business does depend on many different factors, including whether you first need to test and validate your idea or your product, whether manufacturing is involved and to what extent, how complicated the product is, and how the product can reach the market. Starting a fruit juice shop in San Luis Obispo required enough funds to lease a shop, buy blenders, furniture, materials, a sign, and other odds and ends, as well as incorporation of the business, hiring an accounting firm, and consulting the authorities to make sure all laws are followed. Starting a new business making toys with a low-power laser light and mirrors attached to plastic prisms required funding for ordering parts and manufacturing small parts and assembling them in a garage and directly selling to the customers. No shop or furniture was needed, but more money was spent on buying parts and assembly. Microsoft, Hewlett-Packard, and Google all started in a garage as well. How much is needed for your enterprise depends on these factors and must be determined before an attempt is made to secure them.

15.13.1 The Difference between Debt and Equity

If you were to buy a house, but lacked adequate resources, you would get a loan from a bank with which you would buy the house. So long as you pay your monthly payments of (usually) interest and principal, the bank will be happy, and you keep the house until the loan is paid. If you sell the house, you pay back the loan and keep all proceeds from the appreciation of the house. However, even if you lose money due to falling prices, the bank expects to get its money back in full.

Now imagine that instead, you would go to someone and propose that she or he pay you ½ of the cost of buying the house, and together with your own funds, you buy a house with equal shares of ownership in it (equity). Whether you live in the house or you rent it, you would pay your co-owner ½ of the proceeds. When you sell the house, the partner will get ½ of the sale price too. If prices fall dramatically, you both lose money.

This is the difference between a loan (debt) and equity. Similarly, if you get a loan from a bank to start a business, as long as you pay the interest to the bank, the bank will not interfere with the way you run the business. However, even if the business fails, you will have to pay the bank (except in relation to declaring bankruptcy). However, when you have other investors, they will require equity in the company and will share in the profits and appreciation of its value if it takes off, but they will lose their money if the business fails. Due to this higher level of risk, you should expect that the investors will take a much more active role in running the business and its affairs.

15.13.2 Where Does the Money Come From?

The required funds may come from a variety of sources, as outlined below, depending on the stage of the development and how much is needed. ''Bootstrapping'' is a common word in this stage. You have to try to accomplish an almost unimaginable amount with almost nothing, mostly from your own resources, but this resourcefulness is a good sign, especially when you seek future funding from more elaborate sources. You will most likely have to go from one funding type to another as you move from one level of business to another. Few investors will invest in any business that does not have at least some personal financial commitment from the entrepreneur. So keep in mind that you may have to start with personal savings, and as you exhaust one funding level and attain some result, you may move to another level. As long as verifiable progress is made, attaining additional funding is more plausible. Keep in mind that you should expect that anyone who invests in your venture (and that includes your parents) will expect some equity in your enterprise; they will expect to enjoy from your future profits. However, this is a complicated matter and you should consult with experts about how much equity in your company should be given away. In more advanced situations, this becomes a crucial matter and you must be extremely careful about the distribution of your equity and its structure. You must distribute equity willingly (share success) but very cautiously (undesirable partners).

As an example, assume your family members invest in your business with a few thousand dollars for a 50% share of the business. As it grows a bit and its value increases, another person wanting to invest in it with the same amount as your family will not get the same 50% share, but much less. Similarly, as the business grows, or as the stakes in it grow, additional funding may come to you through giving up more and more of your equity. Always ensure that you understand the consequences of relinquishing equity to investors. Now consider the following. Imagine that when you start your business, you register to eventually issue 100,000 shares (stocks). At the beginning, you only issue 50,000 stocks to three other investors plus yourself, each share worth $10. This raises $500,000 (valuation), equally owned by four individuals, each having 25 percent equity in the company for initial investment of $125,000 for 12,500 shares. Later, trying to raise additional funds to grow, you issue another 25,000 shares, each worth $20 ($1,000,000 total valuation). This will raise another $500,000, but now, each of the original investors will have 1/6 part equity (12,500 shares out of a total of 75,000, or 16.7%) in the company. But simultaneously, their shares are worth $20 each, for a total of $250,000. This means that their worth is doubled, even though their equity is less than before (this is referred to as dilution). As more stocks are issued, so long as the worth of each share increases, their equity reduces, but their total worth increases. However, similarly, if the price per share falls, their worth reduces proportionally (selling stocks at a reduced price is referred to as a "down round" which is a major investment disappointment). This gets more complicated when venture capitalists are involved who may require substantial equity or stock options. You must consult knowledgeable business and legal experts to ensure that you do the right thing.

Funding for the start of a new business or venture may be secured from the following sources:

- **Personal savings:** Personal savings, including cash, equity, or other personal resources, is the first level of investment, and is generally required by anyone else who may get involved. Without a personal stake, no one else will give you credit for caring about the business (your "skin in the game"). If you are not prepared to invest your own resources in the business, you must not believe that it is viable; no one else will dare invest their resources.

- **Taking a loan or credit card charges:** Many individuals start small businesses by charging their credit card or taking a loan. Doing this will allow you to keep full control over your enterprise and keep 100 percent of the equity, but it also puts a larger burden on you to succeed. If the required funds are relatively small and you know that you have enough resources and other income to pay back the loan, and so long as you stand a good chance to succeed, this may be a viable option, especially if other sources of funds are not available. Otherwise, especially with credit card loans which generally have the highest interest rates in the industry, you may put your financial future in jeopardy if the loan is large and you do not have other income to pay it back.

- **Grants from government, local authorities, and the like:** In many cases, local governments, state authorities, and the Federal government assist new and small businesses, especially if they fall under certain programs that are targeted. One example is the Department of Defense's Small Business Innovation Research (SBIR) and Small Business Technology Transfer (STTR) programs. At the present time, the SBIR Program provides up to $850,000 in early-stage R&D funding directly to small technology companies (or individual entrepreneurs who form a company). The STTR Program provides up to $850,000 in early-stage R&D funding directly to small companies working cooperatively with researchers at universities and other research institutions. Although these programs require much effort and proposal preparation and the chances of receiving the grants are low, they are great. You may receive a large grant to start your business without giving up any equity (other than giving the government agencies right of use).

Entrepreneurship and new small businesses are a favorite method of growing the economy, and for this reason, governments issue special loans, tax relief, legal assistance, and grants. You should try to find out if your enterprise falls under any of these programs and if you qualify for any assistance or special loans. The Small Business Administration (SBA) is an invaluable asset as well.

- **Borrowing from or giving equity to family, friends, faculty, or fools (the 4-Fs):** This is usually the next step of fundraising for your business. Obviously, not everyone goes to family and friends, but many do. Parents and other family members are usually interested in your success and believe in you more readily than others. As a result, they may provide you an opportunity to succeed with fewer requirements. No doubt they must be given equity in the enterprise for their risk too. Do take advantage of the opportunity, but do not take advantage of them. Be honest about the risks and rewards, the product and the market, and your assessment of chances.

- **Contract with customers, advanced payments from clients:** If you have a product that is ready for use by customers, you may be able to sell your product based on a contract and/or receive advanced payment in order to manufacture the product. This payment is credit, and either the product has to be delivered on time, or the payment (and possible fines and interest) must be paid back to the customer. However, it may help you get off the ground, especially if you do not have other sources of income or funds. Having contracts can also be tremendously helpful in attracting other funding from investors who realize that there are real customers willing to buy the product or service offered by the enterprise.

- **Investors such as angels and angel groups:** Angel investors are wealthy individuals who help entrepreneurs start businesses by investing in their business. After personal and family resources, they are the next logical level of funding. Their investment usually propels the new business into the next level. Angels judge the entrepreneur and the business based on the innovative idea, the team, and the opportunity the business offers. Angels usually invest before venture capitalists. They may expect larger returns than venture capitalists as well.

  There are both individual angels and angel groups. Individuals are usually industry veterans who invest their own money and, if successful, receive full return. Angel groups are usually large networks of industry experts with staff that do most of the preliminary work before the group decides whether to invest or not.

- **Venture capitalists:** Venture capitalists are organizations that develop investment funds used by individual investors who invest in companies in anticipation of receiving generous returns if the company is successful. Their funding level is generally much larger than angels or individual investors. Generally, by the time you seek this level of funding you must have shown the competitiveness of your innovative idea, the worth of your business, and the resourcefulness of your team. In most cases, a short presentation is the sole opportunity you have to prove yourself, your team, and your business. Although getting to this point requires much work, networking, knowing people, and chance, your success or failure at this stage is due to your abilities and worth, and not due to chance or who you know. This is why it is crucial to have an effective, truthful, realistic, and informative presentation. Funding from venture capitalists is perhaps the hardest to get and should be the last resort, as they also seek much control over the affairs of the business.

Investors look into proposals for investment and decide based on the possibility of getting about 10 times their invested money in 5 years. However, out of any 10 new ventures (not new businesses) started, 1 will do well, 3–4 will manage (1–2 times rate of return), and 4–5 will fail. Please refer to www.nasvf.org (National Association of Seed and Venture Funds) for information about venture capitalists.

- **Corporate investors:** Sometimes, corporations agree to invest in promising small companies if the intellectual property is important to them, if they see a large market for the product or strategic value for their own products, or if the product complements their own products. In that case, there is a chance that you may get additional buy-in from larger corporations as the first step to eventual acquisition of your company.

- **License fees from licensees:** If you license your intellectual property or your innovation, wholly or in part, to others, you will receive license fees, which in turn may help you in starting your operations. You may also be able to license your intellectual property to others in exchange for fees to help you start the business with them.

- **Revenues and profits from other products if already in business:** If you already have a business and produce income, but have additional innovative ideas and products, you may be able to invest your own assets into the new adventure, at least initially. Many large corporations do this wholly. As long as they can devote part of their revenues to research and development, they will not need to raise funds through borrowing, selling more stocks, or licensing.

The Funding ladder, coined by Joe Boeddeker[10] is a term that describes the steps that one might follow to secure funds from angels and venture capitalists. It includes:

- **Flyers:** Flyers are used to communicate your idea to others who may become interested in investing. No one expects to receive investment money directly from a flyer, but this is the simplest and easiest way to inform others about the possibility. You may send or post your flyers in strategic locations in hopes that someone with the right connection will see it.

- **2-min "elevator pitch":** This is a simple chance one may get to talk to someone about an idea in 2 minutes (or less). It must be professional, eye catching, informative, and interesting. You need to accomplish everything in an extremely short time.

- **Discussion:** This is when you get a chance to discuss your innovative idea with someone who can invest, or knows others who might be interested. This is usually a few minutes, at a table while eating, during a drive, or in an office meeting. It must be concise, quick, informative, and effective. The presentation should make an adequate impression on the prospective investor to remain interested or to pursue it.

- **Executive Summary:** This is a short, concise summary of your innovative idea, your business and value proposition, the customers, and what you need, perhaps in one or two pages. Boring, data-filled, and long executive summaries are mostly ignored. Important information that is buried in attachments will never be read.

- **Presentation:** This is a chance to present your innovation and your business proposition to a group of investors in a meeting, perhaps 20 minutes long. You must be concise and to the point, speak well, be informative, and although formal, make a positive impression. To get to this point is a significant achievement. However, if you do not succeed here, the investors will no longer be interested. They probably have other businesses to attend to.

- **Business plan:** A business plan includes all the necessary information about a new enterprise, the customers, and what is proposed. It is required by investors if they are interested in your proposition.

- **Due diligence:** If there is an interest in your enterprise, the investor(s) investigate the details of the business plan, the income, cash flow, assets, and other detail prior to investing.

- **Negotiations:** Negotiations are needed in order to arrive at an agreement about the details of a business enterprise and securing a contract and eventual funding. For the most part, attorneys ensure that all aspects of the business are covered. You must seek the assistance of legal experts in this step.

- **Funding:** After a contract is signed and executed, funding is available to you to start your business. In fact, this is when the real work for the technologist starts. Have fun.

15.14 EXECUTION OF A BUSINESS PLAN

When a business plan is developed, and the enterprise is funded, the plan must be executed. Usually, this is among the most intensive and hard-working times in the life of an entrepreneur, where countless hours are spent, many decisions are made, and much heartache is borne until the plan is executed. You will need to work on the technology, hire employees, pay them, provide for their rights (insurance, taxes, safety, etc.), create your management team, develop and test and evaluate your product and service, create a brand-name, advertise, establish a sales force, institute a distribution system, and sell your product or service. As mentioned earlier, most probably, you will need to hire expert managers and financiers and engineers to help you in this endeavor, as no one can do everything alone. Your team will be crucial in your success.

15.14.1 Product Development

As part of your business plan, and as a basis of your enterprise, you must develop your product or service for sale. It is one thing to have a prototype product that was tested in a lab or other controlled environment; it is another to have a real product that will be sold to customers. All issues we have discussed throughout the book, from the safety and reliability of the product to its aesthetics, and from its human interface to its price, and everything else in between, becomes an important issue to consider. You must develop your manufacturing process, as well as your final product, in accordance with all engineering and design skills you have, hopefully together with other designers, engineers, or experts. With due respect for the importance of finances, management, state of the economy, and customer requirements, if your final product does not eventually live up to its desired requirements, it will not be purchased. An innovative product that is needed by consumers and delivers what is promised is central to the success of an enterprise and in creating a long-lasting brand name.

15.14.2   Market Development

The next requirement for a business is to develop the market in which the product or service is offered for sale. The potential for the market for a product may or may not already exist. In other words, there may already be a need for a product, so the market exists, but it does not necessarily mean that the market is available. However, it is also possible that although you have an innovative product or service, the need may not be apparent to the consumers. In that case, you will have to make the consumers realize their need. For example, what is the real need for soda? However, a need is created by the companies for the product through advertising, creating trends, or other means. Did anyone realize that they might need to have access to thousands of songs in the palm of their hands until i-Pods were introduced?

A market, whether in existence or not, must be developed. There are countless methods available to do this, from simple advertising to door-to-door sales to celebrity endorsements and to questionable methods of enticing customers (there was a time when subliminal messages were legal). Obviously, an important factor in this is your financial resources. However, in your business plan, you must identify what methods you will use to reach your customers. Marketing is a fundamentally important part of any business, a subject that requires expertise and knowledge in the subject. You must include marketing experts in your team, early on, in order to develop your market. Otherwise, to whom will you sell your product?

15.14.3   Management Development

Another important factor in any enterprise is the development of the management team. Like most other aspects of this work, your management team is a crucial element in the success of your business. Many start-ups may be able to function adequately when they are extremely small and when they are at their infancy. But as the organization grows and more employees are hired, a management team will be necessary to continue operating efficiently. An organizational chart will also be helpful as the enterprise grows.

15.14.4   Infrastructure Development

This stage involves the development of the actual practice of the business, the staff, the systems, and the service. It also includes the development of accounting practices, a sales and distribution force, and reaching customers and consumers.

15.15   END GAME PLANNING

The ultimate goal of a business enterprise is success. However, success may take on many different meanings. It can be financial success and recognition, it may mean the value of the lessons learned, or it may mean the satisfaction of helping others, the country, or the economy. Whatever your goal, you must think about your end game. What is your ultimate goal? If it is financial success, do you have any particular goal in mind? And what will you do when you achieve it?

In fact, you do not need to answer this question at the beginning, or even if you do, it can be changed at any stage of the game. Certainly you would not expect that the founders of Google had guessed what kind of financial success they might achieve when they started. However, there has to be some understanding of the end game.

Many venture capitalists intend to get involved in start-up enterprises, hoping that they will realize a large rate of return on their investment, but do not intend to remain in it forever. Their goal is to sell the company to other corporations, or offer it to the public. As a result, most start-up companies are either sold or are offered to the public. Corporations purchase the promising companies and either absorb them to acquire their technology, keep them as subsidiaries, or close the business in order to reduce competition. If you involve yourself with venture capitalists, you may run into these issues, even if you would like to keep the company under your control. At some point, others may decide it is time to sell it. This is one reason why it is important to consider what your ultimate goal is when you start a business. As mentioned earlier, there is always a chance that this ultimate goal, or its magnitude, may change. But having a goal in mind may prevent later headaches.

I would like to acknowledge the assistance of Mr. Joe Boeddeker, President and CEO of Venture Growth Alliance in Saratoga, California, and Mr. Dennis J. Fernandez, former Executive Vice-President of FP International Inc. in the preparation of this chapter.

REFERENCES

1. EVANS, HAROLD, "They Made America," Little, Brown, and Company, New York, 2004.
2. KELLEY, TOM, and J. LITTMAN, "The Art of Innovation: Lessons in Creativity from Ideo," Currency, New York, 2001.
3. KELLEY, TOM and J. LITTMAN, "The Ten Faces of Innovation," Currency Doubleday, New York, 2005.
4. COLLINS, JIM, "Good to Great," Harper Business, New York, 2001.
5. YUBAS, MATTHEW, "Product Idea to Product Success: A Complete Step-by-Step Guide to Making Money from Your Idea," Broadword Publishing, New Jersey, 2004.
6. KAWASAKI, GUY, "The Art of the Start: The Time-Tested, Battle-Hardened Guide for Anyone Starting Anything," Portfolio, Published by Penguin Group, New York, 2004.
7. "Getting Started as an Entrepreneur: A Guide for Students," The National Collegiate Innovators and Inventors Alliance (NCIIA), 2002.
8. "The Entreclub Handbook: An Operating Manual for Student Entrepreneurship Clubs," The National Collegiate Innovators and Inventors Alliance (NCIIA), 2002.
9. "The Difference and Distance Between Invention and a Business," *NASA Tech Briefs*, June 2006, pp. 94–96.
10. BOEDDEKER, JOSEPH C., "The Entrepreneurial Bridge," Presentations at the University of Illinois, October 7, 2003, University of Santa Maria, Valparaiso, Chile, May 19, 2006, and University of Queensland, Australia, November 29, 2006.
11. COVELLO, JOSEPH, and B. HAZELGREN, "Your First Business Plan," 3rd Edition, Sourcebooks, Inc., Illinois, 1998.
12. CAGAN, JONATHAN and CRAIG M. VOGEL, "Creating Breakthrough Products," Prentice-Hall, New Jersey, 2002.
13. "Innovate America," National Innovation Initiative Summit and Report, Council on Competitiveness, 2005, p. 8.

CHAPTER 16

DESIGN ANALYSIS OF MACHINE COMPONENTS

16.1 INTRODUCTION

It is entirely impossible to treat the design analysis of machine components to any significant depth in one chapter. Engineers take many years to learn what is covered in this chapter, requiring significant number of courses and books. However, many students and novice designers design products and machines before they finish their engineering studies. It is for these individuals that this chapter is presented. It is meant to be an introduction to the issues of design analysis as it pertains to a novice or a designer who has not yet taken undergraduate analysis courses. The intention is to help students and designers understand how their decisions can affect their designs, and how the components in their design are affected by external loads to which they are subjected. It is expected that those individuals will learn this material in much more depth in their undergraduate and graduate studies, as well as through experience. Appropriate references are mentioned at the end of this chapter for further reading on this material.

As you may remember from Chapter 5, one of the major steps in the design process is the analysis, detail design, and planning of the selected solution. This means that although a general solution has been devised for the problem at hand, the details of the solution must now be analyzed in order to ensure that everything will function properly. The engineering knowledge of the designer provides for this analysis, and this is what differentiates a design engineer from other designers. In mechanical systems and structures, this analysis includes ensuring that each component is strong enough to carry the loads it is supposed to and it can last for the life of the product. The conceptual design takes care of the function of the system, while component design and analysis ensures that the system will function as expected. This can be an extremely important step for products that have critical loads and strength issues related to their components, and engineers spend years studying how to perform this analysis. Courses such as statics, strength of materials, dynamics, stress analysis, and machine design are related to the subject of recognizing the loads, analyzing their effects, and designing the components of the product for strength and long life.

As an example, consider a new robot or a building. The robot is a moving machine with components that rotate and move relative to each other. The building is a static structure that does not move. Still, in both cases, the components of the two systems are subjected to loads, and they must be designed properly in order to carry the loads and be able to function as expected. For instance, the building will be subjected to loads stemming from the weight of the building materials that constitute the structure (concrete, steel, trusses, flooring, etc.), the dead loads such as furniture, cabinets, tables and desks and appliances, live loads such as human occupants, as well as external loads such as wind, earthquake, and movements of the foundations. The robot will experience both static loads coming from its own weight and the weight of the parts it carries (and in some cases, these weights are very large and significant), dynamic loads caused by accelerations, as well as loads that the robot may apply to other machines (such as in assembling an alternator shaft). In both cases, it is extremely important that these loads be determined, analyzed, and calculated and the system be designed to carry the loads without damage.

The sequence of this type of analysis requires that the engineer first estimates or calculates the loads that the product may experience or to which it will be exposed. This includes all types of loads such as static loads, dynamic loads, live loads, dead loads, thermal loads, environmental loads, as well as applied loads. In most engineering programs, this is taught in statics, dynamics, thermodynamics, vibrations, fluid mechanics, and other similar courses. Next, the designer should consider the range of desirable materials that can be used. This requires the knowledge of material properties and material behavior which is taught in material property courses. For example, one needs to know the yield strength of steel in order to be able to design a part without excessive deflection or yielding. By comparing the material strengths and material properties with load requirements and stresses caused by the loads the designer can design the proper shape of the parts and their dimensions.

In this chapter we will follow the same approximate flow. We will discuss forces and moments, free-body diagrams, stresses and strains, strengths, and fatigue loading. We will also discuss common characteristics of power sources. For more detailed information please see References 1–11.

16.2 LOADS: FORCES, MOMENTS

Most products and systems are subjected to forces and moments that can be critical. If the components of the product or system are not designed properly, these loads have the capability to break the components or damage them and fail the product.

For example, consider the page-turner for the disabled, discussed in Chapter 5. For the sake of discussion, assume we picked a mechanical solution such as Figure 5.9, repeated here, with a pair of two-arm linkages and rollers that lift a page and turn it from one side to the other. The bars that turn the page are powered by a motor. Obviously, a motor with adequate torque to operate the arm and turn the page will be needed. Subsequently, all other parts that are operated by the motor will be subjected to the same torque and must be able to carry the load without failing. As a

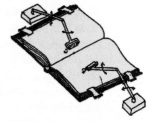

FIGURE 5.9, repeated here.

result, we need to follow the torque (or the load) as it goes from one component to another and make sure that all components have the capability to withstand it. For instance, the bar must be thick and rigid enough to not bend (deflect) under the load. The connection between the first bar and the motor must be capable of carrying this load without breaking, excessive deflection, or sliding. The connection between the two bars must also be strong enough to not break or wear out. The same is true for the second arm and the rollers, etc. Most of this analysis also requires that you first decide how much friction you need, and what material will generate this friction. Therefore, we need to estimate our loads, we need to calculate the applied loads, and we need to determine the load path in order to design all the parts that are subjected to these loads.

In many applications where loads are critical, empirical data may be collected and used in order to estimate these loads. For example, a car is subjected to countless road conditions with varying severities. One way manufacturers and designers use to estimate the applied loads is to record these variations in real tests. For this, a car equipped with sensors and data collection systems is driven in all conceivable conditions, sometimes for long periods of time, and average and peak loads are recorded for design purposes. Loads on airplanes are even more crucial and must be estimated correctly. In other situations, e.g., a building or mechanical devices, loads are estimated based on industry standards, but a healthy safety factors is also used in order to ensure that the product or system remains safe under real loading conditions. This is because we can never really estimate what the maximum loads may be, or whether the material used is as strong as claimed or not.

Static and dynamic load analysis enables a designer to estimate the loads on the product and system and their components. It also enables the designer to determine how these components might fail and what needs to be analyzed (such as modes of failure, tensile loads, compressive loads, bending and torsional loads, strength of the component versus deflections, non-varying loads versus fatigue loading, etc).

16.3   FREE-BODY DIAGRAMS

Free-body diagrams are used extensively in static and dynamic analyses in order to calculate reaction loads as well as loads in each component. For example, to design a truss, it is imperative that not just the reaction forces, but the forces in all truss members be calculated in order to design and specify each individual member. Similarly, if a mechanism is used to accomplish a task, e.g., in the page-turner, all accelerations, and thus dynamic loads, must be calculated. This is always accomplished using a free-body diagram. A free-body diagram or FBD is a representative of what the system or one of its components see when isolated from the environment. For example, when you draw an FBD of a truss, you isolate it from its environment (its support system). In order for the truss to remain at equilibrium, there must be equivalent forces and moments applied to it at the supports. The FBD allows you to see these reaction forces and calculate them. The following examples are used in order to demonstrate this application. You will have to learn about the detail of how FBDs are drawn and used for other situations in a statics course.

EXAMPLE 16.1

Draw the free-body diagram of the truss in Figure 16.1a. L is the load at each joint. Find the reaction forces at A and N. Find the force carried by members DE and IJ. The length of all horizontal and vertical members is d.

Solution   This is a typical problem in which FBDs are used. The FBD of the truss consists of the truss with the external loads applied to it and the reaction loads at the two ends which support it. The FBD is used to find the reaction forces at the supports as follows. As shown in Figure 16.1b, the reaction force at A is broken into two components along x-and y-axes. The reaction at point N only has a vertical component along the axis because the joint is allowed to move along the x-axis.

The second free-body diagram shown in Figure 16.1c is for a portion of the truss, cut along the members that are of interest to us. Since the totality of the truss is at equilibrium, any portion of

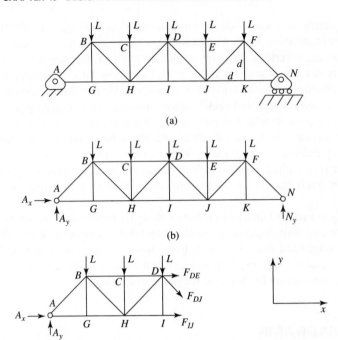

FIGURE 16.1 Free-body diagrams of the truss and a portion of it used to find reaction forces and forces in individual members.

it must be at equilibrium. Therefore, this part, showing the forces on the members *DE*, *IJ*, and *DJ* is also at equilibrium. Using these two FBDs we get

$$\Sigma F_x = 0 \quad \text{or} \quad A_x = 0$$
$$\Sigma F_y = 0 \quad \text{or} \quad A_y + N_y - 5L = 0$$
$$\Sigma M_A = 0 \quad \text{or} \quad N_y(6d) - L(d) - L(2d) - L(3d) - L(4d) - L(5d) = 0$$
$$\text{thus } A_y = N_y = 2.5L$$

For the second part:

$$\Sigma M_D = 0 \quad \text{or} \quad F_{IJ}(d) + L(d) + L(2d) - A_y(3d) = 0$$
$$\text{thus } F_{IJ} = 4.5L$$
$$\Sigma M_J = 0 \quad \text{or} \quad L(3d) + L(2d) + L(d) - F_{DE}(d) - A_y(4d) = 0$$
$$\text{thus } F_{DE} = 4L \qquad \blacksquare$$

EXAMPLE 16.2

Draw the free-body diagrams of all the major components of the floor jack in Figure 16.2.

Solution The FBDs of all the major components are shown. From static equilibrium equations you can find all the forces in all members in order to design these components. $\blacksquare$

16.4 STRESSES, STRAINS, MATERIAL STRENGTHS

Imagine that you apply a tensile (pulling) force to a helical spring and then remove it. So long as the force is relatively small, the spring will stretch under the force, then return to its original length when the force is removed. In this case, the spring is called elastic,

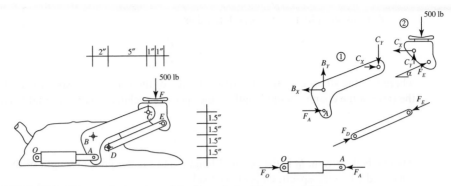

FIGURE 16.2 Schematic drawing of the floor Jack of Example 16.2.

meaning that it returns to its original length without any permanent deformation when the load is removed.

This characteristic is not unique to springs alone; practically, all materials in nature have a certain level of elasticity in them, whether metals, woods, plastics, or even liquids and gases. So, for instance, you may push inward the body panel of a dishwasher or a car just a bit, and as soon as you remove the force, the body panel bounces back. But if you apply a large enough force, whether to the above-mentioned spring or the dishwasher or the car panel, there may be a permanent deformation of the material which will not be recovered when the force is removed; the spring elongates beyond its original length, and the body panel deforms with a permanent dent.

In order to be able to properly design a component that is meant to withstand assumed external loads, engineers and designers need to know the characteristic behavior of the materials under consideration. For this reason, a relatively standard test is performed on most materials which presents the behavior of the particular material when subjected to a load, usually a tensile load (Figure 16.3). Using the graph of Figure 16.4, the component may be designed not just for tensile loads, but for many other loads as well, including fatigue (variable) loads, torsional loads, bending loads, and their combinations. This graph shows the relationship between a load (stress) and deformations (strain) and is usually performed on a sample of the material with standard sizes and/or shapes. Figure 16.3 shows a typical sample made for alloys in tension tests.

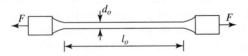

FIGURE 16.3 A typical specimen for testing materials in tension.

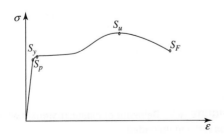

FIGURE 16.4 A typical stress-strain curve for ductile materials.

For this sample, the stress is defined as:

$$\sigma = \frac{F}{A_0} \tag{16.1}$$

where σ is the stress and A_0 is the original area of the neck of the specimen. The strain is the ratio of the change in the length of the specimen relative to the original length, or:

$$\varepsilon = \frac{l - l_0}{l_0} \tag{16.2}$$

where ε is the strain, l_0 is the original length, and l is the present length of the specimen, measured while the specimen is under load.

Figure 16.4 shows a typical stress-strain curve for ductile materials such as low-carbon steel. The first portion of the curve up to S_p is proportional. This means that the stress and strain follow a proportional relationship. The ratio between these two variables is called Modulus of Elasticity (E) or Young's Modulus. Therefore,

$$E = \frac{\sigma}{\varepsilon} \tag{16.3}$$

Additionally, the stress and strain follow an elastic relationship up to S_e, where, as soon as the load is removed, the strain is recovered and the sample returns to its original length (or shape). The value of the stress right before the maximum elastic relationship is called Yield Strength or S_y. This strength indicates how far a sample made from the particular steel may be loaded (in tension, pulled) without permanent deformation or strain. As soon as the load is increased beyond this point, while the stress level is still increasing, you should expect to have permanent deformation or strain in the material. If the load is removed at any point, the same level of elastic recovery may be expected, but with the corresponding permanent deformation (Figure 16.5). The maximum strength level S_u is called Ultimate Strength and indicates the highest level of stress that the sample may resist. Beyond this point, at least for ductile (soft, easily deformable) materials, the cross-sectional area of the material decreases significantly, and although the actual stress still increases due to decreased area, the apparent stress levels drop quickly and the material fails (breaks).

Although this figure is a very simple representation of the material behavior, it is very useful. Based on this behavior, we may also discuss the behavior of the material under varying (fatigue) loads, fractures, as well as with stress concentrations.

Behavior of other materials generally follows the same pattern to some extent. For example, brittle (hard, unforgiving, resisting deformations) materials lack (or have small) inelastic portion of the graph, and under increased stresses, directly go to the ultimate strength and break (Figure 16.6). Whenever you decide to use a particular material, you should find its specific characteristic and use the information in your design. Most common materials are included in tables that are readily available in reference books and from

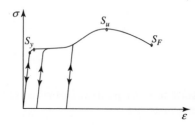

FIGURE 16.5   Behavior of elastic materials when loaded and unloaded.

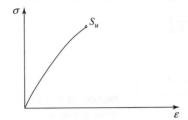

FIGURE 16.6 The behavior of a brittle material indicates lack of (or reduced) inelastic deformation.

FIGURE 16.7 Two steel specimens broken under a tensile load. The more ductile sample deformed more, and as a result, its neck area was smaller.

manufacturers. Please also refer to Chapter 10 for more information about material selection.

Figure 16.7 shows two specimens that failed under tensile loading. As can be seen, one sample is more ductile than the other. The more ductile material deforms more and therefore, breaks with a smaller neck area.

16.5 STRESS ANALYSIS

The ratio of a load to the magnitude of the material devoted to carrying it is the real measure of whether a part or a component will last or not. This is called stress, and it comes in many forms, including tension (tensile stress), compression (compressive stress), and shear stress.

Tensile stress is defined as the ratio of a tensile load to the cross sectional area of the material that carries the load. Similarly, compressive stress is the same ratio for a compressive load, or

$$\sigma = \frac{F}{A_0}$$
(16.1 repeated)

where F is the load (force) and A_0 is the area of the cross section (Figure 16.8).

Obviously, the larger the area, the lower the stress will be. Tensile and compressive stresses are *normal* stresses, because they are normal or perpendicular to the load. Shear stress is parallel to the load, but is still the ratio of the load to the area that carries it, and is expressed as

$$\tau_{ave} = \frac{F}{A_0}$$
(16.4)

Shear stresses have a tendency to rotate, slide, or "smear" a surface. For example, as is shown in Figure 16.8b, one part of the bar tends to slide relative to another. If you

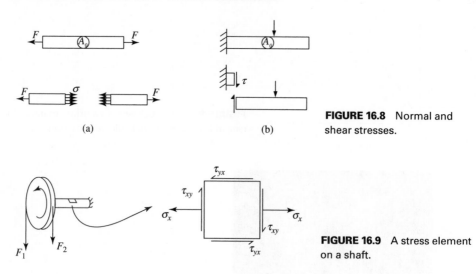

FIGURE 16.8 Normal and shear stresses.

FIGURE 16.9 A stress element on a shaft.

apply a torque on a bolt with a wrench, you will eventually break the bolt head in shear. And if you erect a structure without walls, the structure may collapse, whereas if there are walls in it, the shear load will be taken by the walls (called shear-walls) and it will remain upright.

In general, when a machine element (such as a part of a product or system) is subjected to loads, it may experience a combination of normal and shear stresses at the same time. Figure 16.9 shows an example of this, where the shaft is subjected to both bending and torsional loads (we will discuss how to calculate these stresses shortly). A stress element on the shaft is shown as well, with both normal and shear stresses on the element. However, as you may have guessed, the chosen orientation of the stress element is arbitrary. If we change the orientation of the element, the magnitude of these stresses will change. In reality, the material responds to the maximum stresses regardless of the orientation of the stress element we choose. Since different materials respond differently to normal and shear stresses, it is vitally important for us to calculate the maximum possible stresses and their orientation for designing the part.

Figure 16.10 shows two stress elements on the same machine component. There is no difference between these two elements except that they are at different orientations. This means that, under the same load, the part experiences shear and normal stresses to remain at equilibrium. As you may remember from your physics and static courses, a force may be divided into many different components in different arbitrary directions, but the summation is the same. Here too, the stresses end up being the same, but may be expressed differently in different directions. Therefore, it should be possible to find a particular orientation in which we either have maximum normal or maximum shear stresses. Depending on how the material responds to normal or shear stresses, we will use these corresponding stresses for design purposes.

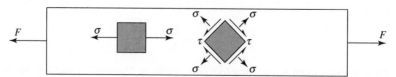

FIGURE 16.10 Two stress elements on a shaft at different orientations.

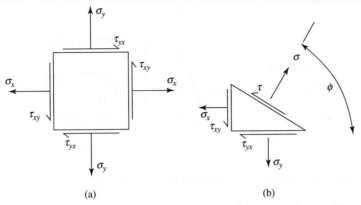

FIGURE 16.11 A stress element and a portion of the same element. For equilibrium, the summation of all forces in all directions must be zero.

Figure 16.11a shows a stress element with normal and shear stresses on it. Figure 16.11b is a portion of the same element cut at an arbitrary angle of ϕ. For this element, through equilibrium equations, we can derive the following:

$$\sigma = \frac{\sigma_x + \sigma_y}{2} + \frac{\sigma_x - \sigma_y}{2}\cos(2\phi) + \tau_{xy}\sin(2\phi)$$
$$\tau = -\frac{\sigma_x - \sigma_y}{2}\sin(2\phi) + \tau_{xy}\cos(2\phi)$$

(16.5)

By taking the derivative of the normal and shear stress equations and setting them equal to zero, one can find the maximum normal stresses (also called principal stresses) and the maximum shear stress and their orientation as follows:

$$\sigma_1, \sigma_2 = \frac{\sigma_x + \sigma_y}{2} \pm \sqrt{\left(\frac{\sigma_x - \sigma_y}{2}\right)^2 + \tau_{xy}^2}$$
$$\tau_{max} = \pm\sqrt{\left(\frac{\sigma_x - \sigma_y}{2}\right)^2 + \tau_{xy}^2}$$
$$\tan(2\phi) = \frac{2\tau_{xy}}{\sigma_x - \sigma_y}$$

(16.6)

These are parametric equations of a circle which represents the relationship between normal and shear stresses, as well as the orientation at which the maximum principal stresses and maximum shear stress occur. This is called *Mohr's circle*. It should be pointed out that at the orientation where the normal stresses are maximum, the shear stress will be zero. However, at the orientation where the shear stress is maximum, the normal stresses will not be zero. The maximum shear stress is always at 45° from the principal axes. This becomes important when we discuss material behavior later.

Please refer to other sources for a complete discussion about Mohr's circle in three dimensions.

EXAMPLE 16.3

A stress element is shown in Figure 16.12 with the corresponding normal and shear stresses. Determine the maximum shear stress as well as the principal stresses and their orientation.

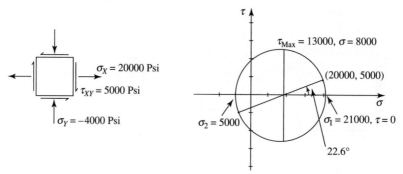

FIGURE 16.12 A stress element and its corresponding Mohr's Circle.

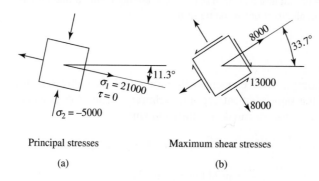

Principal stresses

(a)

Maximum shear stresses

(b)

FIGURE 16.13 The principal stresses, maximum shear stress and the principal axes. As you notice, the maximum shear stress is at an angle of 45° from the principal stress axes.

Solution   Figure 16.13 shows the stress elements with the maximum stresses. These values can be found both from the Mohr's circle in Figure 16.12 or from Equations 16.6., as follows:

$$\sigma_1, \ \sigma_2 = \frac{20000 + (-4000)}{2} \pm \sqrt{\left(\frac{20000 - (-4000)}{2}\right)^2 + 5000^2}$$

$$= 8000 \pm 13000 = 21000 \text{ and } -5000 \text{ psi}$$

$$\tau_{max} = \pm\sqrt{\left(\frac{20000 - (-4000)}{2}\right)^2 + 5000^2} = \pm 13000 \text{ psi}$$

$$\tan(2\phi) = \frac{2(5000)}{20000 - (-4000)} \rightarrow \phi = 11.3°$$

■

Different types of material fail in different modes when subjected to different types of loads. For example, ductile (soft, deformable) materials like low-carbon steel, rubber, and chocolate-caramel bar fail due to shear stresses exceeding their ultimate shear strength when subjected to a tensile load, whereas brittle materials such as chalk, glass, bones, and hard candy, break due to normal stresses if subjected to the same tensile loads. So, imagine that you have a soft rubber bar at hand and you pull the two sides. The bar will break in a cone-like section with about 45° angle. However, if you pull a chalk bar, it will break in a flat plane. Figure 16.14 shows a steel specimen which failed under a tensile load. As you can see, the bar originally started to fail in shear, at a conical angle of 45° all around. However, as is a characteristic of metals, ductile materials may harden when cold worked. (This can be seen in Figure 16.5 too. As a specimen is pulled beyond its yield point and released, it bounces back, but with some permanent deformation. The next time it is pulled, its strength is increased and its total available strain-to-failure is reduced, thus a

FIGURE 16.14   A steel specimen's failure under tensile load.

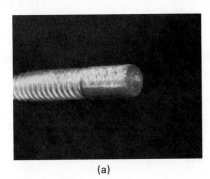

(a)

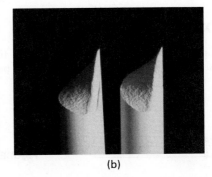

(b)

FIGURE 16.15   Failure of (a) a ductile steel specimen (a bolt) and (b) a brittle material (chalk) under torsional load.

little more brittle. Every time this is repeated, the material becomes stronger, but also more brittle. This is called cold working). This means that as the specimen was pulled and thus deformed, it became more and more brittle. As a result, the last part of the section failed in a flat plane, as expected for brittle materials.

Similarly, if a ductile material is twisted (torsional load), it will fail in shear, as shown in Figure 16.15a, where a ductile bolt was twisted with a wrench until the bolt head severed. However, a piece of chalk twisted the same way would fail due to principal normal stresses, which is 45° from this plane, as shown in the sample in Figure 16.15b. You will see that the chalk is broken at a spiraling 45° angle. A simple detection of parts may reveal much information to you if you look for clues about the type of material and the type of loads.

Wood, as well as many other materials used in industry (such a composite materials) have different characteristics in different directions (anisotropic materials). They are not uniform in all directions and therefore, behave differently. As a result, you must be careful about the direction of their grain and their varying strengths.

16.5.1   Axial Loads

As shown in Figures 16.3 and 16.5, when a machine element is subjected to an axial load, normal stresses will develop perpendicular to the load as

$$\sigma = \frac{F}{A_0}$$

(16.1 repeated)

EXAMPLE 16.4

A bar of 0.25-in. diameter carries a tensile load of 3000 lb. Calculate the normal stress in the bar.

Solution Substituting these values into Equation 16.1 will yield

$$\sigma = \frac{F}{A_0} = \frac{3000}{\pi(0.125)^2} = 61100 \, \text{psi}$$

∎

16.5.2 Shear Loads

As shown in Figure 16.7, when a machine element is subjected to a shear load, shear stresses will develop parallel to the load as

$$\tau_{ave} = \frac{F}{A_0} \qquad\qquad \text{(16.4 repeated)}$$

EXAMPLE 16.5

A piece of $0.25 \times 1 \times 3$ inch steel carries a 1000 lb load as shown in Figure 16.16. Calculate the average shear load in the cross section of the piece.

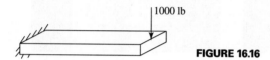

1000 lb

FIGURE 16.16

Solution Substituting these values into Equation 16.4 yields

$$\tau_{ave} = \frac{F}{A_0} = \frac{1000}{0.25 \times 1} = 4000 \, \text{psi}$$

(As you notice, this is referred to as τ_{ave}. You will see the difference between the maximum value and the average value later).

∎

EXAMPLE 16.6

Calculate normal and shear stresses that may be present in different parts of the joined pieces in Figure 16.17. The two pieces of 0.125-in. thick sheet metal are $\frac{3}{4}$ in. wide and are joined together with a square pin of 0.25-in. width. The load F is 100 lb.

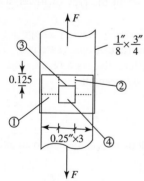

FIGURE 16.17

Solution There are a number of different places where the joint may fail as shown. Referring to number 1, the sheet metal may sever in two places under normal stress as shown. It may also fail as in number 2 in two areas under shear. The insert may also fail under normal stress at number 3, or in shear at the cross section of the pin at number 4. The stresses for these four cases are

$$\sigma_1 = \frac{F}{A_0} = \frac{100}{2 \times 0.25 \times 0.125} = 1600 \, \text{psi}$$

$$\tau_2 = \frac{F}{A_0} = \frac{100}{2 \times 0.25 \times 0.125} = 1600 \, \text{psi}$$

$$\sigma_3 = \frac{F}{A_0} = \frac{100}{0.25 \times 0.125} = 3200 \, \text{psi}$$

$$\tau_4 = \frac{F}{A_0} = \frac{100}{0.25 \times 0.25} = 1600 \, \text{psi}$$

These numbers can be used later to determine the safety factor of the joint based on the strength of the materials used. ∎

16.5.3 Shear and Bending Moment Diagrams

Before we can discuss bending loads and stresses caused by them, we need to first look at the shear diagram and bending moment diagrams. These diagrams are used to calculate the maximum moments that occur when bending loads are present, and are used in the calculation of stresses due to bending. We will use Free-Body-Diagrams (FBD) to draw these. Shear diagram shows the magnitude of shear force at any point along a beam which is loaded by a variety of forces and moments. The bending moment diagram shows the magnitude of the moment at any point along the beam.

Referring to Figure 16.18, you notice that there is a cantilever beam loaded by a force at the end. The FBD of this beam at an arbitrary point along the length of the beam is also shown. For equilibrium, it is necessary that summation of forces along the x- and y-axes, as well as the summation of moments about any point be zero. From these equations we can calculate the magnitude of the shear force and bending moments at any point. The shear and bending moment diagrams show these values along the entire length of the beam.

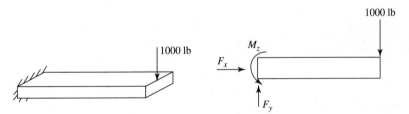

FIGURE 16.18   Free-body diagram of a cantilever beam.

In this particular example, the maximum bending moment is clearly at the end of the beam at the wall. However, in real-life examples, where there are many different forces acting on the beam, some distributed, some variable in intensity, the point of the maximum moment is not necessarily clear. In these cases, it is essential that the shear and bending moment diagrams are plotted in order to find the maximum values of these two variables for stress calculations.

Assuming that there is a continuous load with intensity of $w(x)$ on a beam, it can be shown that the change in shear force between any two points is equal to the integral of the load intensity (or the area under the plot of the intensity) between these two points on the beam. Similarly, it can be shown that the change in bending moment between any two points is equal to the integral of the shear force (or the area under the plot of shear force) between these two points on the beam, or

$$dV = wdx \rightarrow \int_{V_1}^{V_2} dV = \int_{X_1}^{X_2} wdx = V_2 - V_1 \qquad (16.7)$$

$$dM = Vdx \rightarrow \int_{M_1}^{M_2} dM = \int_{X_1}^{X_2} Vdx = M_2 - M_1 \qquad (16.8)$$

These equations can be used in order to calculate and draw the shear and bending moment diagrams.

In drawing these diagrams, the following conventions are used for positive and negative values:

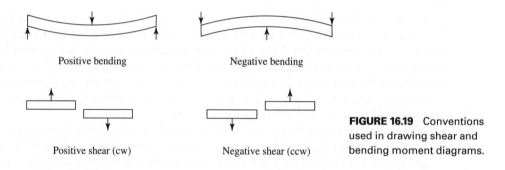

Positive bending   Negative bending

Positive shear (cw)   Negative shear (ccw)

FIGURE 16.19 Conventions used in drawing shear and bending moment diagrams.

EXAMPLE 16.7

The beam shown in Figure 16.20 is loaded with a distributed force at 2000 lb/ft. Draw the shear and bending moment diagrams.

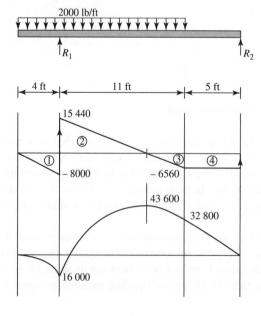

FIGURE 16.20 Free-body diagram, shear diagram, and bending moment diagram for Example 16.7.

Solution First we will draw the free-body diagram and calculate the reaction forces. Then we will draw the shear and bending moment diagrams (Figure 16.20).

As you see, the shear force starts at zero at the left end of the beam, and changes with slope w. At the point where there is a reaction force R_1 there is an abrupt change in shear force equal to this value, then continuing with slope w until the distributed load ends. From this point onward, there is no change to the shear force until the end of the beam where the reaction force brings the total back to zero. Similarly, the moment is the integral of the shear force, which ends up being second order parabola for this case. A positive shear adds to the moment, negative shear reduces it. The moment goes back to zero at the end. As you can see, the area under the load intensity diagram is equal to the change in shear force, and the area under the shear diagram is equal to the bending moment diagram between any two points. Using these, we can calculate all the necessary values as follows:

$$\begin{cases} \Sigma F_y = R_1 + R_2 - (2000)(15) = 0 \\ \Sigma M_{R_2} = (2000)(15)(12.5) - R_1(16) = 0 \end{cases} \rightarrow \begin{cases} R_1 = 23\,440\,\text{lb} \\ R_2 = 6560\,\text{lb} \end{cases}$$

$$Area\,1 = \frac{8000 \times 4}{2} = 16\,000$$

$$Area\,2 = \frac{15440 \times 7.72}{2} = 59\,600$$

$$Area\,4 = 6560 \times 5 = 32\,800$$

The largest bending moment diagram is 43 600 lb.in. This value is used later in order to calculate the maximum normal stresses due to bending. ∎

16.5.4 Bending Loads

Bending loads occur when a machine element is bent as a result of a force and/or a bending moment, for example, in cantilever beams (balcony), or simply supported beams (a bridge). In bending, both normal and shear stresses may be present.

For simplicity of this discussion, let's assume that the cross section of the part is symmetrical. As such, the material in the middle of the cross section (called neutral axis) will remain unaffected by bending; it will not see any stresses. However, as the distance from the middle increases, bending stresses increase too. The maximum stress will be at the outer fibers of the beam, either in tension, or in compression. This is shown schematically in Figure 16.21.

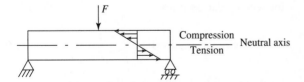

FIGURE 16.21 Normal stresses in bending.

The maximum normal stress in a beam under the bending moment of M is

$$\sigma_{max} = \frac{MC}{I} \tag{16.9}$$

where C is the distance of the farthest fibers from the neutral axis (or in symmetrical cross sections, $\frac{1}{2}$ of the height of the beam) and I is the moment of inertia of the cross section. Obviously, the larger the moment of inertia, the smaller the stress. The same is also true for nonsymmetrical cross sections, except that the maximum tension and compression will not be the same, and the neutral axis will not be in the middle of the section.

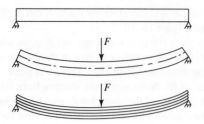

FIGURE 16.22 Shear loads in bending are caused by the tendency of the material layers to "roll" over each other, but being prevented from doing so.

In addition to normal stresses, bending loads create shear stresses too. To better understand this, let's imagine a phone book and how easily it can be bent, even with hundreds of pages. Then imagine that the pages of the phone book are all glued together. Obviously, it is significantly more difficult to bend the book (such as in a thick cardboard). The difference between the two cases is that in the former, the pages are free to roll (slide) over each other, and since their length changes insignificantly, they offer little resistance to bending, whereas in the latter, as the book is bent, all pages other than in the middle (the neutral axis) have to either elongate or shorten, therefore resisting bending. As a result, it is much more difficult to bend a taller object that a shorter one (Figure 16.22). As was mentioned, in the first case, the pages roll over each other in shear. If you prevent this rolling by attaching all pages together, the pages will carry a shear load. Based on this, we expect to also have shear stresses due to bending.

The shear stress is zero at the top or bottom (where normal stresses are maximum), and is maximum on the neutral axis (where normal stresses are zero). The shear stress in bending can be calculated by

$$\tau_{\max} = \frac{VQ}{Ib} \tag{16.10}$$

where V is the shear force at the location of interest, Q is the first moment of the cross-sectional area about the neutral axis, I is the second moment of area of the cross section about the neutral axis, and b is the width of the section at the point of interest. For common cross sections, the following simplified equations may be used:

$$\tau_{\max} = \frac{3V}{2A} \text{ for rectangular sections}$$

$$\tau_{\max} = \frac{4V}{3A} \text{ for solid circular sections} \tag{16.11}$$

$$\tau_{\max} = \frac{2V}{A} \text{ for hollow (thin-walled), circular sections}$$

As you can see, for a rectangular section, the maximum shear stress is 50 percent more than the average (V/A). Consequently, the actual maximum shear stress of Example 16.5 will be 6000 lb instead.

EXAMPLE 16.8

Referring to Figure 16.21, assume that the maximum moment applied to a rectangular beam is 150 lb.in. The beam is 0.75 in high and 0.5 in wide. Calculate the maximum normal stress in the beam.

Solution   Substituting these values into Equation 16.7 we will get

$$I = \frac{1}{12}bh^3 = \frac{1}{12}(0.5)(0.75^3) = 0.01758\,\text{in}^4$$

$$\sigma_{\max} = \frac{MC}{I} = \frac{150(0.375)}{0.01758} = 3,200\,\text{psi}$$
■

The Bones Under Stress   Figure 16.23 shows a simplified schematic of a human arm. As shown, the upper arm is supported by a long bone called humerus. The lower arm has two bones, radius and ulna, which roll over each other at the distal ends providing an opportunity to axially rotate the lower arm relative to the upper arm. At the proximal end (where they touch the humerus bone), they rotate relative to the humerus allowing us to flex and extend the arm. There are no ''pins'' at this joint; the bones roll over each other.

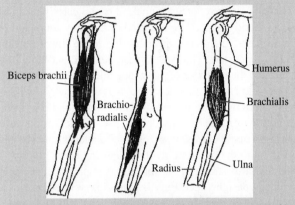

FIGURE 16.23   The schematic of some of the muscles of human arm.

There are a large number of muscles that work in coordination allowing us to flex and extend the lower arm, rotate it, as well as move the wrist and fingers. Since muscles are unidirectional (they can only pull), there are always opposing pairs of muscle to provide opposing motions. Extension of the lower arm (extending the arm outwardly) is accomplished by a large, 3-headed muscle called triceps brachii (not shown). The opposing motion of flexion is accomplished by a combination of a few muscles, most importantly the biceps brachii, the brachialis, and the brachio-radialis. The biceps is the strongest, followed by the brachialis, while the brachio-radialis contributes perhaps only a few percent. Assuming that nature does not have unnecessary redundancies, there must be another reason why these muscles exist.

Looking closely at the detail, you will notice that the biceps muscle is attached to the radius, while brachialis is attached to the ulna. So, although both muscles contribute to flexion, they individually assist in rotation of ulna and radius relative to each other, thus allowing us to rotate our arm. However, the brachio-radialis is also attached to the radius. So what is one of its primary functions?

Let's try to model the lower arm as a cantilever beam attached to the upper arm while lifting a load in flexion, as shown in Figure 16.24a. In this case the biceps muscle acts as a force pulling up, while the load will pull down the arm, creating bending moment on the arm. A close inspection of the sear diagram and the bending moment diagram will reveal that the maximum bending moment will occur where the biceps is attached to the radius, with a very large value that is capable of inflicting pain, if not enough to break the brittle bone. In fact, the role of the brachio-radialis is to significantly reduce this moment by pulling up the distal side of the radius, as a cable would do in a structure (Figure 16.24b). Its other job is to push the radius into the humerus at the joint, thus keeping them together against the action of the load; otherwise, the joint might fail and the bones may separate under the load.

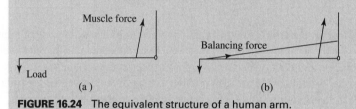

FIGURE 16.24   The equivalent structure of a human arm.

The Horse and the Rider   A horse, other animals, a parent carrying a child on the back while playing horsy, and a bridge can all be modeled as a simply supported beam, as shown in Figure 16.25. In all cases, the load creates tension in the lower portion of the beam and compression in the upper part. The type of material used and how it reacts in tension or compression determine the behavior of the bridge under the load. If the material is good in carrying both tensile and compressive loads, such as mild steel, it will be able to withstand the load. However, certain brittle material such as bones and cast iron are not very strong in tension; in both cases, the beam will collapse. In fact, at the turn of the twentieth century, many bridges made of cast iron collapsed under the newly developed heavy locomotives. The solution to the problem involves devising a way to carry the tensile load with another material that is strong in tension, for example, steel cables and tendons. In the case of the bridge, a design called Fink Bridge, ribs were connected together with steel cables that carry the tensile load. In the case of the horse and the parent it is the tendons and muscles that connect the ribs in the chest cavity that carry the tensile load.

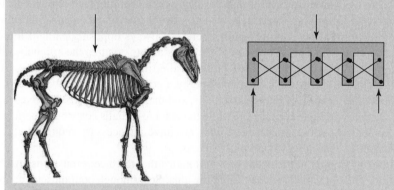

FIGURE 16.25   A horse carrying a rider is similar to a simply supported bridge.

16.5.5   Torsional Loads

Torsional loads occur as a result of twisting actions, for example, when a bolt is tightened. The twisting action is generally produced by a torque. However, the torque may be created as a result of the application of a force. As a result, there may be a combination of loads present, including forces, torsion, and bending (see Figure 16.9). Torsional loads create shear stresses in the cross section of the material.

Although any cross section may be used for carrying torsional loads, circular sections are the most common. For a circular section, the shear stress is zero at the center and maximum at the surface, varying linearly in between. The magnitude of the shear stress is

$$\tau = \frac{T\rho}{J} \quad \text{and} \quad \tau_{max} = \frac{Tr}{J} \tag{16.12}$$

where T is the torque, J is the polar moment of inertia of the surface area, and ρ is the radius at which point the stress is calculated. The maximum stress occurs at maximum ρ, which is the outer surface.

Since the shear stress in the inner fibers of the material is very low, the material is generally wasted; it does not carry much of the load. As a result, it is very common to use a hollow shaft to carry torsional loads. For example, the drive-shaft of rear-wheel drive cars is a hollow shaft. The same amount of material (made into a hollow shaft of a larger diameter) can thus carry much more load. The polar area moment of inertia for a solid circular shaft and a hollow shaft is

$$J = \frac{\pi d^4}{32} \quad \text{and} \quad J = \frac{\pi(d_o^4 - d_i^4)}{32} \tag{16.13}$$

Torsional loads rotate the part and impart an angular displacement in it. The total rotation for a circular shaft is

$$\theta = \frac{Tl}{GJ} \tag{16.14}$$

where l is the length of the part, and G is modulus of rigidity (shear modulus of elasticity).

EXAMPLE 16.9

A circular shaft is subjected to a constant 100 lb-in. torque. The diameter of the shaft is 0.25 in., and it is 3 in, long. The modulus of rigidity for the material is 11×10^6 psi. Find the maximum shear stress and the angular displacement of the shaft. Calculate the increase in maximum shear if the same shaft were hollow with $d_i = 0.125$ in.

Solution Substituting these numbers into Equations 16.10–16.12, we will get

$$J = \frac{\pi d^4}{32} = \frac{\pi(0.25)^4}{32} = 0.0003835 \, in^4$$

$$\tau_{max} = \frac{Tr}{J} = \frac{100(0.125)}{0.0003835} = 32595 \, psi$$

$$\theta = \frac{Tl}{GJ} = \frac{100(3)}{(11 \times 10^6)(0.0003835)} = 0.071 \, rad$$

Calculating J for the hollow shaft we get

$$J = \frac{\pi(d_o^4 - d_i^4)}{32} = \frac{\pi(0.25^4 - 0.125^4)}{32} = 0.0003595 \, in^4$$

$$\tau_{max} = \frac{Tr}{J} = \frac{100(0.125)}{0.0003595} = 34770 \, psi$$

$$\%difference = \frac{34770 - 32595}{32595} = 6.7\%$$

As you see, even though the material is 25 percent less, the maximum stress is only 6.7 percent more. ∎

16.6 STRESS VS. STRENGTH: SAFETY FACTORS

The next step in the design of a machine element is to match the stresses with strengths. Whether a simple analysis or an elaborate one, after stresses are calculated, they must be compared with the strength of the material. If the stresses are below the strength of the material, it will be able to carry the load without permanent deformation or breaking. In reality, in most cases the dimensions of the machine component are not known; instead, based on the strength of the material, a dimension is chosen (designed) for the part such that the stresses *will* be below the strength of the material. In fact, in most cases, a safety factor is included to ensure that even if unknown factors are present, the component will remain safe.

The safety factor can be written as

$$S.F. = \frac{S_y}{\sigma_{max}} \quad \text{or} \quad S.F. = \frac{S_{sy}}{\tau_{max}} \tag{16.15}$$

Designers choose safety factors based on the accuracy of their assumptions and the cost of failure. For example, imagine that you are designing a building. In order to design the correct size of the load-bearing elements (such as the beams and columns), you will need to know the extent of the loads to which the building will be subjected. Let's say that you assume that there will be so many people and so much dead load (furniture, etc.) in the building in addition to the construction materials. Can you imagine that there might be a case where an additional 10 people may come into the building? Should the building collapse? Or imagine that a car is designed for five passengers and 300 pounds of load in the trunk. Would it be possible that there may be more people (or heavier ones) in the cab or more weight in the trunk, or that the car may go through a pothole larger than what you thought? Should it fail? And what if a gust of wind larger than usual hits an airplane? Should it crash? On the other hand, if a building is a little heavier but safer, it may not be a bad trade-off. But can you make your airplane 30 percent heavier to ensure safety? As a result, engineering societies, trade organizations, companies, and regulatory agencies have come up with certain standards that are used as safety factors. These numbers are very different for different situations and different elements. You will need to consult the proper resources for these standards. For example, building codes specify minimum safety factors for construction projects. Airplanes structures are designed with relatively low safety factors (only a few percent), but with much care and elaborate analyses, in order to ensure that assumptions are accurate and loads are accurately represented.

EXAMPLE 16.10

For the hollow shaft used in Example 16.9, assume that the yield strength in shear for the material is 40000 psi. Calculate the safety factor for the shaft.

Solution   Substituting these values into Equation 16.15 we will get

$$S.F. = \frac{S_{sy}}{\tau_{max}} = \frac{40000}{34770} = 1.15$$

■

EXAMPLE 16.11

For the beam of Example 16.8, calculate the width of the beam if we desire a safety factor of 1.5 and the yield strength of the material is 20000 psi. Due to space constraints, the height cannot be any more that 0.75 in.

Solution   Substituting these values into Equations 16.15 and 16.9 we will get

$$I = \frac{1}{12}bh^3 = \frac{1}{12}(b)(0.75^3) = (0.03516)b \text{ in.}^4$$

$$\sigma_{max} = \frac{MC}{I} = \frac{150(0.375)}{(0.03516)b} = \frac{1600}{b} \text{ psi}$$

$$\text{S.F.} = \frac{S_y}{\sigma_{max}} = \frac{20000(b)}{1600} = 1.5 \rightarrow b = 0.12 \text{ in.}$$

In reality, it is most probable that the material will come in a standard thickness of 0.125 in. In that case, obviously we will not cut the thickness to 0.12 in. Rather, we will continue with a slightly higher safety factor resulting from 0.125 in. width.   ∎

16.7   DESIGN FOR STRENGTH VS. DESIGN FOR STRAIN

Imagine that you are designing parts of a robot manipulator, and that the specifications set for the product indicate that the robot will have an accuracy of 1/1000th of an inch. If you were to design the links and the shafts based on strength, it is possible that there may be elastic deformations in the parts due to the applied loads that will exceed the specified accuracies, rendering the robot totally inaccurate. In this case, although the strength requirements are satisfied, the strain requirements are not. Therefore, it is important to also consider allowable strains and ensure that displacements are not larger than allowed or desired. In many cases, although the strength requirements are met, deflections are excessively large and unacceptable. Then, the design of the elements must be based on strain requirements and not just strength requirements.

Robot manipulators are open loop systems. This means that as the robot is subjected to external, as well as dynamic, loads, if there are any displacements in the robot links (arms) or the joints, the end-effector (the last part of the robot that constitutes a tool or grabber) may deflect, but the robot controller will not know about it, and thus, all accuracies are lost. There have been attempts at solving this problem by either adding sensors to the manipulator (such as lasers and vision systems) or by trying to close the open loop through novel combination of linkages. However, most of these systems are either too slow or severely limit robot's utility. The common solution for this problem is to significantly overdesign the robot arm in order to practically eliminate all deflections. As a result, robot arms are very heavy for the loads they carry and the stress levels are very low. A typical small industrial robot may weigh 150 lb with a payload of only 5 lb, even though it can carry many tens of pounds, albeit with larger deflections. An Olympian weight-lifter of the same 150-lb range may life several hundreds of pounds.

16.8   STRESS CONCENTRATIONS

One way to initiate cutting or tearing a material is to inflict a small incision in it and then apply the cutting forces. For example, if you make a small incision in a stack of papers, it will be much easier to tear the stack. Similarly, if there are any cracks or other material flaws in a machine element, it may be in danger of failing. This is due to a phenomenon called stress concentration.

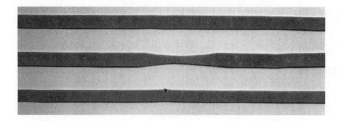

FIGURE 16.26 The effect of stress concentration in a rubber band. As the change in width becomes severe and quick, the stress concentration increases significantly.

As an example of this, take a ¼-inch-wide rubber band and cut it into a long strip, then pull the two ends. You will notice that the rubber band can be pulled significantly without any damage. Next, using a pair of scissors, reduce the width of the rubber band in the middle by gradually cutting out on the sides, as shown in Figure 16.26. You will notice that you can still pull the ends significantly without damage, even if not as much as before. Next, inflict a small incision in the middle and try to pull the ends again. You will break the rubber band with a small force. This shows that stress concentration is severely affected by geometry and how severe the change in geometry is.

Now consider a flat bar with uniform cross section, as in Figure 16.27a. The stress in the bar will be $\sigma_1 = F/A$. Next, assume that the cross section is gradually reduced to $(2A/3)$ as in Figure 16.27b. Then the average stress in the cross section will be $\sigma_2 = f/(2A/3) = 1.5\sigma_1$. If we then do the same, but with a sharp, pointed incision in the bar such that the remaining area remains the same $2A/3$, the average stress will also be the same $(1.5\sigma_1)$, but the actual maximum stress, somewhere in the cross section, will be much larger than that.

To better understand the meaning of stress concentrations consider the traffic flow in a freeway. As long as the number of lanes in the freeway does not change, barring accidents and other mishaps, the traffic will flow smoothly, with uniform speed. However, when there is a gradual reduction in the number of lanes, you can expect to see congested traffic flow, and although vehicles still move relatively fast, there is a reduction in speed, especially if anyone tries to rapidly swerve into the flow. This can be compared to a gradual reduction in the cross section, to an increase in the stress level, and to a corresponding low stress concentration. However, as you can imagine, if there is an accident (simulating a sharp reduction in the remaining cross section) or if there is a sudden loss of a lane, there will be slowed traffic, long lines, and high concentration of vehicles as drivers try to move into another lane or stop, similar to a large stress concentration. In materials too, stress lines can bend somewhat and follow the load when the change is slow. But when the change is dramatic and sudden, stress lines will swerve in and out, and although the average stress level over the entire cross section remains the same, there will be areas of concentrated stresses that can be much larger than the average value (see Figure 16.28). When the concentrated stress level in a portion of the material exceeds the yield strength

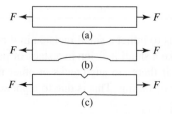

FIGURE 16.27 Stress concentration in a flat bar due to changes in the cross section. As the cross section is reduced with sharper changes, the stress concentration factor increases.

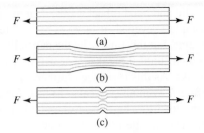

FIGURE 16.28 Variations in stress levels due to changes in the cross section are similar to a traffic flow. The sharper the change, the worst the effect.

of the material, even if the average value is below the yield strength, the material either yields locally or fails locally. If the material in question is ductile, it generally yields first, somewhat relaxing the concentrated stresses. However, if the stress level is too much, it will eventually yield enough to fail. If the material is brittle, it generally fails locally, and since the remaining area will be reduced even further, it will lead to failure. Thus, in general, brittle materials are more susceptible to the effects of stress concentration. You should pay special attention to stress concentration factors for brittle material, and as we will discuss later, especially if the load is varying.

The ratio between the maximum stress level in a cross section and the average stress level is called stress concentration factor k. Reference books on stress analysis, machine design books, and mechanical engineering handbooks have multiple tables that include many different shapes, cross sections, types of loads, and variations in the sections, and indicate the corresponding stress concentration factors[1]. You should refer to these references when necessary. According to these tables, stress concentration factors as large as 5 are possible. Whenever possible, try to avoid sources of stress concentration. However, it should be expected that in countless situations it is necessary to have grooves, bends, and cuts of different shapes. In that case, ensure that the stress concentration factors are taken into account.

Please also see Chapters 7, 9, and 13 for additional discussion on this subject regarding the relationship between this issue and the aesthetic design of products, design factors, and product liability. Please also see the effect of stress concentration factor in fatigue life as follows.

16.9 FATIGUE ANALYSIS

16.9.1 What is Fatigue

The stresses in a static, non-moving, machine part or system with nonvarying loads are constant. They remain the same or vary very little as time goes by. One example of this is the dead loads in a building, where the weight of the structure does not change, or that the change in it is small compared to the weight of the structure. On the other hand, machines are meant to move, and as such, loads (forces, moment, and torques) vary in time; they are not constant. These are called varying loads, and they are very common in machines, bridges, airplanes, automobiles, and most other products. Similarly, even under constant loads, if the machine element rotates, the stresses will vary. Designing for varying stresses is somewhat different and more involved than designing for static, nonvarying stresses. This is because the strength of the material under these loads is generally less than the static yield or ultimate strength, and since loads vary, it requires more analysis to find average loads, maximum loads, as well as equivalent loads (for example, when loads vary randomly).

Now do the following test: take a paperclip and bend it backwards, then return, and repeat until it breaks, and count how many complete reversals it survives before breaking. In most cases, it is just a couple. Next do the same, but this time bend the paperclip half as much, and bend it back again. Repeat (and count) until it breaks. You will notice that it takes many more reversals before the clip breaks. As this is reduced further, the clip will last longer and will survive more cycles before it eventually breaks. In fact, for certain materials like steel, it is possible to reduce the load to a certain level, called endurance limit, at which point the part will essentially have infinite life and will not break due to load variations. This is the essence of fatigue loading and fatigue analysis, that if loads are cyclic, we need to consider fatigue loads and design the machine element based on fatigue (endurance) strength.

First, let's see how we end up with cyclic, variable, fatigue loads. One example is the wing of an aircraft. When the aircraft is not moving, the wings are basically two cantilevered beams hanging from the fuselage. In this case, the weight of the wings due to gravity is downward, creating a moment at the fuselage that opposes the moment created by the weight. As we have already seen in the previous sections, the bending moment will create a normal stress in the wing at all points in relation to the moment. In this case, the normal stress will be tensile at the top of the wing and compressive at the bottom layers. However, as soon as the airplane gains speed, the aerodynamic lift forces will push up the wing, first cancelling the weight, and then increasing until the airplane become airborne. In this case, the stress in the wing will become smaller, zero, and eventually will change direction, such that the tensile stress at the top layers will become compressive while compressive stress at the bottom layers will become tensile. In fact, this is a complete reversal of the state of the normal stresses in the wings. Additionally, during the flight, the lift forces vary due to changes in the wind speed, shear wind forces, changes in speed, and other factors. As a result, as the plane is airborne during the flight, the stresses in the wings keep varying, even if not completely reversed, until it lands again and the stresses revert back to the initial state. Every time an airplane takes off and lands, this process repeats. Should we expect that the wings come off and break like the paperclip did? Figure 16.29 shows a representation of the variations in the stress at some point on the wing.

Now consider a rotating cantilevered shaft, loaded with a downward force, such as the shaft of a motor, an axle in a car, or similar applications, as shown in Figure 16.30. When the shaft is loaded this way, the layers at the top will experience a tensile stress while the layers below the neutral axis will be in compression. However, since the shaft is rotating, the same points on the shaft will move as the shaft rotates. During this process the tensile stress on an element that is at the top will go to zero, to a compressive stress when at the bottom, and back to zero, and finally, to the tensile stress when it returns to the top. All points in the shaft will experience the same shifts in stress values every time the shaft rotates. This is a cyclic, sinusoidal change in the stress that repeats for every complete rotation of the shaft. If the shaft is on a motor that

FIGURE 16.29 A representation of the variable and cyclic nature of stresses in a wing of an airplane during take-off, flight, and landing.

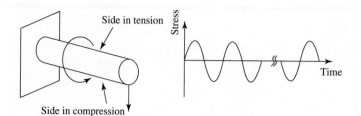

FIGURE 16.30 Fatigue loads on a rotating shaft.

rotates 1700 revolutions per minute (rpm), it will experience 1700 state changes in its stress values.

In the above example, the load may be constant, but the shaft rotates. However, another example of fatigue loading is when the load varies, even if the load-bearing member does not rotate. In addition to the airplane wing example, think about a nail clipper. Every time the nail clipper is pressed, its components experience a stress, which goes back to zero as soon as the load is removed. The same is true for the contact mechanism of a blinker or a thermostat. They move back and forth under thermal forces and experience a fatigue loading. Other examples of parts with varying loads are springs and bolts. The load on these parts may either occasionally or continuously change. So why do they not fail as badly as the paperclip?

In order to understand why parts fail under fatigue loading, even if the stress levels are below the yield strength, let alone ultimate strength, we need to look at the fatigue failure mechanism. This relates to what we discussed regarding material behavior, stress concentrations, stress-strain curve, and material properties.

Imagine that any one of the above-mentioned examples is subjected to the stresses we discussed, and that at one point, there is some imperfection, a crack, a cut, or any other mechanism, that creates a relatively large stress concentration factor. In that case, at the microscopic level of the root of the crack or the imperfection there will be a large stress value that may exceed the yield (or even the ultimate) strength of the material. In that case, the material will yield or fail locally. If it is ductile, the yielding may relieve some of the stress, but due to cold working, the material may harden just a bit at the same location. At the next occurrence of a large load with large stress concentration factor, the material will either yield again and harden a bit more or if it is brittle, it may break microscopically and the crack or the imperfection may grow a bit more. With every application of the load, the crack may grow slightly more and the material may harden microscopically. Generally, since maximum loads are not always constant anyway, larger occurrences of stress cause larger crack growth. These larger crack growths can be clearly seen on a failed cross section, looking like the marks that water leaves behind on a sandy beach. These marks are called beach marks or chevrons (the relatively smooth areas between these chevrons have very small bumps and valleys called striations). The growth continues, sometimes to a vary large extent, until there come a time when one more application of a large load is encountered, when the remaining cross section will no longer be able to take the load and will completely and drastically fail. This is called fatigue failure, and it requires a point or a cause for stress concentration as well as cyclic, varying loads to propagate a crack or imperfection to eventual failure. Figure 16.31 shows a few examples of parts that have failed in fatigue. You can clearly see the chevron marks as well.

In most cases, a clear starting point can also be identified. For example, Figure 16.31a shows a bicycle pedal that failed in fatigue (can you see why a bicycle pedal is subjected to fatigue loading?) exactly at the root of the corner of the stamped brand name, the word Super Mighty (b). The crack initiated at this point (as can be seen from the dirty

FIGURE 16.31 Fatigue failure examples in machine parts. (a) A bicycle pedal broken due to stress concentrations at the root of the name stamped on it as shown in (b). Chevron marks can be seen on the surface as well as the rough final failure area. (c) The shifter of a car with huge fatigue cracked areas. (d) The crankshaft of an automobile. (e) A nail clipper broken where the two members were spot welded together.

area where dirt and oil accumulated for a long time. Chevron marks can be seen clearly, with striations in between. At one point, when one last large force was applied, the remaining area could not withstand the stress and catastrophically broke. The rough area in the bottom is the result. It should be mentioned here that you should never clean the broken area of a failed part, try to put it back together, or otherwise disturb the area, at least until it is completely studied. Every clue will indicate what may have gone wrong and how to improve the design. For example, this surface indicates that the pedal had a safety factor of about 1.8 when new. This is indicated by the ratio of the initial cross section to the final area. It also shows the weak area created by the sharp stamping of the name on the stem. In fact, if the name was not stamped so deeply, the pedal might have just been really mighty. Figure 16.31c shows the shifter of a car. The bar is subjected to fatigue loading every time you shift. However, it should be clear that the force to engage the first gear is larger than the force to disengage it or to engage second gear. As a result, one side of the cross section has a larger fatigue crack than the other. At the bottom of the bar where the cracks initiated there is a groove cut in the material for attaching it to the shifter mechanism. Exactly at this stress-concentration-causing groove cracks started and grew in opposite directions, one larger than the other, until the shifter suddenly broke during the last attempt at shifting gears. The very small area left before the final failure indicates that the part was in fact much stronger than needed if it were not subjected to crack initiation, that it experienced small stresses compared to its strength, and yet, as a result of the groove, cracks developed. In fact, there were enough failures in this model that the company recalled all the cars after about 10 years of service and changed the bars. Figure 16.31d is the crankshaft of a car with similar chevron marks. The rough area of the final failure is very small, indicating the importance of fatigue stresses compared to strength of the material. The crankshaft was not overloaded at all since it still worked even with a huge crack in it. It failed due to stress concentration and a small crack or imperfection in the material. Figure 16.31e is a nail clipper. It failed due to fatigue loading at the point where the two members were spot welded together. Spot welding caused both a stress concentration as well as changes in the material behavior which initiated a crack and resulted in eventual failure. We will discuss surface treatment and its effects on fatigue strength later.

Have you noticed that paperclips, fasteners used in folders, and other similar parts are usually made of very ductile and flexible materials? This is to increase their fatigue life. Obviously, a stronger material has higher yield and ultimate strengths, and as a result, can carry larger loads. But more brittle materials are also more susceptible to fatigue loading due to crack propagation. So, there is a trade-off involved in how strong the material should be. Factors such as size, quality of the material and how well it is made, temperature, and other factors have a negative effect on fatigue life and will be discussed shortly.

No material in nature is perfect. We always expect that materials have imperfections, cracks, and other stress-concentration-causing problems. Material imperfections can arise even at the crystalline level, where there may be inclusions or voids in the material, as shown in Figure 16.32. Additionally, when alloys are formed, it is possible that foreign particles may be present in the mix, causing microscopic imperfections in the material. It is also possible that in many materials, as the mixture cools down, many nuclei form simultaneously, growing into sections that reach each other. As a result, the alloy is made up of many smaller sections joined together. Any of these imperfections may become a nucleus for fatigue crack growth and failure.

Now imagine that a large number of specimens are tested under different fatigue loading conditions until they fail (called Standard Rotating-Bending Fatigue Strengths). Plotting the life of the specimen against the stress to which it was

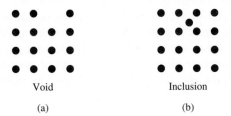

Void

(a)

Inclusion

(b)

FIGURE 16.32 A void and an inclusion in a crystalline structure.

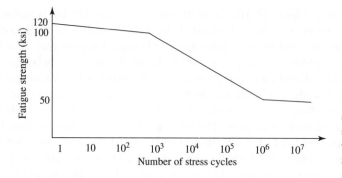

FIGURE 16.33 Fatigue strength for UNS G41300 steel, Normalized. (Data from NACA Technical Note 3866, December 1966.)

subjected reveals the behavior of the material in fatigue loading. From thousands of tests on common materials (such as steels of different grades and aluminum) to exotic and specialty material (such as the ones used in turbines and rotors) a pattern of behavior can be detected for the material at hand. The most common materials tested are steel and aluminum. It has been found that most ferrous specimens follow a pattern similar to Figure 16.33 (which shows the fatigue strength of UNS G43100 type steel specimen), in which, if the stress level is below approximately 50 percent of the ultimate strength of the material, it essentially does not fail in fatigue. This is called *endurance limit* or S'_e, at which point, the material will have infinite life if it is subjected to stresses below this level. Above this level, the specimen will have a limited life, usually expressed in a logarithmic scales $(1, 10^1, 10^2, 10^3$, etc.). However, the endurance limit is approximate; it is affected by many factors and is not necessarily true for all ferrous materials. For example, since cast iron is weaker in tension than in compression, its approximate endurance limit is estimated at $S'_e = 0.4 S_u$. Oddly enough, nonferrous materials such as aluminum do *not* have a knee at 10^6 or any other point. The life increases as the stress level is reduced, but there is no infinite life. As a result, designers use a high life cycle number for their calculation, usually 10^8 or 5×10^8. Isn't this interesting that most of our airplanes are made of aluminum which does not have an infinite life? As a result, it is eventually necessary to replace aluminum parts.

16.9.2 Endurance Limit for Ferrous Materials

Figure 16.34 shows the average strength ratio of S_f / S_{ut} (fatigue strength over the ultimate strength in tension) for a large number of ferrous material samples. These ratios vary much from sample to sample, but the curve shows the average values for all samples. As shown, for low-cycle fatigue, the fatigue strength to ultimate tensile strength can be as high as 0.8–0.9. This means that for low cycle fatigue (about 10^3) the fatigue strength can be as high as 80–90 percent of the ultimate strength of the material. This ratio decreases as the number of cycles increases. Please note that the number of cycles is shown in orders of magnitude (meaning that it is a logarithmic scale), as is the strength ratio. For ferrous materials, when the strength ratio is about 0.5, the number of cycles before fatigue failure becomes infinite. This means that if we subject a ferrous material to stresses that are below 50 percent of its ultimate strength in tension, it may last indefinitely. This endurance limit is used for design of machine elements in fatigue. However, as we will see, this limit is further affected by many other factors.

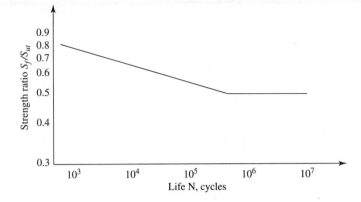

FIGURE 16.34 Strength ratio for ferrous materials. Endurance limit is approximately at 50 percent of ultimate strength for these materials.

16.9.3 Low Cycle Fatigue, High Cycle Fatigue

Based on the information available from Figure 16.34, it is in fact possible to design a machine element for finite life or for infinite life. This is especially useful when the weight of the part and its cost is very important and we want to reduce the weight as much as possible, and when we know that the part will be subjected to limited number of cycles. As we discussed in Chapter 7, the life span of the machine or product is also affected by this design variable. However, as discussed earlier, in the presence of factors that cause stress concentration, the life may be much shorter than we expect.

To design for limited life, you may estimate the fatigue strength of the part based on Figure 16.34 or similar information and use the value for deciding on the dimensions of the part. However, in most cases, and unless we have an acceptable level of certainty about the number of cycles, we should design for infinite life.

EXAMPLE 16.12

The ultimate strength for a steel specimen is 200 000 psi. Estimate the fatigue strength for 10^3 cycles, 10^5 cycles, and infinite life.

Solution From Figure 16.34, we estimate that the ratio of fatigue strength to ultimate strength in tension for 10^3 cycles, 10^5 cycles, and infinite life are 0.85, 0.6, and 0.5, respectively. Thus the fatigue strengths for the specified number of cycles for the sample are approximately:

$$S_{f(10^3)} = 0.85 \times 200\,000 = 170\,000 \text{ psi}$$
$$S_{f(10^5)} = 0.6 \times 200\,000 = 120\,000 \text{ psi}$$
$$S'_e = 0.5 \times 200\,000 = 100\,000 \text{ psi}$$ ∎

16.9.4 Endurance Limit Modifying Factors

The endurance limit found from standard tests is an approximate number and relates only to a standard specimen tested under controlled conditions. In real life, machine parts differ greatly from the standard specimen and the conditions under which it is tested, and therefore, must be modified for a number of different factors. These modifying factors are used to reduce the endurance limit for the variations between the standard test and specimen and the real machine element. The modified endurance limit may be expressed as

$$S_e = S'_e(k_s k_z k_r k_t k_c k_l k_m) \qquad (16.16)$$

where

S_e is the modified endurance limit

S'_e is the endurance limit for the material

k_s is surface factor

k_z is size factor

k_r is reliability (quality) factor

k_t is temperature factor

k_c is stress concentration modifying factor

k_l is load-type factor

and k_m is other miscellaneous factors that may arise.

k_s **or surface factor** is used to introduce the effect of the surface of the material. A rough surface increases the chance of surface cracks and imperfections, thus increasing the chance of fatigue crack initiation and growth. The smoother the surface, the better it is for fatigue life. A polished surface ($k_s = 1$) is the best, followed by ground ($k_s = 0.9$), machined or cold drawn ($k_s = 0.83$ to 0.63), hot rolled ($k_s = 0.7$ to 0.3), and forged ($k_s = 0.55$ to 0.2) for tensile strengths ranging between 60 and 240 kpsi.

k_z **or size factor** is used to incorporate the effects of the size of the part into the endurance limit. As was described briefly earlier, one of the main factors causing fatigue failure is the presence of imperfections and cracks in the material that cause microscopic stress concentrations. The larger a part is the higher are the chances that it may have imperfections. As a result, the endurance limit for larger-sized parts should be somewhat reduced to accommodate the increased chance. Please refer to other resources for elaborate equations that estimate this modifying factor for different ranges, all developed based on experimental results. However, for shafts of up to about 2 inches, the size modifying factor may be as large as 0.8. The endurance limit for tensile loads is not affected by this factor.

k_r **or reliability (quality) factor** relates to the desirable reliability of the part. Since most of the loads for which a machine is designed are estimates, there is always a statistical distribution of stress. Similarly, materials vary from sample to sample, and therefore, there is a statistical distribution of material strength. As a result, there is always a possibility that the strength of the material may fall below the stress caused by the load, which will lead to failure. This is used to calculate the modifying factor for reliability. Desiring high reliability for a part means that we should assume that the endurance limit is lower, and therefore, increase the chance of longer life. For 50 percent reliability, the modifying factor is 1. For 90 percent reliability, it is about 0.9, for 99 percent reliability, it is 0.87, and for 99.9 percent reliability, it is 0.75. Unless otherwise specified, you should assume 50 percent reliability.

k_t **or temperature factor** is used to include the effects of temperature in fatigue life. Use $k_t = 1$ for temperatures below 450°C.

k_c **or stress concentration modifying factor** is used to incorporate the effect of stress concentration factors in the fatigue life. As was discussed earlier, this is a very important factor, as stress concentration is a major element in initiating an eventual fatigue failure. Assuming that the theoretical stress concentration factor for different shapes, sizes, and load types is K_t and q is Notch Sensitivity factor (which can be found in reference tables and is a representation of the shape of the notch), then

$$k_c = \frac{1}{1 + q(K_t - 1)} \qquad (16.17)$$

q is small for ductile materials (0.4–0.8 for notches ranging from 0.01 to 0.15 inches), and increases for more brittle materials (0.8–0.95). Please refer to references at the end of this chapter for values of K_t and q.

k_l or load factor relates the type of loads present. Since most tests on materials are performed on a rotating shaft in bending the results should be modified for other loads. Use $k_l = 1$ for bending loads, $k_l = 0.85$ for axial loads, and $k_l = 0.59$ for torsional loads.

k_m or miscellaneous factors is used to represent the effects of other factors that may arise (residual stresses, directional characteristics, surface treatments, corrosion, etc.). For example, a part that is shot-peened will have better fatigue strength because shot-peening reduces surface cracks and imperfections.

It should be mentioned here that *all* of these factors are experimental and approximate. Their accuracy should always be questioned as applied to any particular application, and to remember that the information is statistical and as such, although true in general, they are not necessarily true in particular. Additionally, different references use different names, factors, or values for these factors, some even completely ignore many of them. You should not be surprised to see discrepancies between these factors in different references.

EXAMPLE 16.13

A beam, loaded in bending, is machined from UNS G10150 hot-rolled steel. The beam has a diameter of 0.75 in., with a groove that is 1/8 in. deep and has a radius of 1/16-in., as shown. We desire a reliability of 90 percent. Calculate its estimated modified endurance limit for fatigue.

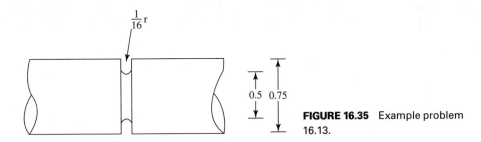

FIGURE 16.35 Example problem 16.13.

Solution The ultimate strength for UNS G10150 hot-rolled steel is 50 000 psi. The endurance limit for this material before modifications is $S'_e = 0.5 \times 50 = 25$ kpsi. The modifying factors can be estimated based on the information provided as follows:

- For machined part with $S_{u_t} = 50$ kpsi, $k_s = 0.83$.
- For sizes up to 2 inches, $k_z = 0.8$.
- For 90 percent reliability, $k_r = 0.9$.
- Since no temperature is mentioned, we assume $k_t = 1$.
- From tables provided in other references, a theoretical stress concentration factor can be found for $D/d = 0.75/0.5 = 1.5$ and $r/d = (1/16)/(0.5) = 0.125$ as $K_t = 1.8$. The

notch sensitivity for the same notch is $q = 0.75$. From these values we can estimate the stress concentration factor to be:

$$k_c = \frac{1}{1 + q(K_t - 1)} = \frac{1}{1 + (0.75)(1.8 - 1)} = 0.625$$

- For bending loads, $k_l = 1$.
- Since no other miscellaneous effects are mentioned, $k_m = 1$.

$$\begin{aligned} S_e &= S_e'(k_s k_z k_r k_t k_c k_l k_m) \\ &= 25 \times 0.83 \times 0.8 \times 0.9 \times 1 \times 0.625 \times 1 \times 1 \\ &= 9.34 \, \text{kpsi} \end{aligned}$$ ∎

16.9.5 Effects of Mean Stress and Amplitude on Fatigue Life

Figure 16.36 shows two fatigue loading situations in which one has an average value $\sigma_m = 0$ while the other has a positive average value, even though the amplitude of the variations in loading σ_a is the same in both cases. The allowable fatigue strength for the machine element is affected by σ_m and σ_a and their ratio. This relationship is shown as in the Modified Goodman Diagram, as shown in Figure 16.37.

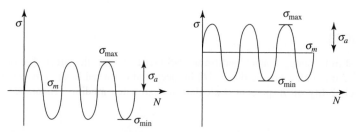

FIGURE 16.36 Fatigue loads may have a zero or a nonzero mean value. Accordingly, their allowable fatigue strength will be influenced by this difference even if their amplitudes are the same.

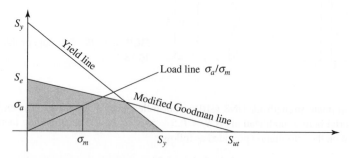

FIGURE 16.37 Modified Goodman Diagram.

To realize this relationship, assume that $\sigma_a = 0$. This means that the load will be static, and as such, the maximum strength allowable for a static load will be S_{ut}. Next assume that the average value of the fatigue load $\sigma_m = 0$. In this case, the load is completely reversed, and the allowable strength for the part can be S_e. However, in all cases, we do not want to load the part beyond yield strength. For all other combinations of loads, the allowable strength will fall in a region that is formed by these limits. To form the Modified Goodman Diagram, mark the values of S_e, S_{ut}, and S_y as shown, and connect.

If the values of σ_a and σ_m are known, draw the load line with the slope of σ_a/σ_m. If these values are below the Goodman line, you can calculate the safety factor. On the other hand, if the values of σ_a and σ_m are not known (for example, if you are designing the part and are calculating the size of the part), then draw the load line based on the ratio of σ_a/σ_m. You may find the allowable strength from the Modified Goodman Line where it intersects with the load line and use this number to calculate the size of the part.

There are other concerns associated with this diagram, such as type of loading and random loading. You can also draw a similar Modified Goodman Diagram for shear loads. Please refer to the list of references for more detail information about these issues.

EXAMPLE 16.14

A machine element is subjected to bending loads that create variable stresses ranging between 7 000 and 13 000 psi. The estimated endurance limit for the part is 13 000 psi, the ultimate strength is 40 000, and the yield strength is 20 000 psi. Find the safety factor associated with this part.

Solution   The Modified Goodman Diagram for the part is drawn in Figure 16.38. The mean and amplitude of the stress can be calculated as

$$\sigma_m = \frac{\sigma_{max} + \sigma_{min}}{2} = \frac{13,000 + 7,000}{2} = 10\,000\,psi$$

$$\sigma_a = \frac{\sigma_{max} - \sigma_{min}}{2} = \frac{13,000 - 7,000}{2} = 3\,000\,psi$$

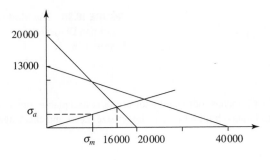

FIGURE 16.38   The Modified Goodman Diagram for the machine part in Example 16.14.

From Figure 16.38, we can estimate that the safety factor for the part is about $16\,000/10\,000 = 1.6$. ∎

EXAMPLE 16.15

A round cantilever beam, 40 inches long, is loaded with a variable force of 100 to 300 lb at the end. The beam is machined from UNS G10180 hot-rolled steel with yield strength of 32 000 psi and ultimate strength of 58 000 psi. For 80 percent reliability ($k_r = 0.93$) and a safety factor of 1.5, estimate the necessary diameter of the beam.

Solution The maximum moment is at the end of the cantilever beam. The mean value and amplitude of the variable stresses can be calculated as

$$M = Fl \quad \text{and} \quad I = \frac{\pi d^4}{64}$$

$$\sigma = \frac{MC}{I} = \frac{Fl(d/2)}{\pi d^4/64} = \frac{32Fl}{\pi d^3}$$

$$\sigma_{max} = \frac{32(300)(40)}{\pi d^3} = \frac{122}{d^3} \text{ kpsi}$$

$$\sigma_{min} = \frac{32(100)(40)}{\pi d^3} = \frac{41}{d^3} \text{ kpsi}$$

$$\sigma_m = \frac{1}{2}\frac{(122+41)}{d^3} = \frac{81.5}{d^3} \text{ kpsi}$$

$$\sigma_a = \frac{1}{2}\frac{(122-41)}{d^3} = \frac{40.5}{d^3} \text{ kpsi}$$

The endurance limit for the material, based on the provided information is

$$S_e = S'_e(k_s k_z k_r) = 0.5 \times 58 \times 0.83 \times 0.8 \times 0.93 = 17.9 \text{ kpsi}$$

From the Modified Goodman Diagram in Figure 16.39, with the load line drawn, σ_m should be 21 kpsi, but with the required safety factor, the calculated diameter will be

$$\frac{81.5}{d^3} = \frac{21}{1.5} \quad \rightarrow \quad d = 1.8 \text{ in.}$$

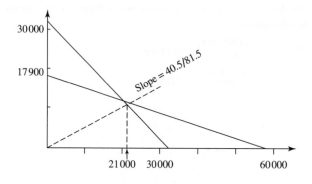

FIGURE 16.39 The Modified Goodman Diagram for Example 16.15.

Since the diameter is less than 2 inches, our size factor of 0.8 is acceptable; otherwise, we would have had to recalculate the endurance limit based on a larger size factor and check again. ∎

16.9.6 Fatigue-Life Design Criteria

The following criteria may be used in order to decide the proper fatigue life for a product: **Infinite Life Design** relates to products and machine elements that are subjected to millions of cycles of fatigue loads. Engine rotors and crankshafts are good examples of this criterion. In this case, the endurance limit for 10^6 or 10^7 (or even higher for nonferrous materials) is used.

Safe Life Design or limited life design criterion is used for machine parts that will be exposed to limited number of cycles. An airplane wing is exposed to large number of fluctuating loads during flight, but its load only reverse completely once per landing or take-off. A pressure vessel may not be exposed to more than between 10^4 to 10^5 cycles. In this case, fatigue strengths for limited number of cycles may be used in order to save in weight and cost of the material.

Fail Safe Design requires that there be enough safeguards that even if a part fails, it will still be safe. This criterion recognizes that there is always a possibility that cracks may grow and a part may fail in fatigue. However, the design includes safeguards to prevent a total failure of the system. For example, garage door openers use springs for counter-balancing the weight of the door. The springs are exposed to fatigue loading, and it is expected that they may fail. However, these springs have a pair of U-shaped bars, engaged together inside the spring and attached to the two ends. Even if the spring fails completely, the two parts remain attached to the door and will keep it in place. Although the door may no longer be operable, it is safe. A multi-engine airplane is safer than a single-engine aircraft. Even if one engine fails, the others will safely fly and land the plane, even if with lower performance and lower speed. In machine parts where stress concentration is inevitable due to geometry, the remedy is to include stress relievers and crack stoppers in the design to reduce the chances of crack propagation until it can be detected. And finally, the part must be checked for cracks on a regular basis, before the crack grows too much.

Damage Tolerant Design is similar to the fail-safe criterion, recognizing that cracks do exist and fatigue life may be limited. Fracture mechanics analysis is used to check whether such cracks will grow large enough to be dangerous before they are detected by periodic inspections. The interval of crack growth between easily detectable size of cracks and a critical size must be longer than the interval between inspections. In this scenario, it is expected that the crack will be detected before it becomes a danger.

16.10 SELECTION OF POWER SOURCES AND ACTUATORS

There is a large array of assorted actuators on the market, from traditional devices such as motors, to more sophisticated types such as brushless DC motors, and to novel devices such as piezoelectric actuators, muscle fibers, and electroactive polymers (EAP). However, in most general applications, electric motors, hydraulic devices, and pneumatic actuators are the devices of choice. They are relatively easier to use, less expensive, easier to incorporate, repair, use, and replace, and are common. Still, a designer should consider the best choice for the application at hand, and in fact, may design even newer, more novel devices for the application.

Before we compare the different types of actuators and their characteristics, it is useful to consider a few factors that can play an important role among all actuators.

16.10.1 Accelerations vs. Actuating Forces and Torques

If there is a net force or torque acting on a body, the body will accelerate linearly or angularly:

$$\overline{F} = m\overline{a} \tag{16.18}$$

and

$$\overline{T} = I\overline{\alpha} \tag{16.19}$$

This means that as the weight or the mass moment of inertia of a body increases, the force or the toque must also be increased in order to create the same acceleration. In most cases, the designer should know (or be able to estimate) the weight and the moment of inertia of the part. Based on desired specifications set for the device, accelerations can also be calculated or estimated. The designer can then choose proper actuating systems that can generate and deliver the required force or torque. This is an important task because if the actuating source is not powerful enough, the system may not perform satisfactorily, or in some cases, not perform at all. On the other hand, excessively powerful actuating sources waste energy and are not desired. Additionally, due to differences between static and kinetic coefficients of friction, starting a system requires more actuating force or torque than continuing to move it. Consequently, this must also be considered in selecting proper actuating systems.

EXAMPLE 16.16

A fan with a moment of inertia of 0.1 kgm^2 is to be accelerated at the rate of 10 rad/s^2. The moment of inertia of the rotor is 0.01 kgm^2. Estimate the required torque needed by the motor.

Solution   The total inertia felt by the motor is the summation of the two moments of inertia. The required torque will be

$$\overline{T} = I\overline{\alpha} = (0.1 + 0.01)(10) = 1.1 \text{ N.m}$$

Based on this number, one can choose an appropriate motor that delivers this much torque at this rate of rotation.   ∎

16.10.2   Back Electromotive Force (back-emf)

Electromotive force is the basic principle on which all motors and transformers operate. This principle states that if a conductor, carrying an electric current, is placed within a magnetic filed, a force will be induced in the conductor. If the conductor is placed on a rotating axis, it will tend to rotate due to the force (or torque). All motors rotate due to this principle. However, the opposite is also true. If a conductor is moved within a magnetic field so as to cross its lines, a current will be induced in it, called back-emf. This is true even for a rotor within a motor. Thus, as a rotor is rotated within a motor, it will generate a current. All generators operate under this principle. Therefore, as the rotor is forced by the electromotive force to rotate due to the supplied current in the windings, a back-emf current in the opposite direction will be induced in it as well.

Now imagine that a motor starts from rest. Due to the electromotive force (torque) the rotor will accelerate. Theoretically, the rotor's speed should continue to increase to infinity due to this acceleration, but we know that as it approaches its nominal speed, it tends to rotate at this speed until disturbed. The reason is that as the rotor's angular speed increases, so does the back-emf. Since the back-emf is a current in the opposite direction of the applied current, its effective current reduces as its speed increases until the effective current at that speed reduces to zero, and therefore, it ceases to accelerate any further. At this point, the output torque of the rotor is obviously zero since it does not accelerate any further. The motor will continue to rotate at this speed until disturbed. Now imagine that a load is applied to the rotor. As soon as the load is applied, the rotor will slow down, reducing its back-emf, resulting in a positive net current which counteracts the applied load. Consequently, as the load on the rotor changes, its speed will change too. In order to control the speed of a DC motor, the current supplied to the motor windings should

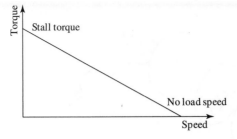

FIGURE 16.40 A typical torque/speed relationship for a DC motor.

change correspondingly, or its speed will change until the generated toque equals the applied load.

This is a very important matter. It means that DC motors will *not* deliver a constant speed under varying loads unless their current is controlled (as in a servomotor). Figure 16.40 shows typical response of an electric DC motor. As you notice, the effective output torque delivered by the motor decreases as its speed increases. The maximum torque is delivered at zero speed (stall torque). However, at this point, the back-emf is zero, and as a result, the current in the rotor generates much heat that can damage the motor. At the no-load maximum speed, the output torque is zero. The designer must consult the manufacturer's information in order to select a proper motor that can deliver the required torque at the desired speed. This is usually accomplished with speed reduction systems, as discussed below.

16.10.3 Speed Reduction

Different actuators produce their actuating force or torque at different rates. For example, a hydraulic ram can make even very small motions at full force, and thus may not need any speed reduction. An electric motor usually runs at its nominal speed that can be several thousand rpm, and as such, must be coupled with a speed reduction system for low speed applications such as in a robot arm or a windshield wiper. A direct drive electric motor can also move a slight angle at full torque and may not require speed reduction at all.

Speed reduction systems increase the cost, weight, space requirements, friction, and backlash in the system. They also reduce the overall reliability. However, they have a positive effect on the effective inertia to which the motor is subjected. To analyze this, assume that, through a set of reduction gears with a reduction ratio of N, a load with inertia I_l, is connected to a motor with inertia I_m (including the inertia of the reduction gears), as shown in Figure 16.41. The torque and speed ratio between the motor and the load will be:

$$T_l = N T_m \tag{16.20}$$

and

$$\omega_l = \frac{1}{N} \omega_m \tag{16.21}$$

The effective inertia of the load felt by the motor is conversely proportional to the square of the speed reduction ratio, or

$$I_{\text{Effective}} = \frac{1}{N^2} I_l \quad \text{and} \quad I_{\text{Total}} = \frac{1}{N^2} I_l + I_m \tag{16.22}$$

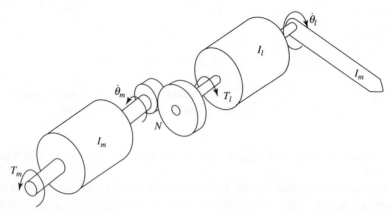

FIGURE 16.41 Inertia and torque relationship between a motor and a load.

This means that speed reduction systems reduce speed and increase torque, but they also reduce the effective inertia that the motor will "feel" by a factor of N^2. As an example, suppose that a reduction ratio of 10 is used in conjunction with a motor. The total effective inertia on the motor will only be 1/100th of the actual inertia of the load, and therefore, the motor can accelerate quickly. In direct drive systems, both electric and hydraulic, the motors are exposed to the full inertial loads.

EXAMPLE 16.17

A motor with rotor inertia of 0.01 Kgm^2 and maximum torque of 10 Nm is connected to a uniformly distributed arm with a concentrated mass at its end, as shown in Figure 16.42. Assume that the inertia of a pair of reduction gears can be ignored. Calculate the total inertia felt by the motor and the maximum angular acceleration it can develop if the speed reduction ratio is (a) 5, (b) 50.

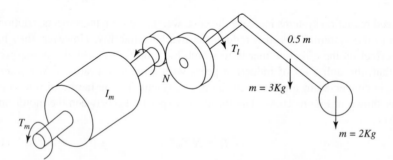

FIGURE 16.42 Schematic drawing of the system of Example 16.17.

Solution   The total moment of inertia of the arm and the concentrated mass at the center of rotation is

$$I_{total} = I_{arm} + I_{mass} = \frac{1}{3}m_{arm}l^2 + m_{mass}l^2$$
$$= \frac{1}{3}(3)(0.5)^2 + (2)(0.5)^2 = 0.75 \, Kgm^2$$

From Equation (16.22),

(a) $I_{\text{Total}} = \dfrac{1}{N^2} I_l + I_m = \dfrac{1}{25}(0.75) + 0.01 = 0.04 \, \text{Kgm}^2$

(b) $I_{\text{Total}} = \dfrac{1}{2500}(0.75) + 0.01 = 0.0103 \, \text{Kgm}^2$

As you can see, the total inertia with the higher gear reduction ratio is practically the same as the rotor inertia of the motor. The maximum angular accelerations for the rotor will be

(a) $\alpha_m = \dfrac{T_m}{I_{\text{total}}} = \dfrac{10}{0.04} = 250 \, \text{rad/sec}^2$

(b) $\alpha_m = \dfrac{T_m}{I_{\text{total}}} = \dfrac{10}{0.0103} = 970 \, \text{rad/sec}^2$ ∎

The following comparison between the common actuators is provided as a guide only. The comparisons allow you to choose a more appropriate actuator for your design. Simple analysis will also allow you to choose the range of actuator size and power for your application. However, in all cases, a more thorough analysis will be needed for a final choice.

16.10.4 Hydraulic Actuators

Hydraulic actuators are strong, reliable, simple, readily available, and with proper valves and controls, their position and speed can be easily controlled. Since they are powerful and provide relatively small displacements, they usually do not need additional speed reduction. This translates into lower costs, fewer parts, lower weight, less friction and backlash, and better reliability. They are responsive and very stiff systems too. However, they are generally expensive and require many parts, from a hydraulic pump and motor, to heat removal devices (radiator or fan), filters, sump, holding valve, servo valve, electronics, etc. Their power to weight ratio, as far as the actuator itself is concerned, is the highest compared to electric devices and pneumatics, but not very high if the pump and associated hardware is included in the ratio (such as in a moving platform that carries all these with the actuator). Hydraulic devices require an umbilical cord for transferring the high pressure fluid to the actuator, plus electricity to run the pump (sometimes 3-phase or 220V power). They are noisy and very susceptible to dirt and foreign material in the fluid. They are also sensitive to the changes in the viscosity of the fluid. They generally leak, and therefore, should not be used for clean-room applications. The high pressure fluid (several thousands of psi are very common) can be a safety risk. The actuator motions can be well controlled if a servo valve is used, but servo valves are expensive and require control circuitry. In general, you should consider hydraulic actuators for systems that require large forces. The force developed by a hydraulic actuator is

$$F = p \times A \tag{16.23}$$

where F is the force, p is the hydraulic pressure, and A is the effective area of the cylinder. For example, for a hydraulic ram with a fluid pressure of 3000 psi, every square inch of the cylinder will develop 3000 lb force.

The flow rate and volume of oil needed in a hydraulic system are

$$d(\text{Vol}) = \frac{\pi d^2}{4}(dx) \tag{16.24}$$

and

$$Q = \frac{d(\text{Vol})}{dt} = \frac{\pi d^2}{4}\frac{dx}{dt} = \frac{\pi d^2}{4}v \tag{16.25}$$

where dx is the desired displacement, and v is the desired velocity of the piston. By controlling the volume of the fluid entering the cylinder the total displacement can be controlled. By controlling the rate at which the fluid enters the cylinder, the velocity can be controlled. It can be controlled either by hand through an operator (such as in a truck), or by a servo valve with control circuitry along with feedback position and speed sensors (such as in a robot).

16.10.5 Electric Actuators

Electric actuators include electric motors (DC, AC, AC reversible, synchronous motors, brushless DC, disk motors), servomotors, stepper motors, Biometal™ wires, direct drive motors, voice coils, piezoelectric drives, and many other novel devices. They are mostly reliable, common, and useful devices and can be used in endless applications. Electric motors come in many different shapes, sizes, and power ranges, and you should expect that you will find a motor appropriate for your applications, from very small to humongous sizes. With proper care, electric motors can last a long time without much difficulty, and if appropriate controls are used, they can be accurate and reliable. Common motors usually run at high speeds (up to several thousand rpm) and therefore, may require speed reduction if lower speeds are needed.

The following are some general characteristics of different types of electric motors and some general applications for them. This is neither an exhaustive list of their characteristics nor their applications, but only a general guideline:

DC electric motors are reversible. Changing the polarity of the current will reverse the direction of rotation. This can be accomplished by switching manually, by control circuits, or by a microprocessor and an H-bridge. By controlling the current, their speed can also be controlled. This can also be accomplished manually, by control circuits (servomotors) or by a microprocessor and pulse-width-modulation. Their life is affected by the life of the commutators that are used to internally switch the current. Since the heat generated in the rotor must pass through the air gap between the rotor and stator or rotor and the casing, and since air is a very good heat insulator, DC motors are prone to overheating and damage. Otherwise, in DC applications (such as cars) and when variable motor speed is needed, DC motors are very common.

AC electric motors are rugged, powerful, and reliable. Their main disadvantage is that since their speed is a function of the frequency of the power source, it cannot be controlled unless the power source frequency is changed by external circuitry. They are generally not reversible unless their stator winding is center-tapped. This means that only half of the winding is used for power generation at any time, but the current can flow in two opposite directions for reversible motions. Unlike DC motors, since the heat is generated in the stator and can be dissipated quickly, AC motors are less prone to damage by heat.

Brushless DC motor is an attempt to operate an AC-type motor in DC mode. In order to accomplish this, the rotor of a brushless motor is a permanent magnet, as is an AC motor. However, a circuit is used to switch DC current between three sets of coils thus creating a rotating magnetic field that will rotate the rotor with it. Sensors are used to detect the position of the rotor and to switch the current to the coils. Since the circuit is in control of the speed at which the current is switched to the coils, the speed of the rotor is controlled. Brushless DC motors can be controlled by microprocessors as well, and therefore, are common in applications with mechatronic devices (VCR, disk drives, etc).

Disk motors are like other motors, but the rotor is made into a flat plate with very small moment of inertia. Disk motors are very thin compared to other motors and are used in applications where vertical space is limited (such as disk drives, VCRs, DVD players, etc).

Servomotors are usually DC motors with a feedback device that controls the position and speed of the motor. As was mentioned earlier, the speed of a DC motor changes as the load on it varies. Additionally, there is no control over how much it rotates. However, in a servomotor, a position feedback sensor such as a potentiometer or encoder is attached to the shaft (or to the output shaft of the speed reduction unit attached to it) which measures the rotational displacement of the shaft. The speed of the shaft can also be measured by a sensor such as tachometer, or the position feedback is differentiated electronically to estimate the angular speed of the shaft. This information is used to control both the position and the speed of the motor. If the speed is lower than desired, the current is increased; if the speed is larger than desired, the current is reduced. Additionally, when the desired position is achieved, the current is cut off. Since this is accomplished with a circuit, it is easily possible to incorporate PID control into the circuit as well and control the motor reliably. An industrial servomotor is generally expensive, but powerful and reliable.

Stepper motors are also an attempt to operate a DC servomotor in AC mode. Unlike regular AC and DC motors, if you connect a stepper motor to a power source (like a battery), it will not rotate; in fact, its highest resistance to movement is at this state, called *holding torque*. Stepper motors have a permanent magnet rotor like AC motors. The motion of the rotor is controlled by a rotating magnetic field that is generated through an external control circuit. A microchip, a microprocessor, or dedicated circuits, switch a sequence of either four or eight coils that create a rotating magnetic field. The rotor follows this field. Every switching of the field causes the rotor to move one step. The total displacement of the motor and its speed can be controlled by how many times the current is switched within the sequence, and how fast.

Stepper motors are reliable and sturdy. Their total displacement and speed can be controlled with microprocessors and dedicated circuits, and therefore, are desirable in many mechatronics-related projects and devices. However, their output torque suffers significantly from increased speeds, even more than regular DC motors. This is due to the added burden of accelerating/cruising/decelerating between every two steps, as well as to the added back-emf caused by currents being turned on and off. Therefore, you should be extremely cautious of their torque generation capability.

Biometal^{TM} **wire,** a patented alloy, shortens about 4 percent when it reaches a transition temperature. The transition temperature can be designed into the material by changing the composition of the alloy, but standard samples are set for about 90° C. At around this temperature, the crystalline structure of the alloy makes a transition from martensitic to austenitic state, and hence, shortens. It will once again switch back to martensitic state when it is cooled down. This process can continue for hundreds of thousands of cycles if the loading on the wire is low. The common source of heat for this transition may be an electric current flowing through the metal itself, which warms due to its electrical resistance. As a result, a piece of Biometal wire can easily be shortened by an electric current from a battery or other power sources. The major disadvantage of the wire is that the total strain occurs within a very small temperature range, and thus, except in on-off situations, it is very difficult to accurately control the strain, and therefore, the displacement. However, it can be a novel actuation source for small forces.

Voice coil motors are common in computer hard drives. They are very fast, and their position can be controlled very accurately. A voice coil is essentially a permanent magnet and a coil, similar to the gauges used in automobiles and voltmeters. As the current in the coil changes, the repelling force (torque) between the two moves the arm accordingly.

16.10.6 Pneumatic Actuators

Pneumatic actuators are relatively simple, easy to use, safe, and reliable. They require a pressurized air source, and as a result, may not be a good choice if pressurized air is not readily available. However, it is still much easier to generate pressurized air than it is to pressurize fluids for hydraulic units. Besides, for simple applications, pressurized air canisters may be utilized. Pneumatic components are mostly off-the-shelf items, and therefore, are relatively inexpensive and can be found in a large variety of sizes and types. A significant characteristic of pneumatic devices is their susceptibility to compression; as the load changes, the air compresses or expands and the actuator moves. This compliance can be a blessing when it is needed and a curse if stiffness is desired. For example, it is very difficult to control the exact movement or position of pneumatic cylinders. A technique called *differential dithering* can be used to control the position of the actuator, in which, a feedback sensor provides signals to either increase or decrease the pressure in the cylinder by adding or removing puffs of air as the actuator moves under the varying load. Differential dithering is expensive and requires additional control circuits. But for on-off applications, when the actuator is all the way to one side or the other, the air pressure can be set high enough to decrease the movements of the actuator under varying loads. Additionally, for operations that are inaccurate, for assembly operations, and for grippers, the extra compliance that pneumatic devices provide can be very useful.

Since air is neutral, even if they leak, pneumatic devices are safe and clean. Utilizing solenoid valves, a simple signal can be used to control the motion of the cylinder. The signal may come from a microprocessor or simple sensors, and therefore, pneumatic devices can be used in many control and automatic applications too.

16.11 SELECTION OF ACTUATORS

In order to choose an appropriate actuator, first you must decide what type of actuator is satisfactory. This decision is based on the available resources, space requirements, power requirements, speed and accuracy requirements, cost, and many other factors. For each system, you also need to estimate or calculate static and dynamic loads on the actuator. The combination of toque (or force) and speed requirements can then be used to select an actuator from among thousands of manufacturers' catalogues. Nowadays, the Internet has made this task much easier since you may search large databases of manufacturers' catalogue quickly. Certain manufacturers also provide their own equations for design calculations and selection of proper actuators. As an example, see http://www.orientalmotor .com/support/mtr_sizing_formulas.htm.

Casual selection of an actuator for a class project, although important, is not as critical as choosing an actuator for mass production. You must look into all the requirements a manufacturer may have placed on their products and ensure that they are met. Each product has specific characteristics and requirements. You must follow manufacturers' recommendations for proper operation.

16.12 DESIGN PROJECT

Consider a design solution for a problem of your choice (perhaps a problem that may have been assigned to you in your class or at work), or one of the solutions presented in Chapter 5 for the page-turner, or even your own solution. Perform a complete analysis of the design to the best of your ability and knowledge level. Obviously, the level of analysis is related to your standing in regards to your curriculum. Even if you do not know how to do detail design analysis, list what type of analyses you believe would be needed on different parts of your solution.

REFERENCES

1. PETERSON, R. E., "Stress Concentration Design Factors," Wiley, New York, 1953.
2. JUVINAL, ROBERT C. and K. M. MARSHEK, "Fundamentals of Machine Component Design," 3rd Edition, Wiley New York, 2000.
3. SHIGLEY, JOSEPH E. and L. MITCHELL, "Mechanical Engineering Design," 4th Edition, McGraw-Hill, New York, 1983.
4. SHIGLEY, JOSEPH E, C. MISCHKE, and R. BUDYNAS, "Mechanical Engineering Design," 7th Edition, McGraw-Hill, New York, 2004.
5. FUCHS, H. O. and R. I. STEPHENS, "Metal Fatigue in Engineering," Wiley, 1980.
6. COLLINS, J. A., "Failure of Material in Mechanical Design," 2nd Edition, Wiley, New York, 1993.
7. "DC Motors, Speed Controls, Servo Systems," Electro-Craft Corporation, 1980.
8. ORTHWEIN, W. C., "Estimating Fatigue Due to Cyclic Multi-axial Stress," American Society of Mechanical Engineers paper # 86-WA/DE-8, 1986.
9. GORDON, J. E., "Structures, or Why Things Don't Fall Down," Da Capo Press, New York, 1978.
10. PARR, ROBERT E., "Principles of Mechanical Design," McGraw-Hill, New York, 1970.
11. www.environmental-robots.com for information on electro active polymers.

HOMEWORK

16.1 Draw the free-body diagram of the truss and calculate the reaction forces.

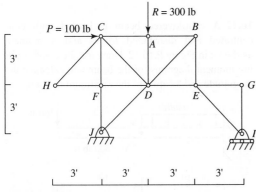

FIGURE P.16.1

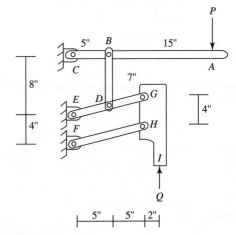

FIGURE P.16.2

16.2 Draw the free-body diagram of the different parts of the machine in Figure P.16.2.

16.3 Using the free-body diagram of the truss at the indicated section, calculate the forces in members *DE*, *FC*, and *FD*.

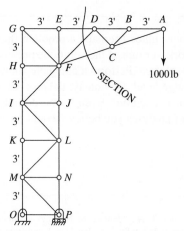

FIGURE P.16.3

16.4 The calculated shear stress and normal stresses on a machine element are 12 000 psi, 22 000 psi, and −5 000 psi, respectively. Calculate the principal normal stresses, maximum shear stress, and the angle associated with the principal stresses.

16.5 A bar is loaded with a force, as shown in Figure P.16.5. Find maximum shear and principal stresses at A. Ignore the transverse shear caused by the force.

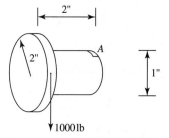

FIGURE P.16.5

16.6 Calculate the normal and shear stresses at the end of the beam in Figure P.16.6.

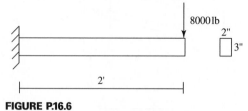

FIGURE P.16.6

16.7 Recalculate the stresses for the beam of Figure P.16.6 if the beam is half as wide or half as tall.

16.8 Determine the areas and the type of stresses that are critical to the analysis of the joint in Figure P.16.8.

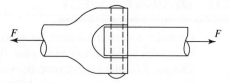

FIGURE P.16.8

16.9 Determine the areas and the type of stresses that are critical to the analysis of the coupling shown in Figure P.16.9. The bolt pattern is shown too.

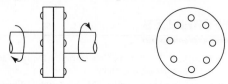

FIGURE P.16.9

16.10 A 10-ft long beam, simply supported at its two ends, is subjected to a 1000 lb force in the middle. Draw the shear and bending moment diagram and calculate the maximum normal and shear stresses on it. The beam has a rectangular cross section and is 2 inches wide and 3 inches tall.

16.11 A beam is loaded with a concentrated moment and distributed forces, as shown in Figure P.16.11. Draw the shear and bending moment diagrams for the beam and calculate the magnitude and location of the maximum moment.

FIGURE P.16.11

16.12 A cantilevered beam is loaded with concentrated and distributed forces and a moment as shown in Figure P.16.12. Draw the shear and bending moment diagrams for the beam and calculate the magnitude and location of the maximum moment.

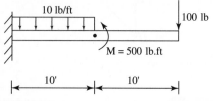

FIGURE P.16.12

16.13 A 2-in. diameter shaft is subjected to a 1000 lb-ft torque. Calculate the maximum shear stress in the shaft.

16.14 Repeat Problem 16.13, but assume it is a hollow shaft with $d_i = 1$ in. Compare the results.

16.15 A 0.25-in. diameter bar is loaded with a 300-lb normal force. Due to a small groove at the attachment point, it is expected that a stress concentration of $K_t = 1.5$ is present. Assume that the effective diameter can still be considered as 0.25 in. For $S_y = 40\,000$ psi, calculate the safety factor for the bar.

16.16 Calculate the endurance limit for a machine part made of UNS G10350 hot-rolled steel. The approximate diameter of the part is 1.5 in., it operates in temperatures below 300 °C, the desired reliability is 99 percent, and it is subjected to tensile loads. The yield strength is 39 kpsi and the ultimate tensile strength is 72 kpsi.

16.17 Draw the Modified Goodman diagram for the sample in Problem 16. Assuming that the maximum and minimum stresses on the part are 8000 psi and 3000 psi, calculate the safety factor for the part.

16.18 Calculate the safety factor for a part, loaded in bending, with maximum stress of 5000 psi and minimum stress of 2000 psi. The part is made of UNS G10100 cold-drawn steel with $S_y = 44\,000$ psi and $S_{ut} = 53\,000$ psi.

16.19 Look online for a motor manufacturer. Determine what is required by the manufacturer in order to specify one of their products and see what you will need to know in order to purchase an AC motor with continuous duty. Could you perform the same for a DC motor? Could you do this for your project?

INDEX